T0342409

# Seashells of Southern Florida

Bivalves

# SEASHELLS
## of Southern Florida

LIVING MARINE MOLLUSKS OF THE

FLORIDA KEYS AND ADJACENT REGIONS

## Bivalves

### Paula M. Mikkelsen and Rüdiger Bieler

**Princeton University Press**

Princeton and Oxford

Copyright © 2008 by Princeton University Press

Published by Princeton University Press, 41 William Street, Princeton, New Jersey 08540
In the United Kingdom: Princeton University Press, 3 Market Place, Woodstock, Oxfordshire OX20 1SY

All Rights Reserved

Library of Congress Cataloging-in-Publication Data

Mikkelsen, Paula M.
Seashells of southern Florida : living marine mollusks of the Florida keys and adjacent regions. Bivalves / Paula M.
Mikkelsen and Rüdiger Bieler.
p.   cm.
Includes bibliographical references and index.
ISBN-13: 978-0-691-11606-8 (hardcover : alk. paper)
ISBN-10: 0-691-11606-7 (hardcover : alk. paper)   1.  Bivalvia—Florida—Florida Keys.   I.  Bieler, Rüdiger.
II.  Title.
QL430.6.M546 2007
594′.40975941—dc22        2007001051

British Library Cataloging-in-Publication Data is available

This book has been composed in Goudy Text and Present for display

Printed on acid-free paper. ∞

nathist.princeton.edu

Composition by Bytheway Publishing Services

Printed in Italy by Eurografica

1 3 5 7 9 10 8 6 4 2

# Contents

# Preface

The crescent-shaped archipelago of the Florida Keys, a chain of low-lying islands at the southernmost tip of Florida, stretches southwestward from Miami into the Gulf of Mexico. In addition to the major string of 30 inhabited limestone islands connected by road bridges and causeways from Key Largo to Key West, the Keys are composed of many hundreds of mangrove mud islands, sandy islands, and rocks of the reef tract.

The Florida Keys lie in the subtropics between 24 and 25 degrees north latitude and their near-tropical climate and environment are closer to that of the Caribbean than to peninsular Florida. Recognized as one of the great recreational and environmental resources of the United States, the region is visited by more than three million visitors per year. A significant part of the attraction is sportfishing, diving, snorkeling, and boating around the coral reefs of the "oceanside" and in the mangrove-fringed back country of the "bayside." This region, where the Gulf of Mexico, Atlantic Ocean, and Caribbean Sea meet, is home to the only shallow-water, tropical coral reef ecosystem on the continental shelf of North America. Together with the numerous other marine habitats such as extensive seagrass meadows and hard-bottom regions, the Keys are host to an amazingly diverse marine fauna.

Although not as visible as corals and fish, the various groups of mollusks, particularly gastropods and bivalves, make up a large part of this diversity. None of the many clams, mussels, oysters, or other bivalves has come close to the iconic status of the region's most well-known gastropod, the Queen Conch. There are no commercial clam or mussel fisheries in the area. If it were not for the ubiquitous mangrove oysters on intertidal roots, docks, and seawalls, few casual observers would even think about bivalves as a substantive part of the Key's faunal diversity.

Although the Keys have long been a popular region for shell collectors, there have only been occasional (and largely unillustrated) attempts to document the bivalve fauna. With the installation in 1990 of the Florida Keys National Marine Sanctuary, which forms a marine protected area encompassing all of the marine waters surrounding the islands of the Keys, a more formal interest was generated in recognizing—and ultimately monitoring—marine biodiversity of the region. Intensive survey and taxonomic work over the past decade by the authors has resulted in the recognition of nearly 400 Florida Keys bivalve species in water depths of less than 300 m. Many of these results were summarized and discussed in a recent technical volume of *Bivalve Studies in the Florida Keys* (Bieler & Mikkelsen, 2004a), which also provides a critical species catalog and an annotated bibliography to regional bivalve literature. Still missing was an *illustrated* monograph for this part of the American coastline that showed the entire bivalve diversity, from intertidal oysters to deepwater "little white clams," in readily comparable photographs, regardless of shell size, habitat accessibility, or rarity.

This book is thus foremost an illustrated monograph of the bivalves of southernmost Florida. Here we attempt to be as complete as possible and set ourselves various parameters to follow. These included (1) to illustrate, in color, each bivalve species that has a verified record (or strongly supported indication) of occurrence in the region; (2) to use only, and clearly identify, specimens that have been deposited in accessible and permanently maintained museum collections, therefore allowing future verification; (3) for wide-ranging species to illustrate specimens that were actually collected in the Florida Keys to highlight regional morphologies, and (4) to update the taxonomy to the most current level. The latter obviously is a "moving target," as identifications are reviewed, systematic research continues, and taxonomic allocations change.

Although deliberately local at its core, the book should prove useful beyond the immediate and literally fluid boundaries of the Keys. Most marine bivalves have large geographic ranges and the majority of the treated species extend north along the Florida coastline and/or into the Caribbean and the Gulf of Mexico, often reaching into South American waters. We have indicated such ranges as known to us and occasionally, as in the case of the Caribbean venus clam *Chione cancellata* (regionally one of the most often cited bivalves that, however, does not occur in the Keys), provide additional images of easily confused extralimital species.

The Florida Keys bivalves are diverse not only in terms of the total number of living species. The fauna also has a surprisingly broad representation of evolutionary lineages, with more than half (59) of the currently recognized (100+) bivalve families represented by at least one species. This allowed us to pursue a second major goal—to provide an introduction to bivalve families with directly comparable descriptions and illustrations of characteristic shell-morphological and anatomical features. For each of the families we selected an exemplar species from the Keys fauna and illustrated its shell and anatomical features in detail, plus giving habitat and other relevant information. In instances where we were unable to base anatomical information on local species because live-collected material was not available, we drew upon specimens and data from elsewhere. In a few cases we realized that no such material or information was available for *any* member of the group, pointing to dire needs for future research.

Bivalve taxonomy is far from settled and higher-level classification in Bivalvia is in a phase of active debate and reorganization. Several names, dates, and hypothesized relationships adopted herein differ from conventional treatments, but for the most part we have left discussions of higher-level relations for more technical venues and have concentrated on highlighting the featured species within their currently assigned family placement.

This work includes more than 1,500 color shell photographs, anatomical and hinge detail drawings, scanning electron micrographs, in situ field photographs, and other illustrations, the vast majority of which were specifically produced for this project. With these we have attempted to showcase the enormous diversity of bivalves—in form, function, and evolutionary lineages—based on representatives from the beautiful, unique, and fragile marine habitats of the Florida Keys.

The Authors

# Seashells of Southern Florida

## Bivalves

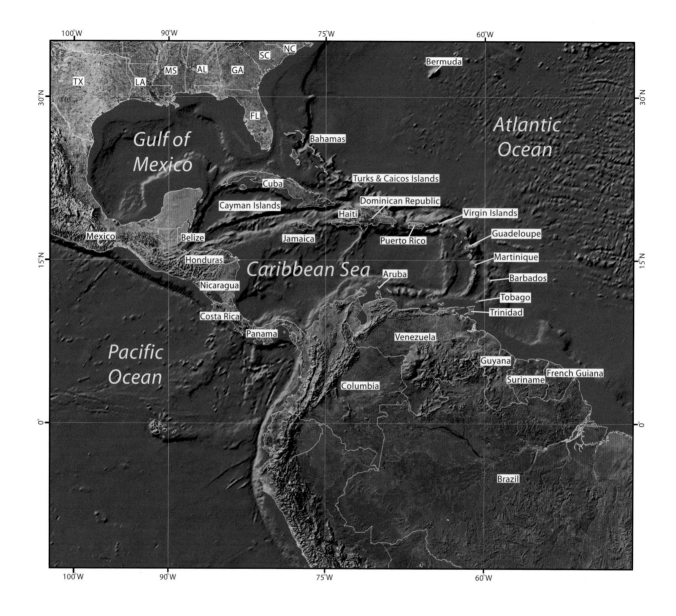

The Florida Keys (arrow) in relation to the Gulf of Mexico and the Caribbean Sea. Many of the species treated in this work have wide distributional ranges in the western Atlantic. A few even extend much further north (e.g., Canada) or south (Uruguay and Argentina) along the eastern coast of the Americas.

# Introduction

## Florida Keys

### Natural History

The Florida Keys form a chain of islands that separate Florida Bay from the Florida Straits at the southern tip of peninsular Florida, extending about 380 km from Soldier Key (15 km south of Miami) to Key West and westward to the Dry Tortugas archipelago (about 24°20′ to 25°21′N and 80° to 83°W). Numerous passes through the ridge of the Florida Keys permit exchange of bay waters with waters of the outer shelf to the south and east. The elevated rocky Keys, many of which nowadays are connected by a roadway leading from Key Largo to Key West, are made of limestone. Of these, the slender arc of Upper and Middle Keys (extending westward to Bahia Honda) are an emergent crest of Key Largo Limestone, the remnants of a Pleistocene coral reef. The Lower Florida Keys from Big Pine Key to Key West, with their dramatically different comblike arrangement, are made of oolitic Miami Limestone. They represent a reef-and-bar/channel complex that also developed about 125,000 years ago when the sea level was 6–8 m higher than today. This formation consists of ooids (a particular kind of near-spherical calcareous sand grain that precipitates in areas of strong reversing tidal currents), which were cemented together when the sea level fell at least 120 m during the advance of glaciers in the northern latitudes and the sand was exposed to slightly acidic rain. In contrast to both of these formations, the islands in Florida Bay and those west of Key West (including the Marquesas Keys and the Dry Tortugas archipelago) are mostly accumulations of modern sediment, carbonate mud, sand, and mangrove peat.

The Florida Reef Tract, a series of offshore bank reefs, marks the southern edge of the Floridian Plateau, 7–13 km from shore. Between the reef tract and the island chain lies Hawk Channel, a wide, V-shaped basin of 5–12 m depth that contains various shoals and patch reefs. South of the reef tract begins the Florida Straits, which is dominated by the fast-flowing Florida Current, a subsystem of the Gulf Stream. Its western portion, responsible for the mild winter temperatures permitting extensive coral reef development in the Keys, is composed of water from the Gulf of Mexico. Tidal range is less than 1 m throughout the Keys. Although the reef tract is exposed to normal salinities (24–37 parts per thousand), enclosed bays and ponds often experience widely fluctuating conditions resulting from heavy rainfall or evaporation.

Few marine environments in the United States can compete with the Florida Keys in terms of natural beauty and natural resources. The region is of particular note for its extensive offshore coral reefs. However, this subtropical region also sustains many other interdependent habitats including fringing mangroves, hypersaline ponds, seagrass meadows, mud banks and tidal channels, sandbars, deep sand plains, and hard pavement below a thin veneer of sand. Florida Bay and its islands serve as nesting, nursery, and/or feeding grounds for numerous marine animals including American crocodiles, West Indian manatees, loggerhead turtles, bottlenose dolphins, and a variety of bird species. Florida Bay serves as nursery grounds for many economically important fish species, pink shrimp, and the Caribbean spiny lobster or crayfish. Taken as a whole, the complex marine ecosystem of the Keys supports one of the most unique and diverse assemblages of plants and animals in North America and is the foundation for the commercial fishing and tourism-based economies that are vital to South Florida.

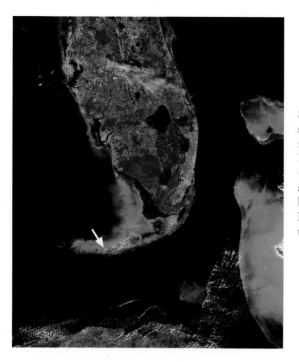

Satellite image of South Florida showing the deep Florida Straits separating Florida from Little and Great Bahama Banks (east), Cay Sal Bank (southeast), and Cuba (south). Corals and nearshore sediments reflect brightly. The 145-km distance from Key West (arrow) to Cuba is shorter than that to Miami.

## Need for Protection

About three million annual visitors join the fewer than 100,000 permanent residents of the Keys, most of them attracted by various water-based activities, including offshore fishing and scuba diving. The heavy use of marine resources over the decades has contributed to environmental problems, particularly for the coral reefs that are rapidly declining. Major issues include water-quality degradation, increase in coral diseases and coral bleaching (which have devastated once-magnificent stands of elkhorn and staghorn corals), as well as invasion of macroalgae into seagrass beds and coral reefs. The stressors on the reef are numerous and those identified range from overfishing, physical damage from ship groundings and individual bottom/coral contact by anchoring boaters, snorkelers, and divers, to pollution and nutrient overload from cesspools and other land-based sources, artificially modified water-flow regimen out of the Everglades, as well as thermal stress caused by climate change. Whatever the full range of causes, the marine habitats of the Keys are under heavy stress and deserve all the protection and nurturing they can get. Formal protection of local marine habitats began in 1960 with the establishment of the world's first underwater park, the John Pennekamp Coral Reef State Park off Key Largo. Today the Keys have numerous preservation areas that have been established for conservation and management of local ecological (and some historical) resources, all now encompassed by the

Satellite image of the Lower Florida Keys. A 50-km (30-mile) stretch of the Overseas Highway (U.S. Route 1) is visible, connecting Ramrod Key on the far right to the densely populated island of Key West. The line of outer bank reefs marks the drop-off into the deeper waters of the Florida Straits.

Detail satellite image of Florida Bay, a shallow inner-shelf lagoon at the southern end of the South Florida watershed. This estuarine area, where freshwater from the Everglades mixes with the ocean waters from the Gulf of Mexico and Florida Straits, is characterized by shallow interconnected basins, with an average depth of only 1 m (3 feet). The basins are lined by a network of mangrove islands and mud banks.

largest, the Florida Keys National Marine Sanctuary (FKNMS), established by the U.S. government in 1990. This vast preserve covers about 10,000 km², including the waters surrounding the island chain out to the 91-m (300-foot or 50-fathom) isobath, including Dry Tortugas National Park at the western end and bordering on the Everglades National Park and Biscayne National Park in the east. The FKNMS employs a system of sanctuary-wide regulations combined with special marine zones (about 6% of the sanctuary is set aside as fully protected zones known as ecological reserves, sanctuary preservation areas, and special use areas). In addition, there are 27 wildlife management areas protecting especially sensitive habitats within the FKNMS, as well as numerous areas managed by other agencies. These include a variety of national parks, national wildlife refuges, state parks, and aquatic preserves.

# Florida Keys Bivalves

## Habitats

Bivalves occur in nearly all of the highly diverse aquatic habitats of the Florida Keys. Marine bivalve species are present from the intertidal to great depths and often are characteristic faunal elements of particular habitats, as components of the coral reef and octocoral communities (e.g., *Chama*, *Pteria*, and *Dendostrea*), the mangrove fringe (*Isognomon*), seagrass beds (*Pinctada* and *Argopecten*), sand flats and beaches (Tellinidae), hard ground (*Glycymeris*), sandy and muddy bottom (*Varicorbula*), or intertidal rocks (*Barbatia* and *Ar*-

Long sandy beaches (top), common and popular with collectors in much of peninsular Florida and other nearby regions, are relatively rare in the Keys. The best examples are part of the Long Key and Bahia Honda State Parks and the Dry Tortugas National Park (photo). Empty shells (especially of lucines, tellins, *Modiolus*, etc.) readily accumulate on such beaches by wave action, as do those attached to uprooted seafans (e.g., arks and wing oysters).

This photograph (middle), taken from Ohio Key toward Bahia Honda Bridge, illustrates the three

*copsis*). As a result of massive development activities over the past 100 years, man-made structures are providing ubiquitous settlement opportunities on hard substrata otherwise rarely available in the region and have shifted relative abundance in favor of cementing or byssally attaching forms such as oysters on bridges, causeways, docks, pilings, and lobster traps. Deliberately scuttled vessels that provide artificial reefs for the scuba diving industry are being settled by large-shelled species of *Spondylus* and *Hyotissa*, which otherwise are rare (or even unknown) in the region. Species tolerant of low-salinity, brackish-water conditions (*Mytilopsis, Anomalocardia*) occur in those parts of Florida Bay that receive substantial freshwater runoff from the Everglades, and others (*Polymesoda*) that can cope with low-saline and extremely hypersaline conditions have found a home in occasionally flooding ponds with conditions hostile to most other marine animals. Assemblages of larger empty shells often provide a three-dimensional substratum for other animals (a reason given by management authorities for the ban on shell collecting in some of the Keys' protected areas), and bivalve shell parts, together with those of other mollusks, calcareous algae, foraminiferans, and coral debris, form the vast majority of the "sand" in the Keys. Permanent surface freshwater is limited to a few ponds on the larger rocky islands and the few freshwater bivalves of the region are not treated in this volume.

## History of Exploration and the Florida Keys Molluscan Diversity Project

The relative ease of access to the Keys has made it a popular shell-collecting site and Florida Keys specimens have become widely distributed in private and museum collections. Many scientific and informal publications on mollusks, including the many popular shell books by R. Tucker Abbott, have included Keys taxa. But despite its rich history of popular and professional mollusk collecting, formal scientific inventories have been few. William Stimpson of Chicago's Academy of Sciences made the first comprehensive attempt in the late 1860s. He accumulated all available records and specimens on loan from various institutional collections. His study material included the extensive holdings of the Smithsonian Institution in Washington, D.C., and original collections from the Florida Straits and the Pourtales Terrace obtained by Louis Francois de Pourtales (of the Museum of Comparative Zoology at Harvard University) during the U.S. Coast Survey expeditions of the 1860s. Tragically, all the specimens amassed by Stimpson were lost in the Great Chicago Fire of 1871. Stimpson also lost his nearly completed manuscript and never attempted to recreate the research. In 1883, William Healey Dall renewed the inventory effort, beginning by discussing the results of collecting by amateur conchologist Henry Hemphill and by analyzing the works of British collector James C. Melvill (1881; who reported on material obtained mainly in Key West during 1871–1872) and of amateur conchologist William W. Calkins (1878; who collected in the Keys during the late 1870s). Botanist/conchologist Charles Torrey Simpson (1887–1889) then produced the first effective Florida Keys checklist by including a separate column in his tabulation of Florida mollusks. This included 98 bivalve species names, of which 86 are currently rec-

---

common hard substrata available to attaching bivalves: limestone "beach rock," mangroves, and man-made structures (e.g., bridge abutments and seawalls).

Seagrass beds are home to numerous, mostly infaunal, bivalve species, including pen shells, lucines, and venerids (bottom). Others, such as mussels, arks, and pearl oysters, attach to the plants themselves and to scattered rocks and empty shells among the blades. Individual larger shells can provide a home for other organisms, such as this empty *Atrina* shell used as shelter by a triggerfish.

ognized as valid taxa. Concurrently, Dall published the results of renewed dredging efforts off southern Florida by the U.S. Coast Survey (Dall, 1886, 1889b), culminating in a preliminary species catalog (Dall, 1889a, revised in 1903) that tabulated 225 species from the Florida Keys (plus 15 species now regarded as synonyms and 34 species out of the range of this survey). The next major collecting effort was made by amateur conchologist John B. Henderson, Jr., who sampled the molluscan fauna of the Florida Keys with his private yacht *Eolis* from 1910 to 1916. These annual cruises resulted in massive collections that were donated to the National Museum of Natural History. No comprehensive taxonomic treatment of the *Eolis* expeditions was ever published, although numerous *Eolis* specimens have been cited in scattered scientific papers (an annotated station list was published by Bieler and Mikkelsen in 2003). In 1936, a privately issued checklist by Norman Wallace Lermond, one of New England's foremost naturalists who also explored Florida's mollusks, reported 247 nominal bivalve species for the Keys, 214 of which are here considered valid. No other comprehensive attempt at summarizing the Keys fauna was made until the inception of the Florida Keys National Marine Sanctuary (FKNMS) in the 1990s, when the FKNMS Draft Management Plan provided a listing of 163 Keys bivalve species compiled by William Lyons and James Quinn (1995).

In 1994, the present authors began the Florida Keys Molluscan Diversity Project, an effort to explore the identity, distribution, and biology of all molluscan species in the region. This research involves new field collections as well as critical study of existing literature and museum collections, with the latter two resources allowing the reconstruction of past distribution of species in the Keys. The previously variously delineated "Florida Keys" region was explicitly defined for this project as the waters surrounding the entire island chain from Broad Creek (about 25°21′N, 80°15′W) at the northern end of Key Largo (including Card and Barnes Sounds but not Biscayne Bay, southwest of but not including Old Rhodes Key) to west of the Dry Tortugas (at 83°30′W). The southern half of Florida Bay is included (with a northern border at the levels of, from east to west, the northern end of the Nest Keys, Russell Key, and the northern limit of Rabbit Key Basin), eliminating that part of Florida Bay that is more properly considered the southern extent of the Florida Everglades. Two annotated species lists were published, listing 325 (Mikkelsen & Bieler, 2000) and, after additional research, 389 bivalve species (Bieler & Mikkelsen, 2004b). The latter publication also set a formal depth limit at the 300-m (= 164-fathom or 984-foot) isobath, which eliminated some exclusively deepwater species included in earlier listings. This book treats and illustrates 377 species, reflecting a slight decrease from our earlier census, based on the elimination of, for example, those only living in freshwater and those for which the only records were based on undocumented species lists (for which no supporting specimens were located).

## Using This Book

This book focuses on the bivalves of the Florida Keys, and specimen photographs have been drawn from there whenever possible. The most frequent sources of this material were the American Museum of Natural History (AMNH) and Field Museum of Natural History (FMNH) mollusk collections, including many specimens specifically collected for this project. A "featured species" was selected for each family, and details of hinge, anatomy, and living appearance are provided specifically for that species whenever possi-

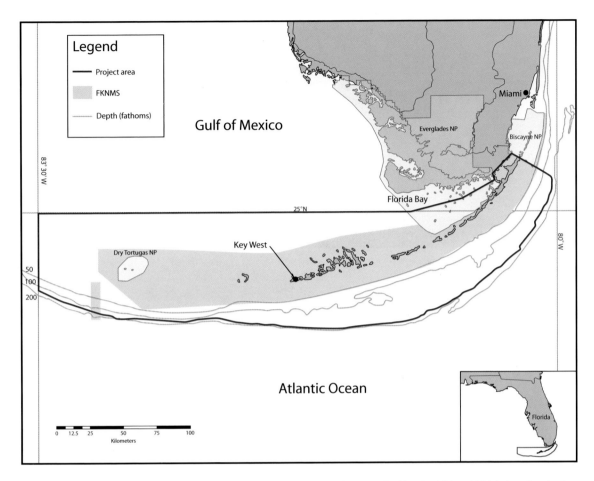

**Legend**

— Project area

▨ FKNMS

-------- Depth (fathoms)

Gulf of Mexico

83° 30' W

25°N

50
100
200

Dry Tortugas NP

Key West

Florida Bay

Everglades NP

Miami ●

Biscayne NP

80° W

Atlantic Ocean

Florida

0   12.5   25        50        75        100
Kilometers

The Florida Keys Molluscan Diversity Project area. Its southern border is marked by the 300-m (164-fathom) isobath. The area of approximately 28,000 km² corresponds to the combined areas of the U.S. states of Massachusetts, Delaware, and Rhode Island, and is only slightly smaller than the country of Belgium. In addition to the Florida National Marine Sanctuary (FKNMS), three national parks (Dry Tortugas, Everglades, and Biscayne) are marked.

ble; in some cases, where these details were not available for the featured species, they are based on another species in the family. Following the featured species, additional species found in the Florida Keys are presented. In a few cases, for which we have credible data attesting to a Florida Keys record, but no specimens readily available, an image from the literature has been reproduced. Image data and photo credits in the back of the book provide detailed information for each set of photographs.

Family descriptions are constructed to be as complete and comparable as possible, especially with regard to soft-body morphology. Some taxa discussed in the family descriptions are not found in the Florida Keys; in the case of species, their geographical location is provided. For some families, scanning electron micrographs are included to better emphasize features not readily seen in light photographs (often with that feature artificially highlighted in green). Technical terms (in SMALL CAPITAL font) used in the descriptions are defined in the Glossary (pp. 415-443).

Most bivalves of the coral reef habitats are cryptic and, when alive, are rarely encountered by snorkelers and divers. Among few exceptions are cementing species of the family Chamidae, such as this coral-embedded specimen of *Chama macerophylla* at Hens and Chickens Patch Reef off Plantation Key (top left).

Most field studies and collecting for this project were done from a small boat, by wading, snorkeling, and scuba

Each set of specimen images includes an external view of the right valve and an internal view of the left valve, unless otherwise noted by RV (right valve) or LV (left valve) designations. Hinge drawings and "transparent clam" anatomical drawings are each presented with the anterior end of the animal to the right. In species where external coloration is demonstrably different on both valves (e.g., Pectinidae) an additional external view of the left valve is provided. Inequivalve species are accompanied by a view of the articulated valves with the smaller valve toward the reader. Other special cases (e.g., Pholadidae and Teredinidae) include extra images of accessory shell plates and other "hard parts." Each set of images is accompanied by a diagnosis that should be sufficient to distinguish the species from similar forms, its known western Atlantic distribution, and size given as approximate anteroposterior length (= "width"), unless otherwise noted, in millimeters of the photographed specimen. If the maximum recorded size from the literature is substantially larger than the photographed specimen, this is indicated by the phrase "to xx mm." Many species that look similar differ substantially in adult size; the reader is encouraged to attend to this information when attempting to identify specimens from this book. Synonyms and former names given here are restricted to those commonly found in popular and regional literature. Comparative statements are provided for species pairs that have been frequently confused in collections and the literature.

Each species account is also accompanied by three icons, reflecting its abundance, depth, and most typical habitat in the Florida Keys:

Abundance: abundant (>50 records); common (11–50 records); rare (<11 records). These categories are based on the records of each species' occurrence in our geographically arranged database of Florida Keys species. A "rare" indication does not necessarily mean that this is an overall rare species, it merely shows that this species in rarely encountered in the Keys (it could be very common elsewhere).

Depth: shallow (known from depths not greater than sampled by conventional scuba diving, i.e., maximally 35 m or 100 feet); deep (greater than conventional scuba depth).

Habitat: sand, mud, or gravel; seagrass, algae, or salt marsh (usually overlying sand or mud); mangrove; seafan (octocorals); rock; wood. These designations point to a typical habitat situation of the species; it is not necessarily exclusive (e.g., a species that usually lives on mangrove roots can alternatively be found on other hard substrata such as rock surfaces).

diving. Deeper-water work involved sampling from larger fishing and research vessel vessels, such as the R/V *Coral Reef II* shown here (top right).

Within the boundaries of the Florida Keys National Marine Sanctuary, sampling was accomplished with small hand tools and vertical grabs, to minimize impact on habitats and living communities. Some dredging on deep sandy bottoms outside the sanctuary boundaries (such as here with a triangle dredge off the Dry Tortugas) provided distribution data for deepwater species (bottom left).

Bycatch (accidentally collected specimens not targeted by primary fishing efforts) of Key West's "pink shrimp" fishing fleet provides many of the commercially traded specimen shells that come from "off Key West" or "off the Dry Tortugas" (middle right).

Just as its northern cousin, the Zebra Mussel (*Dreissena polymorpha*), has done throughout many of the freshwater systems of the United States, the Dark False Mussel (*Mytilopsis leucophaeta*) can be transported attached to boats that are trailered between bodies of water (bottom right). The species is native to the region, but only tolerates low-salinity water. Note the heavy "infestation" of byssally attached mussels under the outboard engine of this boat photographed in the Lower Keys.

Geographic distribution includes: (1) Greenland or Iceland; (2) the eastern North American coastline, beginning with eastern Canada, expressed as northern limit to Florida (e.g, North Carolina to Florida); (3) Bermuda; (4) the Bahama Islands; (5) West Indies (Greater and Lesser Antilles and Windward Islands); (6) Gulf of Mexico, including from western Florida to Texas; (7) Caribbean Central America, including from Mexico (including Yucatan) to Panama; (8) South America ("to" southern limit of range, or individual countries); and (9) extralimital localities (e.g., eastern Pacific). The southernmost extensions here given are in particular need of revisionary work. Ongoing work by South American malacologists indicates that many species deemed to have such wide ranges might in fact be misidentified, unrecognized, cryptic species in South America.

# Bivalve Morphology

The Bivalvia [= Pelecypoda, Lamellibranchiata] are bilateral mollusks lacking a distinct head (also jaws, radular teeth, and cephalic sense organs known from other groups such as gastropods) and usually protected by two shelly valves connected dorsally by a horny ligament. These valves are closed by contraction of muscles attached to the inner surface of the valves, opened by the ligament, and held in position by the hinge teeth and marginal denticles. The dorsal side of bivalves is recognized by location of the hinge and umbones. The mouth, labial palps, and the tip of the foot are anterior, whereas the rectum and siphons (if present) are posterior. Anterior and posterior (along with left and right) are more difficult to determine from empty shells. Posterior can be easily determined in bivalves that have a pallial sinus (on the inner surface) or rostrum associated with siphons; in bivalves without these features, anterior and posterior designation often requires knowledge of the animal.

Several aspects of bivalve morphology have been broadly categorized across the families, and some have been used to define taxonomic groups. The best known of these are the hinge teeth, ligaments, gills, labial palps, siphons, statoliths, and stomachs (e.g., Morton, 1985; Pelseneer, 1891; Purchon, 1987; Ridewood, 1903; Stasek, 1963; Yonge, 1957). Details of these classification systems and other technical terms (below in SMALL CAPITAL font) can be found in the Glossary (pp. 415-443).

Overall bivalve morphology can perhaps be best introduced by examining one of the best understood members of the class, *Mercenaria mercenaria* (Linnaeus, 1758) (Veneridae), the Hard-shelled Clam or Northern Quahog. This species has been harvested or raised in aquaculture for food for hundreds of years within its native range, prompting a very large body of published literature, including much on its morphology and life history (e.g., Jones, 1979; Kellogg, 1903, 1915; Kraeuter & Castagna, 2001).

The adult shell is rounded trigonal in shape, solid, inflated, chalky grayish white, with the umbones anterior of the midline, resulting in a shorter anterior end and longer sloping posterior end (rather ROSTRATE in some specimens). The UMBONES or "beak" (the oldest part of the shell, including the larval shell or PRODISSOCONCH) are PROSOGYRATE over a distinct LUNULE. Exteriorly, shell sculpture consists of crowded, coarse commarginal growth lines that are elevated as erect lamellae anteriorly and posteriorly and lowest, often smooth, at the center of the valve. Posterodorsally, another demarcated near-smooth feature, the ESCUTCHEON, gently slopes toward the posterior end. The LIGAMENT is external and extends from the lunule to approximately half the distance from the umbo to the posterior corner of the valves. The PERIOSTRACUM is thin and generally nonpersistent in the adult. Structurally, the shell is ARAGONITIC calcium carbonate, composed of three layers: an outer composite PRISMATIC layer, a middle CROSSED LAMELLAR layer, and an inner HOMOGENOUS layer.

Internally the shell is porcellaneous white, usually flushed with dark purple near the margins. Muscle attachments to the shell leave distinct scars on the interior. The two largest, the anterior and posterior ADDUCTOR MUSCLE scars, are oval but rendered teardrop-shaped by the adjacent anterior and posterior pedal retractor muscle scars at

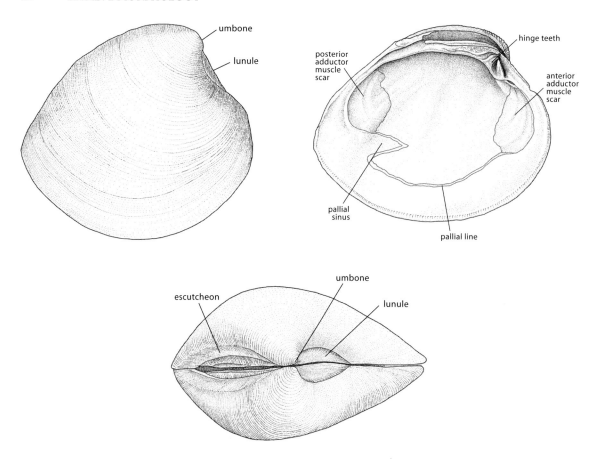

their dorsal edges. The slightly impressed PALLIAL LINE (caused by attachment of the mantle and/or pallial muscles to the shell) extends from the posteroventral points of each adductor muscle scar, paralleling the ventral shell margin. The siphonal retractor muscle (which retracts the short siphons of the clam) creates a triangular, anteriorly pointed PALLIAL SINUS adjacent to the posterior adductor muscle scar. Minute denticles adorn the interior margin of the shell for most of its extent, excluding the siphonal area.

The strong HINGE PLATE at the dorsal side of each valve is HETERODONT, bearing three radiating CARDINAL TEETH just below the umbo. LATERAL TEETH, in other groups often paralleling the shell margin at a distance from the umbones, are absent in *Mercenaria*. A commonly used hinge formula, developed by Bernard (1895) based on ontogenetic development of the teeth, is still in widespread use. In *Mercenaria*, the anterior, middle, and posterior cardinal teeth of the left valve are numbered 2a, 2b, and 4b, respectively. These interlock with the anterior, middle, and posterior cardinal teeth of the right valve, numbered 3a, 1, and 3b, respectively. In *Mercenaria*, the left middle, right middle, and right posterior cardinal teeth are bifid or radially grooved. Paralleling the elongated posterior cardinal tooth in each valve is the NYMPH or attachment area of the ligament.

The soft body of *Mercenaria* shows the prominent muscles that have produced the scars present on the internal shell valves, including the anterior and posterior adductor

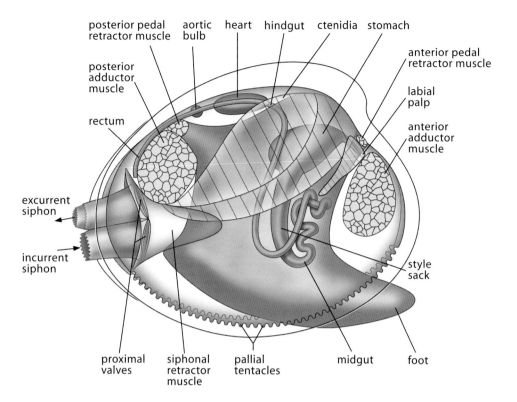

muscles, the pedal retractor muscles, and the trigonal siphonal retractor muscle. MANTLE tissue lines the shell and produces new shell at the margin as the animal grows. In *Mercenaria*, its edges are free (not fused) anteriorly to the ventral point of the siphonal retractor muscle, just posterior to the heel of the foot. The mantle edge is thickened by muscle fibers and has four folds: (1) the outer (which secretes the shell and periostracum), (2–3) the middle, divided in this family into two folds (a smaller outer-middle, which is sensory, and a larger inner-middle, which controls water flow), and (4) the inner (which possibly directs foreign particles out of the MANTLE CAVITY). Posterior EXCURRENT and INCURRENT SIPHONS are short and are formed by fusion of folds 3 and 4 (Yonge's siphonal fusion TYPE B). The larger ventral incurrent and smaller dorsal excurrent siphons are separated for their entire length. The excurrent siphon is equipped with a conical VALVULAR MEMBRANE, and the tip of each siphon is fringed with simple tentacles. The siphons are retracted by trigonal siphonal retractor muscles.

The largest part of the soft body is the visceral mass, which contains the digestive and reproductive systems of the clam and ventrally terminates in the muscular foot. The visceral mass is suspended in the mantle cavity by the anterior and posterior pedal retractor muscles that originate in the muscles of the foot. The FOOT is wedge-shaped, laterally compressed, keeled ventrally, and extended slightly posteriorly as a "heel." Serving as the main tool for burrowing into soft sediment, the foot is composed primarily of muscle fibers. The muscle fibers are irregularly distributed vertically and horizontally, and are interspersed with hemocoelic spaces, which allow expansion when filled. A PEDAL GLAND lies in the midline of the foot, producing the elastic BYSSUS in juvenile *Mercenaria* but becoming inactive in adults.

*Mercenaria mercenaria* is a SUSPENSION or filter feeder, filtering suspended organic particles from the water column through the CTENIDIA (gills). These plicated organs, which also serve in respiration, occupy much of the mantle cavity on either side of the visceral mass. Each side consists of two demibranchs, or double lamellae. Each outer and longer inner demibranch extends from between the LABIAL PALPS, along the ventral edge of the pericardium and kidney, to the base of the siphons. The smaller and shorter outer demibranchs overlap most of their respective inner demibranchs, and together they form a W-shape in cross section on each side of the visceral mass. In addition, supra-axial extensions of the outer demibranchs cover the pericardial space anteriorly to the posterior adductor muscle. Posteriorly the ctenidia connect to the SIPHONAL SEPTUM at the base of the siphons, creating an incurrent INFRABRANCHIAL CHAMBER below the gills and an excurrent SUPRABRANCHIAL CHAMBER above the gills. Structurally, the gill filaments are connected by tissue junctions (EULAMELLIBRANCH gills) and interlamellar septa across the space between lamellae define water tubes within the ctenidia. As water laden with food particles passes through the gills, ciliary currents on the gill surfaces move particles to the food groove at the free edge of each demibranch and then forward to the palps. The anterior ends of the inner demibranchs are fused to the oral groove of the palps (Stasek's labial palp–ctenidia association CATEGORY II). The labial palps, near the anterior adductor muscle at the anterior limit of the gills on either side of the visceral mass, are elongated trigonal structures. Each side comprises two palps, the inner surfaces of which are folded into fine ciliated ridges that further sort and channel food particles collected by the gills into the mouth opening at their center.

The mouth opening leads to a short esophagus, which passes posterodorsally to the STOMACH. The globose stomach lies in the dorsal part of the visceral mass, surrounded anteriorly by lobes of the DIGESTIVE GLAND where assimilation of edible food particles takes place. Internally the stomach includes ciliated sorting areas, ducts to the digestive gland (here mainly concentrated into two outpockets or caeca; Purchon's TYPE V stomach), and an enzyme-packed CRYSTALLINE STYLE secreted by a ventral extension of the stomach, the STYLE SACK. The crystalline style rotates against a chitinous gastric shield covering the interior roof and part of the walls of the stomach. The action of the style against the gastric shield breaks up and begins digestion of edible food particles. A series of ciliated tracts and typhlosoles shuttle small particles into the ducts to the digestive gland, and large inedible particles into the MIDGUT (first portion of the intestine), which is fused to the style sack in this species. Ventral to the style sack, the midgut creates a somewhat variable series of loops in the visceral mass, then runs posterodorsally to exit the visceral mass as the HINDGUT. The hindgut passes through the ventricle of the heart (presumably acting as a stabilizing structure for the contracting ventricle) and past the aortic bulb (an expansion of the posterior aorta), then dorsally over the surface of the posterior adductor muscle to terminate as the rectum near the inner opening of the excurrent siphon.

The heart surrounded by a large pericardium lies at the dorsal midline, posterior to the visceral mass and anterior to the posterior adductor muscle. It consists of a single ventricle and two auricles connected laterally to the ventricle by valved openings. The anterior aorta passes dorsal to the hindgut to supply anterior hemocoelic sinuses. The posterior aorta continues ventral to the hindgut to supply posterior sinuses, interrupted by the muscular, spongy AORTIC BULB posterior to the pericardium. These vessels ultimately feed into numerous hemocoelic spaces throughout the body of the clam. The hemolymph or "blood" contains several types of blood cells capable of phagocytosis. The kidneys lie along the ventral side of the pericardium, communicating with the auricles and emptying into the suprabranchial chamber. Ducts from the gonads within the visceral mass also empty into the suprabranchial chamber in this area.

The nervous system (not shown in figure) is not concentrated: it consists of three main pairs of ganglia, widely separated and joined by long connectives. Fused cerebro-pleural ganglia lie anteriorly near the anterior pedal retractor muscles and innervate the labial palps, the anterior part of the mantle, the visceral mass, pedal retractor muscles, and anterior adductor muscle. Visceral ganglia, lying posteriorly on the anterior face of the posterior adductor muscle near the rectum, innervate the posterior mantle (including the siphons), the ctenidia, kidney, heart, pericardium, and posterior adductor muscle. Pedal ganglia, lying ventrally in the visceral mass near the distal end of the style sack, innervate the foot. The complex system of nerves in *Mercenaria mercenaria* was detailed by Jones (1979).

*Mercenaria mercenaria* is a PROTANDRIC HERMAPHRODITE, with smaller individuals being male and transforming into females as they increase in size. Gametes are spawned freely into the water column. Larval development is PLANKTOTROPHIC, with VELIGER larvae settling after about 12 days. The PRODISSOCONCH or larval shell ranges in diameter from 170 to 240 μm in length. The early benthic juvenile is byssate, with relatively longer siphons and a thinner, more heavily sculptured shell. Growth rate is variable with environmental conditions; growth to marketable length ("littleneck" or 48 mm) can take 15 months in warm southern waters but 4 years in cooler northern waters. Shell layers in cross section show that growth occurs constantly through warm periods, punctuated by growth stoppages in winter. Very large specimens (ca. 15 cm) have been estimated to be at least 40 years old.

*Mercenaria mercenaria* inhabits intertidal to shallow subtidal mud, sandy, or seagrass habitats. It is a relatively rapid burrower, and its adult size and shell thickness aid in deterring most fish and crustacean predators. It is tolerant of a wide range of salinity and water temperature, but requires oceanic conditions for spawning.

The anatomical descriptions and illustrations in this volume illustrate the wide variety of ways that bivalves differ from the *Mercenaria* bauplan, and the reader is referred to those sections for details.

Bivalve shells have provided the primary features for classification, and copious taxonomically valuable shell characters are used in bivalve systematics. Although shape of the shell is the most obvious difference among bivalve taxa, the hinge is of paramount importance, existing in configurations other than the HETERODONT (cardinals + laterals) condition seen in *Mercenaria*. Named hinge types are numerous (see Glossary); notables include the TAXODONT hinge (Arcidae) and the ball-and-socket–like ISODONT hinge (Spondylidae). Many groups have additional structures associated with the hinge, including spoonlike CHONDROPHORES, on which an internal ligament (RESILIUM) is set (see Mactridae), MYOPHORES, on which muscles are attached (see Pholadidae), and CHOMATA or interlocking tubercles (see Gryphaeidae). Other accessory structures are often, although not exclusively, present in EDENTATE or toothless forms. Accessory calcareous structures are present in wood-boring families, such as the PALLETS in Teredinidae, which protect the siphons, and the various accessory shell plates of Pholadidae.

Most marine bivalves are composed of the aragonitic form of calcium carbonate arranged in three layers; details of shell mineralogy vary greatly among bivalve families. CALCITE occurs along with aragonite in some families, including Ostreidae, Pinnidae, and

Limidae, although the two types are almost always segregated. Seven main categories and more than a dozen named shell microstructural varieties have been identified in bivalves (Carter, 1980). The most readily observable variation from *Mercenaria* lies in the innermost layer, which in many families is composed of aragonitic mother-of-pearl or NACRE, most famously present in the pearl oysters (Pteriidae).

Muscle scars on the interior of the shells reflect the position and relative sizes of the muscles of the living bivalve. The most common configuration is as in *Mercenaria*, with anterior and posterior adductors (a condition called DIMYARIAN) of near-equal size (ISOMYARIAN) positioned toward the anterior and posterior extremities of the valves. However, one of the two muscles can be greatly reduced (known as HETEROMYARIAN = differently sized muscles), as in *Pinna* and *Limopsis*, in which the posterior adductor is much larger than the anterior, or in *Nucinella* where the opposite is true. In other bivalves, a central muscle mass (a condition called MONOMYARIAN) can actually be composed of various components. In *Spondylus*, the anterior adductor is absent and the central muscle is derived entirely of the posterior adductor muscle (both pedal retractors are absent in *Spondylus*). In *Malleus* and *Pinctada*, the anterior adductor muscle is minute and the near-central muscle mass is composed of a kidney-shaped posterior adductor with an enlarged posterior pedal retractor muscle nestled in its concavity. In byssate species, literature accounts variably refer to the pedal retractor muscles as byssal or pedo-byssal retractor muscles, implying different origin or function; opinions differ and this distinction is not made here.

The mantle margin is largely free (not fused) in *Mercenaria* anterior of the siphons, leaving a large pedal gape. In other taxa (see Periplomatidae and Hiatellidae), the pedal gape is small and the mantle edges are otherwise fused. In some species of adult Pholadidae, the mantle margins completely fuse and the originally suckerlike foot atrophies. Posterior excurrent and incurrent siphons also exist in a variety of conditions, from short (as in *Mercenaria*) to long and separate (see Tellinidae) to long and united (see Mactridae). Some taxa that appear to have long united siphons actually have an expanded mantle cavity past the posterior limits of the shell (called a POSTVALVULAR EXTENSION, see Teredinidae and Hiatellidae). Other taxa (see Astartidae and Solemyidae) lack siphons, but have excurrent or both incurrent and excurrent apertures controlling water flow. Still other taxa (see Gryphaeidae and Pteriidae) entirely lack defined incurrent or excurrent passages.

The shape of a bivalve's foot is often reflective of its lifestyle. *Mercenaria* is a rather sedentary INFAUNAL burrower, and its wedge- or hatchet-shaped foot is indicative of this task. The byssus is present in most juvenile clams, but is lost in adults of those groups that live in soft sediment. The byssus is retained in species that live associated with hard substrata, and the foot is often correspondingly reduced or modified (for examples, see Isognomonidae and Philobryidae). Bivalves in the subclass Protobranchia, in families such as Nuculidae and Solemyidae, have a broad foot with a planar sole with papillate edges used for rapid movement through or across the substratum. In other sedentary forms, such as cemented Ostreidae or wood-dwelling Teredinidae, the foot is either severely reduced or entirely absent.

The gills of *Mercenaria* are eulamellibranch, defined as having tissue junctions connecting adjacent filaments. Of the several other gill types, the FILIBRANCH gill, found in families such as Arcidae, Pteriidae, and Pectinidae, is also composed of filaments but these are interconnected by cilial junctions only. The PROTOBRANCH gill consists of a close series of flattened leaflets (instead of filaments) connected to one another by cilial junctions. This gill is used primarily for respiration, being less intimately associated with the palps than in clams such as *Mercenaria*. The labial palps in some protobranchs (e.g., *Yoldia* and *Nucula*) are enlarged and have assumed the role of primary feeding organs. Here the palps are equipped with elongated posterior extensions called PALP PROBOSCIDES that agitate the organic-laden surface of the substratum, stirring up particles that are then transported to the mouth along cilial tracts. This process is called DEPOSIT FEEDING (as opposed to suspension feeding in *Mercenaria*), and is also accomplished by the long siphons of tellins. SEPTIBRANCH gills, in Poromyidae and Cuspidariidae, are highly modified sets of filaments restricted to windowlike OSTIA symmetrically positioned in a muscular branchial SEPTUM, which in turn is suspended from the dorsal shell by retractor muscles. The septum divides the mantle cavity into a ventral INFRASEPTAL CHAMBER and a dorsal SUPRASEPTAL CHAMBER (analogous to the infrabranchial and suprabranchial chambers, respectively, of *Mercenaria*). Like the protobranch gill, septibranch gills are used exclusively for respiration—members of the Septibranchia are carnivorous, capturing small crustaceans or polychaete worms using an enlarged scooplike incurrent siphon, and passing them to the trumpetlike mouth (with highly reduced labial palps; for illustration, see Poromyidae). The muscular septum controls internal pressure of the mantle cavity, assisting in prey capture.

Also characteristic of the carnivorous septibranchs is the highly modified Type II stomach defined by Purchon. Many of the features present in the other stomach types are reduced in size or number in Type II, including the sorting areas, ducts to the digestive gland, and style sack. The walls of a Type II stomach are thickened and muscular, and the interior is nearly fully coated with a chitinous lining; both of these modifications facilitate the crushing of prey. Stomach Types I, III, and IV overall similar to *Mercenaria*'s Type V stomach, minus some of the more specialized features. Type I, typical of protobranch bivalves, produces a PROTOSTYLE instead of a crystalline style (differing mainly in lacking enzymes).

Other aspects of morphological variation across the bivalves exist in the path of the hindgut (passing through the heart or not), the aortic bulb (present or absent), PALLIAL EYES (present or absent), and FOURTH PALLIAL APERTURE (present or absent).

# Recent Bivalve Families of the World

The classification of Bivalvia is far from settled. Over time, hundreds of family names have been introduced and various systems have been proposed, often differing widely in composition and arrangement (see Schneider, 2001). The differences and uncertainties stem from various sources: (1) The absence of a well-researched and agreed-upon taxonomic list for bivalves, often resulting in some of the most basic problems (e.g., differences of opinion of which is the older, valid name for a group); (2) the lack of morphological and molecular data for many groups, often resulting in extrapolation of species-specific features to much larger taxonomic units; and (3) different approaches to analyzing available data (e.g., by placing particular weight on individual organ systems, or by particular focus on paleontological or molecular data). The past 10 years have seen a dramatic increase in bivalve research addressing these issues, triggered in part by the new analytical methods of phylogenetic analysis, as well as by the ever-increasing application

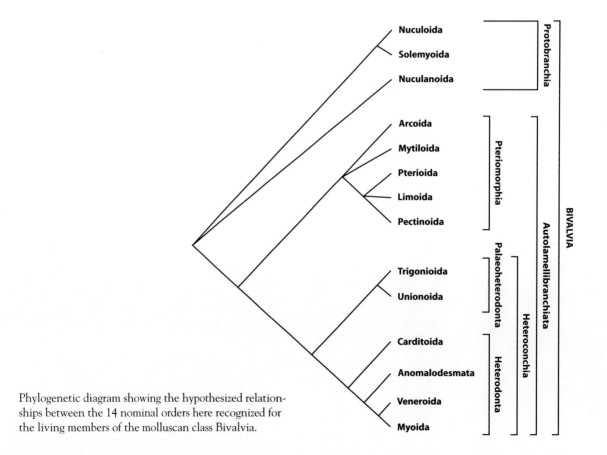

Phylogenetic diagram showing the hypothesized relationships between the 14 nominal orders here recognized for the living members of the molluscan class Bivalvia.

of molecular methods (e.g., Adamkewicz et al., 1997; Campbell, 2000; Campbell et al., 1998; Giribet & Distel, 2003; Giribet & Wheeler, 2002; Steiner & Hammer, 2000). At the same time, efforts are underway to produce a bivalve catalog of names equivalent to that recently published for the Gastropoda (Bouchet & Rocroi, 2005), which will help to provide stability in labeling recognized units.

We leave discussion of the intricacies of evolutionary analyses to other, more technical venues. However, in order to place the Florida Keys fauna into the context of worldwide bivalves, and to clearly delineate our concept of family-level units at this point, we need to present our data within a larger classificatory framework. The system that we are using is a snapshot of our current understanding of bivalve organization and relationships (as outlined by Bieler & Mikkelsen, 2006). It differs from previous systems in various technical details (e.g., recognition of particular authorships or dates of taxa) and composition (e.g., the reorganization of the orders and families belonging to the Heterodonta, or the placement of former family Petricolidae within the Veneridae).

In the following list, we have marked by bold font the families that are represented in the Florida Keys. Author and date citations are as we currently understand them.[1] We have indicated some taxonomic names between class and ordinal levels, but have not assigned ranks (e.g., superorder) to them.

CLASS BIVALVIA Linnaeus, 1758
  PROTOBRANCHIA Pelseneer, 1889
    Order Nuculoida Dall, 1889
      Superfamily Nuculoidea J. E. Gray, 1824
        **Family Nuculidae J. E. Gray, 1824** (see p. 24)
      Superfamily Pristiglomoidea Sanders & Allen, 1973
        Family Pristiglomidae Sanders & Allen, 1973
    Order Solemyoida Dall, 1889
      Superfamily Solemyoidea J. E. Gray, 1840
        **Family Solemyidae J. E. Gray, 1840** (see p. 30)
      Superfamily Manzanelloidea Chronic, 1952
        **Family Manzanellidae Chronic, 1952** (see p. 35)
    Order Nuculanoida Carter, Campbell, & Campbell, 2000
      Superfamily Nuculanoidea H. Adams & A. Adams, 1858 [1854]
        **Family Nuculanidae H. Adams & A. Adams, 1858 [1854]** (see p. 38)
        Family Malletiidae H. Adams & A. Adams, 1858 [1846]
        Family Neilonellidae Schileyko, 1989
        **Family Yoldiidae Dall, 1908** (see p. 44)
        Family Siliculidae Allen & Sanders, 1973
        Family Phaseolidae Scarlato & Starobogatov, 1971
        Family Tindariidae Verrill & Bush, 1897
AUTOLAMELLIBRANCHIATA Grobben, 1894
  Pteriomorphia Beurlen, 1944
    Order Arcoida Stoliczka, 1870

---

[1] Full literature citations are not provided in this context. A project to assemble and publish a fully documented catalog is underway by a team of international collaborators. Nomenclatural decisions herein were guided by the arguments provided by Bouchet and Rocroi (2005: 5–12). Dates given in square brackets indicate that this name has taken the priority of an earlier, replaced name (ICZN, 1999: Art. 40.2.1.).

Superfamily Arcoidea Lamarck, 1809
   **Family Arcidae Lamarck, 1809** (see p. 48)
   Family Cucullaeidae Stewart, 1930
   **Family Noetiidae Stewart, 1930** (see p. 58)
   **Family Glycymerididae Dall, 1908 [1847]** (see p. 62)
Superfamily Limopsoidea Dall, 1895
   **Family Limopsidae Dall, 1895** (see p. 68)
   **Family Philobryidae F. Bernard, 1897** (see p. 74)
Order Mytiloida Férussac, 1822
   Superfamily Mytiloidea Rafinesque, 1815
      **Family Mytilidae Rafinesque, 1815** (see p. 78)
Order Pterioida Newell, 1965
   Superfamily Pterioidea J. E. Gray, 1847 [1820]
      **Family Pteriidae J. E. Gray, 1847 [1820]** (see p. 92)
      **Family Isognomonidae Woodring, 1925 [1828]** (see p. 98)
      **Family Malleidae Lamarck, 1818** (see p. 104)
      Family Pulvinitidae Stephenson, 1941
   Superfamily Ostreoidea Rafinesque, 1815
      **Family Ostreidae Rafinesque, 1815** (see p. 108)
      **Family Gryphaeidae Vyalov, 1936** (see p. 114)
   Superfamily Pinnoidea Leach, 1819
      **Family Pinnidae Leach, 1819** (see p. 120)
Order Limoida Waller, 1978
   Superfamily Limoidea d'Orbigny, 1846
      **Family Limidae d'Orbigny, 1846** (see p. 126)
Order Pectinoida H. Adams & A. Adams, 1857
   Superfamily Pectinoidea Rafinesque, 1815
      **Family Pectinidae Rafinesque, 1815**[2] (see p. 134)
      Family Entoliidae Teppner, 1922
      **Family Propeamussiidae Abbott, 1954** (see p. 148)
      **Family Spondylidae J. E. Gray, 1826**[2] (see p. 154)
   Superfamily Plicatuloidea J. E. Gray, 1857
      **Family Plicatulidae J. E. Gray, 1857** (see p. 158)
   Superfamily Anomioidea Rafinesque, 1815
      **Family Anomiidae Rafinesque, 1815**[2] (see p. 162)
      Family Placunidae Rafinesque, 1815
   Superfamily Dimyoidea P. Fischer, 1886
      Family Dimyidae P. Fischer, 1886
HETEROCONCHIA Hertwig, 1895
   Palaeoheterodonta Newell, 1965
      Order Trigonioida Dall, 1889
         Superfamily Trigonioidea Lamarck, 1819
            Family Trigoniidae Lamarck, 1819
      Order Unionoida Stoliczka, 1870 [freshwater]

---

[2] Some recent authors have credited an article by Wilkes in the *Encyclopaedia Londinensis* (1810) with introducing this family name. However, that work does not consistently apply binominal nomenclature and thus does not fulfill the conditions of ICZN (1999) Art. 11.4; the new names therein are here deemed unavailable.

Superfamily Unionoidea Rafinesque, 1820
    Family Unionidae Rafinesque, 1820
    Family Margaritiferidae F. Haas, 1940
Superfamily Etherioidea Deshayes, 1830
    Family Etheriidae Deshayes, 1830
    Family Hyriidae Swainson, 1840
    Family Mycetopodidae J. E. Gray, 1840
    Family Iridinidae Swainson, 1840

Heterodonta Neumayr, 1883
  Order Carditoida Dall, 1889
    Superfamily Crassatelloidea Férussac, 1822
      **Family Crassatellidae Férussac, 1822** (see p. 166)
      ? Family Cardiniidae Zittel, 1881
      **Family Astartidae d'Orbigny, 1844 [1840]** (see p. 172)
      **Family Carditidae J. Fleming, 1828** (see p. 176)
      **Family Condylocardiidae F. Bernard, 1896** (see p. 182)
  Order Anomalodesmata Dall, 1889
    Family Pholadomyidae King, 1844
    Family Parilimyidae Morton, 1982
    **Family Pandoridae Rafinesque, 1815** (see p. 186)
    **Family Lyonsiidae P. Fischer, 1887** (see p. 192)
    Family Clavagellidae d'Orbigny, 1843
    Family Laternulidae Hedley, 1918
    **Family Periplomatidae Dall, 1895** (see p. 196)
    **? Family Spheniopsidae Gardner, 1928** (see p. 200)
    **Family Thraciidae Stoliczka, 1870 [1830]** (see p. 204)
    Family Myochamidae P. P. Carpenter, 1861
    Family Cleidothaeridae Hedley, 1918
Septibranchia Pelseneer, 1888
    **Family Verticordiidae Stoliczka, 1870** (see p. 208)
    **Family Poromyidae Dall, 1886** (see p. 214)
    **Family Cuspidariidae Dall, 1886** (see p. 220)
  Order Veneroida H. Adams & A. Adams, 1856
    Superfamily Lucinoidea Fleming, 1828
    **Family Lucinidae Fleming, 1828** (see p. 228)
Incertae sedis
    **Family Ungulinidae H. Adams & A. Adams, 1857** (see p. 240)
    **Family Thyasiridae Dall, 1900 [1895]** (see p. 246)
    Family Cyrenoididae H. Adams & A. Adams, 1857 [1853] [freshwater]
    Superfamily Chamoidea Lamarck, 1809
    **Family Chamidae Lamarck, 1809** (see p. 250)
    Superfamily Galeommatoidea J. E. Gray, 1840
      In addition to name-bearing Galeommatidae J. E. Gray, 1840, this poorly re-
      solved group includes numerous nominal family-group names, e.g., Lasaeidae
      J. E. Gray, 1842; Leptonidae J. E. Gray, 1847; Kelliidae Forbes & Hanley,
      1849; Montacutidae W. Clark, 1855; Chlamydoconchidae Dall, 1884;
      Galatheavalvidae Knudsen, 1970; Ephippodontidae Scarlato & Starobogatov,
      1979; Vasconiellidae Scarlato & Starobogatov, 1979; Borniinae F. R. Bernard,

1983; Mysellinae F. R. Bernard, 1983; Orobitellinae F. R. Bernard, 1983; and
Thecodontinae F. R. Bernard, 1983
Florida Keys species herein treated under
**Family Lasaeidae J. E. Gray, 1842** (see p. 258)
Superfamily Hiatelloidea J. E. Gray, 1824
**Family Hiatellidae J. E. Gray, 1824** (see p. 264)
Superfamily Gastrochaenoidea J. E. Gray, 1840
**Family Gastrochaenidae J. E. Gray, 1840** (see p. 268)
Superfamily Arcticoidea R. B. Newton, 1891
Family Arcticidae R. B. Newton, 1891
**Family Trapezidae Lamy, 1920 [1895]** (see p. 274)
Superfamily Glossoidea J. E. Gray, 1847 [1840]
Family Glossidae J. E. Gray, 1847 [1840]
Family Kelliellidae P. Fischer, 1887
Family Vesicomyidae Dall & Simpson, 1901
Superfamily Cyamioidea G. O. Sars, 1878
Family Cyamiidae G. O. Sars, 1878
**? Family Sportellidae Dall, 1899** (see p. 279)
Superfamily Sphaerioidea Deshayes, 1854 [1820]
**Family Corbiculidae J. E. Gray, 1847 [1840]** (see p. 284)
Family Sphaeriidae Deshayes, 1854 [1820] [freshwater]
Superfamily Cardioidea Lamarck, 1809
**Family Cardiidae Lamarck, 1809** (see p. 288)
? Family Hemidonacidae Scarlato & Starobogatov, 1971
Superfamily Veneroidea Rafinesque, 1815
**Family Veneridae Rafinesque, 1815** (see p. 300)
Family Glauconomidae J. E. Gray, 1853
? Family Neoleptonidae Thiele, 1934
Superfamily Tellinoidea Blainville, 1814
**Family Tellinidae Blainville, 1814** (see p. 322)
**Family Donacidae J. Fleming, 1828** (see p. 340)
**Family Psammobiidae J. Fleming, 1828** (see p. 344)
**Family Semelidae Stoliczka, 1870 [1825]** (see p. 350)
**Family Solecurtidae d'Orbigny, 1846** (see p. 358)
Superfamily Solenoidea Lamarck, 1809
Family Solenidae Lamarck, 1809
**Family Pharidae H. Adams & A. Adams, 1856** (see p. 364)
Superfamily Mactroidea Lamarck, 1809
**Family Mactridae Lamarck, 1809** (see p. 368)
Family Anatinellidae Deshayes, 1853
Family Cardiliidae P. Fischer, 1887
Family Mesodesmatidae J. E. Gray, 1840
Superfamily Dreissenoidea J. E. Gray, 1840
**Family Dreissenidae J. E. Gray, 1840** (see p. 374)
Order Myoida Stoliczka, 1870
Superfamily Myoidea Lamarck, 1809
**Family Myidae Lamarck, 1809** (see p. 378)
**Family Corbulidae Lamarck, 1818** (see p. 382)

Family Erodonidae Winckworth, 1932
Superfamily Pholadoidea Lamarck, 1809
   **Family Pholadidae Lamarck, 1809** (see p. 388)
   **Family Teredinidae Rafinesque, 1815** (see p. 396)

# The Florida Keys Bivalves

## Family Nuculidae – Nut Clams

Classification
PROTOBRANCHIA Pelseneer, 1889
Nuculoida Dall, 1889
Nuculoidea J. E. Gray, 1824
Nuculidae J. E. Gray, 1824

**Featured species**
*Nucula proxima* Say, 1822 – **Atlantic Nut Clam**

Obliquely oval, off-white to greenish gray, often with diffuse purple band beneath umbo, glossy smooth with prominent brownish commarginal growth lines, inner shell layer with radially ribbed structure that shows on surface as fine radial striations; interior nacreous white, margin minutely denticulate. Eastern Canada to Florida, Bermuda, Gulf of Mexico, Caribbean Central America. Length 3 mm (to 10 mm). Note: Size, shape, and color of this species vary with environmental conditions, and have led to a number of named forms.

## Family description

The nuculid shell is usually small (including the smallest recorded living bivalves [Moore, 1977], but also to 50 mm), obliquely oval to trigonal or wedge-shaped, solid, with the posterior end truncate and the anterior one longer and rounded. It is EQUIVALVE, inflated, not gaping, INEQUILATERAL (umbones posterior), with OPISTHOGYRATE UMBONES. Shell microstructure is ARAGONITIC and three-layered, with a composite PRISMATIC outer layer, a lenticular NACREOUS or HOMOGENOUS middle layer, and a sheet-nacreous inner layer; TUBULES are apparently absent. Exteriorly nuculids are covered by thick, polished, yellow to green to dark brown, adherent PERIOSTRACUM. Sculpture is usually smooth, or in some species sculptured by various combinations of radial and commarginal striae. Both LUNULE and ESCUTCHEON are usually present but variable in strength and sculpture. [The anterior prolongation of the shell and opisthogyrate umbones of nuculids apparently once caused reverse definition of these two structures; the heart-shaped escutcheon lying below the curved beaks at the shorter end of the shell resembles the more typical lunule of a het-

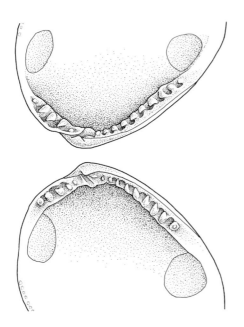

erodont (e.g., Veneridae) and has occasionally been called a "pseudo-lunule" (Schenck, 1934). This is likely the source of the statement by some authors, "true lunule not present."] Interiorly the shell is NACREOUS. The PALLIAL LINE is ENTIRE and obscure. The inner shell margins are usually denticulate, or smooth in some species. The HINGE PLATE is strong and TAXODONT, with an anterior and a posterior series of chevron-shaped teeth separated by a deep, narrow, subumbonal RESILIFER inclined obliquely anteriorly and inwardly projecting. The LIGAMENT is SIMPLE, with an internal portion (RESILIUM).

The animal is ISOMYARIAN; pedal retractor and protractor muscles are well defined. The MANTLE margins are entirely not fused and smooth ventrally; SIPHONS are absent. Large HYPOBRANCHIAL GLANDS are present in the walls of the SUPRABRANCHIAL CHAMBER. The FOOT is large and active with a broad planar sole, a heavily papillate (stellate) margin, and a heel sharply separated from the sole. It is used in locomotion, feeding, and cleansing. A BYSSAL (pedal?) GROOVE is present but the adult is not byssate.

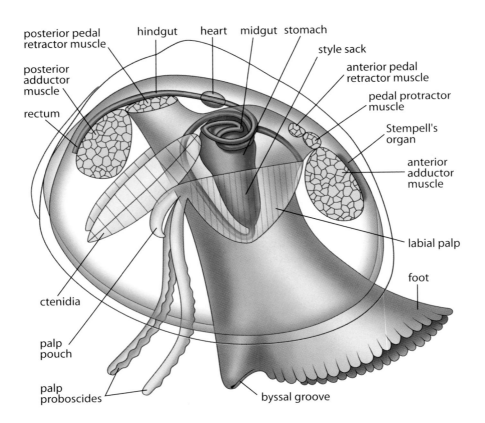

The LABIAL PALPS are large with retractor muscles and elongated ciliated PALP PROBOSCIDES attached posterodorsally. Each palp proboscis sweeps the sediment below the surface, and conveys food particles in mucus strings to the unpaired concave PALP POUCH and thence the mouth via ciliated grooves between the palp lamellae. The palps also par

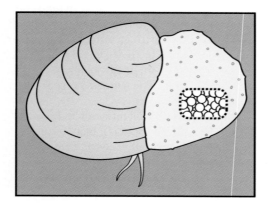

ticipate in sorting food particles brought in by the incurrent water flow. The CTENIDIA are PROTOBRANCH, small to relatively large, somewhat posterior and oblique, and are used mainly for respiration but also to some extent for suspension feeding, passing food particles anteriorly to the palp pouch (there is no direct connection between the ctenidia and palps). Water flow is (primitively) anteroposterior (all other cases of anteroposterior flow in Bivalvia are believed to be secondary). The STOMACH is TYPE I. The MIDGUT is long, wound in numerous irregular coils on the right side of the stomach. The HINDGUT passes either ventral to (in, e.g., *Nucula delphinodonta*) or through (in, e.g., *Nucula proxima*) the ventricle of the heart (which is nearly doubled), and leads to a sessile rectum. Ridges on the interior wall of the rectum in some species produce characteristically grooved fecal pellets. Nuculid blood contains the pigment hemocyanin. Nuculids are GONOCHORISTIC. Most produce PERICALYMMA larvae; some species brood their fertilized eggs in a thin-walled egg case composed of mucus, sand, and other material (diatoms, seagrass fragments, etc.), produced by the hypobranchial glands, attached exteriorly to the shell, and communicating

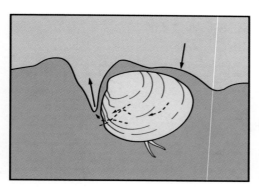

with the mantle cavity. The nervous system is not concentrated, and has separate cerebral and pleural ganglia. STATOCYSTS in adults are open and contain sand grains. ABDOMINAL SENSE ORGANS are absent. A specialized mechanoreceptor (STEMPELL'S ORGAN) is attached to the anterior adductor muscle and detects its contractions.

Nuculids are shallowly INFAUNAL in mud to coarse sand, actively moving through the substrata on the planar foot. Although usually categorized as DEPOSIT FEEDERS, feeding on detritus below the sediment surface using their extensile palp proboscides, nuculids also SUSPENSION FEED, especially as juveniles, resting close beneath the surface with the anterior end of the valves gaping to allow incurrent flow. Experimental removal of the palps in *Nucula sulcata* Bronn, 1831, has shown that adults can also effectively suspension feed.

The family Nuculidae is known since the Ordovician and is represented by 8 living genera and ca. 160 species, inhabiting all seas but most common in the deep sea.

# References

Allen, J. A. 1954. A comparative study of the British species of *Nucula* and *Nuculana*. *Journal of the Marine Biological Association of the United Kingdom*, 33(2): 457–472.

Allen, J. A., and F. J. Hannah. 1986. Reclassification of the Recent genera of the subclass Protobranchia (Mollusca: Bivalvia). *Journal of Conchology*, 32(2): 225–249.

Bergmans, W. 1978. Taxonomic revision of Recent Australian Nuculidae (Mollusca: Bivalvia) except *Ennucula* Iredale, 1931. *Records of the Australian Museum*, 31(17): 673–736.

Drew, G. A. 1899. Some observations on the habits, anatomy and embryology of members of the Protobranchia. *Anatomischer Anzeiger*, 15(24): 493–519.

Drew, G. A. 1901. The life-history of *Nucula delphinodonta* (Mighels). *Quarterly Journal of Microscopical Science*, 44(3): 313–391, pls. 20–25.

Hampson, G. R. 1971. A species pair of the genus *Nucula* (Bivalvia) from the eastern coast of the United States. *Proceedings of the Malacological Society of London*, 39(5): 333–342, pl. 1.

Haszprunar, G. 1985. On the anatomy and fine-structure of a peculiar sense organ in *Nucula* (Bivalvia, Protobranchia). *The Veliger*, 28(1): 52–62.

Heath, H. 1937. The anatomy of some protobranch mollusks. *Mémoires du Musée Royal d'Histoire Naturelle de Belgique*, *Série 2*, 10: 1–26, pls. 1–10.

Moore, D. R. 1977. Small species of Nuculidae (Bivalvia) from the tropical western Atlantic. *The Nautilus*, 91(4): 119–128.

Reid, R. G. B., and D. G. Brand. 1986. Sulfide-oxidizing symbiosis in lucinaceans: implications for bivalve evolution. *The Veliger*, 29(1): 3–24.

Rhind, P. M., and J. A. Allen. 1992. Studies on the deep-sea Protobranchia: the family Nuculidae. *Bulletin of the British Museum (Natural History)*, 58(1): 61–93.

Schenck, H. G. 1934. Classification of nuculid pelecypods. *Bulletin du Musée Royal d'Histoire Naturelle de Belgique*, 10(20): 1–70.

Van de Poel, L. 1955. Structure du test et classification des nucules. *Bulletin de l'Institut Royal des Sciences Naturelles de Belgique*, 31(3): 1–11.

Villarroel, M., and J. Stuardo. 1998. Protobranchia (Mollusca: Bivalvia) Chilenos recientes y algunos fósiles. *Malacologia*, 40(1–2): 113–229.

Verrill, A. E., and K. J. Bush 1897. Revision of the genera of Ledidae and Nuculidae of the Atlantic coast of the United States. *The American Journal of Science*, *Series 4*, 3(13–18): 51–63.

Yonge, C. M. 1939. The protobranchiate Mollusca: a functional interpretation of their structure and evolution. *Philosophical Transactions of the Royal Society of London*, *Series B, Biological Sciences*, 230(566): 79–148.

### *Ennucula aegeensis* (Forbes, 1844) – **Aegean Nut Clam**

Obliquely oval, white, with fine commarginal striae, inner shell layer with radially ribbed structure that shows on surface as fine radial striations; interior margin minutely denticulate. North Carolina, Florida Keys, West Indies, Gulf of Mexico. Length 3 mm.

### *Nucula calcicola* D. R. Moore, 1977 – **Reef Nut Clam**

Obliquely oval, posterior end with nearly vertical truncation, white with thin brownish periostracum, glossy with very weak commarginal ridges crossed by faint radial striae; interior thinly nacreous, margin minutely denticulate. Florida Keys, Bahamas, Caribbean Central America, South America (Colombia). Length < 2 mm.

### *Nucula crenulata* A. Adams, 1856 – **Crenulate Nut Clam**

Obliquely trigonal, yellowish, with numerous fine commarginal riblets, inner shell layer with radially ribbed structure that shows on surface as fine radial striations; interior thinly nacreous, margin finely denticulate. North Carolina to Florida, West Indies, Gulf of Mexico, Caribbean Central America, South America (to Patagonia). Length 5 mm. Note: Also called Atlantic Nut Clam.

### *Nucula delphinodonta* Mighels & C. B. Adams, 1842 – **Dolphin-toothed Nut Clam**

Obliquely trigonal, divided into 3 regions by diverging umbonal slopes, covered by olive brown periostracum, smooth with coarse commarginal growth lines, inner shell layer with radially ribbed structure that shows on surface as fine radial striations; interior nacreous, margin smooth. North Atlantic, eastern Canada to North Carolina, Florida Keys. Length 3 mm.

**A bottom sample,** take by pipe scoop from soft sediment off the Dry Tortugas, is collected, to be sieved and sorted for small infaunal bivalves.

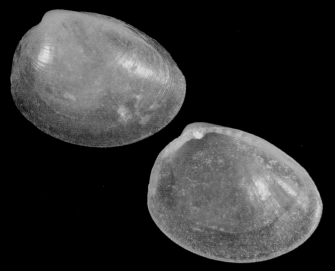

*Ennucula aegeensis*

*Nucula calcicola*

*Nucula crenulata*

*Nucula delphinodonta*

# Family Solemyidae – Awning Clams

Classification
PROTOBRANCHIA Pelseneer, 1889
Solemyoida Dall, 1889
Solemyoidea J. E. Gray, 1840
Solemyidae J. E. Gray, 1840

**Featured species**
*Solemya occidentalis* Deshayes, 1857 – **West Indian Awning Clam**

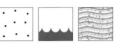

articulated

periostracum removed

Elongated cylindrical with yellowish brown periostracum, smooth with radial rays, internal hinge area with single-pronged buttress directed posteriorly. Florida, Bahamas, West Indies, Gulf of Mexico, Caribbean Central America, South America (Colombia, Brazil). Length 5 mm.

**A living *Solemya occidentalis*** extends its foot, revealing the papillate "stellate disk" of the sole. This species is commonly found in the stomach of its predator, the Bonefish *Albula vulpes* (Linnaeus, 1758).

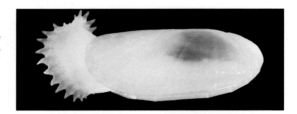

# Family description

The solemyid shell is small to medium-sized (to 100 mm), thin-walled, and anteroposteriorly elongated with parallel dorsoventral margins and rounded ends. It is EQUIVALVE, compressed, and narrowly gaping at both ends. It is INEQUILATERAL (umbones far posterior), with ORTHOGYRATE UMBONES. Shell morphology is highly derived—the shell is weakly calcified especially at the ventral margins, allowing the animal to act as a "flexible tube" that can bend and contract in many directions during burrowing. Shell microstructure is ARAGONITIC and two-layered, with a PRISMATIC outer layer and a HOMOGENOUS inner layer; the inner layer is absent anterior to the umbones, and only the outer layer persists near the margins. Organic content of the shell is very high, comprising two types of conchiolin. TUBULES are apparently absent. Exteriorly solemyids are whitish but covered by a thick, four-layered, glossy, dark brown to greenish PERIOSTRACUM in alternating thin

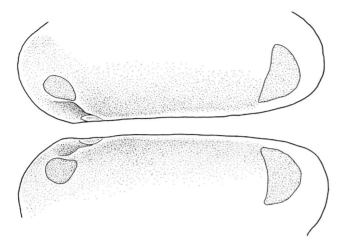

and thick rays or pleats that project beyond the ventral shell margins to form a frill or awning (hence the common name). The periostracum is usually frayed and very brittle when dried. Sculpture is smooth or with irregular, low radial furrows corresponding to the periostracal rays. LUNULE and ESCUTCHEON are absent. Interiorly the shell is non-NACREOUS. The PALLIAL LINE is ENTIRE. The inner shell margins are smooth. The HINGE PLATE is weak and EDENTATE in adults, but thickened buttresses or "horns" are usually present interiorly near the umbones. The LIGAMENT is PARIVINCULAR, set in NYMPHS, and is functionally OPISTHODETIC, but appears AMPHIDETIC by the anterior extension of its outermost layer.

The animal is slightly HETEROMYARIAN (posterior ADDUCTOR MUSCLE slightly smaller), with prominent pedal retractor, elevator, and protractor muscles. The MANTLE margins are extensively fused ventrally, leaving a large anteroventral pedal gape and a small posterior EXCURRENT APERTURE surrounded by tentacles. Crossed mantle margin fibers near the posterior end of the pedal gape resemble (and could be analogous to) the CRUCIFORM MUSCLE of tellinoideans. HYPOBRANCHIAL GLANDS are present in the walls of the SUPRABRANCHIAL CHAMBER. The FOOT is large and active with a broad planar sole and usually a heavily papillate (stellate) margin. A PEDAL GLAND (interpreted as a byssal gland by some authors) is present; the adult is not byssate.

The LABIAL PALPS (which according to some authors are homologous not with true palps but with the palp proboscides of other protobranchs) consist of single, minute, unridged flaps on each side that rest on the anterior edge of the ctenidia near the ends of the anterior food currents. PALP PROBOSCIDES are absent. The CTENIDIA are PROTOBRANCH,

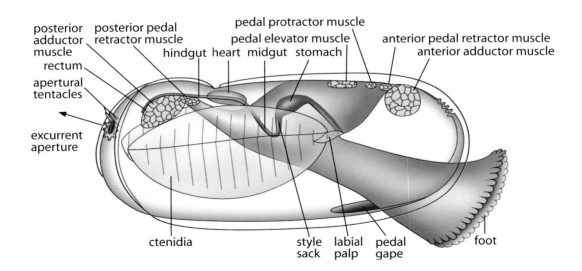

**_Solemya velum_** Say, 1822 – **Atlantic Awning Clam**

Elongated cylindrical with yellowish brown periostracum, internal hinge area with two-pronged buttress. Eastern Canada to Florida. Length 16 mm (to 25 mm).

relatively large, thickened, and are used for both respiration and suspension feeding. Water flow is anteroposterior. Endosymbiotic chemautotrophic (sulfide-oxidizing) bacteria in the fleshy ctenidial filaments provide added nutrition to solemyids, which normally inhabit sulfur-rich anoxic sediments. As a result of at least partial dependence on this supplemental source, the alimentary system is secondarily reduced; a few species (e.g., _Solemya reidi_ Bernard, 1980, from the northeastern Pacific) lack alimentary systems entirely, relying exclusively on chemautotrophy. When present, the small STOMACH is a simplified TYPE I, lacking a gastric shield and sorting areas. The MIDGUT is not coiled, and the HINDGUT, which passes through the ventricle of the large heart, is especially narrow and leads to a sessile rectum. Hemoglobin is found in the ctenidial tissues of some species (e.g., _Solemya velum_). Solemyids are GONOCHORISTIC, with equal cleavage (i.e., the first two cleavages of the egg produce four equal blastomeres—this is the only known family in Bivalvia exhibiting this) and a benthic PERICALYMMA larva developing into a crawl-away hatchling. The nervous system is not concentrated, and has separate cerebral and pleural ganglia. STATOCYSTS in adults are open. ABDOMINAL SENSE ORGANS are absent.

Solemyids are INFAUNAL in soft anaerobic mud or sand with high organic content. They thus thrive in reducing sediments such as sewage outfalls and in mangrove channels in the Florida Keys. Smaller-bodied species live anterior-end downward in relatively deep, closed, U- or Y-shaped burrows, lined by mucus secreted by the foot and without direct communication with the sediment surface; the animal grips the sides of the burrow with its flaring planar foot. Larger species lie at an oblique angle just below the surface, usually in a small depression. Most if not all species have endosymbiotic bacteria upon which they depend to some extent for nutrition; sediment rich in hydrogen sulfide is brought into the anterior MANTLE CAVITY by the foot. As such they can be considered highly modified DEPOSIT FEEDERS. The planar foot also allows active movement through the substrata. "Swimming" movements, reported from laboratory observations, have been interpreted as a relocation mechanism or as mere artifacts. Short forward-directed dashes are achieved by extending the foot to allow water to enter the mantle cavity through the pedal gape, then sealing the mantle edges around the foot and rapidly adducting the valves, expelling a jet of water out the excurrent aperture. Solemyids are also notably water repellent (i.e., they float in the laboratory dish) as a result of a lipoprotein secreted by tubular oil glands at the anterior and posterior limits of the mantle margin; the function of this oil has been suggested as providing the ability to shed the sticky mud of their typical habitat. The unusual physiology of solemyids makes them candidates for studying bivalve–bacteria associations as well as environmental indicators of anoxic conditions (perhaps indicating pollution).

The family Solemyidae is known since the Ordovician and is represented by 3 living genera and ca. 30 species, inhabiting all seas and all depths except the polar regions.

## References

Beedham, G. E., and G. Owen. 1965. The mantle and shell of *Solemya parkinsoni* (Protobranchia: Bivalvia). *Proceedings of the Zoological Society of London*, 144(3): 405–430, pl. 1.

Conway, N. M., B. L. Howes, J. E. McDowell Capuzzo, R. D. Turner, and C. M. Cavanaugh. 1992. Characterization and site description of *Solemya borealis* (Bivalvia; Solemyidae), another bivalve–bacteria symbiosis. *Marine Biology*, 112(4): 601–613.

Dall, W. H. 1908. A revision of the Solenomyacidae. *The Nautilus*, 22(1): 1–2.

Drew, G. A. 1900. Locomotion in *Solemya* and its relatives. *Anatomischer Anzeiger*, 17(15): 257–266.

Gustafson, R. G., and R. A. Lutz. 1992. Larval and early post-larval development of the protobranch bivalve *Solemya velum* (Mollusca: Bivalvia). *Journal of the Marine Biological Association of the United Kingdom*, 72(2): 383–402.

Morse, E. S. 1913. Observations on living *Solenomya*. *The Biological Bulletin*, 25(4): 261–281.

Owen, G. 1961. A note on the habits and nutrition of *Solemya parkinsoni* (Protobranchia: Bivalvia). *Quarterly Journal of Microscopical Science*, 102(1): 15–21, 2 pls.

Pojeta, J., Jr. 1988. The origin and Paleozoic diversification of solemyoid pelecypods. Pages 201–271, in: D. L. Wolberg, comp., *Contributions to Paleozoic Paleontology and Stratigraphy in Honor of Rousseau H. Flower*. New Mexico Bureau of Mines and Mineral Resources, Memoir 44.

Reid, R. G. B. 1980. Aspects of the biology of a gutless species of *Solemya* (Bivalvia: Protobranchia). *Canadian Journal of Zoology*, 58(3): 386–393.

Villarroel, M., and J. Stuardo. 1998. Protobranchia (Mollusca: Bivalvia) Chilenos recientes y algunos fósiles. *Malacologia*, 40(1–2): 113–229.

Vokes, H. E. 1955. Notes on Tertiary and Recent Solemyacidae. *Journal of Paleontology*, 29(3): 534–545.

Yonge, C. M. 1939. The protobranchiate Mollusca: a functional interpretation of their structure and evolution. *Philosophical Transactions of the Royal Society of London, Series B, Biological Sciences*, 230(566): 79–148.

# Family Manzanellidae – Little Nut Clams

**Classification**
PROTOBRANCHIA Pelseneer, 1889
Solemyoida Dall, 1889
Manzanelloidea Chronic, 1952
Manzanellidae Chronic, 1952

**Featured species**
***Nucinella woodii*** (Dall, 1898) – **Wood's Little Nut Clam**

 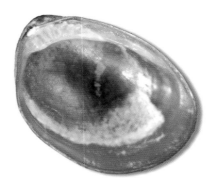

Obliquely oval with swollen umbones, with periostracum orange-yellow blending to green at margin, smooth. Florida Keys. Length 2 mm. Note: The specimens from the Florida Keys are indistinguishable from eastern Pacific specimens of *Nucinella subdola* (Strong & Hertlein, 1937) (P. Valentich-Scott, pers. comm., October 2003). *Nucinella woodii* was originally described from the Florida Pliocene. This genus is in need of taxonomic revision.

## Family description

The manzanellid shell is very small (to 25 mm, usually less than 12 mm), oval to trigonal, usually higher than long, and thin-walled. It is EQUIVALVE, inflated, not gaping, and INEQUILATERAL (umbones posterior), with weakly OPISTHOGYRATE UMBONES. Shell microstructure is ARAGONITIC. TUBULES have not been reported. Exteriorly manzanellids are covered by light olive-yellow to brown, thin, varnishlike PERIOSTRACUM. Sculpture is smooth or sculptured with fine commarginal striae. LUNULE and ESCUTCHEON are absent. Interiorly the shell is non-NACREOUS. The PALLIAL LINE is ENTIRE. The inner shell margins are smooth. The HINGE PLATE is short, strong, and TAXODONT, with a few robust peg- or chevron-shaped teeth (called CARDINAL TEETH by some authors) on each side of the umbo. A prominent anterior LATERAL TOOTH has been reported in some species. The LIGAMENT is submarginal, SIMPLE, OPISTHODETIC, and in some is recessed into FOSSETTES.

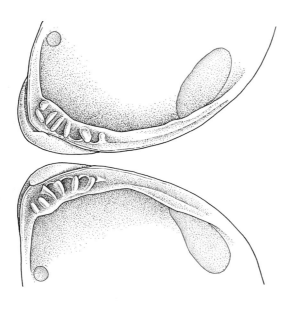

The animal is MONOMYARIAN or extremely HETEROMYARIAN (posterior ADDUCTOR MUSCLE small or absent). This is an unusual configuration for a monomyarian bivalve in which most examples (e.g., Pectinidae and Ostreidae) have retained only the posterior adductor muscle, which has become centralized; neither of these conditions is true for manzanellids. Both anterior and posterior pedal retractor muscles are well defined. The posterior pedal retractor muscles are conjoined just before inserting on the inner shell surface, thus assuming partial function of the absent posterior adductor muscle. The MANTLE margins are not fused, in some cases with a few pallial tentacles; SIPHONS are absent. HYPOBRANCHIAL GLANDS have not been reported. The FOOT is large with a broad planar sole, with a heavily papillate (stellate) margin; the adult is not byssate.

The LABIAL PALPS are minute and unridged; PALP PROBOSCIDES are absent. The CTENIDIA are PROTOBRANCH and relatively large. Water flow is presumed to be anteroposterior. The alimentary system is reduced or absent, and nutrition could be supplemented

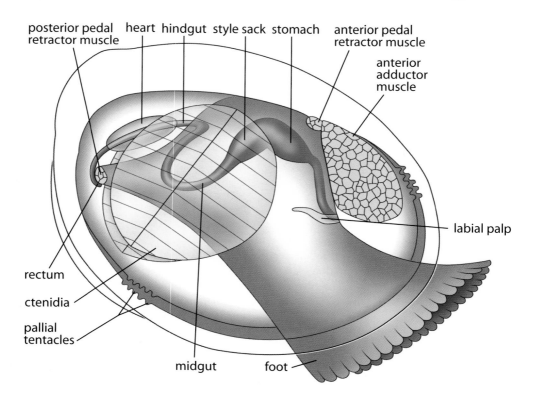

by suspected (but not proven) endosymbiotic chemautotrophic bacteria. The STOMACH is a simplified TYPE I, with only a single aperture to the digestive diverticula and without a dorsal hood. The MIDGUT is not coiled. The HINDGUT passes through the ventricle of the large heart, and leads to a sessile rectum. Nothing is known about the reproductive system or reproduction of manzanellids. The nervous system is not concentrated, and has separate cerebral and pleural ganglia. STATOCYSTS in adults are open. ABDOMINAL SENSE ORGANS are absent.

Manzanellids are active burrowers through soft substrata. Extremely small (1 mm) forms are known from cave habitats. Little else is known about their lifestyle.

The family Manzanellidae is known since the Permian (*Manzanella*) and is represented by 2 living genera (*Nucinella* and *Huxleya*) and ca. 20 species, inhabiting all seas but especially lower latitudes and the deep sea.

### References

Allen, J. A., and H. L. Sanders. 1969. *Nucinella serrei* Lamy (Bivalvia: Protobranchia), a monomyarian solemyid and possible living actinodont. *Malacologia*, 7(2–3): 381–396.

Lamy, E. 1912. Sur le genre *Pleurodon* ou *Nucinella* S. Wood, avec description d'une espèce nouvelle. *Bulletin du Muséum National d'Histoire Naturelle*, 18(7): 429–433.

La Perna, R. 2005. A gigantic deep-sea Nucinellidae from the tropical West Pacific (Bivalvia: Protobranchia). *Zootaxa*, 881: 1–7.

Pojeta, J., Jr. 1988. The origin and Paleozoic diversification of solemyoid pelecypods. Pages 201–271, in: D. L. Wolberg, compiler, *Contributions to Paleozoic Paleontology and Stratigraphy in Honor of Rousseau H. Flower*. New Mexico Bureau of Mines and Mineral Resources, Memoir 44.

Vokes, H. E. 1956. Notes on the Nucinellidae (Pelecypoda) with description of a new species from the Eocene of Oregon. *Journal of Paleontology*, 30(3): 652–671.

# Family Nuculanidae – Pointed Nut Clams

### Classification
PROTOBRANCHIA Pelseneer, 1889
Nuculanoida Carter, Campbell, & Campbell, 2000
Nuculanoidea H. Adams & A. Adams, 1858 [1854]
Nuculanidae H. Adams & A. Adams, 1858 [1854]

**Featured species**
*Propeleda carpenteri* (Dall, 1881) – **Carpenter's Smooth Nut Clam**

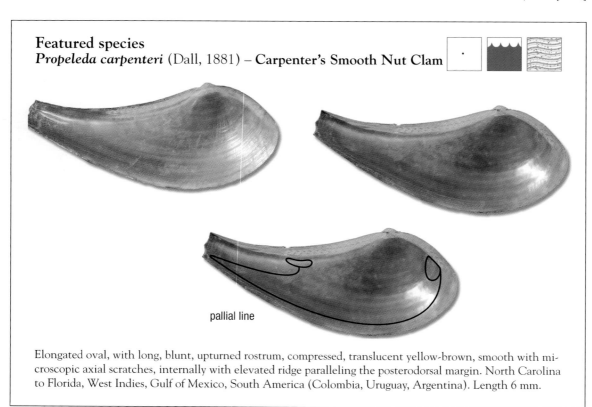

pallial line

Elongated oval, with long, blunt, upturned rostrum, compressed, translucent yellow-brown, smooth with microscopic axial scratches, internally with elevated ridge paralleling the posterodorsal margin. North Carolina to Florida, West Indies, Gulf of Mexico, South America (Colombia, Uruguay, Argentina). Length 6 mm.

# Family description

The nuculanid shell is small to medium-sized (to 70 mm), thin-walled, elongated oval, with the longer posterior end usually ROSTRATE. It is EQUIVALVE, compressed to gaping posteriorly at the siphonal opening, and INEQUILATERAL (umbones anterior), with ORTHOGYRATE UMBONES. Shell microstructure is ARAGONITIC and two-layered, with an irregular PRISMATIC or HOMOGENOUS outer layer, and COMPLEX CROSSED LAMELLAR, CROSSED LAMELLAR, or homogenous inner layer. TUBULES are apparently absent. Exteriorly nuculanids are usually covered by yellow to dark brown, thin, varnishlike PERIOSTRACUM. Sculpture is smooth in some cases, but most often with predominantly commarginal (rarely SCISSU-

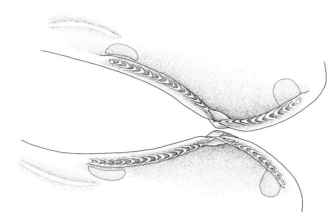

LATE) threads or ridges. LUNULE and ES-CUTCHEON are present or absent. Interiorly the shell is non-NACREOUS. The PALLIAL LINE is weakly impressed with a small to large PALLIAL SINUS. The inner shell margins are smooth. The HINGE PLATE is strong and TAXODONT, arched, with anterior and posterior series of chevron-shaped teeth separated by a subumbonal RESIL-IFER. The LIGAMENT is SIMPLE and submarginal, with an internal portion (RESILIUM).

The animal is ISOMYARIAN or HET-EROMYARIAN (posterior ADDUCTOR MUSCLE smaller). The pedal retractor and protractor muscles are well defined. The MANTLE margins are not fused ventrally. Posterior EXCURRENT and INCURRENT SIPHONS are short, usually united, and either complete or incomplete (i.e., made tubular only by ciliary junctions). A posterior unpaired siphonal tentacle (attached to the right or left mantle margin) and marginal sense organs on the anteroventral mantle edge are present in some species. HYPOBRANCHIAL GLANDS in the walls of the suprabranchial chamber are small or absent. The FOOT is large and active with a broad planar sole, with a heavily papillate (stellate) margin, and in some cases a distinct heel. A BYSSAL (pedal?) GLAND and GROOVE can be present but the adult is not byssate.

The LABIAL PALPS are elongated and narrow with retractor muscles, terminal filaments, and elongated ciliated PALP PROBOSCIDES attached posterodorsally. A PALP POUCH is either small or absent. There is no connection between palps and ctenidia, and any food particle transfer must involve the palp proboscides. The CTENIDIA are PROTO-BRANCH, posterior, oriented horizontally, and can be adjusted to form a SEPTUM separating

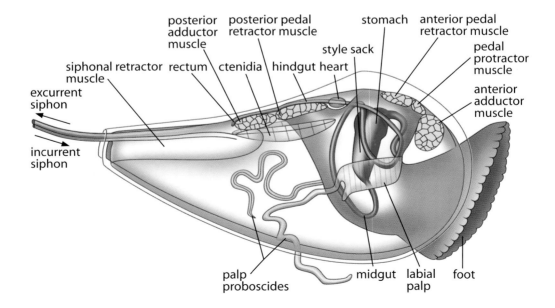

INFRA- and SUPRABRANCHIAL CHAMBERS, perforated by slitlike pores. Incurrent and excurrent water flows are posterior. The STOMACH is TYPE I. The MIDGUT is long and variously convoluted. The HINDGUT passes through the ventricle of the heart, and leads to a sessile rectum. Nuculanids are GONOCHORISTIC; a PERICALYMMA larva is assumed. The nervous system is not concentrated. STATOCYSTS in adults are closed, with STATOLITHS. ABDOMINAL SENSE ORGANS are absent.

Nuculanids are INFAUNAL in mud and sand of high organic content, lying partly buried with the rostrum protruding just above the surface. The planar foot allows plowing and rapid burrowing. Although primarily DEPOSIT FEEDERS through action of the palp proboscides, rhythmic contractions by the gills have been described as setting up a pumping action that can create water flow sufficient for supplemental SUSPENSION FEEDING. One species from hydrothermal vents is known to possess sulfide-oxidizing endosymbiotic bacteria.

The family Nuculanidae is known since the Devonian and is represented by at least 7 living genera and 200–250 species, inhabiting all seas but most common in the deep sea.

*Ledella sublevis* A. E. Verrill & Bush, 1898 – **Polished Nut Clam**

Elongated oval, with short roundly pointed rostrum, whitish, smooth. Massachusetts to Florida, Bermuda, West Indies, South America (Uruguay). Length 4 mm.

## References

Allen, J. A., and F. J. Hannah. 1989. Studies on the deep sea Protobranchia: the subfamily Ledellinae (Nuculanidae). *Bulletin of the British Museum (Natural History), Zoology*, 55(2): 123–171.

Allen, J. A., and H. L. Sanders. 1996. Studies on deep-sea Protobranchia (Bivalvia): the family Neilonellidae and the family Nuculanidae. *Bulletin of The Natural History Museum (London), Zoology Series*, 62(2): 101–132.

Dell, R. K. 1955. A synopsis of the Nuculanidae with check lists of the Australasian Tertiary and Recent species. *Records of the Dominion Museum*, 2(3): 123–134.

Heath, H. 1937. The anatomy of some protobranch mollusks. *Mémoires du Musée Royal d'Histoire Naturelle de Belgique, Série 2*, 10: 1–26, pls. 1–10.

Kilburn, R. N. 1994. The protobranch genera *Jupiteria, Ledella, Yoldiella* and *Neilo* in South Africa, with the description of a new genus. *Annals of the Natal Museum*, 35: 157–175.

Verrill, A. E., and K. J. Bush. 1897. Revision of the genera of Ledidae and Nuculidae of the Atlantic coast of the United States. *The American Journal of Science, Series 4*, 3(13–18): 51–63.

Yonge, C. M. 1939. The protobranchiate Mollusca: a functional interpretation of their structure and evolution. *Philosophical Transactions of the Royal Society of London, Series B, Biological Sciences*, 230(566): 79–148.

### *Nuculana acuta* (Conrad, 1832) – **Pointed Nut Clam**

Elongated trigonal, with sharply pointed rostrum, white with light yellow periostracum, with shallow groove from umbo to ventral margin at both ends of valve, with coarse subequal commarginal ridges extending across posterior ridge. Eastern Canada to Florida, West Indies, Gulf of Mexico, Caribbean Central America, South America (to Brazil). Length 6 mm (to 10 mm).

### *Nuculana concentrica* (Say, 1824) – **Concentric Nut Clam**

Elongated trigonal, with sharply pointed, upturned rostrum, semiglossy yellowish white, with shallow groove from umbo to ventral margin at both ends of valve, with fine commarginal grooves not extending across posterior ridge, smooth at umbo and dorsal center. Florida, Gulf of Mexico, Caribbean Central America, South America (to Brazil). Length 7 mm (to 18 mm).

### *Nuculana jamaicensis* (d'Orbigny, 1853) – **Jamaican Nut Clam**

Elongated trigonal, with sharply pointed rostrum, white, with strong rounded ridge on posterior slope from umbo to tip of rostrum and with radial furrow on anterior slope, with strong regular commarginal sculpture. North Carolina, Florida Keys, West Indies, South America (Suriname). Length 3 mm. Note: The sepia-toned drawings are the original illustrations of the species.

### *Nuculana cf. semen* (E. A. Smith, 1885) – **Seed Nut Clam**

Elongated oval, with short, roundly pointed rostrum that is slightly hooked (indented ventrally), glossy translucent yellowish white, smooth, umbones slightly opisthogyrate. Florida Keys. Length 2 mm. Note: This deepwater species is closest in shape and adult size to *Nuculana semen* (E. A. Smith, 1885), known only from Brazil; its identification has not been confirmed.

### *Nuculana verrilliana* (Dall, 1886) – **Barbed Nut Clam**

Elongated trigonal, with bluntly pointed rostrum, whitish, with widely spaced slightly elevated commarginal ridges, reflexed at anterior end, which is defined by a rounded indistinct ridge and is more strongly sculptured (even when worn) than the remaining shell. North Carolina to Florida, West Indies, Gulf of Mexico. Length 5 mm.

### *Nuculana vitrea* (d'Orbigny, 1853) – **Glassy Nut Clam**

Elongated trigonal, with sharply pointed, upturned rostrum, white, with coarse regular commarginal ridges. Florida, West Indies, South America (Colombia). Length 6 mm. Note: The sepia-toned drawings are the original illustrations of the species.

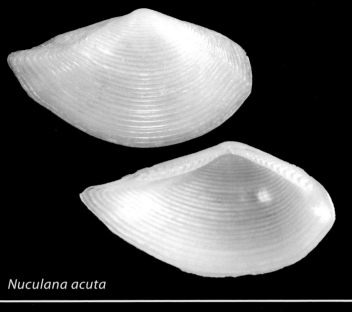

*Nuculana acuta*

*Nuculana concentrica*

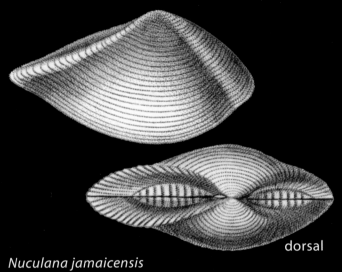

dorsal

*Nuculana jamaicensis*

*Nuculana* cf. *semen*

*Nuculana verrilliana*

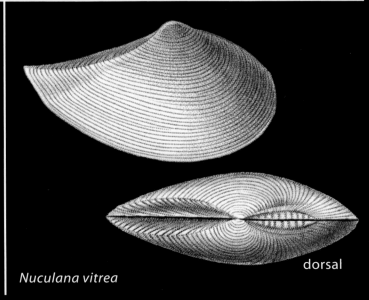

dorsal

*Nuculana vitrea*

# Family Yoldiidae — Yoldia Clams

**Classification**
PROTOBRANCHIA Pelseneer, 1889
Nuculanoida Carter, Campbell, & Campbell, 2000
Nuculanoidea H. Adams & A. Adams, 1858 [1854]
Yoldiidae Dall, 1908

**Featured species**
*Yoldia liorhina* Dall, 1881 – **Smooth-nosed Yoldia**

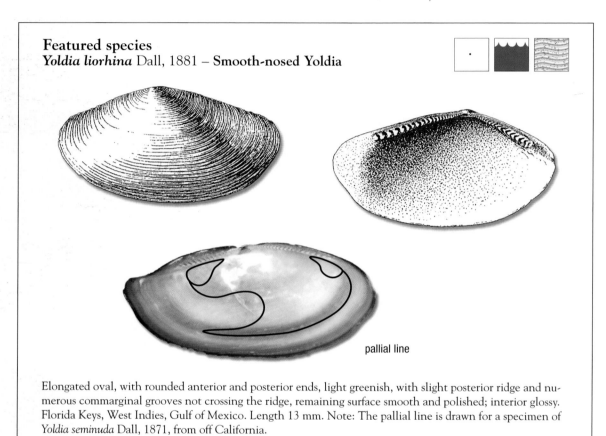

pallial line

Elongated oval, with rounded anterior and posterior ends, light greenish, with slight posterior ridge and numerous commarginal grooves not crossing the ridge, remaining surface smooth and polished; interior glossy. Florida Keys, West Indies, Gulf of Mexico. Length 13 mm. Note: The pallial line is drawn for a specimen of *Yoldia seminuda* Dall, 1871, from off California.

# Family description

The yoldiid shell is small to medium-sized (to at least 60 mm), thin-walled, elongated oval, with the posterior end extended or slightly ROSTRATE. It is EQUIVALVE, usually compressed, usually gaping at both ends, and EQUILATERAL or INEQUILATERAL (umbones slightly anterior or posterior), with PROSOGYRATE UMBONES. Shell microstructure is ARAGONITIC and three-layered, with a PRISMATIC outer layer, a CROSSED LAMELLAR or HOMOGENOUS middle layer, and a COMPLEX CROSSED LAMELLAR or homogenous inner layer (in

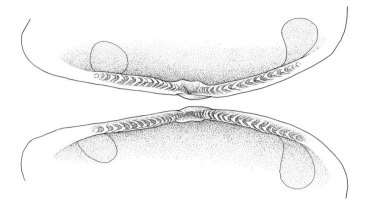

some species also with an innermost prismatic layer). TUBULES are apparently absent. Exteriorly yoldiids are covered by brownish to olive green, varnishlike PERIOSTRACUM. Sculpture is usually smooth and glossy, with fine commarginal growth lines. LUNULE and ESCUTCHEON are present or absent. Interiorly the shell is non-NACREOUS. The PALLIAL LINE has a deep SINUS. The inner shell margins are smooth. The HINGE PLATE is TAXODONT, arched, with anterior and posterior series of chevron-shaped teeth separated by a subumbonal RESILIFER. The LIGAMENT is SIMPLE, AMPHIDETIC, in some species submarginal set on FOSSETTES, and can be extended by periostracum; an internal portion (RESILIUM) is entirely mineralized, and splits in half dorsally with growth.

The animal is ISOMYARIAN; the pedal retractor and protractor muscles are well defined. The MANTLE margins are not fused ventrally. Posterior EXCURRENT and INCURRENT SIPHONS are short and united; the incurrent siphon is either complete or incomplete (i.e., made tubular only by ciliary junctions). A posterior unpaired siphonal tentacle (attached to the right or left mantle margin) and marginal sense organs on the anteroventral mantle edge are present. HYPOBRANCHIAL GLANDS in the suprabranchial chamber are small or

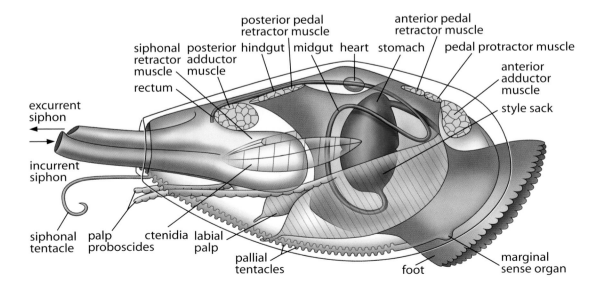

absent. The FOOT is large and active with a broad planar sole, with a heavily papillate (stellate) margin. A BYSSAL (pedal?) GLAND and GROOVE is present but the adult is not byssate.

The LABIAL PALPS are large with retractor muscles, terminal filaments, and narrow ciliated PALP PROBOSCIDES attached posterodorsally. A PALP POUCH is absent. The CTENIDIA are PROTOBRANCH, small to medium-sized, posterior, oriented horizontally, and can

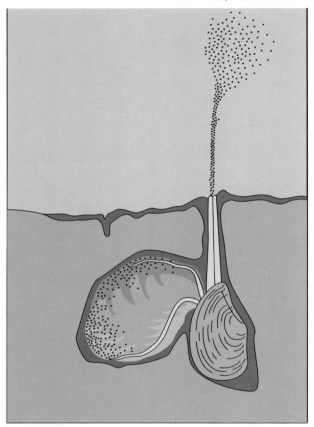

rearrange to form a SEPTUM separating INFRA- and SUPRABRANCHIAL CHAMBERS, perforated by slitlike pores. Incurrent and excurrent water flows are posterior. The STOMACH is TYPE I. The MIDGUT is long and convoluted. The HINDGUT passes through the ventricle of the heart, and leads to a sessile rectum. Yoldiids are GONOCHORISTIC and produce PERICALYMMA larvae. The nervous system is not concentrated, and has fused cerebropleural ganglia (unlike most other protobranch bivalves). STATOCYSTS (closed, with STATOLITHS) have been reported in adult *Yoldia*. ABDOMINAL SENSE ORGANS are absent.

Yoldiids have been biologically investigated more than other protobranchs because they are common, large, and play major roles in the food chains of commercial fish and in bioturbation of the sediment. They are INFAUNAL in mud or sand, and are strong and rapid burrowers. Leaping (under laboratory conditions) was described by Drew (1899b). They DEPOSIT FEED using the palp proboscides predominantly by excavating a subsurface feeding chamber with the labial palps (rejecting feces and PSEUDOFECES through the excurrent siphon into the water column) or less often by sweeping the surface. Yoldiids can also supplement their nutrition by suspension-filtration by the gills, facilitated by the pumping action of the gills.

The family Yoldiidae is known since the Cretaceous and is represented by at least 10 living genera and at least 90 species, inhabiting all seas.

## References

Allen, J. A., and F. J. Hannah. 1986. A reclassification of the Recent genera of the subclass Proto-branchia (Mollusca: Bivalvia). *Journal of Conchology*, 32(4): 225–249.

Bender, K., and W. R. Davis. 1984. The effect of feeding by *Yoldia limatula* on bioturbation. *Ophelia*, 23(1): 91–100.

Drew, G. A. 1899a. Some observations on the habits, anatomy and embryology of members of the Protobranchia. *Anatomischer Anzeiger*, 15(24): 493–519.

Drew, G. A. 1899b. The anatomy, habits, and embryology of *Yoldia limatula*, Say. *Memoirs from the Biological Laboratory of the Johns Hopkins University*, 4(3): 1–37.

Heath, H. 1937. The anatomy of some protobranch mollusks. *Mémoires du Musée Royal d'Histoire Naturelle de Belgique, Série 2*, 10: 1–26, pls. 1–10.

Maxwell, P. A. 1988. Comments on "A reclassification of the Recent genera of the subclass Proto-branchia (Mollusca: Bivalvia)" by J. A. Allen and F. J. Hannah (1986). *Journal of Conchology*, 33(2): 85–96.

Stasek, C. R. 1965. Feeding and particle-sorting in *Yoldia ensifera* (Bivalvia: Protobranchia), with notes on other nuculanids. *Malacologia*, 2(3): 349–366.

Verrill, A. E., and K. J. Bush 1897. Revision of the genera of Ledidae and Nuculidae of the Atlantic coast of the United States. *The American Journal of Science, Series 4*, 3(13–18): 51–63.

Villarroel, M., and J. Stuardo. 1998. Protobranchia (Mollusca: Bivalvia) Chilenos recientes y algunos fósiles. *Malacologia*, 40(1–2): 113–229.

Yonge, C. M. 1939. The protobranchiate Mollusca: a functional interpretation of their structure and evolution. *Philosophical Transactions of the Royal Society of London, Series B, Biological Sciences*, 230(566): 79–148.

# Family Arcidae – Ark Clams

Classification
AUTOLAMELLIBRANCHIATA Grobben, 1894
Pteriomorphia Beurlen, 1944
Arcoida Stoliczka, 1870
Arcoidea Lamarck, 1809
Arcidae Lamarck, 1809

## Featured species
**Barbatia cancellaria** (Lamarck, 1819) – **Red-brown Ark**

Elongated oval to quadrangular, longer posteriorly, solid, dark purplish brown with wide lighter band at mid-valve, covered by radially tufted yellow-brown periostracum, surface cancellate or weakly beaded; interior brownish with white radial ray (reflecting external colors) and narrow byssal gape, margin smooth, cardinal area narrow (see p. 57), hinge line straight, distal teeth nearly parallel to hinge line. North Carolina to Florida, Bermuda, Bahamas, West Indies, Gulf of Mexico, Caribbean Central America, South America (to Brazil). Length 16 mm (to 45 mm).

**The periostracum** of living *Barbatia cancellaria* (here from Indian Key) is thick, fibrous, and occasionally hairy, with projections that extend past the shell margins.

**Barbatia cancellaria** attaches to hard substrata in shallow water, either at the base of other benthic organisms or under rocks (see additional figure on p. 113).

## Family description

The arcid shell is small to medium-sized (to 125 mm), elongated quadrangular to oval, and thin-walled to solid. It is EQUIVALVE to slightly INEQUIVALVE (left valve larger), inflated, and in some species ventrally gaping (or at least indented where the byssus emerges). The shell is EQUILATERAL or INEQUILATERAL (umbones anterior), with PROSO-,

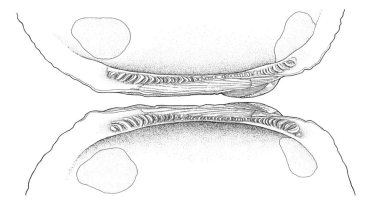

OPISTHO-, or ORTHOGYRATE UMBONES separated by a narrow to wide CARDINAL AREA. Shell microstructure is ARAGONITIC and three-layered, with a PRISMATIC outer layer, a CROSSED LAMELLAR middle layer, and a COMPLEX CROSSED LAMELLAR inner layer. TUBULES are present in some species through all shell layers except the outermost (periostracum). Exteriorly arcids are covered by a thick, fibrous, pilose (occasionally hirsute) PERIOSTRACUM. Sculpture is primarily radial, often with weaker commarginal ridges, or in some species cancellate. LUNULE and ESCUTCHEON are absent. Interiorly the shell is non-NACREOUS; in a few species, the inner margin of the posterior adductor muscle scar is reinforced by a raised MYOPHORIC RIDGE. The PALLIAL LINE is ENTIRE. The inner shell margins are smooth or denticulate. The HINGE PLATE is straight or slightly arched, TAXODONT, with numerous perpendicular to oblique teeth, rarely reduced to a few nearly horizontal ridges, and often diminished in size or absent below the umbones. The LIGAMENT occupies the entire cardinal area and is SIMPLE or more typically DUPLIVINCULAR (with superficial chevron-shaped grooves), and AMPHI-, PROSO-, or OPISTHODETIC.

The animal is ISOMYARIAN or HETEROMYARIAN (anterior ADDUCTOR MUSCLE smaller in, e.g., *Bentharca* and *Bathyarca*); the pedal retractor muscles are elongated and well developed, especially posteriorly where the larger posterior pedal retractors have repositioned the pericardial cavity close to the umbones. Pedal protractor muscles underlie the anterior adductor. The MANTLE margins are not fused ventrally except, in some, for a small posterior excurrent aperture; SIPHONS are absent. In some burrowing forms, INCURRENT and EXCURRENT APERTURES are formed by temporary appression of the mantle lobes. The MANTLE margins are muscular and nontentaculate; simple PALLIAL EYES (cup-shaped,

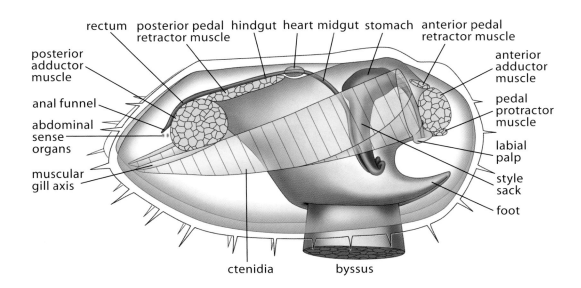

without lens) on the outer folds have been reported for some species, covered by periostracum. Deepwater *Bentharca* and *Bathyarca* possess a pair of prominent posterior mantle flaps and mantle flap glands, presumed to protect the gills and help convey PSEUDOFECES from the MANTLE CAVITY. HYPOBRANCHIAL GLANDS have not been reported. The FOOT is elongated and deeply grooved ventrally. A BYSSUS is present in the adult, is often robust, emanates from a conical process within the BYSSAL GROOVE, and can be shed and resecreted for relocation purposes.

The LABIAL PALPS are relatively small to medium-sized, with ridges restricted to the dorsal part. The CTENIDIA are FILIBRANCH (ELEUTHERORHABDIC), HOMORHABDIC, of about equal size, and not inserted into (or fused with) the distal oral groove of the palps (CATEGORY III association). The posterior third of each gill is attached to a muscular stalk originating on the ventral surface of the posterior adductor muscle. CEPHALIC EYES are present. Incurrent and excurrent water flow is mainly posterior, with a secondary anterior incurrent flow. The STOMACH is TYPE III. The MIDGUT is variable in length and degree of coiling. The HINDGUT passes either dorsal to or through the ventricle of the heart, and leads to a rectum with a free ANAL FUNNEL. In species of *Arca*, the heart includes two lateral pericardia, each with its own auricle and ventricle, and divided by the elongated posterior pedal retractor muscle; the ventricles are medially connected. Intracellular hemoglobin is present in the blood (e.g., the Blood Cockle, *Anadara granosa* (Linnaeus, 1758)) in red blood cells unknown elsewhere in the Mollusca. Arcids are GONOCHORISTIC and usually produce planktonic VELIGER larvae; one species of *Lissarca* is known to brood its larvae. The gonad often extends branches into the mantle tissues. The nervous system is not concentrated. STATOCYSTS in adults are present or absent. ABDOMINAL SENSE ORGANS are present.

Arcids are SUSPENSION FEEDERS and usually marine, rarely inhabiting estuarine or fresh waters (e.g., Indian *Scaphula nagarjunai* Ram & Radhakrishna, 1984). They can be EPIBYSSATE on coral or rock, or (presumably secondarily) ENDOBYSSATE in sand or mud, or less commonly, rock-boring (e.g., Panamic *Litharca lithodomus* (G. B. Sowerby I, 1833)) by a combination of mechanical and chemical means.

The family Arcidae is known since the Jurassic, is represented by 12 living genera and ca. 250 species, and is widely distributed mainly in intertidal or shallow waters. Some species (e.g., *Anadara granosa*) have been exploited for human food throughout Asia and have been cultured since the seventeenth century in China and the mid-nineteenth century in Japan. Members of the superficially similar Noetiidae differ mainly by characters of the ligament.

## References

Heath, H. 1941. The anatomy of the pelecypod family Arcidae. *Transactions of the American Philosophical Society*, 31(5): 287–319, pls. 1–22.

Kusakabe, D., and R. Kitamori. 1949. Anatomical structure of byssus of *Anadara* and *Barbatia* (bivalves). *Bulletin of the Japanese Society of Scientific Fisheries*, 14(5): 223–236. [In Japanese with English abstract.]

Morton, B. 1982. Functional morphology of *Bathyarca pectunculoides* (Bivalvia: Arcacea) from a deep Norwegian fjord with a discussion of the mantle margin in the Arcoida. *Sarsia*, 67: 269–282.

Nicol, D., and D. S. Jones. 1986. *Litharca lithodomus* and adaptive radiation in arcacean pelecypods. *The Nautilus*, 100(3): 105–109.

Nowikoff, M. 1926. Über die Komplexaugen der Gattung *Arca*. *Zoologischer Anzeiger*, 67(11–12): 277–289.

Oliver, P. G., and J. A. Allen. 1980. The functional and adaptive morphology of the deep-sea species of the Arcacea (Mollusca: Bivalvia) from the Atlantic. *Philosophical Transactions of the Royal Society of London, Series B, Biological Sciences*, 291(1045): 45–76.

Oliver, P. G., and A. M. Holmes. 2006. The Arcoidea (Mollusca: Bivalvia): a review of the current phenetic based systematics. *Zoological Society of the Linnean Society*, 148: 237–251.

Reindl, S., and G. Haszprunar. 1996. Fine structure of caeca and mantle of arcoid and limopsoid bivalves (Mollusca: Pteriomorpha). *The Veliger*, 39(2): 101–116.

Reinhart, P. W. 1935. Classification of the pelecypod family Arcidae. *Bulletin du Musée Royal d'Histoire Naturelle de Belgique*, 11(13): 1–68.

Rost, H. 1955. A report on the family Arcidae (Pelecypoda). *Allan Hancock Pacific Expeditions*, 20(2): 177–249.

Scanland, T. B. 1979. The epibiota of *Arca zebra* and *Arca imbricata*: a community analysis. *The Veliger*, 21(4): 475–485.

Sheldon, P. G. 1917. The Atlantic slope arcas. *Palaeontographica Americana*, 1(1): 1–101.

Simone, L. R. L., and A. Chichvarkhin. 2004. Comparative morphological study of four species of *Barbatia* occurring on the southern Florida coast (Arcoidea, Arcidae). In: R. Bieler and P. M. Mikkelsen, eds., *Bivalve Studies in the Florida Keys*, Proceedings of the International Marine Bivalve Workshop, Long Key, Florida, July 2002. *Malacologia*, 46(2): 355–379.

Stanley, S. M. 1972. Functional morphology and evolution of byssally attached bivalve mollusks. *Journal of Paleontology*, 46(2): 165–212.

Tevesz, M. J. S., and J. G. Carter. 1979. Form and function in *Trisidos* (Bivalvia) and a comparison with other burrowing arcoids. *Malacologia*, 19(1): 77–85.

Thomas, R. D. K. 1978. Limits to opportunism in the evolution of the Arcoida (Bivalvia). *Philosophical Transactions of the Royal Society of London, Series B, Biological Sciences*, 284(1001): 335–344.

Waller, T. R. 1980. Scanning electron microscopy of shell and mantle in the order Arcoida (Mollusca: Bivalvia). *Smithsonian Contributions to Zoology*, (313): 1–58.

### *Cucullaearca candida* (Helbling, 1779) – **White Bearded Ark**

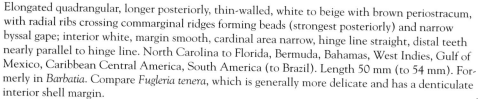

Elongated quadrangular, longer posteriorly, thin-walled, white to beige with brown periostracum, with radial ribs crossing commarginal ridges forming beads (strongest posteriorly) and narrow byssal gape; interior white, margin smooth, cardinal area narrow, hinge line straight, distal teeth nearly parallel to hinge line. North Carolina to Florida, Bermuda, Bahamas, West Indies, Gulf of Mexico, Caribbean Central America, South America (to Brazil). Length 50 mm (to 54 mm). Formerly in *Barbatia*. Compare *Fugleria tenera*, which is generally more delicate and has a denticulate interior shell margin.

### *Fugleria tenera* (C. B. Adams, 1845) – **Delicate Ark**

Elongated oval to quadrangular, longer posteriorly, thin-walled, white with thick brown radial periostracum, with radial ribs crossing commarginal ridges forming beads and narrow byssal gape; interior white, margin denticulate, cardinal area narrow (see p. 57), hinge line straight. Florida, Bermuda, Bahamas, West Indies, Gulf of Mexico, Caribbean Central America, South America (to Brazil). Length 20 mm (to 35 mm). Syn. *balesi* Pilsbry & McLean, 1939. Formerly in *Barbatia*. Compare *Cucullaearca candida*, which has a smooth interior shell margin. Note: See also image of living animal on p. 57. Also known as Doc Bales' Ark.

### *Acar domingensis* (Lamarck, 1819) – **White Miniature Ark**

Elongated quadrangular, skewed anterodorsally, anterior end narrowed, posterior end longer with oblique angular ridge, solid, sandy brown to whitish without noticeable periostracum, surface coarsely cancellate, no byssal gape; interior white, margin denticulate, cardinal area narrow (see p. 57), hinge line slightly arched, teeth more numerous posteriorly. North Carolina to Florida, Bermuda, Bahamas, West Indies, Gulf of Mexico, Caribbean Central America, South America (to Brazil). Length 20 mm. Formerly in *Barbatia*.

### *Lunarca ovalis* (Bruguière, 1789) – **Blood Ark**

Obliquely oval, umbones slightly anterior of center, inflated, solid, white with thick black-brown periostracum, with 26–35 smooth radial ribs, no byssal gape; interior white, margin coarsely denticulate, cardinal area narrow, hinge line arched, teeth more numerous posteriorly, those near umbones coarse. Massachusetts to Florida, West Indies, Gulf of Mexico, Caribbean Central America, South America (to Uruguay). Length 40 mm (to 56 mm). Formerly in *Anadara*.

### *Arca zebra* (Swainson, 1833) – **Atlantic Turkey Wing**

Elongated quadrangular, about twice as long as high, longer posteriorly, with V-shaped indented posterior slope and produced posterior auricle, solid, tan to whitish with ragged stripes of reddish to purple-brown, periostracum yellowish brown, coarse and fibrous, with radial ribs crossing commarginal ridges and narrow byssal gape; interior whitish centrally, mottled to solid reddish brown marginally, margin smooth to weakly denticulate, cardinal area wide, hinge line straight. North Carolina to Florida, Bermuda, Bahamas, West Indies, Gulf of Mexico, Caribbean Central America, South America (to Brazil). Length 60 mm (to 90 mm). Note: See also image of living animal on p. 57.

### *Arca imbricata* Bruguière, 1789 – **Mossy Ark**

Elongated quadrangular, about twice as long as high, longer posteriorly, with bluntly oblique posterior slope, inflated, solid, brown to greenish with thick brown periostracum, with radial ribs crossing commarginal ridges forming beads and wide byssal gape; interior whitish centrally, mottled brown marginally, cardinal area wide (see p. 57), hinge line straight. North Carolina to Florida, Bermuda, Bahamas, West Indies, Gulf of Mexico, Caribbean Central America, South America (to Brazil). Length 50 mm (to 64 mm). Syn. *umbonata* Lamarck, 1819.

*Cucullaearca candida*

*Fugleria tenera*

*Acar domingensis*

*Lunarca ovalis*

*Arca zebra*      LV cleaned

*Arca imbricata*      ventral

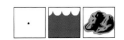

### *Anadara baughmani* Hertlein, 1951 – **Baughman's Skewed Ark**

Elongated oval, skewed anterodorsally, left valve slightly larger, longer posteriorly, solid, white with brown periostracum, with 28–30 flat radial ribs, smooth or beaded commarginally, anteriormost ribs faintly or not cut medially, interrib spaces with close commarginals, no byssal gape; interior white, margin coarsely denticulate, cardinal area wide, hinge line straight. Florida Keys, Gulf of Mexico, Caribbean Central America, South America (to Uruguay). Length 45 mm. Syn. *springeri* Rehder & Abbott, 1951. Compare *Anadara floridana*, which has more radial ribs that are strongly radially grooved. Note: The valves photographed are of the HOLOTYPE specimen of *Anadara springeri* Rehder & Abbott, 1951, a synonym of *A. baughmani*.

### *Anadara floridana* (Conrad, 1869) – **Cut-Ribbed Ark**

Elongated oval, left valve slightly larger, longer posteriorly, inflated, solid, white with light to dark brown thick periostracum, with 30–38 medially cut radial ribs crossing commarginal ridges forming weak beads, no byssal gape; interior white, margin coarsely denticulate, cardinal area wide, hinge line straight. North Carolina to Florida, West Indies, Gulf of Mexico, Caribbean Central America, South America (Venezuela). Length 65 mm (to 90 mm). Compare *Anadara baughmani*, which has fewer radial ribs that are faintly or not radially grooved.

### *Anadara notabilis* (Röding, 1798) – **Eared Ark**

Oval to quadrangular, longer posteriorly with produced posterior auricle, inflated, left valve slightly larger, solid, white with thick brown periostracum, with 25–27 radial ribs crossed by rounded commarginal ridges, no byssal gape; interior white, margin coarsely denticulate, cardinal area wide (see p. 57), hinge line straight, anterior teeth more densely packed. North Carolina to Florida, Bermuda, Bahamas, West Indies, Gulf of Mexico, Caribbean Central America, South America (to Brazil). Length 40 mm (to 92 mm).

### *Anadara transversa* (Say, 1822) – **Transverse Ark**

Elongated oval to quadrangular, left valve slightly larger, longer posteriorly, solid, white with light brown periostracum mainly on margins, with 30–35 radial ribs that are beaded on left valve only, no byssal gape; interior white, margin coarsely denticulate, cardinal area narrow, hinge line straight. Massachusetts to Florida, West Indies, Gulf of Mexico, Caribbean Central America. Length 20 mm (to 38 mm).

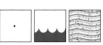

### *Scapharca brasiliana* (Lamarck, 1819) – **Incongruous Ark**

Rounded trigonal, left valve slightly larger, inflated, thin-walled, white with thin light brown periostracum, with 26–28 squared radial ribs crossing barlike beads, no byssal gape; interior white, margin coarsely denticulate, cardinal area trigonal (see p. 57), hinge line straight with slightly arched tooth row. North Carolina to Florida, West Indies, Gulf of Mexico, Caribbean Central America, South America (to Brazil). Length 55 mm. Compare *Scapharca chemnitzii*, which is thicker shelled with fewer radial ribs.

### *Scapharca chemnitzii* (Philippi, 1851) – **Chemnitz's Triangular Ark**

Rounded trigonal, left valve slightly larger, inflated, solid, white with thin light brown periostracum, with ca. 25 squared radial ribs crossed by barlike beads, no byssal gape; interior white, margin coarsely denticulate, cardinal area trigonal, hinge line straight, teeth dense subumbonally. Florida, West Indies, Gulf of Mexico, Caribbean Central America, South America (to Uruguay). Length 30 mm. Compare *Scapharca brasiliana*, which is thinner shelled with more radial ribs.

*Anadara baughmani*

*Anadara floridana*

*Anadara notabilis*

articulated

*Anadara transversa*

*Scapharca brasiliana*

*Scapharca chemnitzii*

### *Bathyarca glomerula* (Dall, 1881) – **Little-Ball Bathyark**

Obliquely oval, inflated, solid, whitish with thin periostracum in radial rows, with fine radial ribs crossed by commarginal ridges forming beads, no byssal gape; interior whitish, margin denticulate, cardinal area narrow, hinge line straight, with <12 teeth. North Carolina, Florida Keys, West Indies, Gulf of Mexico. Length 6 mm.

### *Bentharca sagrinata* (Dall, 1886) – **Shagreen Ark**

Elongated quadrangular, skewed anterodorsally, anterior end narrowed, posterior end longer, solid, whitish with irregular orange rays, with fine radial ribs and few coarse posterior undulations crossed by commarginal ridges, no byssal gape; interior whitish with coarse radial ridges, margin smooth, cardinal area narrow, hinge line straight. Georgia to Florida, West Indies, South America (Colombia). Length 15 mm.

**Dorsal views** of six species in six genera of Arcidae (*Fugleria tenera, Scapharca brasiliana, Acar domingensis, Anadara notabilis, Barbatia cancellaria,* and *Arca imbricata*) illustrate varying configurations of umbones and ligament within the family.

**Living *Arca zebra*** attaches solidly to the surface of hard substrata with its stout byssus, where algae and other fouling biota camouflage it within its environment. In this position, the intraumbonal area of this specimen (on a rock in the backreef area of Looe Key) shows the zebra-stripe pattern characteristic of this species.

**Living *Fugleria tenera*** also byssally attaches to rock surfaces using a strong byssus. When detached, an individual (like this specimen at Looe Key) can crawl to a new attachment site on its colorful foot. Hemoglobin has been shown to be the source of soft-tissue pigmentation in some species of arks.

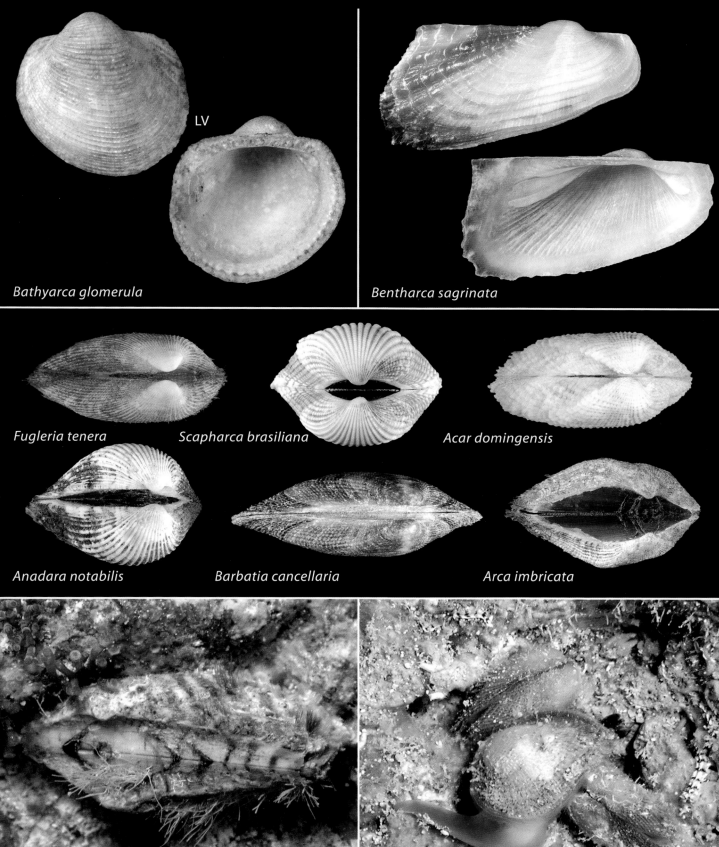

*Bathyarca glomerula*

LV

*Bentharca sagrinata*

*Fugleria tenera*

*Scapharca brasiliana*

*Acar domingensis*

*Anadara notabilis*

*Barbatia cancellaria*

*Arca imbricata*

# Family Noetiidae – False Ark Clams

**Classification**
AUTOLAMELLIBRANCHIATA Grobben, 1894
Pteriomorphia Beurlen, 1944
Arcoida Stoliczka, 1870
Arcoidea Lamarck, 1809
Noetiidae Stewart, 1930

## Featured species
*Arcopsis adamsi* (Dall, 1886) – **Adams' Miniature Ark**

dorsal

Elongated oval to quadrangular, white to cream with inconspicuous periostracum, surface cancellate, no byssal gape; interior white, margin smooth, cardinal area relatively wide, ligament limited to small, trigonal black patch between umbones. North Carolina to Florida, Bermuda, Bahamas, West Indies, Gulf of Mexico, Caribbean Central America, South America (to Brazil). Length 12 mm (to 17 mm). Note: Also known as Adams' Cancellate Ark.

**A group of *Arcopsis adamsi*** crowd closely together on the undersurface of a rock from the bayside of Spanish Harbor Key. When so disturbed, individuals voluntarily release their byssus and relocate away from the light. Discarded byssi show as brown dash-shaped objects on the rock at the lower right and center of this photograph.

# Family description

The noetiid shell is small to medium-sized (to 50 mm), quadrangular to trigonal to oval, and solid. It is EQUIVALVE, inflated, and not gaping. The shell is INEQUILATERAL (umbones slightly anterior), with UMBONES PROSO-, ORTHO-, or OPISTHOGYRATE and separated by a narrow to wide CARDINAL AREA that is flat or V-shaped. Shell microstructure is ARAGO-

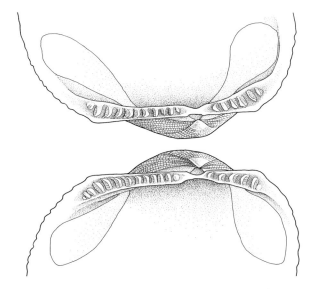

NITIC and three-layered, with a PRISMATIC outer layer, CROSSED LAMELLAR middle layer, and COMPLEX CROSSED LAMELLAR inner layer. TUBULES are present in some species through all shell layers except the outermost (periostracum). Exteriorly noetiids are covered by dense, fibrous, pilose PERIOSTRACUM, in some cases persisting only at the margins. Sculpture is radial or cancellate, usually with dense radial ribs. LUNULE and ESCUTCHEON are absent. Interiorly the shell is non-NACREOUS; one or both adductor muscle scars have a raised inner ridge or shelf (MYOPHORIC RIDGE). The PALLIAL LINE is ENTIRE. The inner shell margins are smooth or denticulate. The HINGE PLATE is straight or weakly arched, TAXODONT, with numerous vertical or rarely oblique teeth, often diminished in size or absent below the umbones. The LIGA-MENT occupies only part of the cardinal area, is PROSO-, AMPHI-, or OPISTHODETIC, and is modified DUPLIVINCULAR, having "vertical" (i.e., transverse, not chevron-shaped) grooves in dorsal view (a SYNAPOMORPHY of the family).

The animal is ISOMYARIAN or HETEROMYARIAN (anterior ADDUCTOR MUSCLE smaller); the pedal retractor muscles are well developed. Pedal protractor muscles underlie the anterior adductor. The MANTLE margins are not fused ventrally; SIPHONS are absent. In some burrowing forms, EXCURRENT and INCURRENT APERTURES are formed by temporary appression of the mantle lobes. The MANTLE margins are muscular and nontentaculate, and usually have well-developed simple PALLIAL EYES (cup-shaped, without lens) on the outer folds, covered by periostracum, and that are restricted to the anterodorsal margin. HYPO-

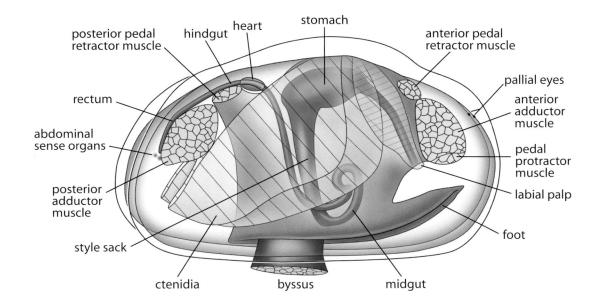

BRANCHIAL GLANDS have not been reported. The FOOT is elongated, heeled, and deeply grooved ventrally, and is usually byssate in the adult.

The LABIAL PALPS are relatively small to medium-sized (large in *Didimacar*). The CTENIDIA are FILIBRANCH (ELEUTHERORHABDIC), HOMORHABDIC, of about equal size (outer demibranch slightly smaller), and not inserted into (or fused with) the distal oral groove of the palps (CATEGORY III association). Incurrent and excurrent water flows are posterior, with a secondary anterior incurrent. The STOMACH is TYPE III. The MIDGUT is coiled. The HINDGUT passes through the ventricle of the heart, and leads to a sessile rectum. The heart frequently has a doubled ventricle. Hemoglobin has been documented in the blood of some species (e.g., *Arcopsis adamsi*). Noetiids are GONOCHORISTIC and usually produce planktonic VELIGER larvae. The gonad often extends branches into the mantle tissues. The nervous system is not concentrated. STATOCYSTS are present in adults. ABDOMINAL SENSE ORGANS are present.

Noetiids are marine, and usually free-living as adults, but some species are EPIBYSSATE on coral or rock.

The family Noetiidae is known since the Cretaceous, is represented by 13 living genera and ca. 40 species, and is widely distributed in shallow waters.

**_Noetia ponderosa_ (Say, 1822) – Ponderous Ark**

dorsal

Rounded trigonal, almost as high as long, slightly skewed anterodorsally, with strong posterior ridge, inflated, white with brown-black periostracum, with 27–31 squared radial ribs each divided by fine incised line, commarginal ridges most apparent between ribs, umbones opisthogyrate, no byssal gape; interior white, margin strongly denticulate, cardinal area wide. Virginia to Florida, West Indies, Gulf of Mexico, Caribbean Central America. Length 50 mm (to 60 mm).

# References

Freneix, S. 1960 ["1959"]. Remarques sur l'ontogénie du ligament et de la charnière de quelques espèces de lamellibranches (Noetidae [*sic*] et Carditidae). *Bulletin de la Société Géologique de France, Série 7,* 1(7): 719–729, pls. 33–34.

Heath, H. 1941. The anatomy of the pelecypod family Arcidae. *Transactions of the American Philosophical Society,* 31(5): 287–319, pls. 1–22.

Oliver, P. G. 1985. A comparative study of two species of Striarciinae [*sic*] from Hong Kong with comments on specific and generic systematics. Pages 283–310, in: B. Morton and D. Dudgeon, eds., *The Malacofauna of Hong Kong and Southern China, II, volume 1.* Hong Kong University Press, Hong Kong.

Oliver, P. G. 1990. Functional morphology and systematics of the genus *Didimacar* (Bivalvia: Arcacea: Noetiidae). Pages 1075–1094, in: B. Morton, ed., *Proceedings of the Second International Marine Biological Workshop: The Marine Fauna and Flora of Hong Kong and Southern China, Hong Kong, 1986.* Hong Kong University Press, Hong Kong.

Oliver, P. G., and J. Järnegren. 2004. How reliable is morphology based species taxonomy in the Bivalvia? A case study on *Arcopsis adamsi* (Bivalvia: Arcoidea) from the Florida Keys. In: R. Bieler and P. M. Mikkelsen, eds., *Bivalve Studies in the Florida Keys.* Proceedings of the International Marine Bivalve Workshop, Long Key, Florida, July 2002. *Malacologia,* 46(2): 327–338.

Rost, H. 1955. A report on the family Arcidae (Pelecypoda). *Allan Hancock Pacific Expeditions,* 20(2): 177–249.

# Family Glycymerididae – Bittersweet Clams or Dog Cockles

**Classification**
AUTOLAMELLIBRANCHIATA Grobben, 1894
Pteriomorphia Beurlen, 1944
Arcoida Stoliczka, 1870
Arcoidea Lamarck, 1809
Glycymerididae Dall, 1908 [1847]

**Featured species**
*Tucetona pectinata* (Gmelin, 1791) – **Comb Bittersweet**

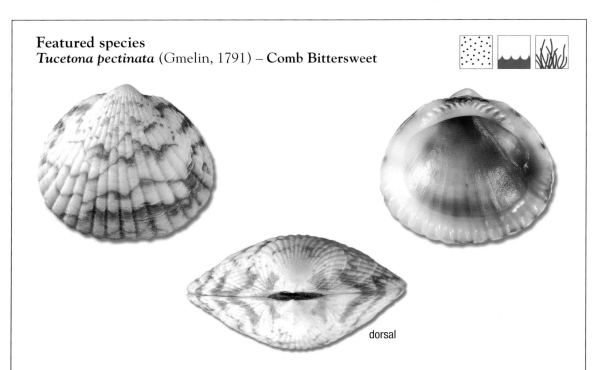

dorsal

Circular, but ventrally flattened, white to gray with brown patches and irregularly commarginal bands, with 20–40 coarse rounded radial ribs crossed by fine commarginal striae, periostracum inconspicuous, umbones orthogyrate; interior white with brown stain. New Jersey to Florida, Bermuda, Bahamas, West Indies, Gulf of Mexico, Caribbean Central America, South America (to Brazil). Length 20 mm (to 31 mm). Formerly in *Glycymeris*. Compare *Tucetona subtilis*, which is ventrally rounded and has more radial ribs.

***Tucetona pectinata* lives** epifaunally on sand or hard-bottom habitats (like these from the bayside of Stirrup Key), with the shell surfaces often fouled by algae.

# Family description

The glycymeridid shell is small to medium-sized (to 100 mm), circular to oval to rounded trigonal, in some cases slightly oblique, and solid. It is EQUIVALVE, moderately inflated,

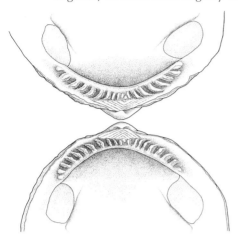

and not gaping. The shell is EQUILATERAL, with the UMBONES central, ORTHO-, occasionally OPISTHO-, or rarely PROSOGYRATE, and separated by a narrow to wide CARDINAL AREA. Shell microstructure is ARAGONITIC and two- or three-layered, with a PRISMATIC outer layer (absent in some species), CROSSED LAMELLAR middle layer, and COMPLEX CROSSED LAMELLAR inner layer. TUBULES have been reported for some species through all shell layers except the outermost (PERIOSTRACUM). Exteriorly glycymeridids are creamy white or brown covered by a usually thick, velvety, dehiscent periostracum that is frequently hirsute in smoother-shelled species. Sculpture is smooth to strongly radial; *Glycymeris* and *Tucetona* genus groups have been recognized based on fine or strong ribs and corresponding hirsute or smooth periostracum, re-

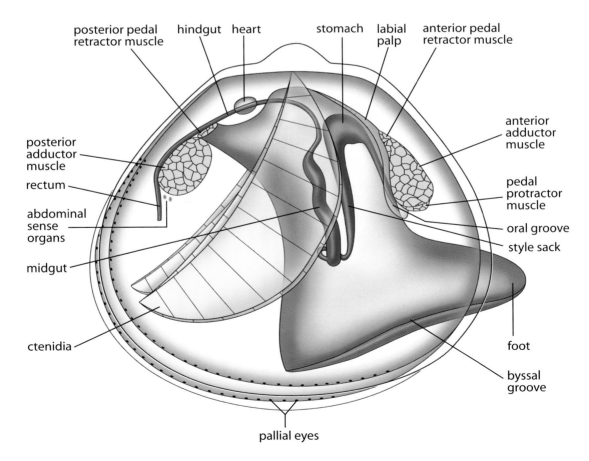

spectively. LUNULE and ESCUTCHEON are absent. Interiorly the shell is non-NACREOUS; the posterior (in some cases also the anterior) adductor muscle scar has a raised ridge (MYOPHORIC RIDGE) on the inner margin. The PALLIAL LINE is ENTIRE. The inner shell margins are denticulate, with denticles corresponding to the external spaces between ribs. The HINGE PLATE is robust and strongly arched, TAXODONT, with numerous radially arranged, in some species chevron-shaped teeth (the larger of which are commonly crenulate), diminished in size below the umbones and laterally. The LIGAMENT occupies the entire CARDINAL AREA, and is PROSO- or AMPHIDETIC, and DUPLIVINCULAR (with superficial chevron-shaped grooves).

The animal is slightly HETEROMYARIAN (posterior ADDUCTOR MUSCLE smaller, a SYNAPOMORPHY of the family); the pedal retractor muscles are well developed. Pedal protractor muscles underlie the anterior adductor. The MANTLE margins are not fused ventrally; SIPHONS are absent. The MANTLE margins are muscular and nontentaculate; the outer folds have simple PALLIAL ("siphonal") EYES (cup-shaped, without lens) on the posterior margin, covered by periostracum. Temporary EXCURRENT and INCURRENT APERTURES can be formed by appression of the mantle edges. HYPOBRANCHIAL GLANDS have not been reported. The FOOT is large, wedge-shaped, and has a deep BYSSAL GROOVE; the adult is nonbyssate.

The LABIAL PALPS are relatively small, and are extended by a long, unridged oral groove to the mouth. The subumbonal CTENIDIA are FILIBRANCH (ELEUTHERORHABDIC), HOMORHABDIC, of about equal size, and not inserted into (or fused with) the distal oral groove of the palps (CATEGORY III association). CEPHALIC EYES are present. Incurrent and excurrent water flows are posterior, with a secondary anterior incurrent. The STOMACH is TYPE III. The MIDGUT is not coiled and is somewhat enlarged in diameter. The HINDGUT passes through the ventricle of the heart, and leads to a freely hanging or sessile rectum. Hemoglobin has been found in the blood of some species. Glycymeridids are generally GONOCHORISTIC, although evidence suggests that at least one species is a PROTANDRIC HERMAPHRODITE, and produce planktonic VELIGER larvae. The nervous system is apparently not concentrated. STATOCYSTS have not been reported in adults. ABDOMINAL SENSE ORGANS are present and asymmetrical.

Glycymeridids are SUSPENSION-FEEDING, shallow INFAUNAL burrowers just below the surface of coarse sand. Some species show intolerance of silt and turbid water. Atlantic species are slow, inefficient burrowers that are believed to be active exclusively at night.

The family Glycymerididae is known since the Cretaceous, is represented by 4 living genera and at least 50 species, and is distributed worldwide except in polar and deep seas. *Glycymeris* is commercially fished in parts of Europe and the Mediterranean; the common name "bittersweet clam" undoubtedly refers to their taste.

## References

Ansell, A. D., and E. R. Trueman. 1967. Observations on burrowing in *Glycymeris glycymeris* (L.) (Bivalvia, Arcacea). *Journal of Experimental Marine Biology and Ecology*, 1(1): 65–75.

Heath, H. 1941. The anatomy of the pelecypod family Arcidae. *Transactions of the American Philosophical Society*, 31(5): 287–319, pls. 1–22.

Matsukuma, A. 1986. Cenozoic glycymerid bivalves of Japan. *Special Papers of the Palaeontological Society of Japan*, 79: 77–94.

Nicol, D. 1945. Genera and subgenera of the pelecypod family Glycymeridae. *Journal of Paleontology*, 19(6): 616–621.

Nicol, D. 1953. A study of the polymorphic species *Glycymeris americana*. *Journal of Paleontology*, 27(3): 451–455.

Nicol, D. 1956. Distribution of living glycymerids with a new species from Bermuda. *The Nautilus*, 70(2): 48–53.

Nicol, D. 1967. How to distinguish between *Limopsis* and *Glycymeris*. *The Nautilus*, 81(2): 45–46.

Thomas, R. D. K. 1975. Functional morphology, ecology, and evolutionary conservatism in the Glycymerididae (Bivalvia). *Paleontology*, 18(2): 217–254, pl. 38.

Thomas, R. D. K. 1976. Constraints of ligament growth, form and function on evolution in the Arcoida (Mollusca: Bivalvia). *Paleobiology*, 2: 64–83.

*Glycymeris americana* (DeFrance, 1826) – **Giant American Bittersweet**

Circular, mottled tan to brown, with numerous radial ribs bearing fine riblets, periostracum velvety, umbones orthogyrate; interior white. North Carolina to Florida, Bermuda, Gulf of Mexico. Length 25 mm (juvenile; to 98 mm).

*Glycymeris decussata* (Linnaeus, 1758) – **Decussate Bittersweet**

Circular, posterior end slightly produced, mottled brown and white with velvety brown periostracum, with fine radial riblets, umbones opisthogyrate; interior white, often stained with brown, cardinal area in front of umbones. Florida, Bermuda, Bahamas, West Indies, Gulf of Mexico, Caribbean Central America, South America (Colombia, Brazil). Length 40 mm (to 55 mm). Compare *Glycymeris undata*, which is orange mottled and less posteriorly rostrate.

*Glycymeris spectralis* Nicol, 1952 – **Spectral Bittersweet**

Circular, but ventrally flattened, uniformly light brown to white with velvety brown periostracum, with numerous radial ribs bearing fine riblets, umbones slightly opisthogyrate; interior white with brown stain. North Carolina to Florida, Gulf of Mexico, Caribbean Central America. Length 25 mm.

*Glycymeris undata* (Linnaeus, 1758) – **Atlantic Bittersweet**

Circular, posterior end slightly produced in some individuals, cream to white with bold orange-brown mottlings and velvety brown periostracum, near-smooth with fine radial ribs bearing fine riblets, umbones orthogyrate; interior white stained with brown. North Carolina to Florida, Bahamas, West Indies, Caribbean Central America, South America (to Uruguay). Length 12 mm (to 50 mm). Compare *Glycymeris decussata*, which is brown mottled and more posteriorly rostrate. Note: Also known as Wavy Bittersweet.

*Tucetona subtilis* Nicol, 1956 – **Bermudan Bittersweet**

Circular, ventrally rounded, white with orange-brown mottlings, with ca. 50 coarse subequal rounded radial ribs crossed by fine commarginal striae, periostracum inconspicuous, umbones orthogyrate or slightly opisthogyrate; interior white. Florida Keys, Bermuda. Length 6 mm (to 12 mm). Formerly in *Glycymeris*. Compare *Tucetona pectinata*, which is ventrally flattened and has fewer radial ribs. Note: The left valve of this specimen has been drilled by a predatory gastropod.

*Tucetona pectinata* **lives** epibenthically on seagrass- and sand-covered hard bottoms in the Florida Keys, such as this site (with an adult Queen Conch, *Strombus gigas* Linnaeus, 1758) at Looe Key back reef.

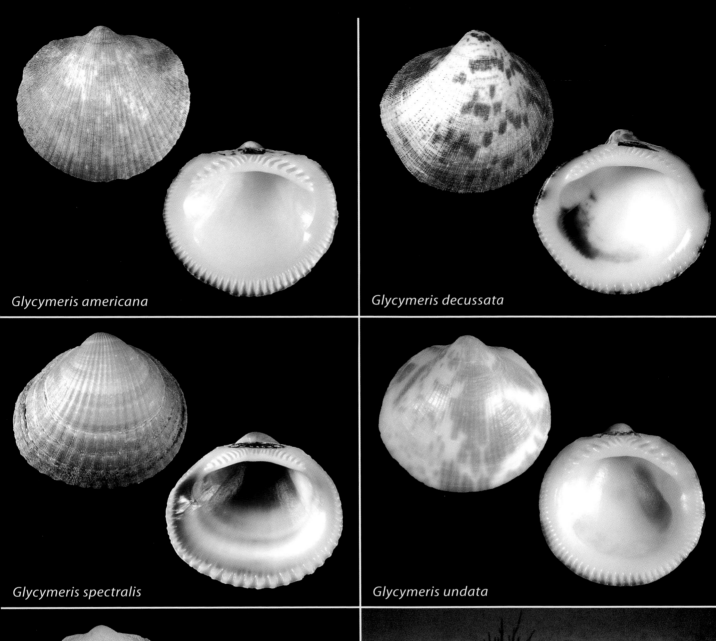

*Glycymeris americana*

*Glycymeris decussata*

*Glycymeris spectralis*

*Glycymeris undata*

*Tucetona subtilis*

# Family Limopsidae – Limops Clams

**Classification**
AUTOLAMELLIBRANCHIATA Grobben, 1894
Pteriomorphia Beurlen, 1944
Arcoida Stoliczka, 1870
Limopsoidea Dall, 1895
Limopsidae Dall, 1895

**Featured species**
*Limopsis cristata* Jeffreys, 1876 – **Crested Limops**

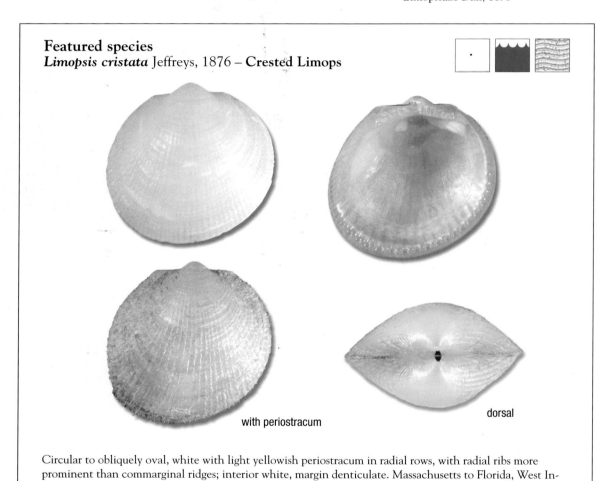

with periostracum

dorsal

Circular to obliquely oval, white with light yellowish periostracum in radial rows, with radial ribs more prominent than commarginal ridges; interior white, margin denticulate. Massachusetts to Florida, West Indies, Gulf of Mexico, also western Europe. Length 4 mm.

# Family description

The limopsid shell is small to medium-sized (to 70 mm), obliquely oval to orbicular to rounded trigonal, and thin-walled to solid. It is EQUIVALVE, compressed to moderately in-

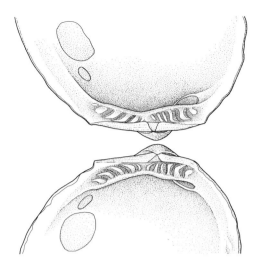

flated, and not gaping. The shell is EQUILATERAL or INEQUILATERAL (umbones anterior), with ORTHO- or PROSO-GYRATE UMBONES separated by a narrow CARDINAL AREA. Shell microstructure is ARAGONITIC and three-layered, with a PRISMATIC outer layer (restricted to the cardinal area), a CROSSED LAMELLAR middle layer, and COMPLEX CROSSED LAMELLAR inner layer. TUBULES have been reported for some species through all shell layers except the outermost (PERIOSTRACUM). Exteriorly limopsids are usually uniformly light-colored, and are covered by a thick, velvety, tufted periostracum that can extend beyond the shell margin. Interlocking periostracal tufts presumably help stabilize the clam in the sediment (analogous to shell spines) and prevent intrusion of sediment into the mantle while the animal is active. Sculpture is smooth or finely radial or commarginal. LUNULE and ESCUTCHEON are absent. Interiorly

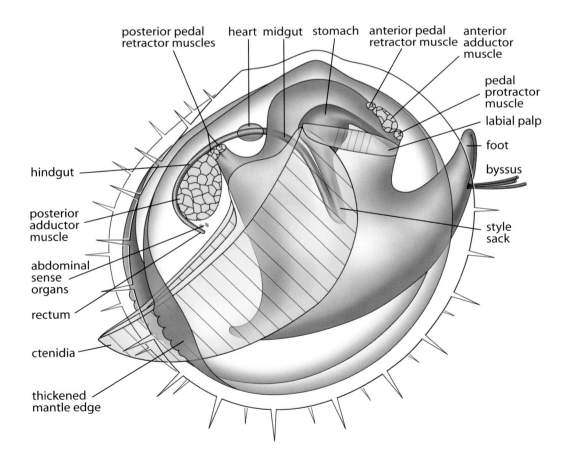

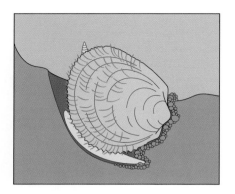

the shell is non-NACREOUS. A radial ridge of varying strength extends from the UMBONAL CAVITY along the posterior edge of the anterior adductor muscle. The PALLIAL LINE is ENTIRE. The inner shell margins are smooth or denticulate. The HINGE PLATE is strong and weakly arched, TAXODONT, with numerous radially arranged teeth that are diminished in size or absent below the umbones. The LIGAMENT is SIMPLE, ALIVINCULAR (interpreted as modified or reduced DUPLIVINCULAR), and AMPHIDETIC; an internal portion (RESILIUM) sits on a deep RESILIFER (especially differing from Glycymerididae by the presence of this feature).

The animal is HETEROMYARIAN (anterior ADDUCTOR MUSCLE smaller and more dorsal); the pedal retractor and protractor muscles are well developed. Pedal elevator muscles have not been reported. The MANTLE margins are not fused ventrally, and in some species have a thickened edge near the termini of the gills; SIPHONS are absent. Temporary EXCURRENT and INCURRENT APERTURES can be formed by appression of the mantle edges and positioning of the gill tips. HYPOBRANCHIAL GLANDS have not been reported. The MANTLE margins are usually nontentaculate, but the outer folds can have a few tentacles and simple PALLIAL EYES (cup-shaped, without lens). The FOOT is elongated, narrow, arcuate, heeled, and has a BYSSAL GROOVE; the adult is byssate in most species. The byssal threads do not have terminal disks and have small sediment particles attached along their lengths.

The LABIAL PALPS are small with few ridges restricted to midpalp. The CTENIDIA are FILIBRANCH (ELEUTHERORHABDIC), HOMORHABDIC, and not inserted into (or fused with) the distal oral groove of the palps (CATEGORY III association). In some species the ascending lamella of the outer demibranchs are absent; the gill axis has been reported as muscular in some species. Incurrent and excurrent water flows are posterior; incurrent flow is also anterior. The STOMACH is TYPE III. The MIDGUT is not coiled. The HINDGUT passes through the ventricle of the heart, and leads to a freely hanging rectum. Limopsids are usually GONOCHORISTIC and produce planktonic VELIGER larvae; deepwater species often are LECITHOTROPHIC. The nervous system is not concentrated. STATOCYSTS and ABDOMINAL SENSE ORGANS are present.

Limopsids are marine SUSPENSION FEEDERS, living epi- or endobyssally in soft sedi-

---

*Limopsis minuta* (Philippi, 1836) – **Minute Limops**

Obliquely oval, flattened posteriorly, white with light yellow-brown periostracum in radial rows, surface cancellate with slightly stronger radial ribs; interior chalky white, margin denticulate. North Atlantic, eastern Canada to Florida, West Indies, Gulf of Mexico, Caribbean Central America, South America (Brazil), also western Europe. Length 13 mm.

ments in deep, cold waters. Living individuals typically plough vertically through the surface of soft sediments, using the byssal surface as a flattened "sole." Inefficient burrowing and the weak byssus make dislodgement from the sediment frequent.

The family Limopsidae is known since the Triassic, is represented by 5 living genera and 25–50 species, and is distributed worldwide mainly in deep, cold, and temperate waters.

### References

Dell, R. K. 1964. Antarctic and subantarctic Mollusca: Amphineura, Scaphopoda and Bivalvia. *Discovery Reports*, 33: 93–250.

Knudsen, J. 1967. The deep-sea Bivalvia. *The John Murray Expedition 1933–34, Scientific Reports*, 11(3): 235–343, pls. 1–3.

Knudsen, J. 1970. The systematics and biology of abyssal and hadal Bivalvia. *Galathea Report*, 11: 1–241, 20 pls.

Nicol, D. 1967. How to distinguish between *Limopsis* and *Glycymeris*. *The Nautilus*, 81(2): 45–46.

Oliver, P. G. 1981. The functional morphology and evolution of Recent Limopsidae (Bivalvia, Arcoidea). *Malacologia*, 21(1–2): 61–93.

Oliver, P. G., and J. A. Allen. 1980. The functional and adaptive morphology of the deep-sea species of the family Limopsidae (Bivalvia: Arcoida) from the Atlantic. *Philosophical Transactions of the Royal Society of London, Series B, Biological Sciences*, 291(1045): 77–125.

Reindl, S., and G. Haszprunar. 1996. Fine structure of caeca and mantle of arcoid and limopsoid bivalves (Mollusca: Pteriomorpha). *The Veliger*, 39(2): 101–116.

Stuardo, J. 1962. Sobre el género *Limopsis* y la distribución de *L. jousseaumei* (Mabille y Rochebrune, 1889) (Mollusca: Bivalvia). *Gayana (Zoologia)*, 6: 1–10, 1 pl.

Tevesz, M. J. S. 1977. Taxonomy and ecology of the Philobryidae and Limopsidae (Mollusca: Pelecypoda). *Postilla, Peabody Museum, Yale University*, 171: 1–64.

### *Limopsis aurita* (Brocchi, 1814) – **Eared Limops**

Obliquely oval, slightly flattened posteriorly, white with light yellow-brown periostracum in radial rows, surface weakly cancellate; interior white, margin denticulate. Florida, Bermuda, Gulf of Mexico, South America (Brazil). Length 9 mm. Note: This species was described as a European fossil; some authors consider the name *L. aurita* a senior synonym of the Recent western Atlantic species *Limopsis paucidentata* Dall, 1886, whereas others (including the present authors) recognize both as valid and extant. The left valve of this specimen has been drilled by a predatory gastropod.

### *Limopsis sulcata* Verrill & Bush, 1898 – **Sulcate Limops**

Obliquely oval, flattened posteriorly, somewhat produced posteroventrally, white with light yellow periostracum in radial rows, with coarse, rounded commarginal ridges notched dorsally by radials; interior white, margin smooth. Eastern Canada to Florida, West Indies, Gulf of Mexico. Length 8 mm.

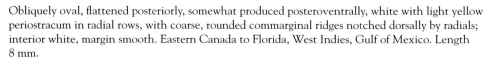

**Seemingly lifeless deepwater sediments** (here at 427 m in the Florida Straits) are home to limopsids and the Blackfin Goosefish.

*Limopsis aurita*

cleaned

*Limopsis sulcata*

# Family Philobryidae – Philobryid Clams

### Classification
AUTOLAMELLIBRANCHIATA Grobben, 1894
Pteriomorphia Beurlen, 1944
Arcoida Stoliczka, 1870
Limopsoidea Dall, 1895
Philobryidae F. Bernard, 1897

## Featured species
*Cratis antillensis* (Dall, 1881) – Antillean Philobryid

Obliquely oval, somewhat auriculate anteriorly and posteriorly, straight or concave posteriorly, solid, white to pink, orange or yellow, with wide flat commarginal ridges covered by periostracum forming narrow radial ridges; interior colored as exterior, margin denticulate. North Carolina to Florida, Bahamas, West Indies, Gulf of Mexico, South America (to Brazil). Length 3 mm (to 7 mm). Formerly in *Limopsis*.

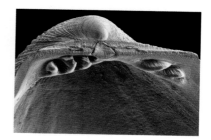

**The hinge plate** of *Cratis antillensis* shows the characteristic central resilium flanked by a short series of taxodont teeth below a row of transverse denticles. The prodissoconch of *C. antillensis* lacks the characteristic caplike appearance of most philobryids.

## Family description

The philobryid shell is small (to 10 mm), solid, and obliquely oval to quadrangular. It is EQUIVALVE, compressed to moderately inflated, and in some species with a distinct anteroventral byssal gape. The shell is INEQUILATERAL (umbones anterior), with an inflated posterior slope and reduced anterior end. The UMBONES are PROSO- or ORTHOGYRATE, and

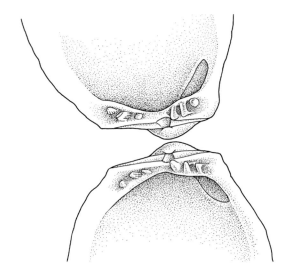

in some cases separated by a narrow CARDINAL AREA. In some species of *Philobrya*, the PRODISSOCONCH is large and caplike, with thickened margins. Shell microstructure is ARAGONITIC and two-layered, with a CROSSED LAMELLAR outer layer and a COMPLEX CROSSED LAMELLAR inner layer. TUBULES are present on the internal surface (except near the margin) and do not open onto the external surface. Exteriorly philobryids are covered by a thick, dehiscent, hirsute PERIOSTRACUM that extends beyond the shell margin. Sculpture is finely radial, in some cases with commarginal striae. LUNULE and ESCUTCHEON are absent. Interiorly the shell is non-NACREOUS. The PALLIAL LINE is ENTIRE. The inner shell margins are smooth or denticulate. The HINGE PLATE is straight to weakly arched, in some cases narrow and reduced TAXODONT teeth (sometimes called PROVINCULAR-like) on either side of a central RESILIFER; *Philobrya* is EDENTATE in adults and has a dreissenid-like myophoral SEPTUM under the umbo. The LIGAMENT is ALIVINCULAR (perhaps reduced DUPLIVINCULAR) and AMPHI- or OPISTHODETIC, with a small internal portion (RESILIUM).

The animal is MONOMYARIAN (anterior ADDUCTOR MUSCLE absent) or extremely HETEROMYARIAN (anterior adductor muscle smaller); the pedal retractor muscles are well developed, with the posterior ones larger. Pedal protractor muscles have not been reported. The MANTLE margins are not fused ventrally; SIPHONS are absent. Hypobranchial GLANDS are present in the SUPRABRANCHIAL CHAMBER along the gill axis. The MANTLE margins are nontentaculate, and have simple PALLIAL ("siphonal") EYES (cap-shaped, without lens) on the outer folds, covered by periostracum. The FOOT is digitiform, laterally compressed, small anteriorly, and with a large, thin posterior heel. It has a BYSSAL GROOVE; the adult is byssate.

The LABIAL PALPS are small. The CTENIDIA are FILIBRANCH (ELEUTHERORHABDIC), HOMORHABDIC, of about equal size, with unusually short filaments that are not inserted into (or fused with) the distal oral groove of the palps (CATEGORY III association). CEPHALIC EYES (cup-shaped, with lens) are present in *Philobrya munita* Finlay, 1930, on each inner demibranch at the junction of the inner labial palp. Incurrent and excurrent water flows are posterior; there can also be an anterior incurrent. The lips are simple; the STOMACH is TYPE III. The MIDGUT is not coiled. The HINDGUT passes either ventral to or through the ventricle of the heart, and leads to a freely hanging rectum. Philobryids are

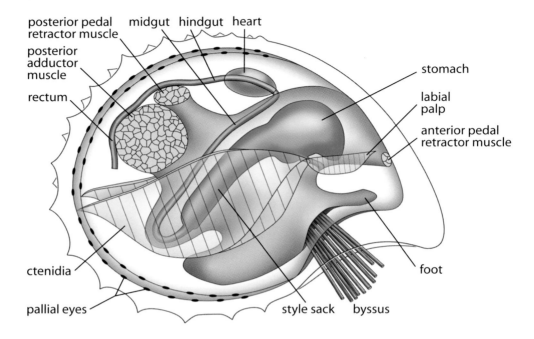

posterior pedal retractor muscle    midgut    hindgut    heart

posterior adductor muscle

rectum

stomach

labial palp

anterior pedal retractor muscle

ctenidia

pallial eyes

foot

style sack    byssus

GONOCHORISTIC and viviparous, producing large, yolky eggs. Brooded larvae are attached to the byssus or gills during incubation. As a result, the gills have short, robust filaments and few interlamellar junctions, and the adult shell can have a large, sharply demarcated PRODISSOCONCH. The nervous system is not concentrated. STATOCYSTS and ABDOMINAL SENSE ORGANS have not been reported.

Philobryids are marine SUSPENSION FEEDERS, living INFAUNALLY or epibyssally, the latter nestling in algae, hydroids, or mussel beds. Efficient EPIFAUNAL crawling has been reported for one species.

The family Philobryidae is known since the Eocene, is represented by 6 living genera and ca. 20 species, and is predominantly distributed in the southern oceans. in warm seas from the intertidal to great depths (1,000 m). Some features of philobryids (e.g., ligament type, adult byssus) suggest that their evolution involves PEDOMORPHOSIS (perhaps NEOTENY).

## References

Bernard, F. 1897. Études comparatives sur la coquille des lamellibranches. II. Les genres *Philobrya* et *Hochstetteria*. *Journal de Conchyliologie*, 45(1): 1–47, pl. 1.

Dell, R. K. 1964. Antarctic and subantarctic Mollusca: Amphineura, Scaphopoda and Bivalvia. *Discovery Reports*, 33: 93–250.

Moore, D. R. 1975. Philobryidae in the Northern Hemisphere. *Bulletin of the American Malacological Union for 1974*, pp. 34–35.

Morton, B. 1978. The biology and functional morphology of *Philobrya munita* (Bivalvia: Philobryidae). *Journal of Zoology*, 185: 173–196.

Prezant, R. S., M. Showers, R. L. Winstead, and C. Cleveland. 1992. Reproductive ecology of the Antarctic bivalve *Lissarca notocadensis* (Philobryidae). *American Malacological Bulletin*, 9(2): 173–186.

Tevesz, M. J. S. 1977. Taxonomy and ecology of the Philobryidae and Limopsidae (Mollusca: Pelecypoda). *Postilla, Peabody Museum, Yale University*, (171): 1–64.

Thiele, J. 1923. Über die Gattung *Philobrya* und das sogenannte Buccalnervensystem von Muscheln. *Zoologischer Anzeiger*, 55(11–13): 287–292.

# Family Mytilidae — True Mussels

**Classification**
AUTOLAMELLIBRANCHIATA Grobben, 1894
Pteriomorphia Beurlen, 1944
Mytiloida Férussac, 1822
Mytiloidea Rafinesque, 1815
Mytilidae Rafinesque, 1815

## Featured species
*Brachidontes exustus* (Linnaeus, 1758) – **Scorched Mussel**

Elongated fan-shaped, often expanded dorsally and/or flattened ventrally, umbones at extreme anterior end, thin-walled, yellow to dark brown, with fine to coarse divaricating radial ribs; interior with purple-brown blotches, with one to four small purplish dysodont hinge teeth, margin denticulate. New Jersey to Florida, Bermuda, Bahamas, West Indies, Gulf of Mexico, Caribbean Central America, South America (to Argentina), also St. Helena. Length 10 mm (to 46 mm). Syn. *domingensis* Lamarck, 1819. Compare *Brachidontes modiolus*, which is larger and more elongate (when adult) and yellow in color.

## Family description

The mytilid shell is small to medium-sized (to 100 mm), thin-walled, pyriform to antero-posteriorly elongated, with the anterior end often narrowly pointed and the posterior rounded. It is usually EQUIVALVE, inflated, and often gaping slightly anteroventrally (but without a byssal gape). Some members (e.g., *Modiolus* and *Trisidos*) are decidedly IN-EQUIVALVE, with the anterior end of the shell twisted relative to the posterior end, as a measure to maintain a low profile in a semi-INFAUNAL posture. The shell is strongly IN-EQUILATERAL (umbones at or near the anterior end), with PROSO- or ORTHOGYRATE UM-BONES. Shell microstructure is ARAGONITIC, CALCITIC, or a mixture of both, and three-lay-ered, with a calcitic PRISMATIC outer layer, an aragonitic NACREOUS (or rarely CROSSED LAMELLAR) middle layer, and an aragonitic (sheet or lenticular) nacreous (or COMPLEX CROSSED LAMELLAR or HOMOGENOUS) inner layer; Lithophaginae differs from this general pattern, and from most other bivalves, in having a nacreous outer layer and a prismatic inner layer. TUBULES are present in some species (e.g., *Lithophaga* and *Modiolus*), but pen-etrate only the inner shell layer. Exteriorly mytilids are covered by an adherent, usually

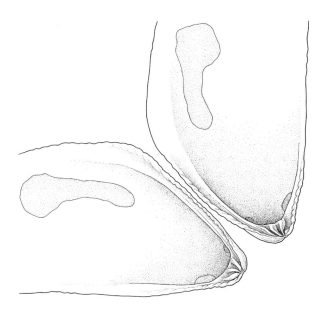

brown or black, frequently hirsute PE-
RIOSTRACUM, which can be partially
calcified; it is three-layered in *Mytilus
edulis*. Sculpture is smooth, radial, or
commarginal, often in several zones.
LUNULE and ESCUTCHEON are absent.
Interiorly the shell is thinly nacreous
(except in lithophagines, which are
non-nacreous). The PALLIAL LINE is
usually ENTIRE and obscure, rarely with
a small posterior concavity, and with a
SINUS in the siphonate lithophagines.
The inner shell margins are smooth or
denticulate. The HINGE PLATE is narrow
and usually weak, EDENTATE in adults or
with small denticles near the beak
(DYSODONT) and vertical crenulations
on one or both sides of the ligament
corresponding to the termini of radial
sculpture and forming a transversely rugose dorsal margin. The LIGAMENT is
PLANI- or ALIVINCULAR and OPISTHODETIC, extending along the dorsal margin about half
the length of the shell, and set on NYMPHS; *Dacrydium* also has an internal portion (RESIL-
IUM) set on a small subumbonal RESILIFER.

Mytilids are the most species-rich family of HETEROMYARIAN bivalves. The anterior
ADDUCTOR MUSCLE is small (absent in *Perna*); the posterior pedal retractor muscles are
large, multiple, and usually insert continuously along a zone anterior to the posterior adductor muscle. The anteriormost muscle bundles of these zones are the pedal elevator
muscles. Pedal protractor muscles are absent. The MANTLE margins are not fused ventrally

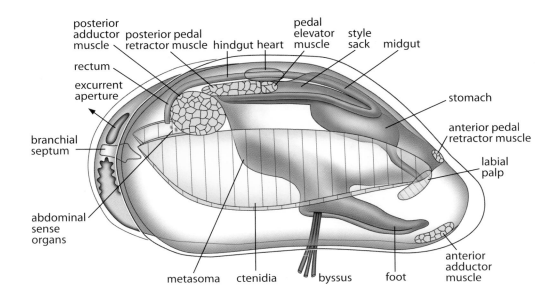

and can be distinctly colored within species. SIPHONS are usually absent (present in Lithophaginae), but the inner folds are fused posteriorly to form a branchial SEPTUM that isolates an EXCURRENT APERTURE and aids in separating the INFRA- and SUPRABRANCHIAL CHAMBERS. HYPOBRANCHIAL GLANDS have not been reported. Marginal tentacles have been reported for some species along the margins of the ventral gape. The rock-boring lithophagines are unusual mytilids in having long, extensible, united or separate siphons equipped with siphonal retractor muscles; the INCURRENT SIPHON is not fused ventrally, becoming functionally complete by apposition of the free edges; the EXCURRENT SIPHON has a basal PROXIMAL VALVE to control water flow. Rock borers also possess specialized dorsal pallial glands that secrete a calcium-binding mucoprotein (once incorrectly characterized as an acid) to excavate burrows. The FOOT is long and digitiform; active and relatively rapid crawling on the highly extensible foot has been recorded for some species. The foot has a BYSSAL GROOVE; the adults are famously byssate often forming extensive "mussel beds" on hard substrata.

The LABIAL PALPS are narrow and curved. The CTENIDIA are FILIBRANCH (ELEUTHER-ORHABDIC), HOMORHABDIC or HETERORHABDIC, and inserted (but not fused) into the distal oral groove of the palps (CATEGORY I association). CEPHALIC EYES are often present (e.g., *Mytilus*). On each side between the inner demibranch and the visceral mass, a vascularized longitudinal series of transverse folds (PLICATE ORGAN) serves as an accessory respiratory organ. Incurrent and excurrent water flows are posterior; incurrent flow is also anterior. The STOMACH is TYPE III. It is simplified in hydrothermal vent–inhabiting bathymodiolines, which supplement their nutrition by endosymbiotic chemautotrophic bacteria in the gills. The MIDGUT is coiled on the right or left side. The HINDGUT passes either dorsal to, ventral to, or through the ventricle of the heart, and leads to a sessile rectum. Mytilids are usually GONOCHORISTIC and produce relatively long-lived planktonic VELIGER larvae; brooding has been reported in several species, including the nest-building *Musculus lateralis*, which retains fertilized eggs within the nest (but outside the parent). The gonad often branches into the mantle tissues and the distensible posterior part of the visceral mass (METASOMA); in *Mytilus edulis*, the presence of gametes causes the mantle to be differently colored in males (beige) and females (reddish). The nervous system not concentrated. STATOCYSTS (open, with STATOCONIA) have been reported in adult *Mytilus*. ABDOMINAL SENSE ORGANS are present.

Mytilids are marine or estuarine (rarely freshwater) SUSPENSION FEEDERS. They are usually EPIBYSSATE, forming dense gregarious beds in shallow water, but show tremendous variation in this regard throughout the family. Lithophagines are (secondarily) chemical burrowers in soft rock or coral, lining their burrows with calcium carbonate derived from their dissolved substratum; diminutive *Musculus lateralis* can be free-living, a nestler that builds "nests" of byssal threads and mucus, or a commensal species embedded within the outer tunic of ascidians; *Modiolus* is ENDOBYSSATE in gravelly sand, attached to small subsurface stones but lying on one side or the other with the shell only partly buried in the sediment. The exposed surfaces of larger mytilid species are often colonized by encrusting algae, bryozoans, barnacles, worm tubes, and other fouling organisms; however, the thick mytilid periostracum apparently deters certain predators such as boring muricid gastropods. Other predators include sea stars, birds, and walrus. Pea crabs (Pinnotheridae) are commensal in the MANTLE CAVITY of some species

***Brachidontes domingensis*** (Lamarck, 1819) is now regarded as a form of *B. exustus*. According to a recent molecular study (Lee & Ó Foighil, 2004), at least four genotypes exist in Florida, suggesting cryptic species originating along the Atlantic and Gulf coasts, Key Biscayne, and the Bahamas. None of these genotypes corresponds morphologically to *B. domingensis*. However, some evidence suggests that this form (characterized by thicker, darker shells with fewer radial ribs) is ecophenotypic; it lives in oceanic, wave-tossed waters, whereas typical *B. exustus* prefers slightly brackish conditions.

of *Mytilus*. Parasitic copepods, polychaete worms, trematodes, and spore-forming protists have been recorded. Small pearls are frequent, naturally nucleated by encysted parasites, and when abundant, often cause closure of mussel beds.

The family Mytilidae is known since the Devonian, is represented by ca. 70 living genera and 250–400 species (depending on author), and is found in all seas from the intertidal zone to the deep sea near hydrothermal vents. Evolution of the heteromyarian "mytiliform" body involved the NEOTENOUS retention of the byssus, followed by loss of locomotory function by the foot and reduction of the anterior (plus expansion of the posterior) muscles. Many mytilids worldwide are harvested or cultured for human food, especially blue mussels (*Mytilus*) and green mussels (*Perna*). Some species of *Lithophaga* and *Botula* present a conservation issue in their habits of boring into living and dead corals, thus undermining the structure of coral reefs. The Asian fresh- or brackish-water Golden Mussel, *Limnoperna fortunei* (Dunker, 1857), has become a serious invasive pest species in South America, fouling natural and artificial hard substrata, competing with native species, and clogging intake pipes of power plants and other water-dependent facilities.

## References

Bertrand, G. A. 1971. The ecology of the nest-building bivalve *Musculus lateralis* commensal with the ascidian *Molgula occidentalis*. *The Veliger*, 14(1): 23–29, 2 pls.

Chanley, P. E. 1970. Larval development of the hooked mussel, *Brachidontes recurvus* Rafinesque (Bivalvia: Mytilidae) including a literature review of larval characteristics of the Mytilidae. *Proceedings of the National Shellfisheries Association*, 60: 86–94.

Fox, R. 2001 (updated 2004). *Invertebrate Anatomy Online*, Mytilus edulis, *Blue Mussel*. http://www.lander.edu/rsfox/310mytilusLab.html, last accessed 28 October 2005.

Gosling, E., ed. 1992. *The Mussel* Mytilus: *Ecology, Physiology, Genetics and Culture*. Elsevier, Amsterdam, The Netherlands, xiii + 589 pp.

Harper, E. M., and P. W. Skelton. 1993. A defensive value of the thickened periostracum in the Mytiloidea. *The Veliger*, 36(1): 36–42.

Jaccarini, V., W. H. Bannister, and H. Micallef. 1968. The pallial glands and rock boring in *Lithophaga lithophaga* (Lamellibranchia, Mytilidae). *Journal of Zoology*, 154: 397–401.

Jukes-Brown, A. J. 1905. A review of the genera of the family Mytilidae. *Proceedings of the Malacological Society of London*, 6(4): 211–224.

Kafanov, A. I., and A. L. Drozdov. 1998. Comparative sperm morphology and phylogenetic classification of Recent Mytiloidea (Bivalvia). *Malacologia*, 39(1–2): 129–139.

Lee, T., and D. Ó Foighil. 2004. Hidden Floridian biodiversity: mitochondrial and nuclear gene trees reveal four cryptic species within the scorched mussel, *Brachidontes exustus*, species complex. *Molecular Ecology*, 13: 3527–3542.

Morton, B. 1982. The mode of life and functional morphology of *Gregariella coralliophaga* (Gmelin 1791) (Bivalvia: Mytilacea) with a discussion on the evolution of the boring Lithophaginae and adaptive radiation in the Mytilidae. *Proceedings of the International Marine Biological Workshop*, 1(2): 875–895.

Pojeta, J., Jr., and T. J. Palmer. 1976. The origin of rock boring in mytilacean pelecypods. *Alcheringa*, 1(2): 167–179.

Savazzi, E. 1985. Adaptive significance of shell torsion in mytilid bivalves. *Palaeontology*, 27(2): 307–314.

Seed, R. 1992. Systematics, evolution and distribution of mussels belonging to the genus *Mytilus*: an overview. *American Malacological Bulletin*, 9(2): 123–137.

Soot-Ryen, T. 1955. A report on the family Mytilidae (Pelecypoda). *Allan Hancock Pacific Expeditions*, 20(1): 1–175.

Turner, R. D., and K. J. Boss. 1962. The genus *Lithophaga* in the western Atlantic. *Johnsonia*, 4(41): 81–116.

Valentich-Scott, P., and G. E. Dinesen. 2004. Rock and coral boring Bivalvia (Mollusca) of the middle Florida Keys, U.S.A., In: R. Bieler and P. M. Mikkelsen, eds., *Bivalve Studies in the Florida Keys*, Proceedings of the International Marine Bivalve Workshop, Long Key, Florida, July 2002. *Malacologia*, 46(2): 339–354.

White, K. M. 1937. *Mytilus*. L. M. B. C. *[Liverpool Marine Biological Committee] Memoirs on Typical British Marine Plants & Animals*, 31: 1–117, 10 pls.

Wilson, B. R., and E. P. Hodgkin. 1967. A comparative account of the reproductive cycles of five species of marine mussels (Bivalves: Mytilidae) in the vicinity of Fremantle, Western Australia. *Australian Journal of Marine and Freshwater Research*, 18(2): 175–203.

Wilson, B. R., and R. Tait. 1984. Systematics, anatomy and boring mechanisms of the rock-boring mytilid bivalve *Botula*. *Proceedings of the Royal Society of Victoria*, 96(3): 113–125.

Yonge, C. M. 1955. Adaptation to rock boring in *Botula* and *Lithophaga* (Lamellibranchia, Mytilidae) with a discussion on the evolution of the habit. *Quarterly Journal of Microscopical Science*, 96(3): 383–410.

Yonge, C. M., and J. I. Campbell. 1968. On the heteromyarian condition in the Bivalvia with reference to *Dreissena polymorpha* and certain Mytilacea. *Transactions of the Royal Society of Edinburgh*, 68(2): 21–43.

Zaixso, H. E. 2003. Sistema nervioso y receptors en la cholga, *Aulacomya atra atra* (Bivalvia: Mytilidae). *Revista de Biología Marina y Oceanografía*, 38(2): 43–56.

### *Brachidontes modiolus* (Linnaeus, 1767) – **Yellow Mussel**

Elongated fan-shaped, expanded winglike dorsally, umbones at extreme anterior end, often stained purple and bordered by brown commarginal bands, thin-walled, whitish with yellow periostracum, with numerous wavy, divaricating radial ribs; interior white with purple stain, posterior margin weakly denticulate, hinge with four small white dysodont teeth. North Carolina to Florida, Bahamas, West Indies, Gulf of Mexico, Caribbean Central America, South America (to Venezuela). Length 30 mm (to 67 mm). Syn. *citrinus* Röding, 1798. Compare *Brachidontes exustus*, which is smaller and brown to black in color.

### *Lioberus castaneus* (Say, 1822) – **Say's Chestnut Mussel**

Elongated oval, flattened ventrally, inflated, umbones near (but not at) anterior end, thin-walled, chestnut to dark brown, anteriorly glossy, posteriorly dull with fine, gray, matted periostracum, smooth; interior bluish white with irregular surface, hinge edentate with slight swelling under umbones, margin smooth. North Carolina to Florida, West Indies, Gulf of Mexico, Caribbean Central America, South America (Colombia, Brazil). Length 17 mm. Compare *Botula fusca*, which is arched and has inrolled umbones.

### *Ischadium recurvum* (Rafinesque, 1820) – **Hooked Mussel**

Elongated fan-shaped, hooked anteriorly, compressed, umbones at extreme anterior end, solid, dark gray-black, with numerous wavy radial ribs; interior purplish to rosy brown with lighter narrow border, margin denticulate, three or four small dysodont hinge teeth displaced ventrally, anterior adductor muscle absent. Massachusetts to Florida, West Indies, Gulf of Mexico, Caribbean Central America. Length 30 mm (to 40 mm). Compare *Geukensia granosissima*, which is yellow to brown, is less anteriorly pointed, and not strongly hooked.

### *Gregariella coralliophaga* (Gmelin, 1791) – **Coral-eating Mussel**

Irregularly cylindrical, posteriorly pointed, with dorsal umbonal ridge from umbo to posterior end, inflated, umbones near (but not at) anterior end and curving anteriorly and inward, fragile, white with thick brown periostracum tufted on posterodorsal slope, with fine, divaricating radial riblets that are absent from center of the valves and that curve posteroventrally from dorsal ridge; interior bluish white, margin finely denticulate, hinge with few teeth; living bored into soft rock or dead coral. North Carolina to Florida, Bermuda, Bahamas, West Indies, Gulf of Mexico, Caribbean Central America, South America (Suriname, Brazil), also eastern Pacific. Length 14 mm (to 19 mm). Syn. *chenui* Récluz, 1842.

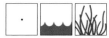

### *Geukensia granosissima* (G. B. Sowerby III, 1914) – **Southern Ribbed Mussel**

Elongated fan-shaped, umbones near (but not at) anterior end, solid, white with yellow to black-brown periostracum, with coarse, beaded, divaricating radial ribs that are weaker anteroventrally; interior bluish white, margin denticulate, hinge teeth obsolete, two anterior muscle scars near umbonal cavity. Florida, Gulf of Mexico, Caribbean Central America, South America (Venezuela). Length 45 mm (to 65 mm). Formerly known as *demissa* Dillwyn, 1817 (eastern Canada to northern Florida). Compare *Ischadium recurvum*, which is gray to black, is more anteriorly pointed, and strongly hooked.

**Some mytilids attach** to small rocks below the sediment surface among salt-marsh vegetation or seagrass, such as these *Geukensia demissa* (Dillwyn, 1817) from the mid-Atlantic coastline of the eastern United States.

*Brachidontes modiolus*

*Lioberus castaneus*

*Ischadium recurvum*

*Gregariella coralliophaga*

*Geukensia granosissima*

**_Modiolus americanus_** (Leach, 1815) – **Tulip Mussel**

Elongated fan-shaped, inflated, umbones near (but not at) anterior end and pink or purple (never white), thin-walled, white to purple with glossy brown periostracum adorned by soft brown strap-like hairs (see inset), with white oblique band at center, occasionally with brown to purple radial bands, smooth; interior white to rose, occasionally purplish, margin smooth, hinge edentate. North Carolina to Florida, Bermuda, Bahamas, West Indies, Gulf of Mexico, Caribbean Central America, South America (to Brazil), also eastern Pacific. Length 55 mm (to 85 mm). Syn. _tulipa_ Lamarck, 1819. Compare _Modiolus squamosus_, which has white umbones and radial bands, and trigonal periostracal hairs.

**_Modiolus squamosus_** Beauperthuy, 1967 – **False Tulip Mussel**

Elongated fan-shaped, umbones white and near (but not at) anterior end, thin-walled, brownish to purple with with glossy brown periostracum adorned by small trigonal hairs (see insets), with oblique whitish band (never with brown to purple radials), smooth; interior purplish to whitish, margin smooth, hinge edentate. North Carolina to Florida, Bahamas, West Indies, Gulf of Mexico, Caribbean Central America, South America (Venezuela). Length 30 mm (to 43 mm). Formerly a subspecies of the Northern Horsemussel, _Modiolus modiolus_ (Linnaeus, 1758). Compare _Modiolus americanus_, which has pink to purplish umbones and radial bands, and narrow straplike periostracal hairs.

**The Indo-Pacific Green Mussel**, _Perna viridis_ (Linnaeus, 1758), introduced to western Florida in 1999, is spreading rapidly in shallow marine habitats but is not yet recorded in the Florida Keys. At a maximum length of 100 mm, Green Mussels grow densely on boats, piers, mangrove roots, water-treatment equipment, and other hard structures.

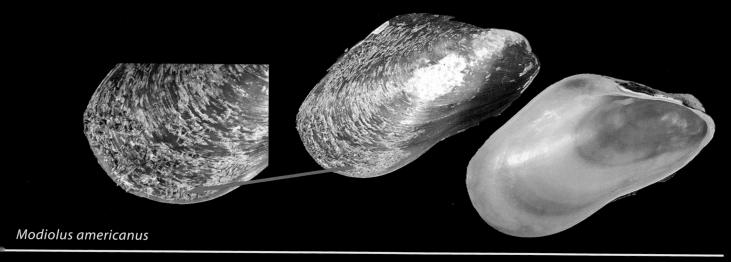

*Modiolus americanus*

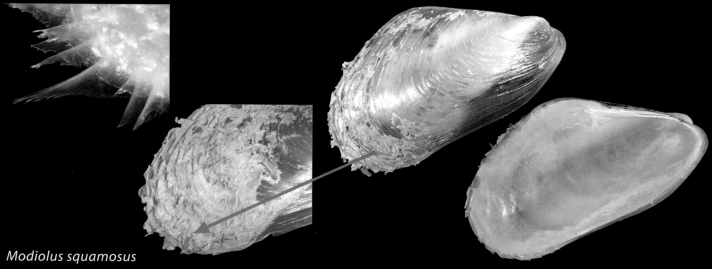

*Modiolus squamosus*

*Perna viridis*

ALERT

GREEN MUSSEL

### *Musculus lateralis* (Say, 1822) – **Lateral Mussel**

Oval, inflated, umbones near (but not at) anterior end, fragile, translucent green, pink, or yellow, occasionally with darker zigzag bands, with radial ribs that are absent from a trigonal area at center of valve, margin denticulate, hinge finely dentate. North Carolina to Florida, Bermuda, Bahamas, West Indies, Gulf of Mexico, Caribbean Central America, South America (Colombia, Venezuela, Brazil). Length 5 mm (to 12 mm).

### *Amygdalum papyrium* (Conrad, 1846) – **Atlantic Paper Mussel**

Elongated cylindrical, wider at posterior end, compressed, umbones near (but not at) anterior end, fragile, white with glistening bluish green to yellowish brown periostracum, smooth; interior iridescent white, margin smooth. Maryland to Florida, Gulf of Mexico, South America (Venezuela). Length 18 mm (to 22 mm).

### *Amygdalum politum* (Verrill & Smith, 1880) – **Polished Paper Mussel**

Elongated oval, wider at posterior end, compressed, umbones near (but not at) anterior end, fragile, white, smooth; interior white, margin smooth. North Atlantic, Florida Keys, West Indies, Gulf of Mexico, South America (Colombia). Length 11 mm (to 42 mm). Compare other *Amygdalum* species, which are either greenish or patterned with cobwebby streaks.

### *Amygdalum sagittatum* (Rehder, 1935) – **Arrow Paper Mussel**

Elongated oval, wider at posterior end, compressed, umbones near (but not at) anterior end, fragile, translucent shiny ivory white, posterior half with cobwebby streaks, smooth; interior with exterior coloration showing through, umbo reinforced by small smooth rib, margin smooth. Florida, West Indies, Gulf of Mexico, South America (to Uruguay). Length 10 mm (to 21 mm). Compare other *Amygdalum* species, which lack the cobwebby radial color pattern.

### *Crenella decussata* (Montagu, 1808) – **Cross-sculptured Crenella**

Oval, inflated, umbones central to slightly anterior, thin-walled, white or tan to yellowish gray, with numerous fine, divaricating radial ribs; interior glossy white and faintly iridescent, margin finely denticulate, hinge finely dentate, dorsal hinge margin finely vertically striate. Arctic circumboreal, eastern Canada to Florida, Bahamas, West Indies, Gulf of Mexico, Caribbean Central America, South America (to Argentina). Length 2 mm (to 6 mm). Syn. *divaricata* d'Orbigny, 1853. Note: Also known as Decussate Crenella.

### *Dacrydium elegantulum hendersoni* Salas & Gofas, 1997 – **Henderson's Glassy Mussel**

Oval to rounded D-shaped, inflated, more compressed posteriorly, umbones near (but not at) anterior end, orthogyrate and touching, fragile, translucent white with iridescent surface, smooth, often with attached silt or sand grains; interior glossy, margin smooth. Florida, Bahamas, Gulf of Mexico. Length 4 mm. Formerly known as *vitreum* Møller, 1842 (North Atlantic, northern Europe, northern Pacific).

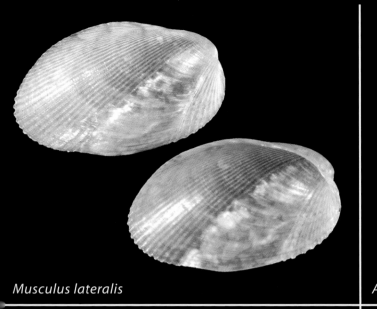

*Musculus lateralis*

*Amygdalum papyrium*

*Amygdalum politum*

*Amygdalum sagittatum*

*Crenella decussata*

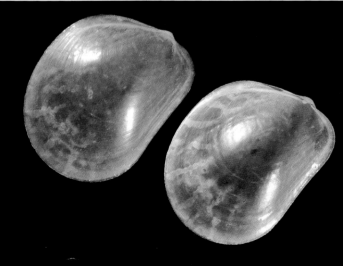

*Dacrydium elegantulum hendersoni*

### *Lithophaga antillarum* (d'Orbigny, 1853) – **Giant Date Mussel**
(nomen protectum, see p. 413)

Elongated cylindrical, inflated, umbones near (but not at) anterior end, thin-walled, cream-white with yellow-brown periostracum, with numerous irregular riblets; interior iridescent cream with purplish tint, margin smooth; living bored into soft rock or dead coral. North Carolina to Florida, Bermuda, Bahamas, West Indies, Gulf of Mexico, Caribbean Central America, South America (Colombia, Brazil), also New Caledonia, Gulf of Siam, India. Length 105 mm (to 110 mm). Compare other *Lithophaga* species (*L. nigra* is black-brown, whereas *L. aristata* and *L. bisulcata* have calcareous posterior extensions) and *Coralliophaga coralliophaga*, which is whiter and thicker shelled with heterodont hinge teeth.

### *Lithophaga aristata* (Dillwyn, 1817) – **Scissor Date Mussel**

Elongated cylindrical, posterior end with scissor-crossed extensions (see inset), inflated, umbones near (but not at) anterior end, solid, whitish with yellow-brown periostracum often with calcareous encrustation, smooth, weakly radially divided by broad rounded keel; interior iridescent yellow-brown with purplish tint, margin smooth; living bored into soft rock or dead coral. North Carolina to Florida, Bahamas, West Indies, Gulf of Mexico, Caribbean Central America, South America (to Venezuela), also eastern Pacific, western Europe. Length 30 mm (to 52 mm). Compare other *Lithophaga* species (*L. antillarum* and *L. nigra* do not have calcareous posterior extensions, whereas those of *L. bisulcata* do not cross scissorlike), and *Coralliophaga coralliophaga*, which is whiter and thicker shelled with heterodont hinge teeth.

### *Lithophaga bisulcata* (d'Orbigny, 1853) – **Mahogany Date Mussel**
(nomen protectum, see pp. 413–414)

Elongated cylindrical, inflated, umbones near (but not at) anterior end, thin-walled, whitish with yellow-brown periostracum and calcareous incrustation that projects bluntly beyond posterior margin, smooth with radial groove dividing valve into two sections, ventral half with periostracal pits (most visible in juveniles); interior yellow-brown with purplish tint, margin smooth; living bored into soft rock or dead coral. North Carolina to Florida, Bermuda, Bahamas, West Indies, Gulf of Mexico, Caribbean Central America, South America (to Uruguay). Length 45 mm. Compare other *Lithophaga* species (*L. antillarum* and *L. nigra* do not have calcareous posterior extensions, whereas those of *L. aristata* cross scissorlike), and *Coralliophaga coralliophaga*, which is whiter and thicker shelled with heterodont hinge teeth.

### *Lithophaga nigra* (d'Orbigny, 1853) – **Black Date Mussel** (nomen protectum, see p. 414)

Elongated cylindrical, inflated, umbones near (but not at) anterior end, thin-walled, cream-white with black-brown periostracum, smooth with coarse ribs on anteroventral third; interior iridescent purplish to bluish gray, margin smooth; living bored into soft rock or dead coral. Florida, Bermuda, Bahamas, West Indies, Gulf of Mexico, Caribbean Central America, South America (Venezuela, Brazil). Length 40 mm (to 65 mm). Compare other *Lithophaga* species (*L. antillarum* is light brown, whereas *L. aristata* and *L. bisulcata* have calcareous posterior extensions), and *Coralliophaga coralliophaga*, which is whiter and thicker shelled with heterodont hinge teeth.

### *Botula fusca* (Gmelin, 1791) – **Cinnamon Mussel**

Elongated oval, slightly arched, inflated, thin-walled, grayish brown with dark brown glossy periostracum, smooth, with coarse commarginal ridges, umbones near (but not at) anterior end, prominent and inrolled; interior purple-brown to white, with small vertical threads on hinge just posterior to ligament, margin smooth; living bored into soft rock or dead coral. North Carolina to Florida, Bermuda, Bahamas, West Indies, Gulf of Mexico, Caribbean Central America, South America (Colombia, Venezuela, Brazil). Length 35 mm. Syn. *cinnamomea* Lamarck, 1819. Compare *Lioberus castaneus*, which is not arched and lacks the inrolled umbones.

**Cracking a piece of dead coral rock** (like this from bayside of Spanish Harbor Key) reveals several species of boring bivalves, here the mytilid *Lithophaga bisulcata* (top, with siphons to right) and the venerid *Petricola lapicida* (bottom, with siphons to left).

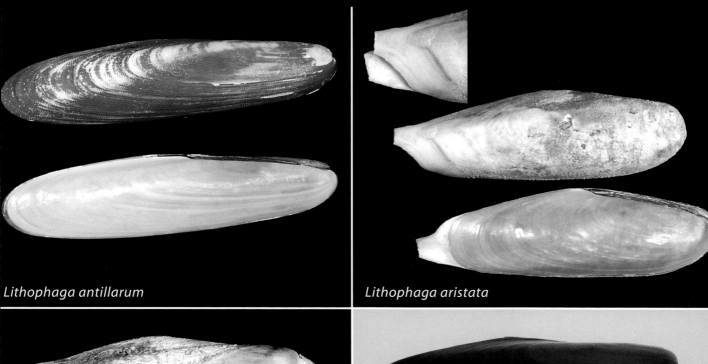

*Lithophaga antillarum*

*Lithophaga aristata*

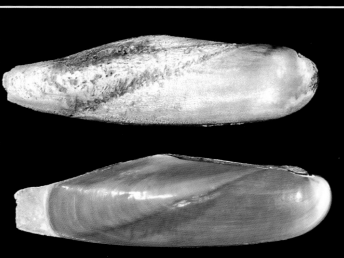

*Lithophaga bisulcata*

*Lithophaga nigra*

*Botula fusca*

# Family Pteriidae — Wing Oysters and Pearl Oysters

**Classification**
AUTOLAMELLIBRANCHIATA Grobben, 1894
Pteriomorphia Beurlen, 1944
Pterioida Newell, 1965
Pterioidea J. E. Gray, 1847 [1820]
Pteriidae J. E. Gray, 1847 [1820]

**Featured species**
*Pinctada longisquamosa* (Dunker, 1852) — **Scaly Pearl Oyster**

Oval to obliquely skewed, posterior auricle elongated by spine(s), compressed, thin-walled, tan to brownish to greenish, with commarginal rows of thin, flattened scales or spines that are often dense, long, and recurved; interior thinly nacreous. Florida, Bermuda, Bahamas, West Indies, Gulf of Mexico, Caribbean Central America, South America (to Venezuela). Length 40 mm. Syn. *xanthia* Schwengel, 1942. Note: The Scaly Pearl Oyster went unrecognized for many years because it was often confused with two other family members common to the western Atlantic—*Pinctada imbricata* because of its spines or *Pteria colymbus* because of its oblique shape. It differs from both of these species by its extremely thin interior nacreous layer, which is thin enough to see through. It is confidently placed in the genus *Pinctada* based on hinge tooth composition, position of the intestine relative to the heart, and molecular sequences. It lives attached by byssal threads to marine vegetation, most commonly on the bayside of the Upper Florida Keys.

**A scanning electron micrograph** of the fractured shell of *Pinctada longisquamosa* shows the large crystals of the outer prismatic layer and the small layered crystals of the inner nacre.

**In the shallow water** of the bayside of Plantation Key, juvenile *Pinctada longisquamosa* attach byssally to epibenthic algae.

A juvenile *Pinctada longisquamosa*, exhibiting two of its common variations, yellow shell color (named *Pteria xanthia* by Schwengel, 1942) and extreme elongation of the posteroventralmost lamella (arrow).

# Family description

The pteriid shell is medium-sized to large (to >300 mm), thin-walled to solid, and obliquely oval to circular, with anterior and posterior AURICLES (posterior usually longer). It is EQUIVALVE to strongly INEQUIVALVE (right valve less convex, byssally notched, and usually resting on it [PLEUROTHETIC on the right valve]), and compressed. The shell is strongly INEQUILATERAL (umbones anterior), with OPISTHOGYRATE UMBONES separated by a narrow CARDINAL AREA. Shell microstructure is a mixture of ARAGONITE and CALCITE, and two-layered, with a calcitic simple PRISMATIC outer layer, and an aragonitic sheet-NACREOUS inner layer. The outer layer often includes scaly lamellae that readily crack and flake, especially in dried shells. TUBULES are absent. Exteriorly pteriids are white to brown to greenish, in some individuals with darker radial stripes, and covered by a nonpersistent PERIOSTRACUM. Sculpture is variable, usually smooth or weakly commarginal; fingerlike lamellae align in radial rows in many species. LUNULE and ESCUTCHEON are absent. Interiorly the shell is thinly to thickly nacreous, with a wide ventral prismatic border. The nacre is widely variable in color, from white to gray to shades of green, blue, yellow, or rose, most famously in the Tahitian populations of the Black-lipped Pearl Oyster, *Pinctada margaritifera*. The PALLIAL LINE is ENTIRE, obscure, and discontinuous. The inner shell margins are smooth. The HINGE PLATE is straight and narrow, EDENTATE in adults (in *Electroma*) or with weak teeth reduced to a single subumbonal denticle fitting into a socket in the opposite valve, plus one set of elongated posterior ridges; teeth are stronger in juveniles, often obsolete in adults. The hinges of *Pinctada* and *Pteria* include identical components but are mirror images of one another. The LIGAMENT is ALIVINCULAR and AMPHIDETIC, extending along the entire dorsal margin; the internal portion (RESILIUM) is set on a subumbonal, posteriorly directed, trigonal RESILIFER.

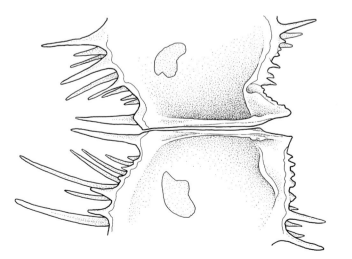

The animal is MONOMYARIAN as an adult (anterior ADDUCTOR MUSCLE absent); the posterior adductor muscle is large and central, and concave to oval in cross section. The posterior pedal retractor muscles are large and insert close to (or within the cavity of) the posterior adductor scar; the anterior pedal retractors are small, and each divides into two portions before insertion at the anterodorsal corner of the shell. Pedal protractors and elevators are absent. The MANTLE margins are not fused ventrally; SIPHONS are absent. The inner mantle folds form a PALLIAL VEIL. Simple PALLIAL EYES (without lens) are present on the outer folds of

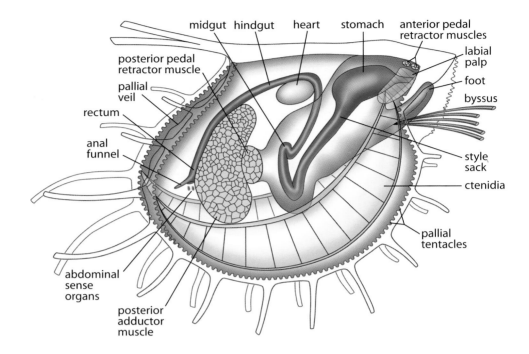

midgut　hindgut　heart　stomach　anterior pedal retractor muscles

posterior pedal retractor muscle

pallial veil

rectum

anal funnel

abdominal sense organs

posterior adductor muscle

labial palp

foot

byssus

style sack

ctenidia

pallial tentacles

some species. HYPOBRANCHIAL GLANDS have not been reported. The FOOT is anterior, small, and digitiform. It has a BYSSAL GROOVE; the adult is byssate.

The LABIAL PALPS are small. The CTENIDIA are large and partially encircle the centralized muscle bundles. They are FILIBRANCH or PSEUDOLAMELLIBRANCH (ELEUTHERORHABDIC), usually HETERORHABDIC, and not inserted into (or fused with) the distal oral groove of the palps (CATEGORY III association). CEPHALIC EYES are present in most species (on both sides in *Pteria,* in some *Pinctada* on the left side only, and absent in *Electroma*). Distally the tips of the ctenidia are attached to the inner mantle fold, separating INFRA- and SUPRABRANCHIAL CHAMBERS; water flow is anteroposterior. The STOMACH is TYPE III. The MIDGUT is not coiled. The HINDGUT usually passes through (dorsal to in *Pinctada*) the ventricle of the heart, and leads to a rectum with an ANAL FUNNEL. Pteriids are usually GONOCHORISTIC (some are PROTANDRIC HERMAPHRODITES) and produce planktonic VELIGER larvae. The nervous system is not concentrated. STATOCYSTS are present in adults, with numerous STATOCONIA. ABDOMINAL SENSE ORGANS are present and asymmetrical.

Pteriids are marine SUSPENSION FEEDERS. They are EPIBYSSATE on hard substrata (rocks, seafans, other shells), often gregariously so. Some are cryptically colored attached to marine plants (e.g., *Pinctada longisquamosa* on seagrass) or heavily coated by epibiotic growth. Commensal pea crabs (Pinnotheridae) or pearlfish (*Onuxodon*, Carapidae) are reported in some species.

The family Pteriidae is known since the Triassic, is represented by 3 living genera and ca. 60 species, and is distributed worldwide in subtropical and tropical seas, from the intertidal zone to deep waters. Species of *Pinctada* and *Pteria* are widely used for mother-of-pearl and for cultured pearl production, most commonly in the Indo-Pacific region. The Atlantic Pearl Oyster (*Pinctada imbricata*) was harvested during the 1500s to near extinction at certain Caribbean localities for its diminutive yellowish natural pearls, but has not

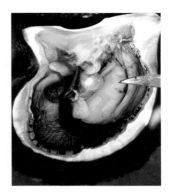

**A five-year-old _Pinctada maxima_** (Jameson, 1901), the Silver-lipped Pearl Oyster, opened after two years in perliculture, reveals its colorful mantle edge, gray-black ctenidia, and a cultured pearl still embedded in the "pearl sack" within the gonad (to left of knife tip).

been successfully used in commercial pearl culturing. The Black-lipped Pearl Oyster (_Pinctada margaritifera_), known for its "black" South Seas or Tahitian cultured pearls, has become established off of southeastern Florida, with reports of single living specimens since the 1990s; its likely source of introduction is ship ballast or fouling. The muscle meat of pteriids is consumed locally in Japan and Australia. Growth rates, diseases, genetics, symbionts, predators, shell and pearl formation, and reproduction are well documented for the commercial species.

## References

Herdman, W. A. 1904. Anatomy of the pearl oyster (_Margaritifera vulgaris_, Schum.). Pages 37–69, pls. 1–9, in: W. A. Herdman, ed., _Report to the Government of Ceylon on the Pearl Oyster Fisheries of the Gulf of Manaar. Part II._ The Royal Society, London.

Landman, N. H., P. M. Mikkelsen, R. Bieler, and B. Bronson. 2001. _Pearls: A Natural History._ Harry N. Abrams, New York, 232 pp.

Mikkelsen, P. M., I. Tëmkin, R. Bieler, and W. G. Lyons. 2004. _Pinctada longisquamosa_ (Dunker, 1852) (Bivalvia: Pteriidae), an unrecognized pearl oyster in the western Atlantic. In: R. Bieler and P. M. Mikkelsen, eds., _Bivalve Studies in the Florida_ Keys. Proceedings of the International Marine Bivalve Workshop, Long Key, Florida, July 2002. _Malacologia_, 46(2): 473–501.

Ranson, G. 1961. Les espèces d'huîtres perlières du genre _Pinctada_ (biologie de quelques-unes d'entre elles). _Institute Royal des Sciences Naturelles de Belgique, Mémoires, Série 2_, 67: 1–95, 42 pls.

Shiino, S. M. 1952. _Anatomy of_ Pteria (Pinctada) martensii _(Dunker), Mother-of-Pearl Mussel._ Special Publication of the Fisheries Experimental Station of Mie Prefecture, 12 pls.

Shirai, S. 1994. _Pearls and Pearl Oysters of the World._ Marine Planning Company, Okinawa, Japan, 108 pp.

Takemura, Y., and T. Kafuku. 1957. Anatomy of the Silver-lip Pearl Oyster _Pinctada maxima_ (Jameson). _Bulletin of Tokai Regional Fisheries Research Laboratory_, 16: 39–40.

Tëmkin, I. 2006. Morphological perspective on classification and evolution of Recent Pterioidea (Mollusca: Bivalvia). _Zoological Journal of the Linnean Society_, 148: 253–312.

Waller, T. R. 1978. Morphology, morphoclines and a new classification of the Pteriomorphia. _Philosopical Transactions of the Royal Society of London, Series B, Biological_ Sciences, 284: 345–365.

### *Pinctada imbricata* Röding, 1798 – **Atlantic Pearl Oyster**

Circular to quadrangular, with subequal auricles, compressed, thin-walled, tan to brownish with mottlings or radial rays of purplish brown or black, with radial rays of overlapping flat scales or spines; interior nacre thick and yellowish, non-nacreous border with alternating light and dark bars. North Carolina to Florida, Bermuda, Bahamas, West Indies, Gulf of Mexico, Caribbean Central America, South America (to Brazil). Length 60 mm. Formerly known as *radiata* Leach, 1814 (Persian Gulf, Arabian Gulf, Mediterranean Sea).

### *Pinctada margaritifera* (Linnaeus, 1758) – **Black-lipped Pearl Oyster**

Circular, with subequal auricles, compressed, solid, light to dark brown to black, with stout commarginal flat scales; interior nacre thick and variably colored from silver to green to blue to gray-black. Native to the Indo-Pacific, introduced to Florida. Length 60 mm (to 300 mm). Note: The photographed specimen, collected in Florida in 1950 (and thus considerably predating the recent discoveries), is a particularly light-nacred specimen.

### *Pteria colymbus* (Röding, 1798) – **Atlantic Wing Oyster**

Obliquely oval, with posterior auricle elongated, solid, dark brown, often with lighter radial rays, smooth or with very fine flat spines, matted periostracum forming radial rows on small invididuals; interior nacre thick and whitish, bordered by wide non-nacreous margin. North Carolina to Florida, Bermuda, Bahamas, West Indies, Gulf of Mexico, Caribbean Central America, South America (to Uruguay). Length 55 mm (to 73 mm).

### *Pteria vitrea* (Reeve, 1857) – **Glassy Wing Oyster**

Obliquely oval, posterior auricle elongated, light brown to white, with very fine flat spines; interior nacre thin. Massachusetts, North Carolina, Florida, Gulf of Mexico. Length 25 mm. Note: The hole in the left valve of this specimen is probably evidence of drilling by a predatory gastropod.

**Living *Pteria colymbus*** byssally attach to upright gorgonians on sand flats and patch reefs, aligning themselves with plankton-laden currents. This specimen was photographed in shallow water near Dove Key off Key Largo. Such vertical attachment is typical of species of the genus *Pteria*.

**A living *Pinctada imbricata***, at the same location as the previous *Pteria colymbus*, is byssally attached to a piece of dead coral on the sand bottom. In contrast to *Pteria*, species in the genus *Pinctada* typically reside near or on the sea bottom, often under rock slabs.

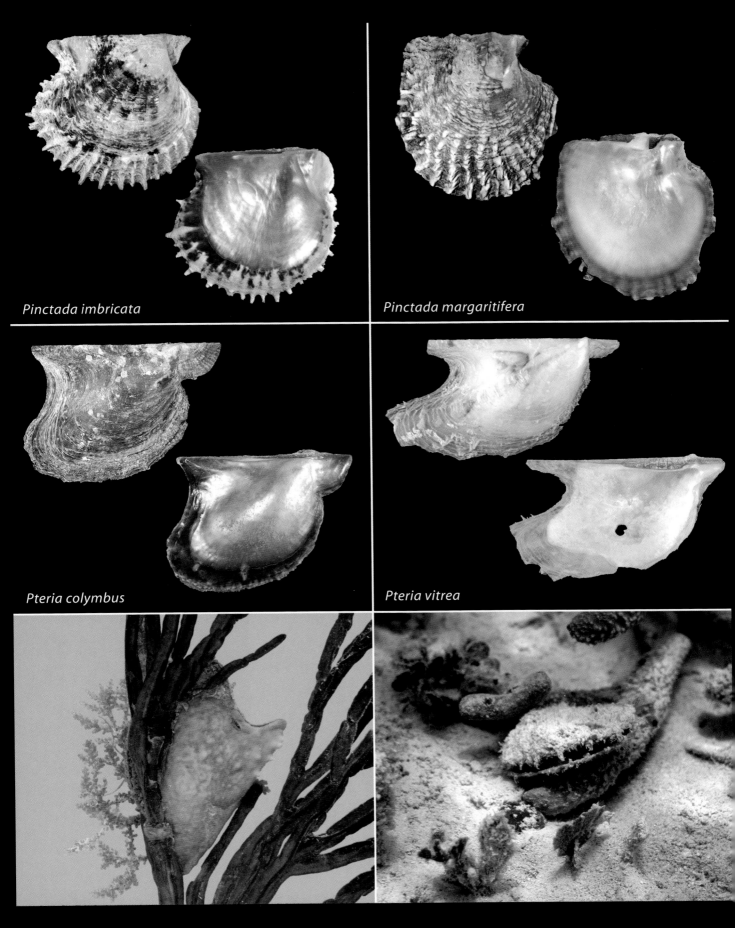

*Pinctada imbricata*

*Pinctada margaritifera*

*Pteria colymbus*

*Pteria vitrea*

# Family Isognomonidae – Tree Oysters or Toothed Oysters

**Classification**
AUTOLAMELLIBRANCHIATA Grobben, 1894
Pteriomorphia Beurlen, 1944
Pterioida Newell, 1965
Pterioidea J. E. Gray, 1847 [1820]
Isognomonidae Woodring, 1925 [1828]

**Featured species**
*Isognomon alatus* (Gmelin, 1791) – **Flat Tree Oyster**

Irregularly obliquely quadrangular or fan-shaped, extremely compressed, thin-walled, drab greenish white to purple-black, with flaky commarginal lamellations; interior nacre yellowish with purple-brown stain, non-nacreous border darker. Florida, Bermuda, Bahamas, West Indies, Gulf of Mexico, Caribbean Central America, South America (to Brazil). Length 75 mm (to 95 mm).

*Isognomon alatus*, also called the Mangrove Oyster, forms large compact colonies or "oyster bars" on dock pilings and cement seawalls in the Florida Keys (here on a PVC piling at Ramrod Key).

## Family description

The isognomonid shell is small to medium-sized (to 150 mm), thin-walled, and irregularly quadrangular to narrowly dorsoventrally elongated, with weakly defined posterior (and rarely anterior) AURICLES. It is EQUIVALVE to strongly INEQUIVALVE (right valve less con-

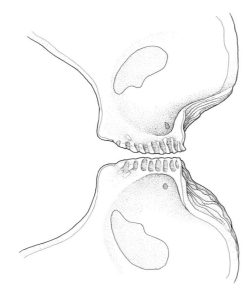

vex), laterally compressed to extremely flattened, and with BYSSAL NOTCH in the less-convex right valve but otherwise not gaping. Most species byssally attach with the less-convex right valve against the substratum (PLEUROTHETIC on the right valve). The shell is INEQUILATERAL (umbones anterior), with OPISTHOGYRATE UMBONES separated by a narrow CARDINAL AREA. Shell microstructure is a mixture of ARAGONITE and CALCITE, and two-layered, with a calcitic PRISMATIC outer layer and an aragonitic NACREOUS inner layer. The outer layer is often lamellar and readily cracks and flakes, especially in dried shells. The inner nacreous layer does not extend to the ventral margin, leaving a broad interior prismatic margin; in *Isognomon bicolor*, this layer is bordered by a thickened VISCERAL RIM that delimits the functional ventral body cavity when the shell is fully closed. TUBULES are absent. Exteriorly isognomonids are covered by a thin, nonpersistent PERIOSTRACUM. Sculpture is smooth or with irregular commarginal lamellae and in some cases with fine to moderate radial sculpture. LUNULE and ESCUTCHEON are absent. Interiorly the shell is nacreous near the umbones, with a wide ventral prismatic border,

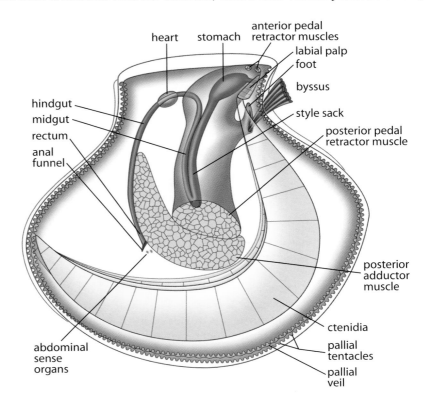

and prismatic inside the AURICLES (when developed). The PALLIAL LINE is ENTIRE and discontinuous. The inner shell margins are smooth. The HINGE PLATE is straight and narrow, EDENTATE in adults (in *Crenatula* and some *Isognomon*) or with weak teeth reduced to a single subumbonal denticle fitting into a socket in the opposite valve and/or one set of elongated posterior ridges; teeth are stronger in juveniles, often obsolete in adults. The LIGAMENT is MULTIVINCULAR and OPISTHODETIC, comprised of numerous, short, parallel internal portions (RESILIA) set in RESILIFERS, which decrease in size posteriorly and increase in number ontogenetically.

The animal is MONOMYARIAN (anterior ADDUCTOR MUSCLE absent); the posterior adductor muscle is large, central, and concave in cross section. The posterior pedal retractor muscles are usually large (absent in *Crenatula*) and insert centrally near or within the cavity of the posterior adductor; the anterior pedal retractors are small, distally divided, and inserted anterodorsally. Pedal protractor and elevator muscles are absent. The MANTLE margins are not fused ventrally and can be darkly pigmented; SIPHONS are absent. The inner mantle folds form a PALLIAL VEIL. Simple PALLIAL ("siphonal") EYES (cap-shaped or cellular, without lens) are present in *Isognomon* (absent in *Crenatula*) on the outer folds under the periostracum. HYPOBRANCHIAL GLANDS have not been reported. The FOOT is anterior, small, and digitiform. It has a BYSSAL GROOVE; the adult is strongly byssate except in sponge-dwelling forms (*Crenatula*).

The LABIAL PALPS are small and trigonal. The CTENIDIA are large and partially encircle the centralized muscle bundles; like the mantle, they are be darkly pigmented in some species. They are FILIBRANCH or PSEUDOLAMELLIBRANCH (ELEUTHERORHABDIC), HOMORHABDIC or HETERORHABDIC, and not inserted into (or fused with) the distal oral groove of the palps (CATEGORY III association). CEPHALIC EYES are present in most species (absent in *Crenatula*). Distally the tips of the ctenidia are attached to the inner mantle fold, separating INFRA- and SUPRABRANCHIAL CHAMBERS; water flow is anteroposterior. The STOMACH is TYPE III. The MIDGUT is not coiled. The HINDGUT passes through the ventricle of the heart, and leads to a rectum with an ANAL FUNNEL. Isognomonids are GONOCHORISTIC and produce planktonic VELIGER larvae. The nervous system is not concentrated. STATOCYSTS are absent in adults. ABDOMINAL SENSE ORGANS are present and asymmetrical.

Isognomonids are marine or estuarine SUSPENSION FEEDERS. Most species are EPIBYSSATE on hard substrata (rock crevices, mangrove roots); *Crenatula* lives obligatorily embedded within sponges, growing in pace with the sponge to maintain water contact. *Isognomon* often forms extensive gregarious colonies on seawalls and other surfaces. Exterior surfaces of the valves can be encrusted by epibionts. The extreme flatness of the valves has been interpreted as an adaptation that minimizes resistance to strong water flow. Predators include boring muricid gastropods. The diet of *Crenatula modiolaris* (Lamarck, 1819) from Hong Kong has been shown to consist of diatoms and dinoflagellates; octopuses prey extensively on this species.

The family Isognomonidae is known since the Permian, is represented by 2 living genera and ca. 20 species, and is distributed worldwide in shallow subtropical and tropical seas; *Isognomon* typically inhabits shallow depths, whereas *Crenatula* is deep to abyssal. *Isognomon alatus* is consumed in Jamaica as a substitute for the dwindling populations of harvestable oysters (Ostreidae). Isognomonids are often called "tooth pearl shells" because of their prominent multivincular ligaments and nacreous interior surfaces.

## References

Domaneschi, O., and C. Mantovani Martins. 2002. *Isognomon bicolor* (C. B. Adams) (Bivalvia, Isognomonidae): primeiro registro para o Brasil, redescrição da espécie e considerações sobre a ocorrência e distribuição de *Isognomon* na costa brasileira. *Revista Brasileira de Zoologia*, 19(2): 611–627.

Fischer, P. 1861. Note sur l'animal du genre *Perna*. *Journal de Conchyliologie*, 9(1) [3rd Série, 1(1)]: 19–28, pl. 4.

Fischer-Piette, E. 1976. Révision des Aviculidées. I. *Crenatula, Pedalion, Foramelina*. *Journal de Conchyliologie*, 113(1–2): 3–42.

Harper, E., and B. Morton. 1994. The biology of *Isognomon legumen* (Gmelin, 1791) (Bivalvia: Pterioida) at Cape d'Aguilar, Hong Kong, with special reference to predation by muricids. Pages 405–425, in B. Morton, ed., *The Malacofauna of Hong Kong and Southern China, III*. Hong Kong University Press, Hong Kong.

Reid, R. G. B. 1985. *Isognomon*: life in two dimensions. Pages 311–319, in: B. Morton and D. Dudgeon, eds., *The Malacofauna of Hong Kong and Southern China, II, volume 1*. Hong Kong University Press, Hong Kong.

Reid, R. G. B., and S. Porteous. 1980. Aspects of the functional morphology and digestive physiology of *Vulsella vulsella* (Linné) and *Crenatula modiolaris* (Lamarck), bivalves associated with sponges. Pages 291–310, in: B. Morton, ed., *The Malacofauna of Hong Kong and Southern China*. Hong Kong University Press, Hong Kong.

Siung, A. M. 1980. Studies on the biology of *Isognomon alatus* Gmelin (Bivalvia: Isognomonidae) with notes on its potential as a commercial species. *Bulletin of Marine Science*, 30(1): 90–101.

Tëmkin, I. 2006. Morphological perspective on classification and evolution of Recent Pterioidea (Mollusca: Bivalvia). *Zoological Society of the Linnean Society*, 148: 253–312.

Trueman, E. R. 1954. The structure of the ligament of *Pedalion* (*Perna*). *Proceedings of the Malacological Society of London*, 30(6): 160–166.

Yonge, C. M. 1968. Form and habit in species of *Malleus* (including the "hammer oysters") with comparative observations on *Isognomon isognomon*. *The Biological Bulletin*, 135: 378–405.

### *Isognomon bicolor* (C. B. Adams, 1845) – Two-toned Tree Oyster

Irregularly oval, with concave anterior byssal notch, very compressed, thin-walled, tan to purple (often both), with coarse commarginal lamellations; interior reflecting external colors, nacreous area strongly demarcated, non-nacreous border thin and wide. Florida, Bermuda, Bahamas, West Indies, Gulf of Mexico, Caribbean Central America, South America (Colombia, Brazil). Length 30 mm (to 35 mm). Note: Also known as Bicolor Purse Oyster.

### *Isognomon radiatus* (Anton, 1838) – Radial Tree Oyster

Irregularly elongated oval, often strongly arched toward left or right, very compressed, thin-walled, tan with pale purple wavy radial rays, more or less smooth, with weak flaky lamellations; interior nacreous area large, non-nacreous border thin. Florida, Bermuda, Bahamas, West Indies, Gulf of Mexico, Caribbean Central America, South America (Colombia, Brazil). Length 40 mm (to 82 mm). Note: Also known as Lister's Purse Oyster.

**A typical habitat** for *Isognomon bicolor* in the Florida Keys, as here at Missouri Key, is on rocks with the cemented wormsnail *Dendropoma corrodens* (d'Orbigny, 1841) (Vermetidae).

***Isognomon radiatus* also byssally attaches** to rocks, often with *Isognomon bicolor*. This group was found near Newfound Harbor (Lower Florida Keys).

**The common name "tree oyster"** is applied to *Isognomon alatus* from its frequent habitat amid mangrove prop roots (here on Crawl Key).

*Isognomon bicolor*

*Isognomon radiatus*

# Family Malleidae – Hammer Oysters and Sponge Fingers

**Classification**
AUTOLAMELLIBRANCHIATA Grobben, 1894
Pteriomorphia Beurlen, 1944
Pterioida Newell, 1965
Pterioidea J. E. Gray, 1847 [1820]
Malleidae Lamarck, 1818

**Featured species**
*Malleus candeanus* (d'Orbigny, 1853) – **Caribbean Hammer Oyster**

Irregularly elongated, compressed, thin-walled, dark purple with areas of yellowish white, early growth stages with regular coarse commarginal lamellae, becoming irregular and smooth ventrally; interior nacreous area small, non-nacreous ventral area thin and dark purple, with elevated midridge. Florida, Bermuda, Bahamas, West Indies, Gulf of Mexico, South America (Colombia, Brazil). Length 55 mm. Note: Some authors consider this species a synonym of *Malleus* (= *Malvufundus*) *regulus* (Forsskål, 1775), from the Indo-Pacific and Mediterranean Sea.

*Malleus candeanus* **in situ** on an algae-covered coral reef in the Florida Keys. When nestled into soft coral rock, only the ventralmost shell tips extend out of the crevice.

## Family description

The malleid shell is small to large (to >250 mm), thin-walled to solid, and irregularly dorsoventrally elongated, usually with both posterior and anterior AURICLES (of variable development according to species and in part to habitat constraints; absent in *Vulsella*). It is EQUIVALVE or INEQUIVALVE (right valve less convex), compressed, with a BYSSAL NOTCH in the less-convex right valve (absent in *Vulsella* and some *Malleus*), and gaping or irregularly undulate posteriorly. The shell is EQUILATERAL to INEQUILATERAL (umbones anterior;

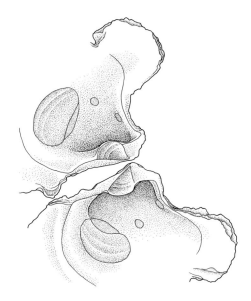

posterior in *Vulsella*), with OPISTHOGYRATE UM-BONES separated by a narrow CARDINAL AREA. Shell microstructure is a mixture of ARAGONITE and CALCITE, and two-layered, with a calcitic simple PRISMATIC outer layer and an aragonitic sheet-NACREOUS inner layer. The outer layer is often lamellar and readily cracks and flakes, especially in dried shells. In *Malleus*, the outer layer comprises most of the shell, whereas the inner layer is restricted to the small dorsal part of the shell above the muscle scar that surrounds the visceropedal mass; in *Vulsella*, the inner nacreous layer extends nearly to the ventral shell margins. TUBULES are absent. Exteriorly malleids can be unicolored or radially banded and are covered by a nonpersistent PERIOS-TRACUM. Sculpture is smooth or with often-irregular commarginal lamellae. LUNULE and ESCUTCHEON are absent. Interiorly the shell is nacreous near the umbones with a wide ventral prismatic border, and prismatic inside the AURICLES. Tightest shell closure in some cases is not at the ventral shell margin, but rather at the thickened margin of the small internal nacreous layer (VISCERAL RIM). A longitudinal median ridge (PALLIAL RIDGE; absent in *Vulsella*) extends either from the visceral rim or from the (more dorsal) insertion of the pallial retractor muscle down the length of each valve, probably functioning to strengthen the thin-walled, elongated shell. The PALLIAL LINE is absent (as is a SINUS). The inner shell margins are smooth. The HINGE PLATE is straight or arched and narrow, completely EDENTATE in adults (in *Vulsella*) or with one set of elongated posterior ridges that are present in juveniles and occasionally retained by adults (subumbonal teeth, as found in pteriids, are absent in all). The LIGAMENT is large, submarginal, ALIVINCULAR, and AMPHIDETIC; an internal portion (RESILIUM) is sunken into a posteriorly directed, trigonal to semicircular, subumbonal RESILIFER.

The animal is MONOMYARIAN (anterior ADDUCTOR MUSCLE absent); the posterior adductor muscle is large and central, and concave in cross section (oval in *Vulsella*). The posterior pedal retractor muscles are large and insert centrally within the cavity of the posterior adductor scar (or are absent in *Vulsella* spp. and *Malleus albus* (Lamarck, 1819)); the anterior pedal retractors are small, in some species distally divided, and inserted anterodorsally or anteriorly (in some species asymmetrically on the left and right sides). Pedal protractor and elevator muscles are absent. The MANTLE margins are not fused ventrally; SIPHONS are absent. The inner mantle folds form a PALLIAL VEIL. Simple PALLIAL ("siphonal") EYES (cellular, without lens) are present or absent on the outer folds. HYPO-BRANCHIAL GLANDS have not been reported. The mantle is capable of great extension and withdrawal through action of the pallial retractor muscles, originating along the gill axes and inserting at a single point (rather than along a PALLIAL LINE) on the median shell ridge near the ventral visceral mass. The presence of a space (PROMYAL PASSAGE) between the posterior adductor and visceral mass (actually between the two posterior pedal retractor muscles), coupled with the absence of the SUPRAMYAL SEPTUM (in other families connecting the right and left mantle lobes on the dorsal side of the adductor muscle), allows enhanced posterior water flow. The FOOT is anterior and divided into two portions. The

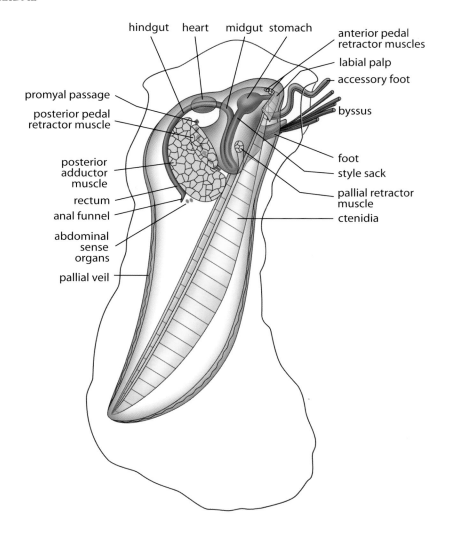

hindgut  heart  midgut  stomach

anterior pedal
retractor muscles

labial palp

accessory foot

promyal passage

posterior pedal
retractor muscle

byssus

posterior
adductor
muscle

foot

style sack

rectum

pallial retractor
muscle

anal funnel

ctenidia

abdominal
sense
organs

pallial veil

foot proper is small and digitiform; it has a BYSSAL GROOVE and the adult is strongly
BYSSATE except in some INFAUNAL or sponge-dwelling species (e.g., *Malleus albus* and
*Vulsella* spp.). Dorsal to the foot proper, an elongated, grooved process (ACCESSORY FOOT),
larger than the foot itself, is used for cleaning the INFRABRANCHIAL CHAMBER.

The LABIAL PALPS are small to large. The CTENIDIA are large and extend the entire
length of the mantle cavity; with the mantle, they are the only organs to occupy the ventral prismatic part of the shell. They are FILIBRANCH or PSEUDOLAMELLIBRANCH (ELEU-
THERORHABDIC), HOMORHABDIC or HETERORHABDIC, and not inserted into (or fused with)
the distal oral groove of the palps (CATEGORY III association). CEPHALIC EYES are present
in most species (absent in *Vulsella*). Distally the tips of the ctenidia are attached to the inner mantle fold, separating INFRA- and SUPRABRANCHIAL CHAMBERS. The mantle and gills
can be withdrawn into the relatively small chamber within the visceral rim (when present). Water flow is anteroposterior. The STOMACH is TYPE III; the MIDGUT is not coiled.
The HINDGUT passes through the ventricle of the heart, and leads to a rectum with an

ANAL FUNNEL. In *Malleus*, the heart is asymmetric and displaced to the left side of the animal because of the presence of the promyal passage. Malleids produce planktonic VELIGER larvae; their reproductive mode has yet to be described. The nervous system is not concentrated. STATOCYSTS have not been recorded in adults. ABDOMINAL SENSE ORGANS are present and asymmetrical.

Malleids are marine SUSPENSION FEEDERS. They are usually EPIBYSSATE on hard substrata such as rocks, coral, or pilings. Some species have been reported to stand more or less erect, umbo down, on the surface of the substratum; a layer of encrusting biota frequently coats the exterior of both valves. Other species nestle in crevices; the larvae of *Malleus candeanus* have been shown to be both photonegative and geonegative, driving them to settle in dark overhanging portions of the coral reef. *Vulsella* lives obligatorily embedded within sponges, growing in pace with the sponge to maintain water contact. Some species in the genus *Malleus* are INFAUNAL, stabilized by their auricles and shell undulations; with growth in the infaunal White Hammer Oyster (*Malleus albus*), the byssus is lost, the byssal notch closes, and the foot proper and byssal retractor muscles are reduced in size. Exposure of the fragile ventral shell ends of malleids leads to frequent breakage; rapid shell repair has been reported. The diet of *Vulsella vulsella* (Linnaeus, 1758) from Hong Kong has been shown to consist of diatoms and dinoflagellates.

The family Malleidae is known since the Cretaceous, is represented by 3 living genera and ca. 15 species, and is distributed worldwide in warm seas. The common name "hammer oyster" is based on the hammer or "T" shape of some of the larger species (e.g., the Black Hammer Oyster, *Malleus malleus* (Linnaeus, 1758)).

## References

Boss, K. J., and D. R. Moore. 1967. Notes on *Malleus* (*Paramalleus*) *candeanus* (d'Orbigny) (Mollusca: Bivalvia). *Bulletin of Marine Science*, 17(1): 85–94.

Johnson, C. W. 1918. The *Avicula candeana* of d'Orbigny, from Bermuda. *The Nautilus*, 32: 37–39, pl. 3.

Kühnelt, W. 1938. Der Anpassungstypus der Hammermuschel. *Palaeobiologica*, 6: 230–241.

Reid, R. G. B., and S. Porteous. 1980. Aspects of the functional morphology and digestive physiology of *Vulsella vulsella* (Linné) and *Crenatula modiolaris* (Lamarck), bivalves associated with sponges. Pages 291–310, in: B. Morton, ed., *The Malacofauna of Hong Kong and Southern China.* Hong Kong University Press, Hong Kong.

Tëmkin, I. 2006. Morphological perspective on classification and evolution of Recent Pterioidea (Mollusca: Bivalvia). *Zoological Society of the Linnean Society*, 148: 233–312.

Waller, T. R., and I. G. Macintyre. 1982. Larval settlement behavior and shell morphology of *Malleus candeanus* (d'Orbigny) (Mollusca: Bivalvia). Pages 489–497, in: K. Rützler and I. G. Macintyre, eds., *The Atlantic Barrier Reef Ecosystem at Carrie Bow Cay, Belize, 1: Structure & Communities.* Smithsonian Contributions to the Marine Sciences, 12.

Yonge, C. M. 1968. Form and habit in species of *Malleus* (including the "hammer oysters") with comparative observations on *Isognomon isognomon*. *The Biological Bulletin*, 135: 378–405.

# Family Ostreidae – True Oysters

**Classification**
AUTOLAMELLIBRANCHIATA Grobben, 1894
Pteriomorphia Beurlen, 1944
Pterioida Newell, 1965
Ostreoidea Rafinesque, 1815
Ostreidae Rafinesque, 1815

**Featured species**
*Dendostrea frons* (Linnaeus, 1758) – **Frond Oyster**

LV

Irregularly oval or narrowly elongated, often with clasper spines, compressed, solid, purple-red to golden brown, radially plicate with saw-toothed margins; interior usually white, chomata restricted to margin near hinge or on entire margin, anal funnel present. North Carolina to Florida, Bermuda, Bahamas, West Indies, Gulf of Mexico, Caribbean Central America, South America (to Brazil). Length 47 mm (to 50 mm). Formerly in *Lopha*. Note: The habitat of this species is oceanic. The right valve of this specimen has been drilled by a predatory gastropod. Also known as "Coon" (i.e., Racoon) Oyster.

*Dendostrea frons*, in its most distinctive elongated form, uses "claspers" on its lower left valve to cling to upright gorgonians on coral reefs and patch reefs (here at Hens and Chickens Reef off Plantation Key).

## Family description

The ostreid shell is medium-sized to large (to 600 mm), usually solid, and rounded to narrowly dorsoventrally elongated or pyriform. It is more or less flattened but often variable, distorted and irregular within a species, and ce-

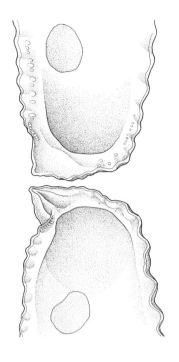

mented by (PLEUROTHETIC on) the left valve by a small to large attachment area (influencing shell shape) or secondarily detached. An exception is the genus *Cryptostrea*, which lives embedded in sponges. The valve margins are often irregularly plicate. It is usually INEQUIVALVE (lower left valve more convex; upper right valve often flattened lidlike), inflated or compressed, and not gaping. The shell is EQUILATERAL with OPISTHOGYRATE, generally eroded, UMBONES separated by a CARDINAL AREA of variable prominence. Shell microstructure is a mixture of ARAGONITE and (primarily) CALCITE, and two-layered, with a calcitic PRISMATIC outer layer and a foliated calcitic inner layer; this structure is characteristic of the upper right valve, whereas the lower left valve is entirely of foliated calcite; the aragonite in the shell is restricted to the MYOSTRACUM (area of muscle attachment). Irregular, porous, white, CHALKY DEPOSITS of unknown cause or function are common on the interior surface. TUBULES are absent. Exteriorly ostreids are (usually) chalky whitish (the Indo-Pacific Cockscomb Oyster, *Lopha cristagalli* (Linnaeus, 1758), is deep purple) and covered by a thin, nonpersistent PERIOSTRACUM. Sculpture is of irregular commarginal lamellae, often with pleated ribs, lamellose scales (especially on the upper right valve), large HYOTE spines, and/or CLASPER SPINES. LUNULE and ESCUTCHEON are absent. Interiorly the shell is non-NACREOUS and usually whitish, but can have a metallic darker color in Lophinae; the adductor muscle scars are usually darkly pigmented in crassostreines. The PALLIAL LINE is ENTIRE and obscure. An angulate shell margin (COMMISSURAL SHELF) is usually poorly defined (well defined in *Planostrea*), and is usually denticulated by elongated or pustular tubercles (CHOMATA) on the upper right valve interlocking with sockets on the lower left valve; these are usually obvious on either side of the ligament (but can continue ventrally) and vary in number, strength, and ontogenetically. The HINGE PLATE is wide and EDENTATE in adults. The LIGAMENT is ALIVINCULAR and AMPHIDETIC, with a large, internal portion (RESILIUM) set on a trigonal, subumbonal RESILIFER.

The animal is MONOMYARIAN (anterior ADDUCTOR MUSCLE absent); the posterior adductor muscle is large, posterocentral or -ventral, and concave to oval in cross section. Pedal retractor, elevator, and protractor muscles are absent. Ctenidial protractor muscles (QUENSTEDT MUSCLES) insert below the umbones. The MANTLE margins are not fused ventrally except at the point of ventral gill attachment; SIPHONS are absent. Black melanin pigment is present in mantle tissues of Ostreinae. The mantle lobes are thickened in some species for glycogen storage. The inner

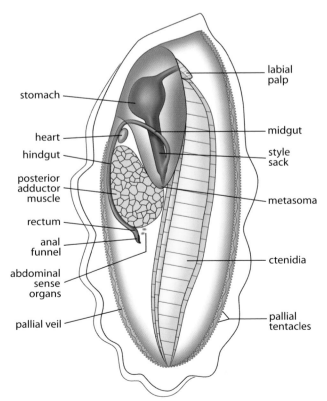

stomach
heart
hindgut
posterior adductor muscle
rectum
anal funnel
abdominal sense organs
pallial veil

labial palp
midgut
style sack
metasoma
ctenidia
pallial tentacles

mantle folds form a PALLIAL VEIL. The mantle is not fused to the visceral mass on the right side of the animal, creating a space (PROMYAL PASSAGE; absent in most Ostreinae and Lophinae, including *Dendostrea*) dorsal to the posterior adductor muscle that allows more efficient posterior flow of water. HYPOBRANCHIAL GLANDS have not been reported. The FOOT and BYSSUS are absent in the adult.

The LABIAL PALPS are small to medium-sized; they are fused medially to form a hood over the mouth. The CTENIDIA are large and partially encircle the centralized adductor muscle. They are PSEUDOLAMELLIBRANCH (SYNAPTORHABDIC), HETERORHABDIC, of about equal size, and not inserted into (or fused with) the distal oral groove of the palps (CATEGORY III association). CEPHALIC EYES are present. Distally the tips of the ctenidia are attached to the inner mantle fold, separating INFRA- and SUPRABRANCHIAL CHAMBERS; water flow is anteroposterior. The STOMACH is TYPE III; the STYLE SACK extends into a tapering sacklike extension of the visceral mass (METASOMA). The MIDGUT is not coiled. The HINDGUT passes dorsal to the ventricle of the heart, and leads to a freely hanging rectum either with or without an ANAL FUNNEL or fleshy collarlike rim. ACCESSORY HEARTS are present. Ostreids are GONOCHORISTIC or PROTANDRIC or SIMULTANEOUS HERMAPHRODITES, and produce planktonic VELIGER larvae or brood them in the gills (or brood for a short period then release veligers). Some species undergo sex reversal annually or more often. Differences in the PRODISSOCONCH hinge have been identified to distinguish ostreid larvae from those of gryphaeids: ostreids have subequal PROVINCULAR teeth and sockets that are interrupted by a median gap, whereas the line is uninterrupted in gryphaeids. The nervous system is not concentrated but pedal ganglia and STATOCYSTS are absent in adults (associated with absence of the foot and active locomotion; although the larvae of *Ostrea edulis* Linnaeus, 1758, have statocysts with STATOCONIA). ABDOMINAL SENSE ORGANS are present and in some species asymmetrical.

Ostreids are marine or estuarine SUSPENSION FEEDERS. They are well known for tolerating a wide range of turbidity due to suspended sediment or plankton, and for remaining tightly closed and healthy (due to the strong pulling force of the adductor muscle) for many days. They cement to hard substrata of many kinds (e.g., rocks, coral, seafans, mangroves, floating buoys), with some species forming dense gregarious beds or reefs in shallow water, and the shape of the shell often reflecting its substratum and surroundings. Lophines are primarily coral reef associates. Exterior surfaces of the valves are often encrusted by epibionts and (in larger species) bored by sponges, polychaete worms, and smaller bivalves. Commensals include pea crabs (Pinnotheridae); *Cryptostrea* is itself considered a commensal with the sponge it inhabits. Natural predators include carnivorous gastropods, sea stars, flatworms, crabs, fish, and birds. As a result of the many environmental factors affecting oysters, ostreids show perhaps more intraspecific variability than any other group of living bivalves.

The family Ostreidae is known since the Triassic, is represented by 15 living genera and >40 species, and is distributed worldwide in tropical and temperate seas

**The American Oyster** (*Crassostrea virginica*) forms dense beds in intertidal waters (here at Cedar Key on Florida's Gulf Coast). Many other organisms find refuge from predators in the interstices of the complex habitat. Such aggregations have formed the basis for easy harvesting and culture practices in the past. Today this species is commercially raised in aquaculture.

in depths to 30 m. There is a conspicuous absence of *Crassostrea* in the Florida Keys because of an almost complete absence of estuarine conditions, although these species are abundant elsewhere in Florida and the Caribbean. Edible oysters are probably the best known mollusks in all respects. Ostreids (especially Crassostreinae) include many species that are harvested or maricultured for human food (e.g., *Crassostrea virginica* of the eastern United States, *Crassostrea gigas* (Thunberg, 1793) of Japan, and *Ostrea edulis* of Europe), and as a consequence, many (along with their associated organisms and diseases) have been transported around the world through commercial trade or intentional transplantation. Because they are suspension feeders, oysters often pick up and concentrate particles and microscopic organisms in the water column; many diseases caused by degraded environmental conditions are reported in commercial oysters; the most serious of these in the United States is "Dermo" disease, caused by the protistan endoparasite *Perkinsus marinus* (Mackin, Owen, & Collier, 1950), which proliferates under conditions of high temperature and salinity. Oyster diseases cause at least physical changes to the oyster flesh, rendering them unmarketable; some diseases cause widespread die-offs of oyster populations, and can be harmful or fatal to humans who ingest the infected oysters.

## References

Galtsoff, P. S. 1964. The American oyster *Crassostrea virginica* Gmelin. *Fishery Bulletin*, 64: 1–480.

Kennedy, V. S., R. I. E. Newell, and A. F. Eble, eds. 1996. *The Eastern Oyster* Crassostrea virginica. Maryland Sea Grant College, College Park, xvi + 734 pp.

Forbes, M. L. 1966. Life cycle of *Ostrea permollis* and its relationship to the host sponge, *Stelletta grubii*. *Bulletin of Marine Science*, 16(2): 273–301.

Forbes, M. L. 1971. Habitats and substrates of *Ostrea frons*, and distinguishing features of early spat. *Bulletin of Marine Science*, 21(2): 613–625.

Fox, R. 2005. *Invertebrate Anatomy Online*, Crassostrea virginica, *American Oyster*. http://www.lander.edu/rsfox/310CrassostreaLab.html, last accessed 28 October 2005.

Harry, H. W. 1985. Synopsis of the supraspecific classification of living oysters (Bivalvia: Gryphaeidae and Ostreidae). *The Veliger*, 28(2): 121–158.

Hopkins, A. E. 1934. Accessory hearts in the oyster. *Science*, 80(2079): 411–412.

Kirkendale, L., T. Lee, P. Baker, and D. Ó Foighil. 2004. Oysters of the Conch Republic (Florida Keys): a molecular phylogenetic study of *Parahyotissa mcgintyi, Teskeyostrea weberi* and *Ostreola equestris*. In: R. Bieler and P. M. Mikkelsen, eds., *Bivalve Studies in the Florida Keys*. Proceedings of the International Marine Bivalve Workshop, Long Key, Florida, July 2002. *Malacologia*, 46(2): 309–326.

McLean, R. A. 1941. The oysters of the western Atlantic. *Notulae Naturae*, 67: 14 pp.

Nelson, T. C. 1938. The feeding mechanism of the oyster. I. On the pallium and the branchial chambers of *Ostrea virginica, O. edulis* and *O. angulata*, with comparisons with other species of the genus. *Journal of Morphology*, 63(1): 1–61.

Nelson, T. C. 1960. The feeding mechanism of the oyster. II. On the gills and palps of *Ostrea edulis, Crassostrea virginica* and *C. angulata*. *Journal of Morphology*, 107(2): 163–203.

Ranson, G. 1967. Les espèces d'huitres vivant actuellement dans le monde, définies par leurs coquilles larvaires ou prodissoconques: étude des collections des quelques-uns des grands musées d'histoire naturelle. *Revue des Travaux de l'Institut des Pêches Maritimes*, 31(2): 127–199, 31(3): 205–274.

Thomson, J. M. 1954. The genera of oysters and the Australian species. *Australian Journal of Marine and Freshwater Research*, 5(1): 132–168.

Waller, T. R. 1981. Functional morphology and development of veliger larvae of the European oyster, *Ostrea edulis* Linné. *Smithsonian Contributions to Zoology*, 328: 1–70.

### *Crassostrea rhizophorae* (Guilding, 1828) – **Root Oyster**

Irregularly elongated oval with undulating margins, deeply cupped, with flat upper right valve fitting well into deep lower left valve, moderately compressed, solid, grayish white often blotched with purple, with irregular commarginal lamellations; interior glossy white with purple elongated muscle scar, margin smooth (chomata absent), anal funnel absent. Florida Keys, West Indies, Gulf of Mexico, Caribbean Central America, South America (to Uruguay). Length 85 mm. Formerly considered a subspecies or form of *C. virginica* (see following). Compare *Crassostrea virginica*, which has a smaller, rounder purple adductor muscle scar, and *Ostrea equestris*, which is greenish internally. Note: The habitat of this species is estuarine.

### *Crassostrea virginica* (Gmelin, 1791) – **American Oyster**

Irregularly elongated oval to rounded trigonal with only slightly undulating or straight margins, deeply cupped, with flat upper right valve fitting well into deep lower left valve, moderately compressed, solid, grayish white, with irregular commarginal lamellations; interior glossy white with purple round muscle scar, margin smooth (chomata absent), anal funnel absent. Eastern Canada to Florida, Bermuda, West Indies, Gulf of Mexico, South America (French Guiana, Brazil). Length 10 mm (to 150 mm). Compare *Crassostrea rhizophorae*, which has an elongated purple adductor muscle scar, and *Ostrea equestris*, which is greenish internally. Note: The habitat of this species is estuarine.

### *Ostrea equestris* Say, 1834 – **Crested Oyster**

Irregularly oval, with flat right valve fitting inside deeper left valve, lower left valve with raised crenulated margins usually higher on one side, moderately compressed, very thin-walled, whitish gray, periostracum easily flaking off; interior dull gray with olive green stain, margin occasionally stained purple, chomata on entire margin or nearly so, anal funnel present. Virginia to Florida, Bermuda, Bahamas, West Indies, Gulf of Mexico, Caribbean Central America, South America (to Argentina). Length 47 mm. Formerly in *Ostreola*. Compare *Crassostrea* spp., which have purple adductor muscle scars. Note: The habitat of this species is oceanic; *Ostrea equestris* is the most commonly encountered oyster in the Florida Keys.

### *Teskeyostrea weberi* (Olsson, 1951) – **Threaded Oyster**

Oval, compressed, thin-walled, apricot-colored, with fine, divaricating radial ribs and thin lamellose extensions; interior glossy apricot with darker margin, muscle scar relatively small, chomata restricted to margin near hinge, anal funnel absent. Florida Keys, Bermuda. Length 27 mm (to 37 mm). Its former status as a free-living ecomorph of the sponge-commensal species *Cryptostrea permollis* (G. B. Sowerby II, 1871), was recently refuted by the results of a molecular phylogenetic analysis (Kirkendale et al., 2004). Note: The habitat of this species is oceanic.

**The flat apricot-colored shells** of *Teskeyostrea weberi*, here with *Barbatia cancellaria* at Grassy Key, are commonly found with *Ostrea equestris* on the underside of rocks in oceanside locations.

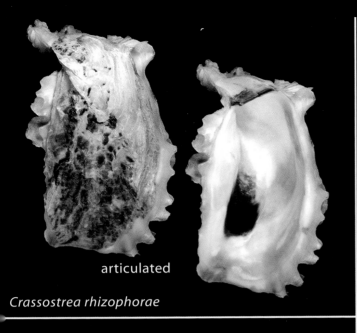

articulated

*Crassostrea rhizophorae*

articulated

*Crassostrea virginica*

articulated

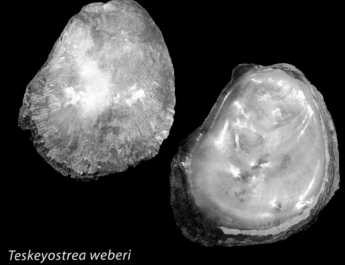

*Teskeyostrea weberi*

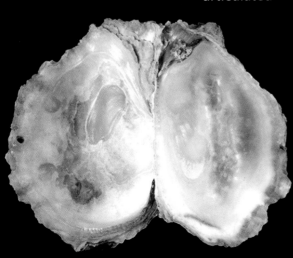

*Ostrea equestris*

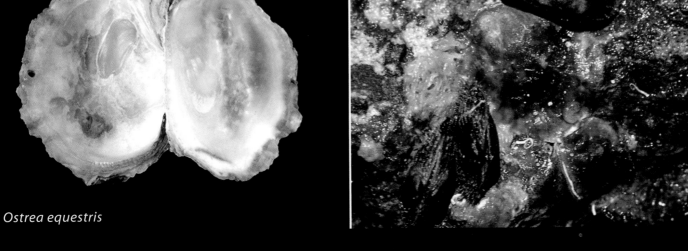

# Family Gryphaeidae – Foam Oysters

**Classification**
AUTOLAMELLIBRANCHIATA Grobben, 1894
Pteriomorphia Beurlen, 1944
Pterioida Newell, 1965
Ostreoidea Rafinesque, 1815
Gryphaeidae Vyalov, 1936

**Featured species**
*Hyotissa mcgintyi* (Harry, 1985) – **Atlantic Foam Oyster**

Circular to D-shaped in profile, margin often irregularly plicated, solid, compressed, cream to lavender, occasionally with short hollow spines; interior cream, pinkish, or light brown, with lightest colored area often near shell margin. North Carolina to Florida, Bermuda, West Indies, Gulf of Mexico, Caribbean Central America, South America (Colombia, Brazil). Length 61 mm (to 10 cm). Formerly in *Parahyotissa*. Note: Inset shows the vesicular shell structure characteristic of this family. Also known as Honeycomb Oyster.

# Family description

The gryphaeid shell is medium-sized to large (to 300 mm), thin-walled to solid, and circular to oval, with the valve margins often plicate. It is cemented by (PLEUROTHETIC on) the left valve, usually by a small attachment area on the left valve, or secondarily detached. Most species are highly INEQUIVALVE (lower left valve more convex, upper right valve lidlike [e.g., *Neopycnodonte*]), whereas some are EQUIVALVE (e.g., *Hyotissa*). The umbonal recess of inequivalve forms usually fills with solid shell matter. The shell is inflated or compressed, not gaping, and EQUILATERAL, with OPISTHOGYRATE UMBONES separated by a wide CARDINAL AREA. Shell microstructure is a mixture of ARAGONITE and (primarily) CALCITE,

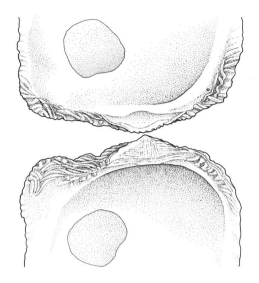

and one- or two-layered, with a thin or absent calcitic PRISMATIC outer layer and a foliated calcitic inner layer; the latter has a characteristic vesicular structure that is most apparent marginally and appears spongy or foam-like under magnification. Irregular, porous, white, CHALKY DEPOSITS of unknown cause or function are common on the interior surface. Fluid-filled cavities can be present between the shell layers. TUBULES are absent. Exteriorly gryphaeids can be distinctly colored (e.g., pink, brown, or purple-black) and are covered by a thin, nonpersistent PERIOSTRACUM. Sculpture is radially undulate or ribbed in many species, in some with scales and/or HYOTE spines; the left valve can have a deep posterior radial groove or flexure. LUNULE and ESCUTCHEON are absent. Interiorly the shell is non-NACREOUS. The PALLIAL LINE is ENTIRE and obscure. An angulate shell margin (COMMISSURAL SHELF) is usually wide and well defined, and is denticulated by an interlocking complex of anastomosing elongated, sinuous ridges and tubercles (VERMICULAR CHOMATA), limited to each side of the ligament. The HINGE PLATE is wide and EDENTATE in adults. The LIGAMENT is ALIVINCULAR and AMPHIDETIC; an internal portion (RESILIUM) is large and sunken into a trigonal, subumbonal RESILIFER.

The animal is MONOMYARIAN (anterior ADDUCTOR MUSCLE absent); the posterior adductor muscle is large, posterocentral or -ventral, and round in cross section or only slightly flattened dorsally. The ventral border of the muscle scar in the left valve is ele-

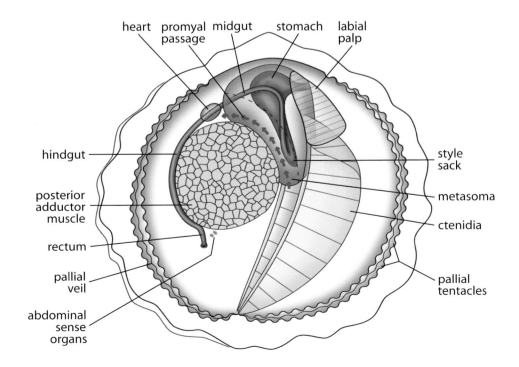

vated above the surface of the valve, whereas its dorsal border is indented. Pedal retractor, elevator, and protractor muscles are absent. Ctenidial protractor muscles (QUENSTEDT MUSCLES) insert below the umbones. The mantle and gills can be darkly pigmented. The MANTLE margins are not fused ventrally except at the point of ventral gill attachment; SIPHONS are absent. The inner mantle folds form a PALLIAL VEIL. The mantle is not fused to the visceral mass on the right side of most species (not fused on both sides in *Hyotissa*), creating a space (PROMYAL PASSAGE) dorsal to the posterior adductor muscle that allows more efficient posterior flow of water. HYPOBRANCHIAL GLANDS have not been reported. The FOOT and BYSSUS are absent in the adult.

The LABIAL PALPS are large; they are fused medially to form a hood over the mouth. The CTENIDIA are large and partially encircle the centralized adductor muscle. They are PSEUDOLAMELLIBRANCH (SYNAPTORHABDIC), HETERORHABDIC, of about equal size, and not inserted into (or fused with) the distal oral groove of the palps (CATEGORY III association). CEPHALIC EYES can be present. Distally the tips of the ctenidia are attached to the inner mantle fold, separating INFRA- and SUPRABRANCHIAL CHAMBERS; water flow is anteroposterior. The STOMACH is TYPE III; the STYLE SACK extends into a tapering sacklike extension of the visceral mass (METASOMA). The MIDGUT is not coiled. The HINDGUT passes through the ventricle of the heart, and leads to a freely hanging rectum with a fleshy collarlike rim. ACCESSORY HEARTS are absent. Gryphaeids are PROTANDRIC or SIMULTANEOUS HERMAPHRODITES and produce planktonic VELIGER larvae. Differences in the PRODISSOCONCH hinge have been identified to distinguish gryphaeid larvae from those of ostreids: gryphaeids have subequal PROVINCULAR teeth and sockets that continue uninterrupted along the entire hinge line, whereas a median gap is present in ostreids. The nervous system is not concentrated but pedal ganglia and STATOCYSTS are absent in adults (associated with absence of the foot and active locomotion). ABDOMINAL SENSE ORGANS are present and asymmetrical.

Gryphaeids are marine or estuarine SUSPENSION FEEDERS. They are cemented to hard subtidal substrata (e.g., rocks, dead coral, shipwrecks), but are generally not gregarious or reef-forming. Exterior surfaces of the valves are often encrusted by epibionts and (in larger species) bored by sponges, polychaete worms, and smaller bivalves. In the Indo-Pacific, a pearlfish (*Onuxodon parvibrachium* (Fowler, 1927), Carapidae) has been reported to inhabit the MANTLE CAVITY of *Hyotissa hyotis*. The latter species has been introduced to Florida, probably via commercial ship traffic (ballast water).

The family Gryphaeidae is known since the Triassic, is represented by 2 living genera and only ca. 5 species, and is distributed in tropical and subtropical seas. The characteristic vesicular shell structure, most visible near the inner shell margins, is the source of the common name "foam oysters" for this family.

## References

Bieler, R., P. M. Mikkelsen, T. Lee, and D. Ó Foighil. 2004. Discovery of the Indo-Pacific oyster *Hyotissa hyotis* (Linnaeus, 1758) in the Florida Keys (Bivalvia: Gryphaeidae). *Molluscan Research*, 24(3): 149–159.

Harry, H. W. 1985. Synopsis of the supraspecific classification of living oysters (Bivalvia: Gryphaeidae and Ostreidae). *The Veliger*, 28(2): 121–158.

Ranson, G. 1941. Les espèces actuelles et fossils du genre *Pycnodonta* F. de W.—I. *Pycnodonta hyotis* (L.). *Bulletin du Muséum National d'Histoire Naturelle, Série 2*, 13(2): 82–92.

Sevilla-H., M. L., F. García-D., and E. Uria-G. 1998. Datos anatómicos de *Hyotissa hyotis* (Linnaeus, 1758), Ostreacea: Gryphaeidae. *Anales de la Escuela Nacional de Ciencias Biológicas*, 43(1–4): 25–32.

Thomson, J. M. 1954. The genera of oysters and the Australian species. *Australian Journal of Marine and Freshwater Research*, 5(1): 132–168.

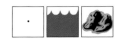

### *Hyotissa hyotis* (Linnaeus, 1758) – **Giant Foam Oyster**

Circular to oval with roundly plicate margins, moderately inflated, solid, purplish black to magenta to violet-brown, with irregular sculpture of pleated ribs bearing hollow spines; interior bluish or yellowish white with black or brown stains, ventral margin of adductor muscle scar elevated in larger specimens. Tropical Indo-Pacific, introduced to Florida. Length 178 mm (to 30 cm). Note: Also known as Coxcomb Foam Oyster.

### *Neopycnodonte cochlear* (Poli, 1795) – **Deepwater Foam Oyster**

Circular to oval, free upper right valve flat, lower attached left valve deeply cupped, margins rising vertically from substratum, inflated, thin-walled, white to pink to orange, surface relatively smooth with few indistinct plications on left valve; interior surface dull. North Carolina to Florida, Bermuda, West Indies, also western Europe and Indo-Pacific from Red Sea, Madagascar, Philippines, Japan, and Hawaii. Length 13 mm (to 7 cm). Note: This species lives at greater depths than any other Recent oyster.

**Artificial hard substrata** are often the habitat of gryphaeids. The encrusted hulls of artificial reefs (here the intentionally sunken wreck of the *Thunderbolt*) camouflage the shells of cementing bivalves such as *Hyotissa hyotis*, making them nearly imperceptible to scuba divers. The smaller *Hyotissa mcgintyi* is one of the most abundant oysters on offshore oil platforms in the Gulf of Mexico.

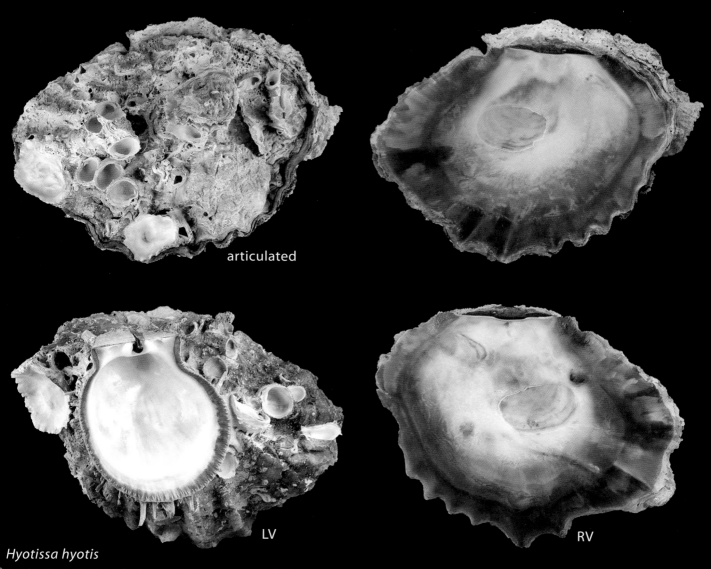

articulated

LV

RV

*Hyotissa hyotis*

articulated

RV

*Neopycnodonte cochlear*

# Family Pinnidae Leach, 1819 – Pen Shells or Fan Shells

**Classification**
AUTOLAMELLIBRANCHIATA Grobben, 1894
Pteriomorphia Beurlen, 1944
Pterioida Newell, 1965
Pinnoidea Leach, 1819
Pinnidae Leach, 1819

**Featured species**

*Pinna carnea* Gmelin, 1791 – **Amber Pen Shell**

Narrowly trigonal, compressed, thin-walled, pale orange to amber, smooth or with 8–12 weak radial ridges occasionally with scattered inrolled spines; interior with central radial ridge most conspicuous anteriorly, ventral lobe of nacreous layer equal to or longer than dorsal lobe. North Carolina to Florida, Bermuda, Ba-

*Pinna carnea* **(left) is known** as the Razor Shell in the Bahamas, reflecting its typical in situ posture with its sharp posterior edges protruding from the sand.

**The prodissoconch** of *Pinna carnea* (right), rarely retained intact on the adult shell, is smooth and trigonal in shape.

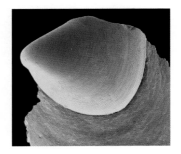

## Family description

The pinnid shell is medium-sized to large (to 900 mm; e.g., *Pinna nobilis* (Linnaeus, 1758), one of the world's largest living bivalves), thin-walled, elongated fan-shaped to trigonal (occasionally distorted), anteriorly pointed, and posteriorly truncate. It is EQUIVALVE, compressed, with a small anterior byssal gape, and widely gaping (but closeable by the flexible shell) posteriorly. It is INEQUILATERAL (umbones at the pointed anterior end), with OPISTHOGYRATE UMBONES. Shell microstructure is a mixture of ARAGONITE and CAL-CITE, and two-layered, with a thick calcitic simple PRISMATIC outer layer and a thin lenticular or sheet-NACREOUS inner layer confined to the anterior part of the shell. The outer layer is high in organic content, very flexible in some species, and readily cracks and flakes, especially in dried shells; the prismatic crystals are the largest recorded for any Recent mollusk. TUBULES are absent. Exteriorly pinnids are covered by nonpersistent PE-RIOSTRACUM. Sculpture is smooth to radially ribbed or spiculose. LUNULE and ES-CUTCHEON are absent. Interiorly the shell is thinly nacreous from the umbones to the posterior adductor muscle, with a very wide ventral prismatic border; the shape of the nacreous layer is taxonomically important. *Pinna* has a distinct median sulcus in the anterior part of the shell (absent in other genera). The PALLIAL LINE is absent (as is a SINUS). The inner shell margins are smooth. The HINGE line is long and straight; the HINGE PLATE is weak and EDENTATE in adults. The anteriormost point of the shell is usually eroded and strengthened internally by a series of transverse septa. The LIGAMENT is PLANIVINCULAR and OPISTHODETIC, set in PSEUDONYMPHS, reaching only as far as the posterior limit of the inner NACREOUS layer; beyond this point, a thinner secondary extension (of prismatic shell, rather than periostracum) bridges the valves along the remaining long, straight hinge line. The nonelastic primary ligament merely holds the valves together, having lost its primary function of opening the shell valves.

The animal is HETEROMYARIAN (anterior ADDUCTOR MUSCLE smaller); the posterior adductor muscle is large and central. The anterior pedal retractor muscle is minute, whereas the posterior pedal retractor is large, well-developed (mainly involved in byssal retraction), and inserts next to the posterior adductor muscle. Pedal elevator and protractor muscles are absent. Pallial retractor muscles, attached to the shell at a single point (rather than along a PALLIAL LINE) anteroventral to the posterior adductor muscle, allow the mantle and gills to be withdrawn to the level of the muscle when disturbed. Mantle color can be species-specific (see figure of living animal in situ, p. 123). The MANTLE margins are greatly expanded posteriorly (resulting in "centralization" of the posterior adductor muscle) and not fused ventrally; SIPHONS are absent. The inner mantle folds form a PALLIAL VEIL. The middle and inner folds lack PALLIAL EYES but have pigmented glandular spots (ORGANS OF WILL) of unknown function (perhaps associated with production of pigment rays on the shell). A posterior muscular stalklike structure (PALLIAL ORGAN, a SYNAPOMORPHY of the family) with a conical glandular head serves to clear the suprabranchial chamber of shell fragments when the exposed edge of the shell is broken

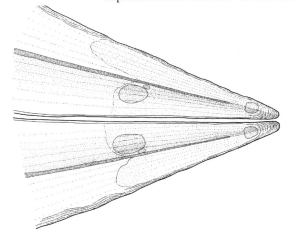

and the mantle withdrawn (an alternative defensive function was proposed for *Atrina pectinata* (Linnaeus, 1758), when the pallial tentacle was discovered to produce acidic secretions and to react to damage but not to introduced debris); extensive marginal repair is common and relatively rapid (*Pinna carnea* can mend a 3- to 5-cm break in 3 days). The INFRABRANCHIAL CHAMBER is continuously cleaned by ciliated waste canals (probably a synapomorphy of Pinnidae) that transport PSEUDO-FECES (largely sand) to the posterior margin. HYPO-BRANCHIAL GLANDS have not been reported. The FOOT is anterior, conical, and elongated. It has a large BYSSAL GROOVE; the adult is byssate, producing profuse silky byssal threads.

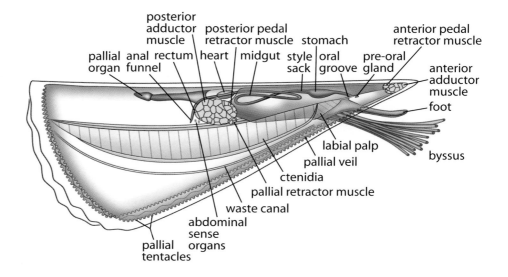

An unpaired PRE-ORAL GLAND (unique to Pinnidae and excretory in function) lies dorsal to the mouth. The LABIAL PALPS are elongated triangular, medium-sized to large, and are extended by a long, unridged oral groove to the mouth. The CTENIDIA are PSEUDOLAMELLIBRANCH (SYNAPTORHABDIC), HETERORHABDIC, of about equal size, and not inserted into (or fused with) the distal oral groove of the palps (CATEGORY III association). Distally the tips of the ctenidia are attached to the inner mantle folds, separating INFRA- and SUPRABRANCHIAL CHAMBERS. Incurrent and excurrent water flows are posterior; incurrent flow can also be ventral but always above the sediment surface. The STOMACH is TYPE III; the CRYSTALLINE STYLE is unusually long and slender in pinnids, extending well into the initial part of the "MIDGUT" (actually combined midgut/STYLE SACK); the midgut is coiled. The HINDGUT passes through the ventricle of the heart, and leads to a freely hanging rectum with an ANAL FUNNEL. Brown pigment in the blood has been called "pinnaglobin" but its oxygen-carrying capacity or other function have not been determined. The kidneys are often very large and produce granules (NEPHROLITHS) probably associated with rapid calcium uptake (which in turn also allows rapid shell repair). Pinnids are GONOCHORISTIC or HERMAPHRODITIC and produce planktonic VELIGER larvae. The nervous system is concentrated and includes cerebropleural ganglia in close proximity to the pedal ganglia; their connectives and commissures form a ring around the esophagus. STATOCYSTS in adults are large, located (unusually for bivalves) near the tip of the foot, variably shaped, and can each contain a STATOLITH (absent in some individuals). ABDOMINAL SENSE ORGANS are present and asymmetrical.

Pinnids are marine SUSPENSION FEEDERS. They are shallowly ENDOBYSSATE, half- to mostly embedded in sand, mud, or shell hash, with the wider posterior end projecting, and the pointed anterior end buried, anchored to small stones or shells. If dislodged slightly, pinnids can reanchor to some extent using a water jet expelled anteriorly from the MANTLE CAVITY to fluidize the substratum (this technique also allows it to burrow deeper into the sediment as it grows); fully dislodged individuals cannot rebury. Predators include sea stars and carnivorous gastropods. Some species support a community of fouling organisms; the shell surface is often bored extensively by sponges and polychaete worms. Commensal pea crabs (Pinnotheridae) and shrimps (Pontoniidae) often live in the mantle cavity.

The family Pinnidae is known since the Carboniferous, is represented by 3 living genera and only ca. 20 species, and is distributed in shallow tropical and subtropical seas. The scalloplike adductor muscles of pinnids are edible, and specimens are harvested in many countries (e.g., Mexico, Japan) for human food. Pinnid pearls (either nacreous or porcellaneous) have held historical interest in Mediterranean countries and India. Mediterranean *Pinna nobilis* was harvested for its golden brown byssal threads in ancient times and as recently as the early twentieth century, for weaving novelty items (e.g., hats, gloves); the term "byssus" was first coined in 1476 in reference to that of *P. nobilis*. This species is endemic and now endangered in the Mediterranean Sea and is legally protected by many surrounding countries. Pinnids are called "razor fish" in Australia because of the sharp posterior shell margin that protrudes from the sand in shallow water.

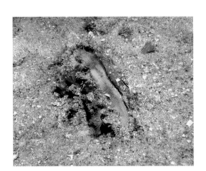

**Mantle color** is an important character in distinguishing Caribbean *Atrina* species. That of *A. rigida* is orange (shown here in a specimen from Lake Worth, Florida), whereas that of *A. seminuda* is beige.

**"Silk"** produced from the golden brown byssus of Mediterranean *Pinna* is first documented from the fourth century and has become the topic of various legends, including speculations on the origin of Jason's "Golden Fleece." Harvesting the byssus of *Pinna nobilis* Linnaeus, 1758, sustained an active "mussel silk" industry in southern Italy from the eighteenth to early twentieth centuries that manufactured knitted, woven, or furlike objects such as caps, gloves, collars, and stoles.

### References

Dall, W. H. 1897. Synopsis of the Pinnidae of the United States and West Indies. *The Nautilus*, 11(3): 25–26.

Grave, B. H. 1911 (1909). Anatomy and physiology of the wing-shell *Atrina rigida. Bulletin of the United States Bureau of Fisheries*, 29(744): 409–439, pls. 48–50.

Liang, X. Y., and B. Morton. 1988. The pallial organ of *Atrina pectinata* (Bivalvia: Pinnidae): its structure and function. *Journal of Zoology*, 216(3): 469–477.

Maeder, F., A. Hänggi, and D. Wunderlin, eds. 2004. *Bisso marino, Fili d'oro dal fondo del mare— Muschelseide, Goldene Fäden vom Meeresgrund*. Naturhistorisches Museum Basel, Switzerland, 127 pp. [In Italian and German.]

Reid, R. G. B., and D. G. Brand. 1989. Giant kidneys and metal-sequestering nephroliths in the bivalve *Pinna bicolor*, with comparative notes on *Atrina vexillum* (Pinnidae). *Journal of Experimental Marine Biology and Ecology*, 126(2): 95–117.

Rosewater, J. 1961. The family Pinnidae in the Indo-Pacific. *Indo-Pacific Mollusca*, 1(4): 175–226.

Turner, R. D., and J. Rosewater. 1958. The family Pinnidae in the western Atlantic. *Johnsonia*, 3(38): 285–326.

Winckworth, R. 1929. Marine Mollusca from south India and Ceylon. III: *Pinna*. With an index to the Recent species of *Pinna. Proceedings of the Malacological Society of London*, 18(6): 276–297.

Yonge, C. M. 1953. Form and habit in *Pinna carnea* Gmelin. *Philosophical Transactions of the Royal Society of London, Series B, Biological Sciences*, 237(648): 335–374.

### *Atrina rigida* (Lightfoot, 1786) – **Stiff Pen Shell**

Broadly trigonal, compressed, thin-walled, dark brown, with 15–25 coarse radial ridges adorned with inrolled spines; interior glossy black-brown, with central muscle scar at or protruding above upper margin of internal nacreous border, mantle golden orange (see p. 123). North Carolina to Florida, Bermuda, Bahamas, West Indies, Gulf of Mexico, Caribbean Central America, South America (Venezuela). Length 90 mm (to 355 mm). Compare *Atrina seminuda*, which has a beige or pale yellow mantle and a muscle scar well within the inner nacreous layer.

### *Atrina seminuda* (Lamarck, 1819) – **Half-naked Pen Shell**

Broadly trigonal, compressed, thin-walled, brown, with coarse radial ridges adorned with inrolled spines on about half of the valve, central muscle scar well within internal nacreous area, mantle beige to pale yellow. North Carolina to Florida, West Indies, Gulf of Mexico, Caribbean Central America, South America (to Argentina). Length 180 mm (to 243 mm). Compare *Atrina rigida*, which has an orange mantle and a muscle scar at or above the edge of the inner nacreous layer.

### *Atrina serrata* (G. B. Sowerby I, 1825) – **Saw-toothed Pen Shell**

Broadly trigonal, compressed, thin-walled, tan to light brown, with ca. 30 fine radial ridges adorned with numerous small inrolled spines, central muscle scar well within internal nacreous area. North Carolina to Florida, West Indies, Gulf of Mexico, South America (Colombia, Suriname). Length 180 mm (to 295 mm).

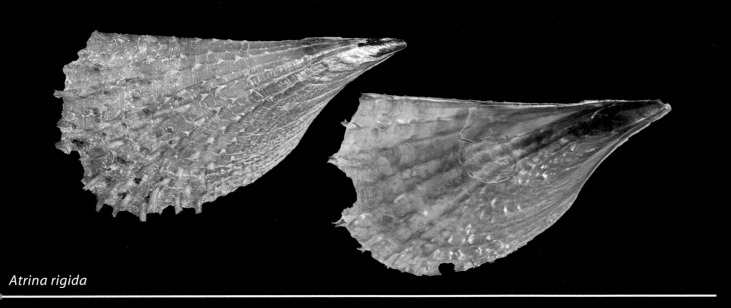

*Atrina rigida*

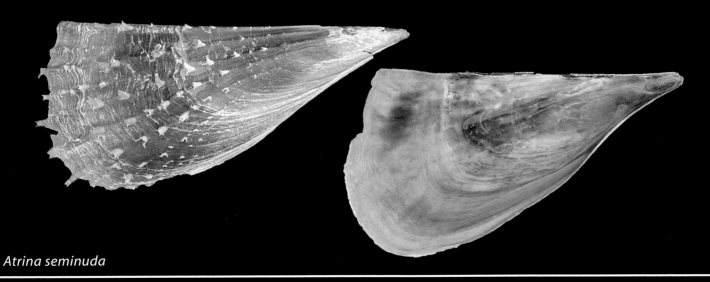

*Atrina seminuda*

*Atrina serrata*

# Family Limidae – Flame Scallops and File Clams

**Classification**
AUTOLAMELLIBRANCHIATA Grobben, 1894
Pteriomorphia Beurlen, 1944
Limoida Waller, 1978
Limoidea d'Orbigny, 1846
Limidae d'Orbigny, 1846

**Featured species**
*Ctenoides mitis* (Lamarck, 1807) – **Smooth Flame Scallop**

anterior

Elongated oval, gaping anteriorly and posteriorly near hinge, compressed, thin-walled, white with light brown periostracum, with ca. 90 fine, scaly, divaricating radial ribs, interspaces transversely grooved; interior white, margin crenulated by exterior ribs; animal with long, white pallial tentacles, and reddish orange mantle, gills, and pallial veil. North Carolina to Florida, Bermuda, Bahamas, West Indies, Gulf of Mexico, Caribbean Central America, South America (Colombia, Venezuela). Height 60 mm (to 83 mm). Syn. *floridana* Olsson & Harbison, 1953, and *tenera* G. B. Sowerby II, 1843. Compare *Ctenoides scabra*, which has ca. 50 coarse scaly ribs. Note: The Smooth Flame Scallop was long considered a form, variety, or subspecies of the more coarsely sculptured Rough Flame Scallop (*C. scabra*). Recent studies based on living populations in the Florida Keys confirmed its status as a separate species based on shell characters; the two species often co-exist on the same reef.

*Ctenoides mitis* **in situ** on a coral reef in the Florida Keys.

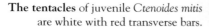

**The tentacles** of juvenile *Ctenoides mitis* are white with red transverse bars.

**The white tentacles** of *Ctenoides mitis*, contrasted here against the red mantle, are in several layers and several lengths. Interspersed among the tentacle bases are minute orange spots indicating the location of pallial eyes. Limid eyes lie below a cornealike surface of the mantle epithelium (not exposed and stalked like those of Pectinidae). In *C. mitis* they are equipped with a lens and can detect shadow and movement of predators; in others they are without lens or are present only in larval stages.

**The prodissoconch** (in green) of *Ctenoides mitis* is relatively small, indicating planktotrophic development, and smooth.

# Family description

The limid shell is small to large (to 200 mm height), thin-walled to solid, dorsoventrally elongated oval to trigonal, frequently oblique, and with small anterior and (often larger) posterior AU-RICLES. It is EQUIVALVE, usually compressed and gaping anteriorly, in some species also posteriorly, but usually without a distinct BYSSAL NOTCH. *Ctenoides* features a distinct marginally thickened byssal gape. The shell is EQUILATERAL to INEQUILATERAL (umbones slightly anterior), with UM-BONES PROSO- or ORTHOGYRATE and well separated by a trigonal CARDINAL AREA. Shell microstructure is a mixture of ARAGONITE and CALCITE, and two- or three-layered, with a foliated-calcite-like outer layer, an aragonitic CROSSED LAMELLAR middle layer (absent in some), and an aragonitic COMPLEX CROSSED LAMELLAR inner layer. TUBULES are absent. Exteriorly limids are usually white to cream-colored, translucent if thin-shelled, and in some species are covered by a thin brownish PERIOSTRACUM. Sculpture is comprised of prominent radial ribs, in some species DIVARICATING and/or scaly. LUNULE and ESCUTCHEON are absent, although the sunken anterior border of *Lima*, *Divarilima*, and *Acesta* has been called a lunule by some authors. Interiorly the shell is non-NACRE-OUS and whitish. The PALLIAL LINE is ENTIRE. The inner shell margins are smooth to strongly denticulated by the exterior ribs. The HINGE PLATE is straight, short, and EDENTATE in adults, but in some taxa with tubercles assuming the appearance of TAXODONT teeth. The LIGAMENT is ALIVIN-

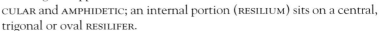

CULAR and AMPHIDETIC; an internal portion (RESILIUM) sits on a central, trigonal or oval RESILIFER.

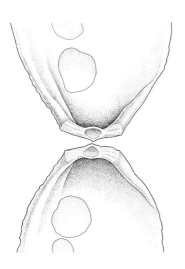

The animal is MONOMYARIAN (anterior ADDUCTOR MUSCLE absent); Limidae is the only monomyarian family that is not PLEU-ROTHETIC, that is, consistently resting on either side with concomitant anatomical asymmetry. The posterior adductor muscle is variable in size and position, but is often posterodorsal or central. One or two pairs of pedal retractor muscles can be present; the posterior pedal retractors attach either to the shell (ventral to the posterior adductor muscle) or to the visceral mass; the anterior pedal retractor muscle is small and inserts dorsally on the shell. In *Ctenoides*, separate anterior pedal and byssal retractor muscles are present. Pedal elevator and protractor muscles are absent. A series of anterior and posterior ctenidial retractors can be present. The MANTLE margins are usually not fused ventrally and can be brightly colored, often in shades of red or orange; SIPHONS are absent. The inner mantle folds form a PALLIAL VEIL (or velum, which is fused anteriorly in some genera) with or without guard tenta-

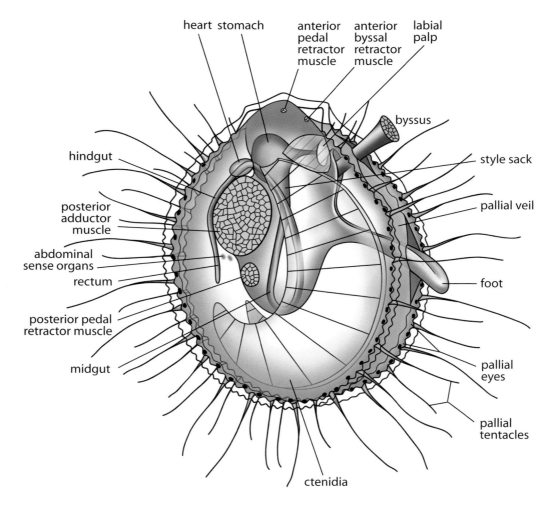

heart stomach     anterior pedal retractor muscle    anterior byssal retractor muscle    labial palp

byssus

hindgut

style sack

posterior adductor muscle

pallial veil

abdominal sense organs

rectum

foot

posterior pedal retractor muscle

midgut

pallial eyes

pallial tentacles

ctenidia

cles. The middle folds bear long, extensible, multiform, (in some reports) glandular sensory tentacles (which vary in length and form in different genera); in *Limatula* and *Limea*, an enlarged muscularized pair of ANAL TENTACLES that might be involved in cleaning the excurrent chamber lies at the posterodorsal margin near the point of mantle fusion. PALLIAL EYES (cap-shaped or invaginated, with lens) have been reported for some species (absent in *Limaria*, *Limea* and *Limatula*). The Indo-Pacific species *Ctenoides ales* (Finlay, 1927) is notable in having a silvery white outer edge of the pallial veil; fluorescence and transmission electron microscopy have shown that inclusions in the cells of this edge reflect light when the animal rapidly rolls and releases the veil, producing a lightninglike "flash" emulating bioluminescence. HYPOBRANCHIAL GLANDS have not been reported. The FOOT is anterior, short to long, slender, byssate in the adult, and is unique among bivalves for being rotated 180° relative to the visceral mass (a larval process that results in the BYSSAL GROOVE facing the mouth instead of ventrally; although somewhat analogous to gastropod torsion, it results only in twisting of the pedal nerves, not of the ganglionic ring); this last feature is recognized as a SYNAPOMORPHY of the order Limoida.

    The LABIAL PALPS are small. The CTENIDIA are large and partially encircle the centralized muscle bundles. They are EULAMELLIBRANCH (SYNAPTORHABDIC), HETERORHABDIC or HOMORHABDIC, and not inserted into (or fused with) the distal oral groove of the palps (CATEGORY III association; however, *Limatula* and *Limea* are CATEGORY II—inserted and fused); some species of *Limea* have only the inner demibranch. Small CEPHALIC EYES are present. Water flow is antero- and ventroposterior. The LIPS are either simple (perhaps secondarily) or (in most) hypertrophied and arborescent, interdigitating, and in some cases fused, leaving only a series of pores over the mouth, allowing exit of excess water without loss of food. The STOMACH is TYPE IV. The MIDGUT is not coiled. The HINDGUT passes through the pericardium but does not pierce the ventricle of the heart (which is double with separate or fused halves), and leads to a usually freely hanging rectum. Limids are GONOCHORISTIC or PROTANDRIC HERMAPHRODITES and produce planktonic VELIGER larvae; brooding in the suprabranchial chamber also has been reported. The nervous system can be concentrated, with cerebropleural and pedal ganglia close together, and in some species also very near the visceral ganglia creating a very small nerve ring. STATOCYSTS have been reported in the adults of some species. ABDOMINAL SENSE ORGANS are present.

Limids are marine or estuarine SUSPENSION FEEDERS. They are usually EPIBYSSATE or nestling on rocks, seagrass blades, or other hard substrata. Living individuals of two shallow-water genera (*Ctenoides* and *Lima*) are usually associated with reefs, living under rocks or in crevices. *Limaria* lives similarly or in seagrass beds, and can build byssal "nests" (when sufficient cover is unavailable) comprised of byssal threads, mucus, and in some cases incorporating small stones and shells. One species of *Limatula* lives in colonies of tube worms. Most species can irregularly swim to relocate or escape predators by clapping the valves (i.e., contraction of the adductor muscle) aided by waving the long pallial tentacles; *Limatula* can "crawl" by repeatedly extending and anchoring its foot then propelling itself to the new anchoring site with its tentacles. Tentacles of some limids can autotomize and include glands that secrete mucus that is apparently distasteful or irritating to predators.

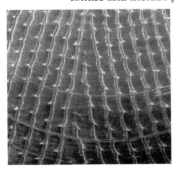

The family Limidae is known since the Carboniferous and is represented by 10 living genera and ca. 130 species, inhabiting all seas from shallow to great depths. In the Florida Keys, limids are well known to scuba divers for their long flamboyant tentacles and (often) bright red coloration. Overcollecting of Flame Scallops, mainly for private and commercial marine aquaria, has led to protective statutes in the State of Florida.

**Divaricating shell ribs**, here at midvalve in *Ctenoides sanctipauli* from Key Vaca, regularly increase the total number of ribs per valve as the animal grows older.

## References

Allen, J. A. 2004. The Recent species of the genera *Limatula* and *Limea* (Bivalvia, Limacea) present in the Atlantic, with particular reference to those in deep water. *Journal of Natural History*, 38: 2591–2653.

Fleming, C. A. 1978. The bivalve mollusc genus *Limatula*: a list of described species and a review of living and fossil species in the southwest Pacific. *Journal of the Royal Society of New Zealand*, 8(1): 17–91.

Gilmour, T. H. J. 1990. The adaptive significance of foot reversal in the Limoida. Pages 249–263, in: B. Morton, ed., *The Bivalvia, Proceedings of a Memorial Symposium in Honour of Sir Charles Maurice Yonge (1899–1986), Edinburgh, 1986*. Hong Kong University Press, Hong Kong.

Merrill, A. S., and R. D. Turner. 1963. Nest building in the bivalve genera *Musculus* and *Lima*. *The Veliger*, 6(2): 55–59, pls. 9–11.

Mikkelsen, P. M., and R. Bieler. 2003. Systematic revision of the western Atlantic file clams, *Lima* and *Ctenoides* (Bivalvia: Limoida: Limidae). *Invertebrate Systematics*, 17(5): 667–710.

Morton, B. S. 1979. A comparison of lip structure and function correlated with other aspects of the functional morphology of *Lima lima*, *Limaria* (*Platilimaria*) *fragilis*, and *Limaria* (*Platilimaria*) *hongkongensis* sp. nov. *Canadian Journal of Zoology*, 57(4): 728–742.

Stuardo, J. R. 1968. *On the Phylogeny, Taxonomy and Distribution of the Limidae (Mollusca: Bivalvia)*. Ph.D. Dissertation, Harvard University, Cambridge, Massachusetts, 327 pp., 37 pls., 24 maps, 44 figs.

### *Ctenoides miamiensis* Mikkelsen & Bieler, 2003 – **Miami Flame Scallop**

Elongated obliquely oval, concave and gaping anteriorly, convex and not gaping posteriorly, compressed, thin-walled, white with light brown periostracum, with ca. 50 coarse, divaricating, scaly radial ribs, interspaces with oblique transverse grooves; interior white to light brown, margin crenulated by exterior ribs. Florida, West Indies, South America (Brazil). Height 10 mm. Note: The valves photographed are of a PARATYPE specimen of *Ctenoides miamiensis*. The habitat of this recently discovered species is thus far unknown.

### *Ctenoides planulata* (Dall, 1886) – **Flat Flame Scallop**

Elongated oval, gaping anteriorly with dorsally flaring anterior auricle, not gaping posteriorly, extremely compressed, solid, white with light brown periostracum, with ca. 150 very fine, closely packed, scaly to smooth, divaricating radial ribs, interspaces very narrow and punctate; interior white, margin smooth and interiorly beveled. Florida, Bahamas, West Indies, Caribbean Central America, South America (Colombia). Height 10 mm (to 17 mm). Note: The valves photographed are of PARALECTOTYPE specimens of *Ctenoides planulata*. Previously listed as *planulatus* (masculine), but the original author of genus *Ctenoides*, Mörch (1853), implied feminine gender of the name.

### *Ctenoides sanctipauli* Stuardo, 1982 – **Brazil Flame Scallop**

Circular to oval, narrowly gaping anteriorly, not gaping posteriorly, compressed, thin-walled, translucent white with light yellowish periostracum, near-smooth, with ca. 115 fine, flattened, divaricating radial ribs, interspaces smooth; interior white to light yellow, margin smooth. South Carolina, Florida, West Indies, Gulf of Mexico, Caribbean Central America, South America (Brazil). Height 19 mm (to 42 mm).

### *Ctenoides scabra* (Born, 1778) – **Rough Flame Scallop**

Elongated oval, gaping anteriorly and posteriorly near hinge, compressed, solid, white with brown periostracum, with ca. 50 coarse, scaly radial ribs, interspaces transversely grooved; interior white, margin crenulated by exterior ribs, animal with long, red pallial tentacles, and reddish orange mantle, gills and pallial veil. North Carolina to Florida, Bahamas, West Indies, Gulf of Mexico, Caribbean Central America, South America (to Brazil). Height 75 mm (to 107 mm). Formerly in *Lima*. Compare *Ctenoides mitis*, which has ca. 90 fine, scaly ribs. Note: Previously listed as *scaber* (masculine), but the original author of genus *Ctenoides*, Mörch (1853), implied feminine gender of the name.

### *Divarilima albicoma* (Dall, 1886) – **Carinate File Clam**

Oval with oblique strongly keeled anterior margin, posterior auricle sharply protruding, compressed, thin-walled, translucent white with yellowish periostracum, with numerous fine, divaricating radial ribs; interior margin smooth. Florida Keys, West Indies, South America (Colombia, Brazil). Height 8 mm. Note: The valves photographed are of the HOLOTYPE specimen of *Divarilima albicoma*.

### *Lima caribaea* d'Orbigny, 1853 – **Spiny File Clam**

Elongated trigonal, strongly sloping anteriorly, rounded posteriorly, white, with 22–34 scaly, nondivaricating radial ribs, interspaces smooth; interior white, margin crenulated by exterior ribs, animal with multiple rows of short pallial tentacles, and white body with varying degrees of orange to purplish red markings. North Carolina to Florida, Bermuda, Bahamas, West Indies, Gulf of Mexico, Caribbean Central America, South America (to Brazil). Height 40 mm (to 56 mm). Formerly known as *Lima lima* Linnaeus, 1758 (Mediterranean Sea).

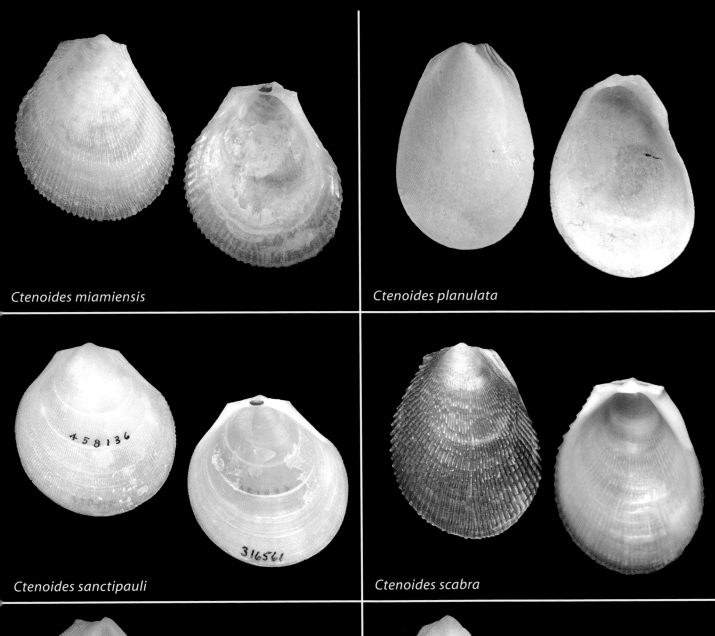

Ctenoides miamiensis

Ctenoides planulata

Ctenoides sanctipauli

458136

316561

Ctenoides scabra

Divarilima albicoma

Lima caribaea

### *Limea bronniana* Dall, 1886 – **Bronn's Dwarf File Clam**

Obliquely oval, not gaping, moderately compressed, solid, white, with 25–30 smooth, rounded radial ribs, interspaces with microscopic commarginal scratches; interior white, hinge line with three or four small teeth, margin crenulated by exterior ribs. North Carolina to Florida, West Indies, Gulf of Mexico, South America (Brazil). Height 3 mm (to 5 mm).

### *Limatula confusa* (E. A. Smith, 1885) – **Confusing File Clam**

Elongated oval with anterior bulge, not gaping, moderately compressed, thin-walled, white, with ribs fine, crossed by numerous commarginal striae thus appearing finely scaled or serrated, margin finely crenulated by exterior ribs. North Carolina to Florida, West Indies, South America (Brazil), also eastern North Atlantic to Azores. Height 3 mm (to 15 mm). Note: The sepia-toned drawing is the original illustration of the species.

### *Limatula setifera* Dall, 1886 – **Bristly File Clam**

Elongated oval, slightly oblique and broader dorsally, not gaping, moderately compressed, thin-walled, white, with spiny radial ribs; interior white, margin crenulated by exterior ribs. North Carolina to Florida, West Indies, Gulf of Mexico. Height 7 mm (to 9 mm). Note: The valve photographed is of a SYNTYPE specimen of *Limatula setifera*.

### *Limatula subovata* (Jeffreys, 1876) – **Subovate File Clam**

Elongated oval, not gaping, moderately compressed, thin-walled, white, with low radial ribs, two larger ribs at midvalve; interior white, margin crenulated by external ribs. North Atlantic to North Carolina, Florida Keys, also eastern North Atlantic to Mediterranean. Height < 2 mm (to 8 mm). Compare other *Limatula* species, which lack larger ribs at midvalve that create a visible "sulcus."

### *Limaria pellucida* (C. B. Adams, 1846) – **Antillean File Clam**

Obliquely oval, anteriorly and posteriorly gaping, compressed, thin-walled, translucent white, with irregular, fine radial ribs; interior white, margin crenulated by exterior ribs. North Carolina to Florida, Bermuda, Bahamas, West Indies, Gulf of Mexico, Caribbean Central America, South America (to Brazil). Height 17 mm (to 26 mm). Note: This species is recorded from a wide variety of environmental conditions, from estuarine mangrove creeks, to shallow-water seagrass beds, to oceanic coral reefs; it is possible that more than one cryptic species make up this complex.

**Living *Limaria pellucida*** show the long, flowing tentacles and wide shell gape characteristic of this genus. Individuals can autotomize their annulated, sticky tentacles and swim rapidly to escape predators such as fish or octopuses.

*Limea bronniana*

*Limatula confusa*

LV

*Limatula setifera*

*Limatula subovata*

*Limaria pellucida*

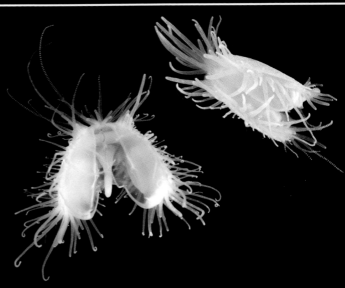

# Family Pectinidae – True Scallops

**Classification**

AUTOLAMELLIBRANCHIATA Grobben, 1894
Pteriomorphia Beurlen, 1944
Pectinoida H. Adams & A. Adams, 1857
Pectinoidea Rafinesque, 1815
Pectinidae Rafinesque, 1815

## Featured species
*Caribachlamys sentis* (Reeve, 1853) – Scaly Scallop

LV

Oval fan-shaped, anterior auricle much larger than posterior, compressed with left valve slightly more so (see p. 147), solid, usually purple to orange-red, with ca. 50 radial ribs on main body, major ribs scaly throughout, 2–4 minor ribs between major ones; interior glossy white or pigmented as exterior. North Carolina to Florida, Bahamas, West Indies, Gulf of Mexico, Caribbean Central America, South America (Colombia, Brazil). Length 25 mm (to 41 mm). Formerly in *Chlamys*. Compare *Caribachlamys ornata*, which has large patches of maroon or orange on the left valve and has radial ribs arranged in groups. Note: Also known as Sentis Scallop.

**A close-up image** of the interior right auricle of *Caribachlamys sentis* shows the characteristic comblike ctenolium at the edge of the byssal notch.

**This orange color form** of *Caribachlamys sentis* was photographed at Pigeon Key.

**Caribachlamys sentis**, like this purple specimen at Looe Key, usually lives in rock crevices. The shells are often fouled by the same organisms that coat the surfaces of their environment.

# Family description

The pectinid shell is small to large (to 300 mm, cemented *Hinnites* to 500 mm), usually thin-walled, orbicular to trigonal to oval fan-shaped, and with distinct anterior and poste-

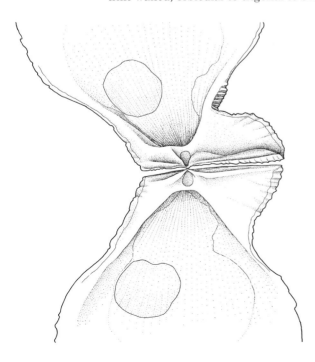

rior AURICLES. It is either EQUIVALVE or INEQUIV- ALVE (right valve more convex and resting on or cemented by it [PLEUROTHETIC on the right valve]), compressed to inflated, usually not gap- ing, and with a BYSSAL NOTCH below the auricle in the right valve (accompanied by a shallow SI- NUS in the left valve). As the animal grows, the byssal notch fills with shell material, leaving be- hind a distinct track (BYSSAL FASCIOLE). A comb- like series of denticles (CTENOLIUM) is found, at least in early growth stages, along the ventral margin of the byssal notch of the right valve. This structure, recognized as a SYNAPOMORPHY of the family, separates the byssal strands, prevents their rotation and breakage, and mechanically strengthens their attachment. In some species, an analogous tooth (PSEUDOCTENOLIUM) is pres- ent, formed from external sculptural elements on the right margin of the byssal notch. The shell is EQUILATERAL, with the UMBONES central and OR- THOGYRATE, but often with one auricle larger than the other. Shell microstructure is a mixture

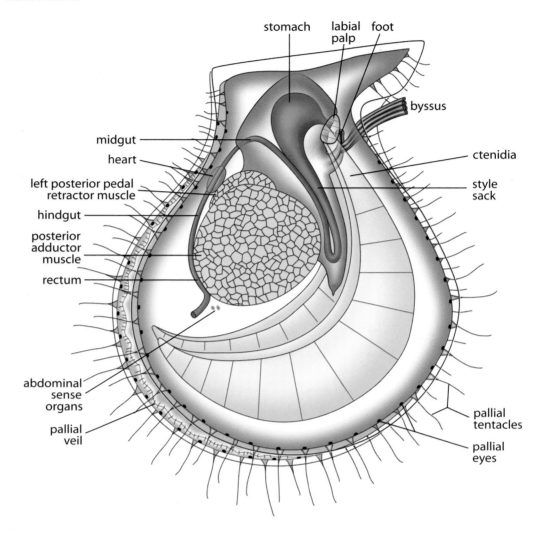

of ARAGONITE and CALCITE, two- or three-layered, with a foliated calcitic outer layer, an aragonitic CROSSED LAMELLAR middle layer (absent in some), and a foliated calcitic inner layer. The right valve of early postmetamorphic juveniles can be of PRISMATIC calcite (i.e., "prismatic stage"), conveying additional flexibility to the right larval shell, but disappears with the onset of radially ribbed sculpture. TUBULES are apparently absent. Exteriorly pectinids are often distinctively and brightly colored, covered by a nonpersistent PERIOSTRACUM. Sculpture is most often composed of radial ribs but can be smooth or cancellate, or differing on the two valves. LUNULE and ESCUTCHEON are absent. Interiorly the shell is non-NACREOUS and often reflects the color(s) of the exterior. The PALLIAL LINE is ENTIRE. The inner shell margins are denticulated by the exterior ribs, in some cases extending onto the inner surface. The HINGE PLATE is straight and EDENTATE in adults, but in some, one to several hinge teeth or auricular ridges (CRURAE) radiate from the umbo on both sides. The dorsal edge of the right valve overlaps that of the left valve in many species. The LIGAMENT is ALIVINCULAR and AMPHIDETIC; an internal portion (RESILIUM) has a nonfibrous core and sits on a central trigonal RESILIFER. A secondary external ligament of fused periostracum unites the valves dorsally.

The animal is MONOMYARIAN (anterior ADDUCTOR MUSCLE absent); the posterior adductor muscle is large and central. Of the pedal retractor muscles, the anterior pair is absent in all and in some only the left posterior is present. Pedal elevator muscles are present, but pedal protractors have not been reported. The MANTLE margins are not fused ventrally; SIPHONS are absent. The inner mantle folds form a PALLIAL VEIL (velum) with guard tentacles; the edges of the velum can be appressed in some species at certain areas to form an EXCURRENT APERTURE. The middle folds bear extensible sensory tentacles and complex PALLIAL EYES (invaginated, with lens). Pectinid eyes are generally blue in color (see p. 138), 1.0–1.5 mm in diameter, and present at the mantle margins of both valves. HYPOBRANCHIAL GLANDS have not been reported. The FOOT is anterior, small to medium-sized, and is suckerlike distally, functioning as a cleansing organ for the adjacent MANTLE CAVITY; it has a BYSSAL GROOVE and in some (e.g., *Caribachlamys*) produces a BYSSUS in the adult. Other pectinids (e.g., *Euvola*) are free-living and one (*Hinnites*) is secondarily cemented (byssally attached as a juvenile) by the right valve.

The LABIAL PALPS are small to medium-sized. The CTENIDIA are large and partially encircle the centralized muscle bundles. They are FILIBRANCH or PSEUDOLAMELLIBRANCH (in both cases ELEUTHERORHABDIC), HETERORHABDIC, and are not inserted into (or fused with) the distal oral groove of the palps (CATEGORY III association). Water flow is antero- and ventroposterior. The LIPS are hypertrophied and arborescent, interdigitating to create a series of pores over the mouth, allowing exit of excess water without loss of food. The STOMACH is TYPE IV; the MIDGUT is not coiled but can form a wide loop. The HINDGUT passes through the ventricle of the heart, and leads to a freely hanging rectum. Pectinids are usually HERMAPHRODITES (mostly PROTANDRIC; GONOCHORISTIC species also occur) and produce planktonic VELIGER larvae. The nervous system is unusually concentrated, with a complex, fused parietovisceral ganglion that is the largest and most intricate of all Bivalvia; in *Pecten*, the pedal ganglia are closely associated with the cerebropleural ganglia. STATOCYSTS (with STATOCONIA) are present in adults and are asymmetrical in some species. ABDOMINAL SENSE ORGANS are present and occasionally unpaired.

Pectinids are marine SUSPENSION FEEDERS. Most species of scallops are EPIBYSSATE on hard substrata (e.g., rocks, kelp, or pilings); some are free-living. Common predators include sea stars, whelks, and octopuses. True scallops are perhaps best known for their swimming behavior, achieved by clapping the valves together (by contraction of the adductor muscle), and in some cases by pursing the margins of the pallial veil together, resulting in a form of jet propulsion. Under normal conditions, water jets from right and left dorsal water channels near the auricles, propelling the scallop "forward" with the ventral edge leading; this behavior is also used for expelling PSEUDOFECES. A more rapid escape response jets water from all around the valve margin, sending the shell "backward," usually with the dorsal edge leading. Some species can intentionally detach the byssus and swim away to relocate when danger threatens. Although most scallops are poor swimmers, using the response solely for escape, others are rapid swimmers; *Amusium pleuronectes* (Linnaeus, 1758) has been clocked at 73 cm/sec. Certain species (e.g., *Argopecten irradians* (Lamarck, 1819), *Placopecten magellanicus* (Gmelin, 1791), and Australian *Pecten fumatus* (Reeve, 1852)) are locally abundant and have been targeted by commercial harvesting, with *P. magellanicus* providing one of the most valuable shellfisheries along the U.S. eastern seaboard. Today, nearly 60 species of true scallops form the basis of successful aquaculture practices in at least 15 countries. Larger species can harbor diverse epizoic and/or boring communities; pea crabs (Pinnotheridae) have been reported as commensal in the mantle cavities. Pectinids have been widely used symbols in art and heraldry throughout history, particularly in Greek, Roman, and Medieval Europe.

The family Pectinidae is known since the Triassic and is represented by ca. 50 living genera and ca. 400 species, inhabiting intertidal to hadal depths (ca. 7,000 m) from the tropics to polar seas. In the Florida Keys, true scallops are common in shallow-water sea-grass beds and under rocks in reefal environments.

**Living *Argopecten gibbus*** from Sanibel Island (western Florida) shows its pallial tentacles and blue pallial eyes. These eyes have structures analogous to those of vertebrates, including a cornea, lens, and retina, with the latter, however, in direct contact with the lens (rather than being separated by a fluid-filled space). "Sight reactions," including swimming, shell orientation, and response to movement, indicate that pectinids can "see" using some form of image formation.

## References

Barber, V. C., and P. N. Dilly. 1969. Some aspects of the fine structure of the statocysts of the molluscs *Pecten* and *Pterotrachea*. *Zeitschrift für Zellforschung und Mikroskopische Anatomie*, 94(4): 462–478.

Cox, I., ed. 1957. *The Scallop: Studies of a Shell and Its Influence on Mankind*. Shell Transport and Trading Company, London, 135 pp.

Dakin, W. J. 1909. *Pecten*. The edible scallop. *Proceedings and Transactions of the Liverpool Biological Society, Memoir 17*, 23: 333–468.

Dakin, W. J. 1928. The eyes of *Pecten*, *Spondylus*, *Amussium* and allied lamellibranchs, with a short discussion on their evolution. *Proceedings of the Royal Society of London, Series B; Containing Papers of a Biological Character*, 103(725): 355–365.

Dijkstra, H. H. 1999. Type specimens of Pectinidae (Mollusca: Bivalvia) described by Linnaeus (1758–1771). *Zoological Journal of the Linnean Society*, 125: 383–443.

Fox, R. 2001 (updated 2004). *Invertebrate Anatomy Online*, Argopecten irradians, *Bay Scallop, with notes on* Placopecten magellanicus. http://www.lander.edu/rsfox/310argopectenLab.html, last accessed 04 November 2005.

Küpfer, M. 1916. *Die Sehorgane am Mantelrande der Pecten-Arten. Entwicklungsgeschichtliche und Neuro-histologische Beiträge mit Anschließenden Vergleichend-anatomischen Betrachtungen*. Gustav Fischer, Jena, Germany, v + 312 pp., 8 pls.

Moir, A. J. G. 1977. On the ultrastructure of the abdominal sense organ of the giant scallop, *Placopecten magellanicus* (Gmelin). *Cell and Tissue Research*, 184(3): 359–366.

Morton, B. 1994. The biology and functional morphology of *Leptopecten latiauratus* (Conrad, 1837): an "opportunistic" scallop. *The Veliger*, 37(1): 5–22.

Rombouts, A. 1991. *Guidebook to Pecten Shells: Recent Pectinidae and Propeamussiidae of the World*. Universal Book Services and Dr. W. Backhuys, Oegstgeest, The Netherlands, 157 pp.

Sastry, A. N. 1962. Some morphological and ecological differences in two closely related species of scallops, *Aequipecten irradians* Lamarck and *Aequipecten gibbus* Dall from the Gulf of Mexico. *Quarterly Journal of the Florida Academy of Sciences*, 25(2): 89–95.

Shumway, S. E., ed. 1991. *Scallops: Biology, Ecology and Aquaculture*. Elsevier, Amsterdam, The Netherlands, 1,095 pp.

Stasek, C. R. 1963. Orientation and form in the bivalved Mollusca. *Journal of Morphology*, 112(3): 195–214.

Waller, T. R. 1978. Morphology, morphoclines and a new classification of the Pteriomorpha (Mollusca: Bivalvia). *Philosophical Transactions of the Royal Society of London, Series B, Biological Sciences*, 284(1001): 345–365.

Waller, T. R. 1984. The ctenolium of scallop shells: functional morphology and evolution of a key family-level character in the Pectinacea (Mollusca: Bivalvia). *Malacologia*, 25(1): 203–219.

Waller, T. R. 1993. The evolution of "*Chlamys*" (Mollusca: Bivalvia: Pectinidae) in the tropical western Atlantic and eastern Pacific. *American Malacological Bulletin*, 10(2): 195–249.

Waller, T. R. 2006. Phylogeny of families in the Pectinoidea (Mollusca: Bivalvia): importance of the fossil record. *Zoological Society of the Linnean Society*, 148: 313–342.

Yonge, C. M. 1981. On adaptive radiation in the Pectinacea with a description of *Hemipecten forbesianus*. *Malacologia*, 21(1–2): 23–34.

***Caribachlamys mildredae*** (F. M. Bayer, 1941) – **Alternate-ribbed Scallop**

Oval fan-shaped, anterior auricle much larger than posterior, compressed with left valve slightly more so, solid, purple to orange-red, with ca. 30 radial ribs, every third or fourth rib on left valve larger with large, erect scales, those of right valve in groups of 2 or 3; interior yellowish with marginal purple stains. Florida, Bermuda, Caribbean Central America. Length 30 mm (to 52 mm). Formerly in *Chlamys*.

***Caribachlamys ornata*** (Lamarck, 1819) – **Ornate Scallop**

Oval fan-shaped, anterior auricle much larger than posterior, compressed with left valve slightly more so, solid, left valve creamy white with large patches of dark maroon to orange, right valve lighter colored, with ca. 18 undercut radial ribs covered by fine radial striae in 18 groups of 3 closely spaced riblets; interior glossy white with exterior pattern visible through shell. Florida, Bahamas, West Indies, Gulf of Mexico, Caribbean Central America, South America (Colombia, Brazil). Length 16 mm (to 40 mm). Formerly in *Chlamys*. Compare *Caribachlamys sentis*, which is more evenly colored and has more evenly arranged radial ribs.

***Caribachlamys pellucens*** (Linnaeus, 1758) – **Knobby Scallop**

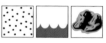

Oval fan-shaped, anterior auricle much larger than posterior, compressed with left valve slightly more so, solid, left valve white to yellow with squarish orange to reddish blotches, right valve lighter colored, with 8–10 major radial ribs that are knobby on left valve, scaly on right valve; interior white stained with purple and yellow. Florida, Bermuda, Bahamas, West Indies, Gulf of Mexico, Caribbean Central America, South America (Colombia, Venezuela). Length 36 mm (to 53 mm). Formerly known as *Chlamys imbricata* (Gmelin, 1791) (see Dijkstra, 1999).

***Spathochlamys benedicti*** (Verrill & Bush, 1897) – **Yellow-spotted Scallop**

Oval fan-shaped, anterior auricle much larger than posterior, the latter meeting dorsal margin at 90° angle, compressed, thin-walled, yellow, orange, pink, or purple with whitish zigzags, with ca. 22 radial ribs adorned with concave scales and alternating with weaker riblets; interior glossy, with exterior pattern visible through shell and distinct yellow patch below umbones. North Carolina to Florida, Bermuda, Bahamas, West Indies, Gulf of Mexico, Caribbean Central America, South America (Suriname, Brazil). Length 13 mm (to 25 mm). Formerly in *Chlamys*.

***Cryptopecten phrygium*** (Dall, 1886) – **Spathate Scallop**

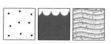

Widely fan-shaped, posterior auricle only slightly smaller than anterior, compressed, solid, dull gray with indistinct pink blotches, with ca. 17 sharp, M-shaped radial ribs, each adorned with three rows of fine, closely packed scales; interior white with pink markings. Massachusetts to Florida, Bermuda, West Indies, Gulf of Mexico. Length 40 mm (to 42 mm).

***Laevichlamys multisquamata*** (Dunker, 1864) – **Many-ribbed Scallop**

Oval fan-shaped, anterior auricle much larger than posterior and with dorsal margin sharply folded over, compressed, thin-walled, bright yellow flushed with purple toward margin or strongly flushed with dark red, umbonal area near-smooth, remainder of shell with 120–170 weak, finely scaled radial ribs; interior whitish, yellow toward umbones. Florida, Bermuda, Bahamas, West Indies, Gulf of Mexico, South America (Brazil). Length 34 mm (to 70 mm). Formerly in *Chlamys*.

*Caribachlamys mildredae*

*Caribachlamys ornata*

*Caribachlamys pellucens*

*Spathochlamys benedicti*

*Cryptopecten phrygium*

*Laevichlamys multisquamata*

*Argopecten gibbus* (Linnaeus, 1758) – **Atlantic Calico Scallop**

Broadly fan-shaped, auricles subequal, valves equally convex, solid, white with orange to pink mottlings, right valve lighter colored, with ca. 19 squared radial ribs, interspaces with fine commarginal threads, posterior auricle with 7–10 ribs, ctenolium with two teeth; interior white with pinkish stain. Maryland to Florida, Bermuda, Bahamas, West Indies, Gulf of Mexico, Caribbean Central America, South America (Venezuela, Brazil). Length 40 mm (to 52 mm). Compare other *Argopecten* species; *A. nucleus* is brown mottled, whereas *A. irradians* is more evenly brown to gray. Note: This species is a target of commercial fisheries in peninsular Florida, with highly fluctuating annual landings.

*Argopecten irradians* (Lamarck, 1819) – **Atlantic Bay Scallop**

Broadly fan-shaped, auricles subequal, moderately inflated with right valve slightly more convex, solid, drab brown to gray, right valve lighter colored, with 19–21 squared radial ribs, interspaces with commarginal threads, ctenolium with 3–5 teeth; interior white. Eastern Canada to Florida, Gulf of Mexico, Caribbean Central America, South America (Colombia). Length 40 mm (to 95 mm). Compare other *Argopecten* species, which are less evenly colored; *A. nucleus* is brown mottled, whereas *A. gibbus* is pink or orange mottled. Note: Three subspecies have been recognized based on geographic range and the number of radial ribs. Once a plentiful commercial species in Florida, commercial harvesting of Atlantic Bay Scallops is no longer allowed (and recreational catches are limited) to enhance natural recovery in Florida waters.

*Argopecten nucleus* (Born, 1778) – **Nucleus Scallop**

Broadly fan-shaped, auricles subequal, moderately inflated with right valve slightly more convex, solid, left valve cream with chestnut brown mottlings and opaque white specks, right valve lighter colored, with ca. 20 squared radial ribs, interspaces with fine commarginal threads, posterior auricle with 4–6 ribs; interior white flushed with yellow. Florida, Bahamas, West Indies, Gulf of Mexico, South America (to Suriname). Length 17 mm (to 45 mm). Compare other *Argopecten* species; *A. gibbus* is usually pink or orange mottled, whereas *A. irradians* is more evenly brown to gray.

*Aequipecten muscosus* (Wood, 1828) – **Rough Scallop**

Broadly fan-shaped, width of auricles equal to that of valves, anterior auricle slightly longer than posterior, moderately inflated (see p. 147), yellow, orange-brown, red, brown, or purple, with white mottlings, with 18–20 rounded radial ribs bearing three to five rows of thin, broad scales; interior white. North Carolina to Florida, Bermuda, Bahamas, West Indies, Gulf of Mexico, Caribbean Central America, South America (to Brazil). Length 40 mm (to 45 mm). Syn. *acanthodes* Dall, 1925. Formerly in *Lindapecten*.

*Aequipecten glyptus* (Verrill, 1882) – **Red-ribbed Scallop**

Broadly fan-shaped, auricles subequal, compressed, thin-walled, left valve with broad, rose rays, right valve lighter colored or pure white, with ca. 17 radial ribs that are prickled near umbones, flattened near margins; interior white. Massachusetts to Florida, Gulf of Mexico. Length 30 mm (to 75 mm).

*Aequipecten lineolaris* (Lamarck, 1819) – **Wavy-lined Scallop**

Broadly fan-shaped, auricles subequal, thin-walled, glossy, left valve rose-tan with numerous thin, wavy, pink-brown commarginal lines, right valve white, near-smooth with ca. 18 low, rounded ribs; interior white. Florida, West Indies, Caribbean Central America, South America (to Suriname). Length 25 mm. Formerly in *Argopecten*.

*Argopecten gibbus*

*Argopecten irradians*

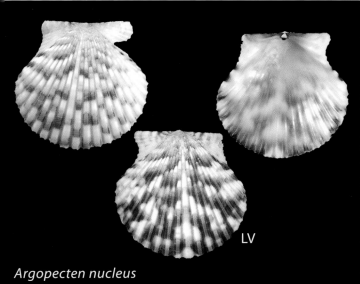

*Argopecten nucleus*

*Aequipecten muscosus*

*Aequipecten glyptus*

*Aequipecten lineolaris*

### *Euvola raveneli* (Dall, 1898) – **Round-ribbed Scallop**

Broadly fan-shaped to circular, left valve flat to slightly concave, right valve deeply convex (see p. 147), auricles sube-qual, inflated, thin-walled, left valve darker, right valve commonly whitish with tan or pinkish grooves, rarely lemon yellow or orange; with smooth radial ribs, those of left valve rounded and widely spaced, those of right valve broad and low and bifurcated by central groove in larger specimens; interior white with brownish stain. North Carolina to Florida, Bermuda, West Indies, Gulf of Mexico, Caribbean Central America. Length 50 mm. Formerly in *Pecten*. Compare *Euvola ziczac*, in which the ribs of the left valve are closely spaced, and *E. chazaliei*, which is smaller, lighter in color, and more delicate.

### *Euvola ziczac* (Linnaeus, 1758) – **Zigzag Scallop**

Broadly fan-shaped to circular, left valve flat to slightly concave, right valve deeply convex, auricles subequal, inflated, thin-walled, left valve mottled purple to yellow to brown with few darker, weak, concentric zigzag bands, right valve brownish red to orange; with smooth radial ribs, those of left valve squared and closely spaced, separated by narrow grooves, those of right valve broad and low, fading near lateral margins; interior white with brownish stain. North Carolina to Florida, Bermuda, Bahamas, West Indies, Gulf of Mexico, Caribbean Central America, South America (to Brazil). Length 70 mm. Formerly in *Pecten*. Compare *Euvola raveneli*, in which the ribs of the left valve are widely spaced, and *E. chazaliei*, which is smaller, lighter in color, and more delicate.

### *Euvola chazaliei* (Dautzenberg, 1900) – **Dwarf Zigzag Scallop**

Broadly fan-shaped to circular, left valve flat to concave, right valve convex, auricles subequal, thin-walled, slightly translucent creamy white to pale brown with pinkish umbones, occasionally with darker brown, zigzag commarginal bands on left valve, right valve white mottled with pink to orange-brown; left valve with narrow radial ribs separated by wide interspaces with commarginal striae, right valve with low, widely spaced roundish ribs; interior white with brownish stain. North Carolina to Florida, Bermuda, Bahamas, West Indies, Gulf of Mexico, South America (to Brazil). Length 30 mm (to 33 mm). Formerly in *Pecten*. Compare *Euvola raveneli* and *E. ziczac*, which are both larger, darker colored, and more solid. Note: Also known as Tereinus Scallop.

### *Euvola laurentii* (Gmelin, 1791) – **Laurent's Scallop**

Broadly fan-shaped to circular, left valve flat to slightly concave, right valve convex, auricles subequal, thin-walled, left valve darker in shades of pink and beige often in radial and chevron pattern, right valve white, smooth; interior white, with fine radial ribs arranged in pairs and extending to margin. Florida Keys, Bermuda, West Indies, Caribbean Central America, South America (Colombia, Venezuela). Length 65 mm. Formerly in *Amusium*. Compare *Euvola "papyracea,"* which is generally darker colored interiorly and exteriorly.

### *Euvola "papyracea"* auctt. – **Paper Scallop**

Broadly fan-shaped to circular, left valve flat to slightly concave, right valve convex, auricles subequal, inflated, thin-walled, right valve whitish usually with yellow to cream margins, left valve light mauve to reddish brown with darker flecks, near-smooth, glossy, with faint radial ribbing; interior yellow to reddish brown, with ca. 22 fine radial ribs arranged in pairs and extending to margin. Florida, West Indies, Gulf of Mexico. Length 80 mm. Formerly in *Amusium*. Compare *Euvola laurentii*, which is generally lighter colored interiorly and exteriorly. Note: True *"Euvola" papyracea* (Gabb, 1873), is a Dominican Republic fossil in the genus *Amusium*; Recent specimens so called are either *E. marensis* Weisbord, 1964 (in northern South America), or this as yet unnamed species (from Gulf of Mexico and eastern Florida) (Waller, 1991).

### *Brachtechlamys antillarum* (Récluz, 1853) – **Antillean Scallop**

Oval fan-shaped, anterior auricle slightly larger than posterior, compressed, thin-walled, pastel yellowish gray to light brown with lines and patches of opaque white, with 11–15 broad, low, rounded radial ribs; interior with radials reflecting exterior pattern. Florida, Bermuda, Bahamas, West Indies, Caribbean Central America, South America (Colombia). Length 17 mm (to 26 mm). Formerly in *Lyropecten*.

*Euvola raveneli*

LV

*Euvola ziczac*

LV

*Euvola chazaliei*

LV

*Euvola laurentii*

LV

*Euvola "papyracea"*

LV

*Brachtechlamys antillarum*

LV

### *Nodipecten fragosus* (Conrad, 1849) – **Northern Lion's Paw**

Broadly fan-shaped, posterior auricles meeting dorsal margin at 90° angle, hinge of right valve overlapping left (so articulated valves appear to have recessed umbo), compressed, solid, dark maroon to red or orange or lemon yellow (see inset), with ca. eight coarse radial ribs covered by numerous riblets and large knobs arranged in commarginal rows; interior whitish, darker at margins, with radial ribs reflecting exterior pattern. North Carolina to Florida, Gulf of Mexico. Length 70 mm (to 150 mm). Formerly known as *Lyropecten nodosus* (Linnaeus, 1758) (Caribbean and South America). *Nodipecten nodosus* has nine radial ribs, and so "looks" wider than *N. fragosus*, with coarser macrosculpture and less tendency to develop knobs on the ribs; the hinge of its right valve does not overlap that of its left valve as is true in *N. fragosus*. Note: This species is common on wrecks and artificial reefs.

**In dorsal view** with left valve uppermost, the articulated shells of *Caribachlamys sentis* (left), *Aequipecten muscosus* (center), and *Euvola raveneli* (right) show the variability of shell inflation within this family.

*Nodipecten fragosus*

LV

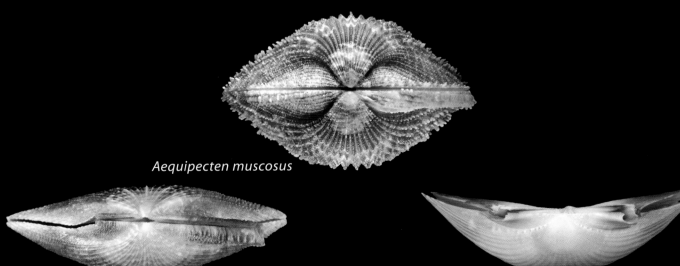

*Aequipecten muscosus*

*Caribachlamys sentis*

*Euvola raveneli*

# Family Propeamussiidae – Glass Scallops

**Classification**
AUTOLAMELLIBRANCHIATA Grobben, 1894
Pteriomorphia Beurlen, 1944
Pectinoida H. Adams & A. Adams, 1857
Pectinoidea Rafinesque, 1815
Propeamussiidae Abbott, 1954

## Featured species
*Similipecten nanus* (Verrill & Bush, 1897) – **Dwarf Glass Scallop**

  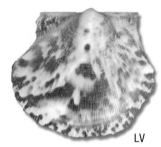

LV

Circular, auricles subequal, compressed (left valve slightly more inflated), thin-walled, translucent grayish white with milky and opaque white mottling, left valve with fine radial lines and cancellate umbonal area, right valve near-smooth with very fine, regular commarginal lamellae, hinge line with transverse striations; interior without ribs, margin smooth. Delaware to Florida, West Indies, Gulf of Mexico, South America (to Brazil). Length 3 mm (to 7 mm). Formerly in *Cyclopecten*.

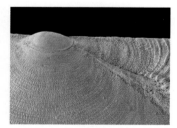

**The right umbo** of *Similipecten nanus* shows a relatively large prodissoconch (in green; suggesting lecithotrophic or direct larval development) and the byssal fasciole or track left by the byssal notch as the animal grows.

## Family description

The propeamussiid shell is small to medium-sized (to 120 mm), thin-walled, often fragile, circular to obliquely ovate, and with distinct anterior and (smaller) posterior AURICLES that can be denticulate on the dorsal margin. It is EQUIVALVE or INEQUIVALVE (right valve flatter), resting on (PLEUROTHETIC on) the right valve, compressed, and usually widely gaping. The ventral margin of the right valve is flexible. A BYSSAL NOTCH (variable in

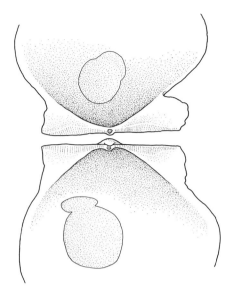

depth among species) lies below the auricle in the right valve; as the animal grows, the byssal notch fills with shell material, leaving behind a distinct track (BYSSAL FASCI-OLE). A comblike series of denticles (CTENOLIUM) is absent in the majority of species (only two exceptions are known) along the ventral margin of the byssal notch, but in some a weak analogous tooth (PSEUDOCTENOLIUM) occurs on the leading edge of the fasciole, formed from external sculptural elements on the right margin of the byssal notch; most Recent species have neither structure. The shell is EQUILATERAL to slightly INEQUILATERAL (umbones slightly posterior or appearing so due to inequal auricles), with ORTHOGYRATE UMBONES. Shell microstructure is a mixture of ARAGONITE and CALCITE, and two-layered, with an outer layer of columnar PRISMATIC calcite (right valve only, imparting its flexibility) and foliated calcite, and a CROSSED LAMELLAR aragonitic inner layer that extends outside the pallial line in some cases nearly to the ventral margin. TUBULES have not been re-

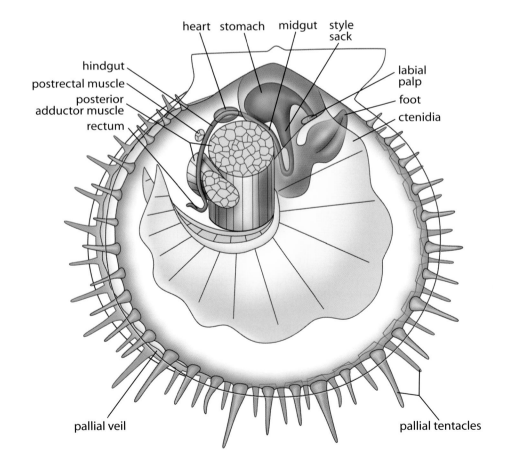

ported. Exteriorly propeamussiids are generally translucent whitish, covered by a nonpersistent PERIOSTRACUM. Sculpture is smooth or with fine radial or commarginal sculpture, or both, differing on the two valves and usually more strongly developed on the left valve (uppermost when in situ). The auricles are radially sculptured in some species. LUNULE and ESCUTCHEON are absent. Interiorly the shell is non-NACREOUS. The PALLIAL LINE is ENTIRE. The inner shell margins are smooth. The inner surface often has noninterlocking internal riblets unrelated to external sculpture. The HINGE PLATE is straight and EDENTATE in adults; the dorsal margins of the auricles are flattened medially and can be transversely striated or roughened. The dorsal edge of one valve can overlap that of the other valve. The LIGAMENT is ALIVINCULAR and AMPHIDETIC; an internal portion (RESILIUM) has a nonfibrous core and sits on a subumbonal rounded RESILIFER, which can have elevated edges (RESILIAL TEETH).

Knowledge about propeamussiid anatomy and life habits is limited. The animal is MONOMYARIAN (anterior ADDUCTOR MUSCLE absent); the posterior adductor muscle is large and central, in some species with its striate or "quick" portion separate and obliquely oriented relative to the smooth or "catch" portion. Adduction of these two asymmetrical parts results in the right valve insetting into the left, facilitated by the flexible margin of the right valve, creating a seal that concentrates expelled water jets to the posterior, auricular shell gape. A POSTRECTAL MUSCLE connects the two valves dorsal to the hindgut. Pedal retractor muscles are absent or present (in byssate species). Pedal elevator and protractor muscles have not been reported. The MANTLE margins are not fused ventrally; SIPHONS are absent. The inner mantle folds form a very large PALLIAL VEIL (velum) usually without guard tentacles; the middle folds bear extensible sensory tentacles and well-developed PALLIAL EYES (secondarily absent in many). Propeamussiid eyes are structurally complex (invaginated, with lens), and are secondarily lost in deepwater species. HYPOBRANCHIAL GLANDS have not been reported. The FOOT is anterior and bulbous, functioning as a cleaning organ for the adjacent MANTLE CAVITY and perhaps also in prey capture; it has a BYSSAL GROOVE and in a few (e.g., *Cyclopecten* and *Catillopecten*) produces a BYSSUS in the adult.

The LABIAL PALPS are small and fused to the simple (unelaborated) LIPS. The CTENIDIA are large and partially encircle the centralized muscle bundles. Some that have been examined structurally appear to lack all unifying junctions between filaments, however, the gills of *Propeamussium lucidum* (Jeffreys, 1873) are FILIBRANCH (ELEUTHERORHABDIC), HOMORHABDIC, and are not inserted into (or fused with) the distal oral groove of the palps (CATEGORY III association). In *Catillopecten vulcani* (Schein-Fatton, 1985), from the Galapagos hydrothermal vent fields, each ctenidium is N-shaped; the outer demibranch consists of the descending lamella only and no endosymbionts are present (in contrast to other vent-associated bivalves). In *P. lucidum*, each V-shaped ctenidium has been interpreted as comprised only of the descending lamella of the outer demibranch plus the ascending lamella of the inner demibranch, these two fused at the ventral edge, resulting in an "inside out" gill; such a gill cannot filter or gather/sort food particles and functions only in respiration. Water flow is antero- and ventroposterior. The esophagus is wide. The STOMACH is small and TYPE II; the MIDGUT is not coiled. The HINDGUT passes through the ventricle of the heart, and leads to a freely hanging rectum. Propeamussiids so far examined are GONOCHORISTIC; all species examined have a large PRODISSOCONCH indicating a reduced or absent planktonic larval stage. The nervous system has not been extensively studied. STATOCYSTS are present in adults. ABDOMINAL SENSE ORGANS are present or absent.

Propeamussiids are marine; most are EPIFAUNAL in fine to coarse soft substrata. Few if any species have been observed alive, due to the abyssal depths inhabited by most species.

The lack of pedal retractor muscles has been interpreted as evidence that propeamussiids are proficient swimmers (probably as an escape mechanism); all species are apparently capable of swimming. Some species, such as *Catillopecten vulcani*, are SUSPENSION FEEDERS. However, the stomach contents of many species include foraminiferans, radiolarians, and copepods, but no detritus; this, the unusual gill structure, and Type II stomach suggest that at least some propeamussiids are carnivorous, capturing individual epifaunal zooplankton in the incurrent water flow. In *Propeamussium lucidum*, the foot has been noticed lodged between the lips, suggesting it might be used in transferring captured prey to the mouth.

The family Propeamussiidae is known since the Carboniferous and is represented by 8 living genera and ca. 200 species, worldwide in distribution, with some inhabiting shallow waters, but most in polar or deep seas (to ca. 5,000 m). The shallow species are cryptic or utilize specialized habitats. One Caribbean species resembles a blade of the calcareous alga *Halimeda* among which it nestles.

## References

Beninger, P. G., S. C. Dufour, P. Decottignies, and M. Le Pennec. 2003. Particle processing mechanisms in the archaic, peri-hydrothermal vent bivalve *Bathypecten vulcani*, inferred from cilia and mucocyte distributions on the gill. *Marine Ecology Progress Series*, 246: 183–195.

Bernard, F. R. 1978. New bivalve molluscs, subclass Pteriomorphia, from the northeastern Pacific. *Venus*, 37: 61–75.

Dijkstra, H. H. 1991. A contribution to the knowledge of the pectinacean Mollusca (Bivalvia: Propeamussiidae, Entoliidae, Pectinidae) from the Indonesian Archipelago. *Zoologische Verhandelingen, Nationaal Natuurhistorisch Museum, Leiden*, 271: 1–57.

Dijkstra, H. H. 1995. Bathyal Pectinoidea (Bivalvia: Propeamussiidae, Entoliidae, Pectinidae) from New Caledonia and adjacent areas. Pages 9–73, in: P. Bouchet, ed., *Résultats des Campagnes MUSORSTOM, 14. Mémoires du Muséum National d'Histoire Naturelle*, 167.

Hayami, I. 1988. Taxonomic characters of propeamussiids from Japan. *Venus*, 47(2): 71–82.

Knudsen J. 1967. The deep-sea Bivalvia. *The John Murray Expedition, 1933–34, Scientific Reports*, 11(3): 237–343.

Knudsen, J. 1970. The systematics and biology of abyssal and hadal Bivalvia. *Galathea Report*, 11: 7–241.

Le Pennec, G., P. G. Beninger, M. Le Pennec, and A. Donval. 2003. Aspects of the feeding biology of the pectinacean *Bathypecten vulcani*, a peri-hydrothermal vent bivalve. *Journal of the Marine Biological Association of the United Kingdom*, 83: 479–482.

Le Pennec, M., A. Herry, R. Lutz, and A. Fiala-Medioni. 1988. Premières observations ultrastructurales de la branchie d'un bivalve Pectinidae hydrothermal profond. *Comptes Rendus de l'Académie des Sciences de Paris, Série 3*, 307: 627–633.

Merrill, A. S. 1959. A comparison of *Cyclopecten nanus* Verrill and Bush and *Placopecten magellanicus* (Gmelin). *Occasional Papers on Mollusks*, 2(25): 209–228.

Morton, B., and M. H. Thurston. 1989. The functional morphology of *Propeamussium lucidum* (Bivalvia: Pectinacea), a deep-sea predatory scallop. *Journal of Zoology*, 218: 471–496.

Rombouts, A. 1991. *Guidebook to Pecten Shells: Recent Pectinidae and Propeamussiidae of the World*. Universal Book Services and Dr. W. Backhuys, Oegstgeest, The Netherlands, 157 pp.

Waller, T. R. 1971. The glass scallop *Propeamussium*, a living relict of the past. *The American Malacological Union, Annual Reports for 1970*, pp. 5–7.

Waller, T. R. 1978. Morphology, morphoclines and a new classification of the Pteriomorpha (Mollusca: Bivalvia). *Philosophical Transactions of the Royal Society of London, Series B, Biological Sciences*, 284(1001): 345–365.

### *Hyalopecten strigillatus* (Dall, 1889) – **File Glass Scallop**

Circular, anterior auricle narrow, posterior auricle arching gradually into posterodorsal shell margin, thin-walled, white, with crowded elevated commarginal ridges on both valves. Georgia, Florida, West Indies, Gulf of Mexico, South America (Brazil). Length 10 mm. Formerly in *Cyclopecten*.

### *Cyclopecten thalassinus* (Dall, 1886) – **Costate Glass Scallop**

Broadly fan-shaped to circular, anterior auricle larger than posterior, compressed, left valve less convex and slightly smaller, thin-walled, white with red, brown, and yellow mottlings on both valves (right valve brighter), left valve with coarse commarginal ridges and radial ribs most noticeable near margin, right valve with elevated commarginal ribs crossed by radial threads; interior with flattened thickened smooth margin (especially in left valve), without internal ribs. Massachusetts to Florida, West Indies, Gulf of Mexico, South America (Brazil). Length 7 mm. Formerly in *Propeamussium*.

### *Propeamussium cancellatum* (E. A. Smith, 1885) – **Cancellate Glass Scallop**

Circular to obliquely oval, auricles small and subequal, posterior one larger than anterior, somewhat denticulate at dorsal margin, posterior end of left valve separated from auricle by deep groove, compressed (left valve slightly more inflated), thin-walled, translucent white or creamy, left valve cancellate, right valve with closely spaced commarginal lirae; interior glossy, with 10-12 white radial ribs that are visible through exterior, somewhat thickened toward outer ends, interspaces with rudimentary marginal riblets. Florida, Bermuda, West Indies, Gulf of Mexico, South America (Brazil). Length 10 mm (to 26 mm).

### *Parvamussium pourtalesianum* (Dall, 1886) – **Pourtales' Glass Scallop**

Broadly fan-shaped to circular, auricles subequal, compressed, thin-walled, translucent, right valve with commarginal threads, left valve near-smooth; interior with ca. 9 rod-like, opaque, white radial ribs. Florida, West Indies, Gulf of Mexico, South America (Colombia, Brazil). Length 9 mm (to 12 mm). Formerly in *Propeamussium*.

### *Parvamussium sayanum* (Dall, 1886) – **Say's Glass Scallop**

Broadly fan-shaped, auricles subequal, compressed, thin-walled, white, left valve with ca. 12 radial ribs (between which additional ribs appear toward margin) crossed by fine commarginal ridges, right valve with fine commarginal ridges and obscure low radials; interior with 10–16 radial ribs. Florida Keys, West Indies, Gulf of Mexico. Length 8 mm (to 15 mm). Formerly in *Propeamussium*.

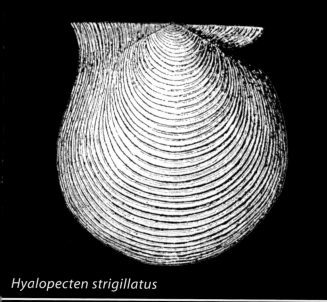

*Hyalopecten strigillatus*

*Cyclopecten thalassinus*

LV

*Propeamussium cancellatum*

LV

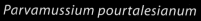

*Parvamussium pourtalesianum*

LV

*Parvamussium sayanum*

LV

# Family Spondylidae J. E. Gray, 1826 – Thorny Oysters

**Classification**
AUTOLAMELLIBRANCHIATA Grobben, 1894
Pteriomorphia Beurlen, 1944
Pectinoida H. Adams & A. Adams, 1857
Pectinoidea Rafinesque, 1815
Spondylidae J. E. Gray, 1826

**Featured species**
*Spondylus americanus* Hermann, 1781 – **Atlantic Thorny Oyster**

articulated

Oval to circular, white with yellow, orange or reddish umbones, with low radial ribs adorned with often-spectacular spines, usually erect nonfrondose spines 5 cm or less in length; interior white, reddish purple at margins, margin denticulate (see also p. 119), juveniles much less spiny and often with irregular lamellae on lower valve (resembling *Chama*). North Carolina to Florida, Bermuda, Bahamas, West Indies, Gulf of Mexico, Caribbean Central America, South America (to Brazil). Height 85 mm (to 150 mm).

**The variably elongated and complex spines of** *Spondylus americanus*, also called the Chrysanthemum Shell, protect its colorful mantle edge from predatory fish and act as framework for camouflaging growth. The bright red color of this specimen, from the *Thunderbolt* wreck off Marathon, is an encrusting layer of sponge, not of the shell itself.

## Family description

The spondylid shell is small to large (to >200 mm), solid, orbicular to dorsoventrally oval, and with weak anterior and posterior AURICLES more prominent in juveniles. It is INEQUIVALVE (lower right valve usually larger and more convex), cemented by (PLEUROTHETIC on) the right valve by a variable area of cementation, inflated, and not gaping

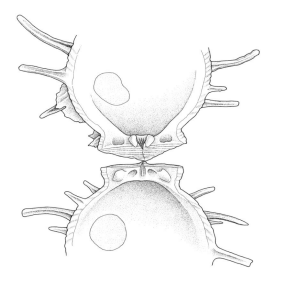

as adults; juveniles (PECTINIFORM STAGE) of *Spondylus* have a BYSSAL NOTCH (and subsequent BYSSAL FASCIOLE) in the lower right valve below the auricle (without CTENOLIUM). The shell is nearly EQUILATERAL, with OR-THOGYRATE UMBONES separated by a wide CARDINAL AREA (larger on the lower right valve). Shell microstructure is a mixture of ARAGONITE and CALCITE, and three-layered, with a foliated calcitic outer layer (also on narrow margin of shell interior; PRISMATIC calcite is secondarily absent) and aragonitic CROSSED LAMELLAR middle and inner layers (the former restricted to the hinge; the latter extending outside the pallial line nearly to the ventral margin). TUBULES penetrate all shell layers and the periostracum. Exteriorly spondylids are often brightly colored, covered by a nonpersistent PERIOSTRACUM. Sculpture is irregular, usually conspicuously radially spinose and/or lamellose (especially on the upper left valve). LUNULE and ESCUTCHEON are absent. Interiorly the shell is non-NACREOUS. The PALLIAL LINE is ENTIRE. The inner shell margins are denticulate. The HINGE PLATE is straight and strong, with strongly interlocking, secondary ISODONT teeth in adults. The cardinal area is steeply trigonal in the lower right valve, and smaller in the upper left valve; both sides have a conspicuous central

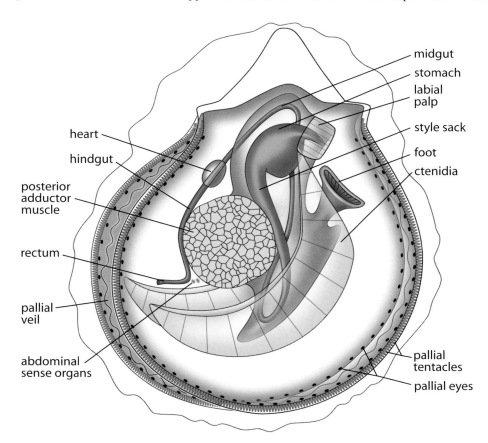

groove reflecting the position of the resilium. The LIGAMENT is internal (RESILIUM) with a nonfibrous core and sits on a deep, subumbonal, trigonal RESILIFER. A secondary external ligament of fused periostracum unites the valves along the posterodorsal hinge margin.

The animal is MONOMYARIAN (anterior ADDUCTOR MUSCLE absent); the posterior adductor muscle is large and posterocentral, with striated and nonstriated portions that are subequal (rather than demonstrably unequal as in most bivalves). Pedal retractor, elevator, and protractor muscles are absent (although a considerable number of muscle fibers pass from the foot into the mantle). The MANTLE margins are not fused ventrally; SIPHONS are absent. The inner mantle folds form an often-colorful PALLIAL VEIL (velum) without guard tentacles; the middle folds bear sensory tentacles and structurally complex PALLIAL EYES (invaginated, with lens) similar to those in Pectinidae. Pallial retractor muscles are present and extensive. HYPOBRANCHIAL GLANDS have not been reported. The FOOT is anterior and suckerlike, with a radially grooved, highly flexible "terminal funnel"; it has a BYSSAL GROOVE (juveniles initially attach by a byssus) but adults are nonbyssate. The foot functions as a cleaning organ for the INFRABRANCHIAL CHAMBER, accumulating waste particles and mucus at its center and discharging them to the exterior or to the inner mantle surface with its own PSEUDOFECES rejection pathways.

The LABIAL PALPS are quadrangular. The CTENIDIA are large and partially encircle the centralized adductor muscle. They are PSEUDOLAMELLIBRANCH (ELEUTHERORHABDIC), HETERORHABDIC, and are not inserted into (or fused with) the distal oral groove of the palps (CATEGORY III association). Water flow is antero- and ventroposterior. The LIPS are hypertrophied and arborescent, interdigitating to create a series of pores over the mouth, allowing exit of excess water without loss of food. The STOMACH is TYPE IV; the MIDGUT is not coiled. The HINDGUT passes through the ventricle of the heart, and leads to a usually freely hanging rectum. Spondylids are GONOCHORISTIC or HERMAPHRODITIC, and produce planktonic VELIGER larvae. The gonad often extends branches into the mantle tissues. The nervous system is not concentrated, but the pedal ganglia are (unusually among bi-

***Spondylus ictericus*** Reeve, 1856 – **Digitate Thorny Oyster**

articluated

Obliquely oval, brick red or dull purple with white mottling at umbones, with coarse, irregular radial ribs adorned with short spines, minor spines with minute prickles, spines of all sizes often digitate at ends; interior white, margin brick red and denticulate. Florida, Bermuda, Bahamas, West Indies, Gulf of Mexico, Caribbean Central America, South America (to Brazil). Height 55 mm (to 95 mm).

**Even at small juvenile stages**, the color and spination of *Spondylus americanus* (left) and *S. ictericus* (right) shells are distinct.

valves) joined via connectives with the visceral rather than the cerebropleural ganglia; the fused parietovisceral ganglia form a very large ganglionic complex (also including part or all of the cerebropleural ganglia according to some authors). STATOCYSTS are present in adults. ABDOMINAL SENSE ORGANS are present and asymmetrical.

Spondylids are marine EPIFAUNAL SUSPENSION FEEDERS. They are most abundant and diverse on rock and coral reefs, and are often heavily encrusted with sponges and other fouling and/or boring organisms. Their shell spines and tight closure generally protect the animal against predators, but one fragile Australian species is preyed upon by boring muricid gastropods through the exposed part of the right (lower) hinge plate. The complex eyes are responsible for the marked shadow reflex reaction of spondylids. High intraspecific variability makes species-level identification difficult in this family.

The family Spondylidae is known since the Jurassic and is represented by 2 living genera and ca. 70 species, distributed worldwide mainly in tropical and subtropical waters; one species inhabits depths to 1,800 m. *Spondylus* shells have a strong presence in the anthropological records of Peru, in the form of beads, inlay, or whole shell, where its distinct orange-red shells first appeared during the Late Preceramic Period (ca. 5000–3800 BP) and by 2200 BP had become the symbol of the Peruvian elite class.

## References

Dakin, W. J. 1928a. The anatomy and phylogeny of *Spondylus*, with a particular reference to the lamellibranch nervous system. *Proceedings of the Royal Society of London, Series B, Containing Papers of a Biological Character*, 103(725): 337–355.

Dakin, W. J. 1928b. The eyes of *Pecten*, *Spondylus*, *Amussium* and allied lamellibranchs, with a short discussion on their evolution. *Proceedings of the Royal Society of London, Series B, Containing Papers of a Biological Character*, 103(725): 355–365.

Lamprell, K. 1987. *Spiny Oyster Shells of the World*, Spondylus. E. J. Brill / Dr. W. Backhuys, Leiden, The Netherlands, 82 pp.

Logan, A. 1974. Morphology and life habits of the Recent cementing bivalve *Spondylus americanus* Hermann from the Bermuda platform. *Bulletin of Marine Science*, 24(3): 568–594.

Sandweiss, D. H. 1999. The return of the native symbol: Peru picks *Spondylus* to represent new integration with Ecuador. *Society for American Archaeology Bulletin*, 17(2): 1, 8–9. Online version http://www.saa.org/publications/saabulletin/17-2/SAA1.html, last accessed 10 November 2005.

Waller, T. R. 1978. Morphology, morphoclines and a new classification of the Pteriomorpha (Mollusca: Bivalvia). *Philosophical Transactions of the Royal Society of London, Series B, Biological Sciences*, 284(1001): 345–365.

Yonge, C. M. 1973. Functional morphology with particular reference to hinge and ligament in *Spondylus* and *Plicatula* and a discussion on relations within the superfamily Pectinacea (Mollusca: Bivalvia). *Philosophical Transactions of the Royal Society of London, Series B, Biological Sciences*, 267(883): 173–208.

# Family Plicatulidae – Kitten's Paw Clams

**Classification**

AUTOLAMELLIBRANCHIATA Grobben, 1894

Pteriomorphia Beurlen, 1944

Pectinoida H. Adams & A. Adams, 1857

Plicatuloidea J. E. Gray, 1857

Plicatulidae J. E. Gray, 1857

**Featured species**
***Plicatula gibbosa*** Lamarck, 1801 – **Atlantic Kitten's Paw**

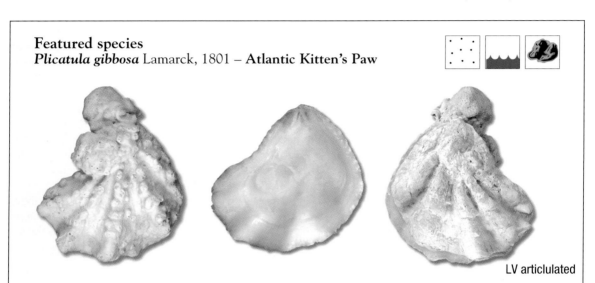

LV articlulated

Trigonal to fan-shaped, compressed, solid, grayish white with orange rays on radial ribs, with central rounded area bordered by 5–12 broad, trigonal, radial undulations forming plicate saw-toothed margins; interior glossy white, muscle scar occasionally raised. North Carolina to Florida, Bermuda, Bahamas, West Indies, Gulf of Mexico, Caribbean Central America, South America (to Argentina). Length 25 mm (to 33 mm).

***Plicatula gibbosa*** is a common species in deepwater (80 m) *Oculina* coral colonies off the eastern coast of Florida.

## Family description

The plicatulid shell is small to medium-sized (usually <100 mm), solid, commonly irregular in outline, dorsoventrally elongated trigonal to spathate, oval, or orbicular, often with undulating valve margins, and occasionally with anterior and posterior AURICLES. It is EQUIVALVE TO INEQUIVALVE, with the right valve usually more convex, cemented by

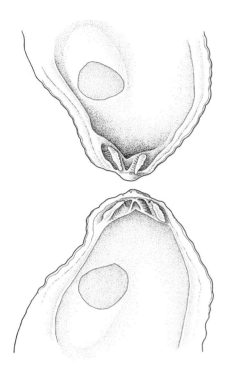

(PLEUROTHETIC on) a small attachment area on either valve, usually compressed, and not gaping. The shell of most species is nearly EQUILATERAL, with ORTHOGYRATE UMBONES. Shell microstructure is a mixture of ARAGONITE and CALCITE, and three-layered, with a foliated calcitic outer layer (PRISMATIC calcite is secondarily absent) and aragonitic CROSSED LAMELLAR middle and inner layers, the latter extending outside the pallial line nearly to the ventral margin. TUBULES have not been reported. Exteriorly plicatulids are often reddish brown covered by a nonpersistent PERIOSTRACUM. Sculpture consists of broad radial ribs that are usually commarginally lamellose. LUNULE and ESCUTCHEON are absent. Interiorly the shell is non-NACREOUS. The PALLIAL LINE is ENTIRE. The inner shell margins are smooth but usually heavily plicate by the broad external sculpture. The HINGE PLATE is short with strongly interlocking, occasionally serrated, secondary ISODONT teeth. The LIGAMENT is internal (RESILIUM) and sits on a deep subumbonal trigonal RESILIFER. A secondary external ligament of fused periostracum unites the valves along the dorsal hinge margin, including dorsal to the resilium.

The animal is MONOMYARIAN (anterior ADDUCTOR MUSCLE absent); the posterior adductor muscle is large and posterocentral. Pedal retractor, elevator, and protractor muscles

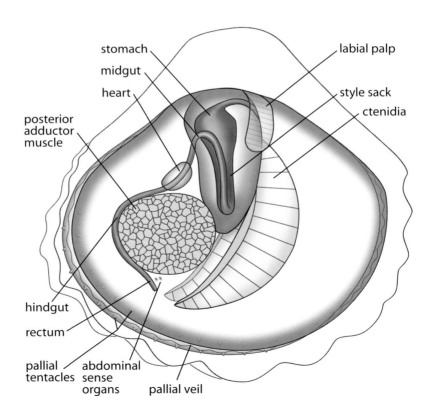

are absent. The MANTLE margins are not fused ventrally; SIPHONS are absent. The inner mantle folds form a narrow PALLIAL VEIL (velum) without guard tentacles; the inner folds bear short tentacles but no PALLIAL EYES. HYPOBRANCHIAL GLANDS have not been reported. The FOOT and BYSSUS are absent in the adult; juveniles apparently do not pass through a byssally attached phase.

The LABIAL PALPS are small with the outer pair fused medially to form a hood over the mouth. The CTENIDIA are large and partially encircle the centralized adductor muscle. They are FILIBRANCH (ELEUTHERORHABDIC), HOMORHABDIC, and are not inserted into (or fused with) the distal oral groove of the palps (CATEGORY III association). The ascending limbs of the inner demibranchs are absent in some species. Water flow is antero- and ventroposterior. The lips are simple. The STOMACH is TYPE IV. The MIDGUT is not coiled. The HINDGUT passes through the ventricle of the heart, and leads to a freely hanging rectum. Plicatulids are GONOCHORISTIC; larval development has not been described. The nervous system is not concentrated. STATOCYSTS are present in adults. ABDOMINAL SENSE ORGANS are present and asymmetrical.

Plicatulids are marine SUSPENSION FEEDERS, attached to hard substrata (e.g., coral rubble). Because of the large isodont hinge teeth, plicatulids cannot open their shell valves much more than a slit.

The family Plicatulidae is known since the Triassic and is represented by 1 living genus and ca. 10 species, distributed worldwide in warm waters, mainly at shallow depths.

## References

Waller, T. R. 1978. Morphology, morphoclines and a new classification of the Pteriomorpha (Mollusca: Bivalvia). *Philosophical Transactions of the Royal Society of London, Series B, Biological Sciences*, 284(1001): 345–365.

Watson, H. 1930. On the anatomy and affinities of *Plicatula*. *Proceedings of the Malacological Society of London*, 19(1): 25–31, pl. 5.

Yonge, C. M. 1973. Functional morphology with particular reference to hinge and ligament in *Spondylus* and *Plicatula* and a discussion on relations within the superfamily Pectinacea (Mollusca: Bivalvia). *Philosophical Transactions of the Royal Society of London, Series B, Biological Sciences*, 267(883): 173–208.

Yonge, C. M. 1975. The status of the Plicatulidae and the Dimyidae in relation to the superfamily Pectinacea (Mollusca: Bivalvia). *Journal of Zoology*, 176: 545–553.

Yonge, C. M. 1977. The ligament in certain "anisomyarians." *Malacologia*, 16(1): 311–315.

# Family Anomiidae – Saddle Oysters or Jingle Shells

**Classification**
AUTOLAMELLIBRANCHIATA Grobben, 1894
Pteriomorphia Beurlen, 1944
Pectinoida H. Adams & A. Adams, 1857
Anomioidea Rafinesque, 1815
Anomiidae Rafinesque, 1815

## Featured species
### *Anomia simplex* d'Orbigny, 1853 – Common Jingle Shell

LV

Irregularly circular to oval, thinner right valve flatter with byssal hole, left valve often convex, compressed, thin-walled, translucent white to yellow or orange (turning black after burial in sand), right valve often lighter, smooth but assuming shape and sculpture of substratum; interior of left valve with two small and one large central muscle scars. Eastern Canada to Florida, Bermuda, Bahamas, West Indies, Gulf of Mexico, Caribbean Central America, South America (to Argentina). Length 30 mm (to 44 mm). Note: Some authors consider this a synonym of *Anomia ephippium* Linnaeus, 1758 (northern Europe).

**Anomia simplex assumes** the shape of the surface upon which it attaches, often becoming ridged or misshapen as a result. This specimen is on the smooth interior umbo of a *Mya arenaria* shell from Massachusetts.

## Family description

The anomiid shell is medium-sized to large (to 150 mm), thin-walled, circular to oval, flat, or folded, usually distorted according to the contour of the substratum. The lower right valve is usually flat, much thinner than the upper left valve, and has a small to large, oval, subumbonal hole (BYSSAL NOTCH or FORAMEN) through which the byssus passes; the dorsal margin of the hole is open in some taxa, fused in others (e.g., *Pododesmus*). It is IN-

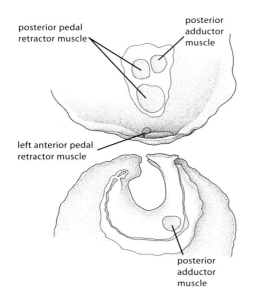

posterior pedal
retractor muscle

posterior
adductor
muscle

left anterior pedal
retractor muscle

posterior
adductor
muscle

EQUIVALVE, with the upper left valve usually more convex, usually byssally cemented on (PLEUROTHETIC on) the right valve, compressed, and not gaping. The shell is INEQUILATERAL, with umbones displaced from the margin by overgrowth of the mantle lobes and shell, especially in the upper left valve. Shell microstructure is a mixture of ARAGONITE and CALCITE, and one- or two-layered, with a foliated calcitic outer layer (also with PRISMATIC calcite in the right valve) and a thin aragonitic COMPLEX CROSSED LAMELLAR inner layer that does not extend outside the pallial line and is absent in some. TUBULES are apparently absent. Exteriorly anomiids are translucent and variably colored, covered by a thin PERIOSTRACUM. Sculpture is weak and irregular, and can reflect that of the substratum. LUNULE and ESCUTCHEON are absent. Interiorly the shell is thinly NACREOUS or hyaline. The PALLIAL LINE is ENTIRE. The inner shell margins are smooth. The HINGE PLATE is short and EDENTATE in adults, but with various large tubercles and ridges. The LIGAMENT is internal (RESILIUM), elongated oval, and sits on a stalked or mushroom-shaped RESILIFER (CRURUM).

The animal, like the shell, is highly asymmetrical due to twisting of the body that laterally repositions the byssus to exit through the foramen in the right valve. It is MONOMYARIAN (anterior ADDUCTOR MUSCLE absent); the posterior adductor muscle is small and posterocentral. The posterior pedal retractor is larger than the adductor, is divided into

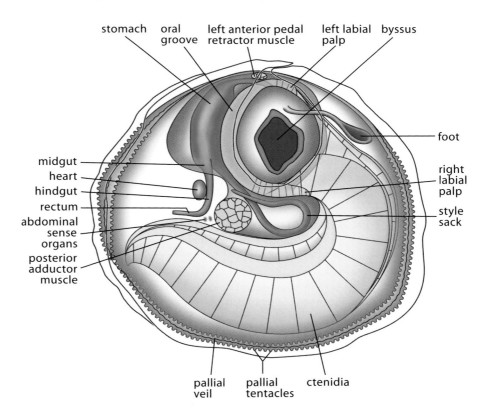

stomach oral groove left anterior pedal retractor muscle left labial palp byssus foot midgut heart hindgut rectum abdominal sense organs posterior adductor muscle right labial palp style sack pallial veil pallial tentacles ctenidia large and small portions in *Anomia*, and inserts on the upper left valve (only) opposite the byssus and foramen; contraction of this muscle produces shell closure. In *Heteranomia*, the scars of the posterior adductor and (larger) posterior pedal retractor muscles are merged into one. The left anterior pedal retractor inserts near the resilium, again on the upper left valve (only); the right anterior pedal retractor is present only in *Pododesmus*. Pedal elevator and protractor muscles are absent. The MANTLE margins are not fused ventrally; SIPHONS are absent. Right and left pallial extensions converge (and can fuse) dorsal to the byssal apparatus, displacing the umbones away from the shell margin. The inner mantle folds form a small PALLIAL VEIL (velum) without guard tentacles; the inner folds bear short tentacles but no PALLIAL EYES. In the Indo-Pacific limpet-like *Enigmonia*, complex pallial eyes (invaginated, with lens) are uniquely present solely under the upper left valve on the outer surface of the left mantle lobe. Large HYPOBRANCHIAL GLANDS are present in the SUPRABRANCHIAL CHAMBER only in *Pododesmus*. The FOOT is anterior, slender-stalked, and spatulate in most anomiids, and is used mainly for cleaning the area around the byssus. The foot of limpetlike *Enigmonia* is secondarily used for locomotion; it is also anterior, but distally bulbous with a thick stalk and a filamentous (probably sensory) terminal flagellum and normally extends through the foramen with the stout byssus emerging from its "heel." The BYSSUS is uncalcified and temporarily attached in *Enigmonia*, whereas in others (more typical of the family) it is short, pluglike, composed principally of CALCITIC calcium carbonate, and irreversibly cemented to the substratum.

The right LABIAL PALP lies at the ventral side of the byssal complex, whereas the left palp lies near the mouth and umbo on the opposite side; this results in a wide separation of the right and left palps, connected by an unridged oral groove. The CTENIDIA are large and partially encircle the centralized muscle bundles. They are FILIBRANCH (ELEUTHERORHABDIC), asymmetric (due to widely separated labial palps), HOMORHABDIC, and are not inserted into (or fused with) the distal oral groove of the palps (CATEGORY III association). In *Anomia* and *Enigmonia*, three of the demibranchs (both left and the inner right) are associated with the left labial palp, and only one demibranch (outer right) with the right palp (representing a unique condition in Bivalvia). CEPHALIC EYES are often present on the first gill filament. The ascending limbs of both demibranchs are absent in *Heteranomia*; the descending lamellae are attached to each other by ciliary junctions, and as a result, the gills assume in cross section a single inverted W shape. Water flow is antero- and ventroposterior; apposition of the mantle margins can narrow the flow to a small anterior incurrent area when conditions are turbid. The alimentary tract is extremely short. The lips are simple, but can be longer on the right side; the mouth is displaced to the right of the midline near the umbo. The STOMACH is TYPE IV; the MIDGUT is very short and not coiled. The STYLE SACK is separate from the midgut, very long (often describing a complete circle), and extends into the right mantle lobe. The HINDGUT passes dorsal to the ventricle of the heart, between the two auricles, and leads to a rectum that is freely hanging or attached to the right mantle lobe. The pericardium is absent (a unique condition in Bivalvia). Anomiids are GONOCHORISTIC and produce planktonic VELIGER larvae; in some, the gonad extends into the mantle lining each valve. The nervous system is asymmetric with the cerebropleural ganglia displaced from midline, and somewhat concentrated with the cerebropleural and pedal ganglia in close proximity. STATOCYSTS (with STATOCONIA) have been reported in adult *Anomia*. ABDOMINAL SENSE ORGANS are present and asymmetrical, or absent.

Anomiids are marine or estuarine SUSPENSION FEEDERS, usually permanently and tightly attached to hard substrata (e.g., shells or rocks) by the calcified byssus. Limpetlike *Enigmonia aenigmatica* (Holten, 1803) is unique in being mobile and semiterrestrial, moving about on mangroves and other hard substrata in damp atmospheres; it is tolerant of long periods of exposure and varies in color with its substratum.

The family Anomiidae is known since the Jurassic and is represented by 4 living genera and ca. 15 species, distributed worldwide mainly in temperate waters, from shallow to deep depths. The common name "jingle shells" is derived from the shell craft industry, wherein empty shells shaken together, as in a mobile, make a jingling sound.

---

*Pododesmus rudis* (Broderip, 1834) – **Atlantic False Jingle**

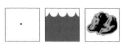

LV

Oval, right valve with byssal hole, compressed, thin-walled, brown, tan, yellowish, white, or purplish, with irregular rounded radial riblets; interior of left valve with one small and one large central muscle scar. North Carolina to Florida, Bermuda, West Indies, Gulf of Mexico, Caribbean Central America, South America (to Argentina). Length 20 mm (to 80 mm).

---

### References

Beu, A. G. 1967. Notes on the Australian Anomiidae (Mollusca: Bivalvia). *Transactions of the Royal Society of New Zealand, Zoology,* 9(18): 225–243.

Bourne, G. C. 1907. On the structure of *Aenigma aenigmatica,* Chemnitz; a contribution to our knowledge of the Anomiacea. *Quarterly Journal of Microscopical Science, New Series,* 51(2): 253–295, pls. 15–17.

Morton, B. 1976. The biology, ecology and functional aspects of the organs of feeding and digestion of the S. E. Asian mangrove bivalve, *Enigmonia aenigmatica* (Mollusca: Anomiacea). *Journal of Zoology,* 179(4): 437–466.

Prezant, R. S. 1984. Functional microstructure and mineralogy of the byssal complex of *Anomia simplex* Orbigny (Bivalvia: Anomiidae). *American Malacological Bulletin,* 2: 41–50.

Seed, R., and D. Roberts. 1976. A study of three populations of saddle-oysters (family Anomiidae) from Strangford Lough, Northern Ireland. *The Irish Naturalists' Journal,* 18(11): 317–321.

Waller, T. R. 1978. Morphology, morphoclines and a new classification of the Pteriomorpha (Mollusca: Bivalvia). *Philosophical Transactions of the Royal Society of London, Series B, Biological Sciences,* 284(1001): 345–365.

Yonge, C. M. 1977. Form and evolution in the Anomiacea—*Pododesmus, Anomia, Patro, Enigmonia* (Anomiidae), *Placunanomia, Placuna* (Placunidae fam. nov.). *Philosophical Transactions of the Royal Society of London, Series B, Biological Sciences,* 276(950): 453–523.

Yonge, C. M. 1980. On *Patro australis* with comparisons of structure throughout the Anomiidae (Bivalvia). *Malacologia,* 20(1): 143–151.

# Family Crassatellidae – Crassatella Clams

**Classification**
AUTOLAMELLIBRANCHIATA Grobben, 1894
HETEROCONCHIA Hertwig, 1895
Heterodonta Neumayr, 1883
Carditoida Dall, 1889
Crassatelloidea Férussac, 1822
Crassatellidae Férussac, 1822

---

## Featured species
*Eucrassatella speciosa* (A. Adams, 1852) – **Beautiful Crassatella**

Rounded trigonal, posteriorly rostrate, with oblique keel from umbones, compressed, solid, yellowish white to light brown with orange-brown markings, with thin nut-brown periostracum and regular rounded commarginal ridges; interior glossy white with tan or pink flush, margin smooth. North Carolina to Florida, West Indies, Gulf of Mexico, Caribbean Central America, South America (Colombia). Length 43 mm (to 65 mm).

---

LV

**The juvenile of *Eucrassatella speciosa*** shows a deceptively different external sculpture and hinge line than the adult.

# Family description

The crassatellid shell is small to medium-sized (to 115 mm), solid, and quadrangular to rounded trigonal, anteriorly rounded, with the posterior end ROSTRATE or truncate. It is EQUIVALVE or INEQUIVALVE (left valve slightly larger), compressed, and not gaping. The shell is EQUILATERAL to INEQUILATERAL (umbones slightly posterior), with PROSO-,

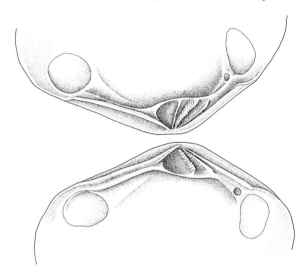

OPISTHO-, or ORTHOGYRATE UMBONES. Shell microstructure is ARAGONITIC and two-layered, with a CROSSED LAMELLAR outer layer and a COMPLEX CROSSED LAMELLAR or HOMOGENOUS inner layer. TUBULES are present penetrating the inner layer (or occasionally both layers). Exteriorly crassatellids are covered by a thin to thick, brown, polished or fibrous PERIOS-TRACUM. Sculpture is smooth or (in most species) with distinct commarginal (occasionally oblique) ribs or folds, and sometimes with an internal layer of very fine radial sculpture. LUNULE and ESCUTCHEON are distinct and often indented; in *Crassinella,* the lunule is narrow whereas the escutcheon is wide and lunulelike. Interiorly the shell is non-NACRE-OUS. The PALLIAL LINE is ENTIRE, although a slight change in curvature marks the position

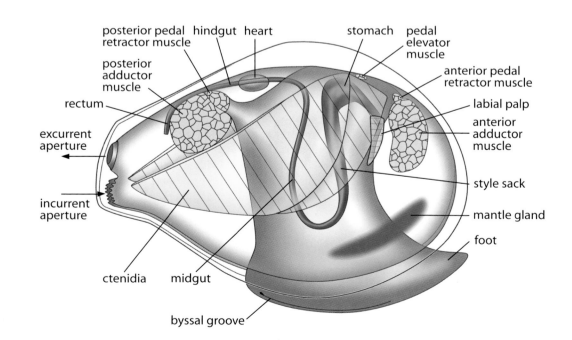

of the posterior excurrent aperture. The inner shell margins are smooth. The HINGE PLATE is strong, flat, wide, and HETERODONT, with to or three CARDINAL TEETH posterior to the resilifer, plus anterior and posterior laminar LATERAL TEETH; all teeth can be finely transversely striate or serrate. The LIGAMENT is ALIVINCULAR and AMPHIDETIC (although mostly OPISTHODETIC), set on NYMPHS posterodorsal to an internal portion (RESILIUM) that is in turn set on a large, trigonal RESILIFER. A secondary external ligament of fused periostracum extends anteriorly and posteriorly.

The animal is ISOMYARIAN; pedal retractor muscles are well developed. Pedal elevator muscles insert in the umbonal recess; pedal protractor muscles have not been reported. The MANTLE margins are not fused ventrally. SIPHONS are absent but the inner folds are fused posteriorly to form a posterodorsal EXCURRENT APERTURE equipped with a conelike extension; the INCURRENT APERTURE is not fused ventrally. Elongated scythe-shaped mantle glands (that have been reported to produce a luminescent secretion in one species of *Crassinella*) are present on the inner surface near the mantle edge. HYPOBRANCHIAL GLANDS have not been reported. The FOOT is large, bladelike, laterally compressed, quadrangular, pointed anteriorly, and heeled posteriorly; it has a BYSSAL GROOVE but the adult

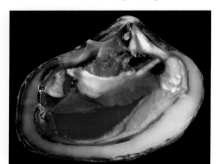

is usually nonbyssate (some smaller *Crassinella* are byssate).

The LABIAL PALPS are small to large, and trigonal. The CTENIDIA are EULAMELLIBRANCH (SYNAPTORHABDIC) and HOMORHABDIC (a few existing HETERORHABDIC reports need confirmation); the outer demibranchs are absent in *Crassinella*. The gills are not inserted into (or fused with) the distal oral groove of the palps (CATEGORY III association), or are inserted but not fused (CATEGORY I). Incurrent and excurrent water flows are posterior; an anterior incurrent also is present in some species. The lips are simple or are enlarged laterally. The STOMACH is TYPE IV; the MIDGUT is not coiled. The HINDGUT passes through the ventricle of the heart, and leads to a freely hanging rectum. Living individuals can be brightly colored because of the presence of extracellular hemoglobin in the blood. Crassatellids are GONOCHORISTIC; some species of *Crassinella* and *Eucrassatella* brood VELIGER larvae in the suprabranchial chamber. The nervous system is not concentrated. STATOCYSTS and ABDOMINAL SENSE ORGANS have not been reported.

Crassatellids are marine SUSPENSION FEEDERS, shallowly burrowing or attaching to hard particles with byssal threads; some are EPIFAUNAL, lying on the substratum. Predators include octopuses, boring gastropods, and shell-crushing sharks and rays. Certain features (e.g., adult byssus, absent outer demibranchs, and brooding) of some small *Crassinella* suggest that their evolution involves PEDOMORPHOSIS.

The family Crassatellidae is known since the Devonian and is represented by 9 living genera and ca. 40 species, distributed worldwide mainly in shallow tropical and subtropical waters. The robust shells of large South Australian *Eucrassatella* have been traditionally used as hand tools by Aboriginal hunter-gatherers.

## References

Allen, J. A. 1968. The functional morphology of *Crassinella mactracea* (Linsley) (Bivalvia: Astartacea). *Proceedings of the Malacological Society of London*, 38(1): 27–40.

Coan, E. V. 1979. Recent eastern Pacific species of the crassatellid bivalve genus *Crassinella*. *The Veliger*, 22(1): 1–11.

Coan, E. V. 1984. The Recent Crassatellinae of the eastern Pacific, with some notes on *Crassinella*. *The Veliger*, 26(3): 153–169.

Darragh, T. A. 1964. A preliminary revision of the living species of *Eucrassatella* (Pelecypoda: Crassatellidae). *Journal of the Malacological Society of Australia*, 1(8): 3–9.

Harry, H. W. 1966. Studies on bivalve molluscs of the genus *Crassinella* in the northwestern Gulf of Mexico: anatomy, ecology and systematics. *Publications of the Institute of Marine Science, Texas*, 11: 65–89.

Lyons, W. G. 1989. An Atlantic molluscan assemblage dominated by two species of *Crassinella* (Bivalvia: Crassatellidae). *American Malacological Bulletin*, 7(1): 57–64.

Taylor, J. D., E. A. Glover, and S. T. Williams. 2005. Another bloody bivalve: anatomy and relationships of *Eucrassatella donacina* from south Western Australia (Mollusca: Bivalvia: Crassatellidae). Pages 261–288, in: F. E. Wells, D. I. Walker, and G. A. Kendrick, eds. *The Marine Flora and Fauna of Esperance, Western Australia [Proceedings of the Twelfth International Marine Biological Workshop: The Marine Flora and Fauna of Esperance, Western Australia, February 2003], volume 1*. Western Australian Museum, Perth.

### *Crassinella dupliniana* (Dall, 1903) – **Round-ridged Crassinella**

Trigonal, umbonal angle hooked (observed from interior), anterior margin straight, posterior convex, with narrow ridges forming grooves along posterior left valve and anterior right valve, solid, creamy white to orange-brown to dark purple-brown, occasionally flushed with white or radially rayed, with rounded commarginal ridges and microstructure resembling snakeskin; interior colors reflecting exterior, margin smooth. Florida, Bahamas, Gulf of Mexico. Length 2 mm (to 4 mm). Note: Originally described as a Pliocene fossil (from Duplin Formation).

### *Crassinella lunulata* (Conrad, 1834) – **Lunate Crassinella**

Trigonal, umbonal angle slightly hooked (observed from interior) and at ca. 90° angle, anterior margin more convex and slightly longer, compressed, solid, white to yellow-brown to dark brown, occasionally radially rayed, with ca. 15–17 coarse commarginal ridges varying from rounded to thin and erect, plus microstructure resembling snakeskin; interior white with brown staining or reflecting exterior colors, margin smooth. Massachusetts to Florida, Bermuda, Bahamas, West Indies, Gulf of Mexico, Caribbean Central America, South America (to Uruguay). Length 5 mm (to 9 mm). Compare *Crassinella martinicensis*, which has an unhooked umbonal angle.

### *Crassinella martinicensis* (d'Orbigny, 1853) – **Martinique Crassinella**

Trigonal, umbonal angle not hooked (observed from interior) and just less than 90°, anterior margin more convex and slightly longer, compressed, solid, white shaded with brown, with 8–15 sharp commarginal ribs covering entire valve, plus microstructure resembling snakeskin; interior color reflecting exterior, margin smooth. Florida, Bahamas, West Indies, Gulf of Mexico, Caribbean Central America, South America (to Brazil). Length 3 mm (to 4 mm). Compare *Crassinella lunulata*, which has a distinctly hooked umbonal angle.

**The Carysfort Reef Lighthouse** is the oldest functioning "iron-pile" lighthouse in the United States, the first of three reef lights that (later Civil War General) George G. Meade would build in Florida. The 34-m-tall structure constructed in 1852 marks the outer reef line located about 10 km off Key Largo.

*Crassinella dupliniana*

*Crassinella lunulata*

*Crassinella martinicensis*

# Family Astartidae – Astarte Clams

**Classification**
AUTOLAMELLIBRANCHIATA Grobben, 1894
HETEROCONCHIA Hertwig, 1895
Heterodonta Neumayr, 1883
Carditoida Dall, 1889
Crassatelloidea Férussac, 1822
Astartidae d'Orbigny, 1844 [1840]

**Featured species**
*Astarte smithii* Dall, 1886 – **Smith's Astarte**

Rounded trigonal to quadrangular, light brown, periostracum finely reticulate, with strong, low, rounded or squared commarginals; interior color reflecting exterior, margin smooth, becoming denticulate with age. Florida, West Indies, Gulf of Mexico. Length 6 mm (to 7 mm).

# Family description

The astartid shell is small to medium-sized (to at least 45 mm), solid, and oval to trigonal to quadrangular. It is EQUIVALVE or INEQUIVALVE with the left valve larger, compressed, and not gaping. The shell is usually EQUILATERAL, with PROSOGYRATE UMBONES. Shell microstructure is ARAGONITIC and two-layered, with a CROSSED LAMELLAR outer layer and a COMPLEX CROSSED LAMELLAR or HOMOGENOUS inner layer. TUBULES are apparently absent. Exteriorly astartids are covered by thick, strongly adherent, yellow to brown to black, varnished or fibrous PERIOSTRACUM, with a microscopically reticulate or pitted or ridged pattern that is species-diagnostic. Sculpture is smooth or with commarginal ribs (always present

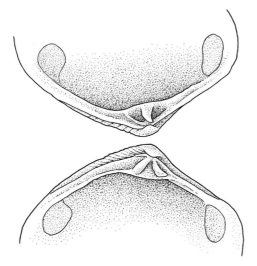

umbonally on early stages), and with internal layers of radial riblets (although external radial sculpture is absent). Lunule and escutcheon are present and usually distinct; the lunule of the right valve overlaps the left and the escutcheon of the left valve overlaps the right. Interiorly the shell is non-nacreous. The pallial line is entire. The inner shell margins are smooth or denticulate. The hinge plate is strong, wide, and heterodont, with two or three prominent cardinal teeth in each valve; lateral teeth are laminar or absent (some authors claim fused to the cardinals). The ligament is parivincular, opisthodetic, and usually set on strong nymphs. A secondary external ligament of fused periostracum extends posteriorly.

The animal is isomyarian or rarely heteromyarian (posterior adductor muscle smaller); pedal retractor muscles are well developed. Pedal elevator and protractor muscles have not been reported. The mantle margins are not fused ventrally. Siphons are absent but the inner mantle folds are fused posteriorly to form a posterodorsal excurrent aperture that has a conelike extension in some species; the incurrent aperture is not fused ventrally. The unusual periostracum is secreted by the middle mantle fold (rather than by the outer mantle fold as in most bi-

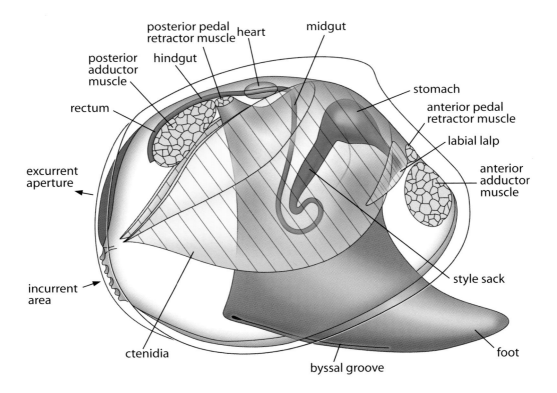

valves). Scattered gland cells are present on the inner surface near the mantle edge (similar in position to the mantle glands of crassatellids). HYPOBRANCHIAL GLANDS have not been reported. Other mantle elaborations have been reported in single species: the mantle edge can be brightly colored (i.e., orange-red in *Astarte undata* Gould, 1841, because of the presence of extracellular hemoglobin in the blood), can bear luminescent organs in the middle mantle fold (*Astarte sulcata* Da Costa, 1778), or can be glandular in the apertural area. The FOOT is large, quadrangular, usually laterally compressed, pointed anteriorly, and heeled posteriorly; it has a BYSSAL GROOVE but the adult is nonbyssate.

The LABIAL PALPS are small to medium-sized and trigonal. The CTENIDIA are EULAMELLIBRANCH (SYNAPTORHABDIC) and HOMORHABDIC; the outer demibranchs are smaller than the inner demibranchs or are absent (in small-bodied species). The gills are inserted into but not fused with the distal oral groove of the palps (CATEGORY I association). Incurrent and excurrent water flows are posterior; an anterior incurrent has not been reported. The STOMACH is a simplified TYPE IV, without sorting area but with a spiral typhlosole; the MIDGUT is loosely coiled. The HINDGUT passes through the ventricle of the heart, and leads to a sessile rectum. Astartids are usually PROTANDRIC HERMAPHRODITES; temperate–tropical species produce planktonic VELIGER larvae, whereas nonpelagic larvae (from large eggs that adhere to the substratum) are typical of cold-water species. Details of the nervous system, including STATOCYSTS and ABDOMINAL SENSE ORGANS, have not been reported.

Astartids are marine SUSPENSION FEEDERS, shallowly INFAUNAL in gravelly mud or sand; some species lie horizontally or at a 45° angle on the right valve at or near the surface. Two Northern Hemisphere species, *Astarte montagui* (Dillwyn, 1817) and *Tridonta borealis* Schumacher, 1817, are known to be sluggish burrowers, but are active at night, leaving their burrows to meander the surface. Food particles consist of coarse particles (retained in the gut because of the absence of sorting areas in the stomach) including diatoms. Predators of cold-water species include fish and walrus.

The family Astartidae is known since the Devonian and is represented by 11 living genera and ca. 50 species, distributed mainly in cold boreal and Arctic waters.

Note: Additional Astartidae species appear after the Carditidae, p. 180.

## References

Dall, W. H. 1903. Synopsis of the family Astartidae, with a review of the American species. *Proceedings of the United States National Museum*, 26(1342): 933–951.

Matveeva, T. A. 1977. Reproduction of bivalves of the family Astartidae. *Issledovaniia Fauny Morei, Akademiia Nauk SSSR, Zoologicheskii Institut*, 14(22): 418–427. [In Russian with English abstract.]

Philippi, R. A. 1839. Einige zoologische Notizen. 7. Ueber das Thier von *Astarte incrassata* De la Jonk. *Archiv für Naturgeschichte*, 5(1): 125–127, pl. 4.

Saleuddin, A. S. M. 1965. The mode of life and functional anatomy of *Astarte* spp. (Eulamellibranchia). *Proceedings of the Malacological Society of London*, 36(4): 229–257.

Saleuddin, A. S. M. 1967. Notes on the functional anatomy of three North American species of *Astarte, A. undata* Gould, *A. castanea* Say and *A. esquimalti* Baird. *Proceedings of the Malacological Society of London*, 37(6): 381–384, pl. 40.

Yonge, C. M. 1969. The functional morphology and evolution within the Carditacea (Bivalvia). *Proceedings of the Malacological Society of London*, 38(6): 493–527.

Zettler, M. L. 2002. Ecological and morphological features of the bivalve *Astarte borealis* (Schumacher, 1817) in the Baltic Sea near its geographical range. *Journal of Shellfish Research*, 21(1): 22–40.

# Family Carditidae – Little Heart Clams

**Classification**
AUTOLAMELLIBRANCHIATA Grobben, 1894
HETEROCONCHIA Hertwig, 1895
Heterodonta Neumayr, 1883
Carditoida Dall, 1889
Crassatelloidea Férussac, 1822
Carditidae J. Fleming, 1828

**Featured species**
*Carditamera floridana* Conrad, 1838 – **Broad-ribbed Carditid**

Elongated quadrangular, lunule small, oblique, and deeply indented, solid, whitish gray with small bars of chestnut brown arranged concentrically on ribs, covered by gray periostracum, with ca. 20 coarse, rounded, beaded radial ribs; interior white with small light brown patches above muscle scars, margin crenulated by exterior ribs. Florida, Gulf of Mexico, Caribbean Central America, South America (Brazil, Uruguay). Length 18 mm (to 38 mm).

*Carditamera floridana* is a typical seagrass bed inhabitant; however, it occasionally settles on floating algal mats (here bayside of Plantation Key).

## Family description

The carditid shell is small to large (to at least 100 mm), solid, and rounded to quadrangular or trapezoidal; members of *Bequina* are mussel-shaped with the umbones at the narrowed anterior end. It is EQUIVALVE, inflated, and not gaping. The shell is INEQUILATERAL (umbones anterior), with PROSOGYRATE, often corroded UMBONES. Shell microstructure is ARAGONITIC and two-layered, with a CROSSED LAMELLAR outer layer and a COMPLEX

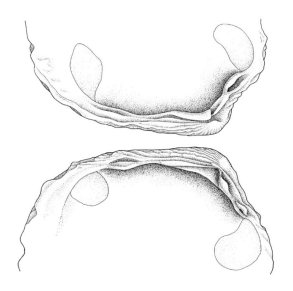

CROSSED LAMELLAR inner layer. TUBULES are present penetrating the inner layer. Exteriorly carditids are dully colored or mottled, covered by a brownish PERIOSTRACUM, often pilose with radial rows of "hairs"; the periostracum is evident only at the margins in *Bequina*. Sculpture consists of strong radial ribs, in some species crossed by commarginal ribs forming nodules or scales. The ventral edge of the posterior region is generally marked by a prominent rib or angulation. The LUNULE is small, deep, and usually bordered by a groove; the ESCUTCHEON is obscure. Interiorly the shell is non-NACREOUS. The PALLIAL LINE is ENTIRE. The inner shell margins are denticulate with elements corresponding to the external radial ribs. The HINGE PLATE is wide and HETERODONT, with two or three oblique CARDINAL TEETH, which can be transversely striated, and anterior and posterior LATERAL TEETH (absent in some species) in each valve. The LIGAMENT is PARIVINCULAR, OPISTHODETIC, and set on strong NYMPHS. A secondary external ligament of fused periostracum covers the primary ligament dorsally and extends anteriorly and posteriorly.

The animal is ISOMYARIAN or HETEROMYARIAN (anterior ADDUCTOR MUSCLE smaller); the posterior adductor muscle can be set on a platform raised anteriorly from the internal shell surface. Anterior and posterior pedal retractor muscles are present. Pedal elevator and protractor muscles have not been reported. The MANTLE margins are not fused ventrally. SIPHONS are absent but the inner mantle folds are fused posteriorly to form a posterodorsal EXCURRENT APERTURE; the INCURRENT APERTURE is usually not fused ventrally

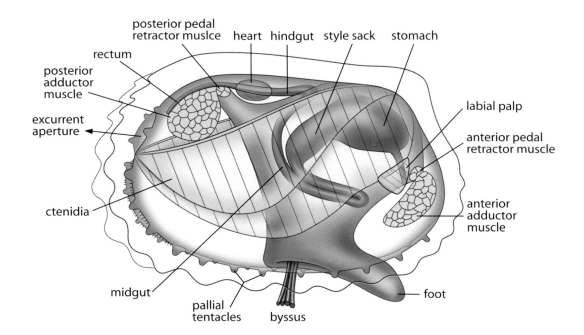

but can be apposed to form a functional incurrent aperture. Extensive mantle glands are present in some species, presumably associated with cleansing the MANTLE CAVITY. HYPO-BRANCHIAL GLANDS have not been reported. The FOOT is wedge-shaped, carinate, pointed anteriorly, heeled posteriorly, and reportedly quite inactive; a BYSSAL GROOVE and BYSSUS are present in the adult.

The LABIAL PALPS are small to medium-sized and trigonal. The CTENIDIA are EU-LAMELLIBRANCH (SYNAPTORHABDIC) and HOMORHABDIC; the outer demibranchs are usu-ally smaller than the inner demibranchs. The gills are united posteriorly to form a di-aphragm separating INFRA- and SUPRABRANCHIAL CHAMBERS. The gills are inserted into and fused with the distal oral groove of the palps (CATEGORY II association). Incurrent and excurrent water flows are posterior; an anterior incurrent is also present. The STOM-ACH is TYPE IV with a spiral typhlosole; the MIDGUT is either coiled or not coiled. The HINDGUT passes through the ventricle of the heart or in some species dorsal to it, and leads to a sessile rectum. Extracellular hemoglobin is present in the blood of some species (e.g., *Cardita affinis* G. B. Sowerby I, 1833). Carditids are usually GONOCHORISTIC and pro-duce either LECITHOTROPHIC or DIRECT-DEVELOPING larvae (planktotrophy is unknown in the family), brooded in the suprabranchial chamber of the inner or both demibranchs; fe-males of Thecaliinae further brood byssally attached young in a MARSUPIUM in the shell or a ventral fold of the mantle (representing one of the few secondary sexually dimorphic characters in Bivalvia). The nervous system is not concentrated. STATOCYSTS have been reported in adult *Cardita*. ABDOMINAL SENSE ORGANS have not been reported.

Carditids are marine SUSPENSION FEEDERS, and include shallowly INFAUNAL species in soft substrata, byssally attached species under rocks, and nestlers in rubble.

The family Carditidae is known since the Devonian and is represented by 16 living genera and ca. 50 species, widely distributed in all seas except in polar regions.

 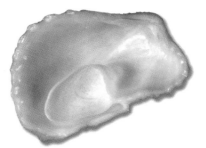

***Thecalia concamerata* (Gmelin, 1791)**, from South Africa, has a peculiar shelly MARSUPIUM on the inside of the shell valves, where young attach by byssal threads for a second round of parental protection (after brooding of larvae in the gills). This structure is present only in females.

## References

Coan, E. V. 1977. Preliminary review of the northwest American Carditidae. *The Veliger*, 19(4): 375–386, 4 pls.

Dall, W. H. 1903. Synopsis of the Carditacea and of the American species. *Proceedings of the Academy of Natural Sciences of Philadelphia*, 54(for 1902)(4): 696–719.

Jones, G. F. 1963. Brood protection in three southern Californian species of the pelecypod genus *Cardita*. *The Wasmann Journal of Biology*, 21(2): 141–148.

Kaspar, J. 1913. Beiträge zur Kenntnis der Familie der Eryciniden und Carditiden. *Zoologische Jahrbücher, Supplement 13, Fauna Chilensis*, 4(4): 545–625.

Schneider, J. A. 1993. Brooding of larvae in *Cardita aviculina* Lamarck, 1819 (Bivalvia: Carditidae). *The Veliger*, 36(1): 94–95.

Yonge, C. M. 1969. Functional morphology and evolution within the Carditacea (Bivalvia). *Proceedings of the Malacological Society of London*, 38(6): 493–527.

## ASTARTIDAE

**Astarte crenata subequilatera** G. B. Sowerby II, 1854 – **Lentil Astarte**

Rounded triangular, umbones often eroded, light to dark brown, periostracum reticulate in fine, wavy-ridge pattern, shell with strong, rounded, regular commarginal ridges; interior white, margin smooth. North Atlantic, Florida Keys. Length 22 mm. Syn. *lens* Verrill, 1872.

**Astarte "nana"** E. A. Smith, 1881 – **Dwarf Astarte**

Rounded trigonal, cream to rose-brown with white (eroded) umbones, periostracum reticulate, with ca. 25 regular, rounded commarginal ridges; interior white to light brown, margin denticulate. North Carolina to Florida, Gulf of Mexico. Length 9 mm. Note: The relationship of the western Atlantic *Astarte "nana"* to the Mediterranean form described by Jeffreys (1864) as *Astarte compressa* var. *nana* has yet to be determined.

## CARDITIDAE

**Glans dominguensis** (d'Orbigny, 1853) – **Santo Domingo Carditid**

Quadrangular to oval, lunule narrow and poorly defined, inflated, solid, whitish with rose tint or orange blotches, with sharply defined, coarse, beaded radial ribs; interior glossy white, margin crenulated by exterior ribs. North Carolina to Florida, Gulf of Mexico, Caribbean Central America. Length 4 mm (to 6 mm). Compare *Pleuromeris tridentata*, which is more trigonal in overall shape, and *Carditopsis smithii*, which is much smaller and has a strong commarginal ridge bordering the prodissoconch.

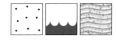

**Pleuromeris tridentata** (Say, 1826) – **Three-toothed Carditid**

Rounded trigonal, lunule small and impressed, solid, white to gray-brown to bright rose, occasionally with patches of orange-brown, with 15–18 low, coarse, beaded radial ribs; interior glossy white or orange-brown, hinge teeth often purplish blue, margin crenulated by exterior ribs. New Jersey to Florida, Bahamas, Gulf of Mexico, Caribbean Central America. Length 4 mm (to 8 mm). Compare *Glans dominguensis*, which is more quadrangular in overall shape.

**Pteromeris perplana** (Conrad, 1841) – **Flattened Carditid**

Obliquely rounded trigonal, lunule poorly defined, compressed, solid, pinkish or mottled brown, with wide, low radial ribs and weaker commarginal ridges; interior white, margin crenulated by exterior ribs. North Carolina to Florida, Caribbean Central America. Length 4 mm (to 7 mm).

*Astarte crenata subequilatera*

*Astarte "nana"*

*Glans dominguensis*

*Pleuromeris tridentata*

*Pteromeris perplana*

# Family Condylocardiidae – Condyl Clams

## Classification
AUTOLAMELLIBRANCHIATA Grobben, 1894
HETEROCONCHIA Hertwig, 1895
Heterodonta Neumayr, 1883
Carditoida Dall, 1889
Crassatelloidea Férussac, 1822
Condylocardiidae F. Bernard, 1896

**Featured species**
*Carditopsis smithii* (Dall, 1896) – **Smith's Tiny Condyl Clam**

Trigonal, with coarse, raised commarginal ridge at edge of large prodissoconch, inflated, solid, orange-brown, with 10 or 11 coarse, beaded radial ribs and scaly surface; interior reflecting exterior color, margin crenulated by exterior ribs. Florida Keys, Bermuda, Bahamas, West Indies, Caribbean Central America, South America (Colombia, Brazil). Length < 2 mm. Compare *Glans dominguensis*, which is larger and lacks the commarginal ridge bordering the prodissoconch.

**The prodissoconch** (in green) of diminutive *Carditopsis smithii* is relatively large and bordered by a strong commarginal ridge.

## Family description

The condylocardiid shell is minute to small (to 9 mm), solid, higher than long, and dorsoventrally oval to orbicular or trigonal, with a few species oval to trapeziform. It is EQUIVALVE, moderately inflated, and not gaping. The shell is EQUILATERAL or slightly IN-

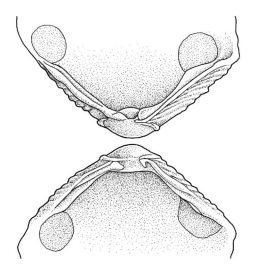

EQUILATERAL (umbones slightly posterior), with PROSO-, OPISTHO-, or (most often) ORTHOGYRATE UMBONES; the PRODISSOCONCH in some species is large, radially ribbed, and saucerlike with a bulging rim and/or "AURICLES." Shell microstructure is ARAGONITIC and two-layered, with a CROSSED LAMELLAR outer layer and a COMPLEX CROSSED LAMELLAR inner layer. TUBULES have not been reported. Exteriorly condylocardiids are covered by thin PERIOSTRACUM, persisting mainly in the interspaces of the external sculpture. Sculpture is rarely smooth, but often consists of radial (DIVARICATING in some *Cuna*) and/or commarginal ribs (ornamented by oblique striae in *Propecuna*). LUNULE and ESCUTCHEON are usually present (absent in, e.g., *Austrocardiella*, *Benthocardiella*, and *Ovacuna*). Interiorly the shell is non-NACREOUS. The PALLIAL LINE is ENTIRE. The inner shell margins are smooth or denticulate. The HINGE PLATE is narrow, straight or arched, and HETERODONT (called "spondyliform" by some authors), with two or three CARDINAL TEETH (the posterior cardinal often weaker) and one or two elongated LATERAL TEETH per valve. The LIGAMENT is internal

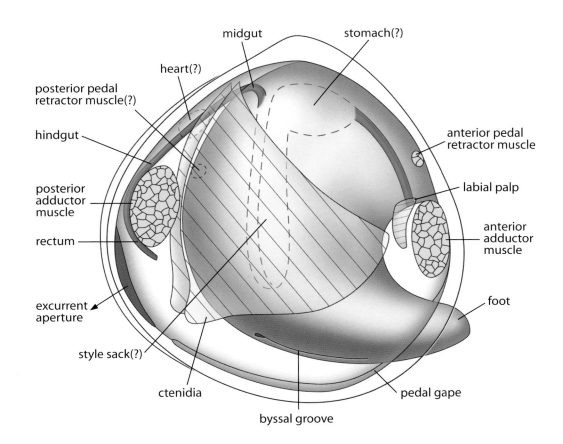

(RESILIUM), small, OPISTHODETIC, and set in a central, trigonal to rounded RESILIFER; it partially covers the cardinal tooth.

The scant knowledge of condylocardiid anatomy is based on incomplete data for *Condylocardia notoaustralis* Cotton, 1930 (Middelfart, 2002a), *Condylocuna io* (Bartsch, 1915) (Oliver & Holmes, 2004), *Cuna concentrica* Hedley, 1902, and *Ovacuna atkinsoni* (Tenison Woods, 1877) (both Middelfart, 2002b). The animal is ISOMYARIAN or weakly HETEROMYARIAN (posterior ADDUCTOR MUSCLE smaller); pedal retractor muscles are present. The MANTLE margins are largely not fused ventrally. SIPHONS are absent but the mantle folds are fused posteriorly to form a posterior EXCURRENT APERTURE; the INCURRENT APERTURE is not fused ventrally. The FOOT is pointed anteriorly and in some cases heeled posteriorly; a BYSSAL GROOVE and BYSSUS are present or absent (in *Ovacuna)* in the adult.

The LABIAL PALPS are small and trigonal. The CTENIDIA are EULAMELLIBRANCH (SYNAPTORHABDIC) and HOMORHABDIC; the outer demibranchs are absent. The association of the anterior gill filaments and the labial palps is unknown. Excurrent water flow is posterior; incurrent flow is probably both anterior and ventral. The STOMACH is probably TYPE IV but has not been confirmed; configurations of the gut loops and heart are unknown; the rectum is freely hanging. Condylocardiids are known to be GONOCHORISTIC and all studied so far are ovoviviparous, incubating larvae in the suprabranchial chamber. The nervous system has not been studied.

Condylocardiids are marine SUSPENSION FEEDERS, and in some cases are abundant on shallow-water sandflats where they can be ecologically important.

The family Condylocardiidae is known since the Tertiary and is represented by at least 21 living genera and at least 65 species, distributed worldwide but mainly in the antiboreal regions. Some features of condylocardiids (e.g., adult byssus, brooding, and absence of outer demibranch) suggest that their evolution involves PEDOMORPHOSIS.

## References

Bernard, F. 1897 ("1896"). Études comparatives sur la coquille des lamellibranches, *Condylocardia*, type nouveau de lamellibranches. *Journal de Conchyliologie*, 44(3): 169–207, pl. 6.

Coan, E. V. 2003. The tropical eastern Pacific species of the Condylocardiidae (Bivalvia). *The Nautilus*, 117(2): 47–61.

Middelfart, P. 2002a. A revision of the Australian Condylocardiinae (Bivalvia: Carditoidea: Condylocardiidae). *Molluscan Research*, 22(1): 23–85.

Middelfart, P. 2002b. Revision of Australian Cuninae *sensu lato* (Bivalvia: Carditoidea: Condylocardiidae). *Zootaxa*, 112: 1–124.

Oliver, P. G., and A. M. Holmes. 2004. Cryptic bivalves with descriptions of new species from the Rodrigues lagoon. *Journal of Natural History*, 38: 3175–3227.

Salas, C., and E. Rolán. 1990. Four new species of Condylocardiidae from Cape Verde Islands. *Bulletin du Muséum National d'Histoire Naturelle, Section A, Zoologie, Biologie et Écologie Animals, Série 4*, 12(2): 349–363.

Salas, C., and R. von Cosel. 1991. Taxonomy of tropical West African bivalves. III. Four new species of Condylocardiidae from the continental shelf. *Bulletin du Muséum National d'Histoire Naturelle, Section A, Zoologie, Biologie et Écologie Animals, Série*, 13(3–4): 263–281.

# Family Pandoridae Rafinesque, 1815 – Pandora Clams

**Classification**
AUTOLAMELLIBRANCHIATA Grobben, 1894
HETEROCONCHIA Hertwig, 1895
Heterodonta Neumayr, 1883
Anomalodesmata Dall, 1889
Pandoridae Rafinesque, 1815

**Featured species**
*Pandora inflata* Boss & Merrill, 1965 – **Inflated Pandora**

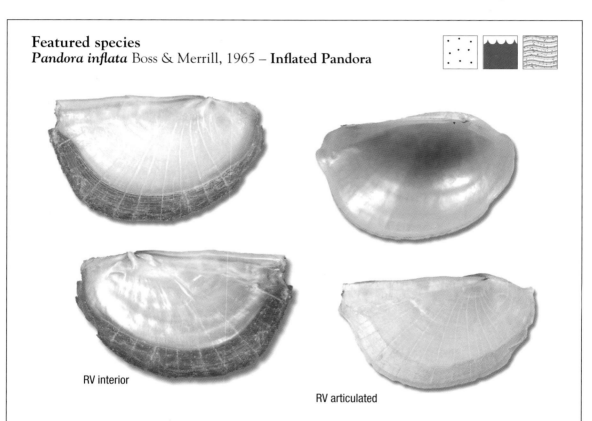

RV interior

RV articulated

Broadly semicircular, ventral margin roundly protruding midventrally, left valve convex and larger than flattened right valve; thin-walled, white, left valve smooth with faint radial scratches, right valve with radial scratches and two radial carinate ridges along posterodorsal margin; interior nacreous with non-nacreous border on right valve, margin smooth. Massachusetts to Florida, Gulf of Mexico. Length 11 mm.

# Family description

The pandorid shell is small to medium-sized (to 60 mm), thin-walled to solid, and lunate or crescent-shaped, with the anterior end rounded and the posterior end bluntly pointed. It is INEQUIVALVE (left valve larger and more convex, overlapping the flat to concave

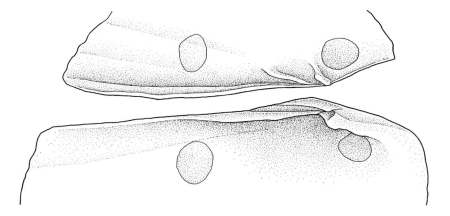

right valve), highly compressed, and not gaping. The shell is INEQUILATERAL (umbones anterior), with PROSO-, OPISTHO-, or ORTHOGYRATE UMBONES. Shell microstructure is ARAGONITIC "PRISMATONACREOUS" and three-layered, with a simple PRISMATIC outer layer, a lenticular NACREOUS middle layer, and a sheet-nacreous inner layer. TUBULES are apparently absent. Exteriorly pandorids are white, often iridescent (when the prismatic outer shell layer erodes away, exposing the underlying nacre), covered by a thin, occasionally radially arranged PERIOSTRACUM. When the valves are closed, the periostracum of the flattened right valve forms a tight seal ventrally against the inner margin of the larger left valve. Sculpture is smooth or with commarginal lirae. LUNULE and ESCUTCHEON are absent. Interiorly the shell is nacreous. The PALLIAL LINE is discontinuous and ENTIRE. The inner shell margins are smooth. The HINGE PLATE is straight and EDENTATE in adults, but with secondary, low, toothlike ridges or CRURAE (often called "CARDINAL TEETH"), one or

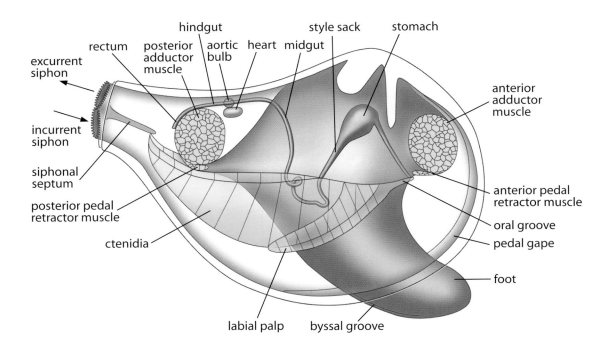

three teeth in the right valve and two weaker ones in the left valve, on either side of a central resilifer. The LIGAMENT is internal (RESILIUM) and OPISTHODETIC, rests on the right posterior CRURUM or a weak RESILIFER, and in some species is reinforced by an elongated LITHODESMA. A secondary ligament of fused periostracum unites the valves dorsally both anterior and posterior to the umbones.

The animal is ISOMYARIAN or slightly HETEROMYARIAN (anterior ADDUCTOR MUSCLE smaller); pedal retractor muscles are small and unusual in being inserted ventral to their respective adductor muscles. Pedal elevator and protractor muscles have not been reported. The MANTLE margins are partially fused ventrally, with a small to large anteroventral pedal gape. A FOURTH PALLIAL APERTURE is absent. Posterior EXCURRENT and INCURRENT SIPHONS are short and united. Siphonal retractor muscles are not concentrated, but individual muscle bundles are strong and often leave distinct blocklike muscle scars along the pallial line on the shell. Periostracum sheaths the base of the siphons, and can include embedded sand grains in this region. HYPOBRANCHIAL GLANDS have not been reported. The FOOT is large, laterally compressed, unheeled, and has a BYSSAL GROOVE; the adult is nonbyssate. The LABIAL PALPS are large, narrow, elongated, and are extended by an unridged oral groove to the mouth. The CTENIDIA are EULAMELLIBRANCH (SYNAPTORHABDIC) and HETERORHABDIC; the much smaller outer demibranchs consist only of reflected descending lamellae. The gills are posteriorly united with the SIPHONAL SEPTUM, separating INFRA- and SUPRABRANCHIAL CHAMBERS. The gills are not inserted into (or fused with) the distal oral groove of the palps (CATEGORY III association). Incurrent and excurrent water flows are posterior. The STOMACH is TYPE IV, with an unusually poorly developed major typhlosole; the MIDGUT is coiled. The HINDGUT passes dorsal to or through the ventricle of the heart and leads to a sessile rectum. An AORTIC BULB is present in some species. Pandorids are SIMULTANEOUS HERMAPHRODITES and produce large, yolky eggs (exuded on mucus strands that cling to the substratum) that hatch LECITHOTROPHIC VELIGER larvae; larvae settle close to the parental population. The nervous system is not concentrated. STATOCYSTS (TYPE B1, with single large STATOLITHS) are known in adult *Frenamya*. ABDOMINAL SENSE ORGANS have not been reported.

Pandorids are marine SUSPENSION FEEDERS, in most cases shallowly INFAUNAL in a horizontal position, usually with the more convex left valve downward; other species (and other individuals within otherwise infaunal species) are EPIFAUNAL in a similar position in coarse muddy gravel or sand; pandorids have also been recorded nestled among oysters. The unusual ventral positions of the pedal retractor muscles and gills have been attributed by one author to the extremely laterally compressed body. Most pandorids live in localized populations, and collections are rare.

The family Pandoridae is known since the Oligocene and is represented by ca. 6 living genera and ca. 25 species, distributed worldwide mainly in the Northern Hemisphere in shallow subtidal waters.

**A living *Pandora gouldiana*** Dall, 1886, from New York has siphons that are short, united, and sheathed by a translucent envelope of periostracum.

## References

Allen, J. A. 1954. On the structure and adaptations of *Pandora inaequivalvis* and *P. pinna*. *Quarterly Journal of Microscopical Science*, 95(4): 473–482.

Allen, J. A. 1961. The development of *Pandora inaequivalvis* (Linné). *Journal of Embryology and Experimental Morphology*, 9(2): 252–268.

Allen, M. F., and J. A. Allen. 1955. On the habits of *Pandora inaequivalvis* (Linné). *Proceedings of the Malacological Society of London*, 31(5-6): 175–185.

Boss, K. J. 1965. Catalogue of the family Pandoridae (Mollusca: Bivalvia). *Occasional Papers on Mollusks, Harvard University*, 2(33): 413–424.

Boss, K. J., and A. S. Merrill. 1965. The family Pandoridae in the western Atlantic. *Johnsonia*, 4(44): 181–216.

Morton, B. 1984. The adaptations of *Frenamya ceylanica* (Bivalvia: Anomalodesmata: Pandoracea) to life on the surface of soft muds. *Journal of Conchology*, 31(6): 359–371.

Stasek, C. R. 1963. Orientation and form in the bivalved Mollusca. *Journal of Morphology*, 112(3): 195–214.

Thomas, K. A. 1994. The functional morphology and biology of *Pandora filosa* (Carpenter, 1864) (Bivalvia: Anomalodesmata: Pandoracea). *The Veliger*, 37(1): 23–29.

Yonge, C. M., and B. Morton. 1980. Ligament and lithodesma in the Pandoracea and the Poromyacea with a discussion on evolutionary history in the Anomalodesmata (Mollusca: Bivalvia). *Journal of Zoology*, 191: 263–292.

### *Pandora bushiana* Dall, 1886 – **Bush's Keeled Pandora**

Narrowly semicircular, left valve slightly inflated, right valve slightly concave, rostrate and demarcated by radial ridge on left valve; anterodorsal margin anteroventrally sloping, anterior end separated by weak radial groove, moderately compressed, thin-walled, white, occasionally tinged with yellow, covered by light brown periostracum, left valve smooth, right valve with radial lines; interior margin smooth. North Carolina to Florida, Bahamas, West Indies, Gulf of Mexico, Caribbean Central America, South America (Colombia, Brazil). Length 11 mm (to 25 mm).

### *Pandora glacialis* Leach, 1819 – **Glacial Pandora**

Semicircular, left valve convex, centrally inflated and larger than right valve, right valve centrally flattened, sharply concave ventrally and overlapping left valve along posterodorsal margin, moderately compressed, thin-walled, white with light brownish periostracum, left valve smooth with irregular commarginal surface, right valve with weak radial scratches; interior margin smooth. North Atlantic, Massachusetts, Florida Keys, also Arctic coasts worldwide. Length 18 mm.

**Deepwater habitats,** such as this featuring sponge and *Stylaster* coral at 173 m on Pourtales Terrace, Florida, also include soft sediments for species of *Pandora*.

*Pandora bushiana*

*Pandora glacialis*

# Family Lyonsiidae – Lyonsia Clams

**Classification**
AUTOLAMELLIBRANCHIATA Grobben, 1894
HETEROCONCHIA Hertwig, 1895
Heterodonta Neumayr, 1883
Anomalodesmata Dall, 1889
Lyonsiidae P. Fischer, 1887

**Featured species**
*Entodesma beana* (d'Orbigny, 1853) – **Pearly Entodesma**

Irregularly oval to quadrangular, often distorted, posterior end elongated and truncated, slightly gaping anteriorly and posteriorly, with byssal notch at anterior end of ventral margin, moderately compressed, thin-walled, translucent pearly with coarse, crinkly, thin, light brown periostracum, with weak, raised radial threads and coarse commarginal growth lines near margin, early part of valves with rasplike surface and radiating lines of small pimples; interior margin smooth. North Carolina to Florida, Bermuda, Bahamas, West Indies, Gulf of Mexico, Caribbean Central America, South America (Colombia, ? Brazil). Length 19 mm (to 23 mm). Note: The relationship to *Entodesma brasiliensis* (Gould, 1850), often considered a synonym, needs further study.

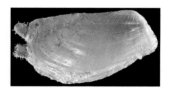

**The tissues of the living animal** of *Entodesma beana* are bright orange. This species commonly lives in sponges.

# Family description

The lyonsiid shell is small to medium-sized (rarely to 100 mm), very thin-walled, and irregularly oblong to orbicular, in some species posteriorly ROSTRATE, with the edges poorly calcified, and usually somewhat distorted due to the nestling habit. It is INEQUIVALVE (left valve larger and more convex, overlapping the right valve slightly), inflated anteriorly,

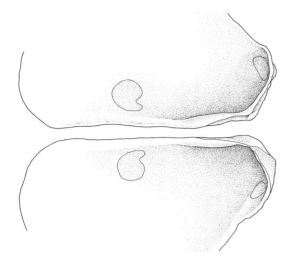

compressed posteriorly, and can be gaping posteriorly for the siphons and in some cases ventrally for the byssus. The shell is INEQUILATERAL (umbones anterior), with PROSOGYRATE UMBONES. Shell microstructure is ARAGONITIC "PRISMATONACREOUS" and three-layered, with a simple PRISMATIC (or granular HOMOGENOUS) outer layer, a lenticular NACREOUS middle layer, and a sheet-nacreous inner layer. TUBULES are present, penetrating the inner and middle shell layers only. Exteriorly lyonsiids are white, often iridescent, covered by a thin to very thick, often radially arranged, dehiscent PERIOSTRACUM that often incorporates sand grains (variously interpreted as for increased camouflage, strength, or stability within soft sediment) and overlaps the free shell margins slightly. Sculpture is weak and variable, often with microscopic spines or nodules, in some cases with rugose commarginal wrinkles. LUNULE and ESCUTCHEON are absent. Interiorly the shell is thinly nacreous. The PALLIAL LINE is ENTIRE but is posteriorly weakly sinuate in some species. The inner shell margins are smooth. The HINGE PLATE is irregular and EDENTATE in adults. The LIGAMENT is internal (RESILIUM), OPISTHODETIC, set in a long, narrow RESILIFER, and reinforced by a short to elongated, more or less spatulate LITHODESMA. A secondary external ligament of fused periostracum unites the valves along the dorsal hinge margin.

The animal is ISOMYARIAN or HETEROMYARIAN (anterior ADDUCTOR MUSCLE smaller and more ventral); the posterior adductor muscle lies mid-dorsally, approximately halfway between the umbones and the posterior end of the shell. Anterior and posterior pedal retractor muscles are small. Pedal elevator and protractor muscles are absent. The MANTLE margins are extensively fused ventrally, usually with a small anteroventral pedal gape. Posterior EXCURRENT and INCURRENT SIPHONS are short and separate, but siphonal retrac-

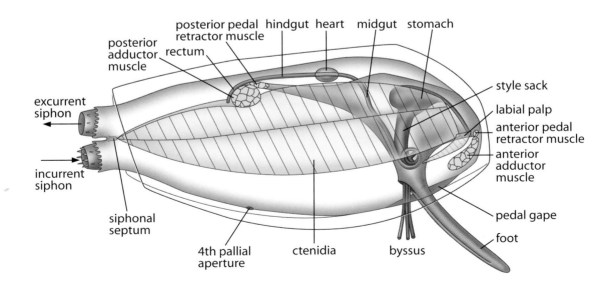

tor muscles are poorly developed. Periostracum covers the base of the siphons and the ventral surface of the fused mantle edges. A FOURTH PALLIAL APERTURE is present posteroventrally. Arenophilic radial mantle glands (involved in adhering sand grains to the external shell surface) are present at the mantle edge (*Lyonsia* and juvenile *Entodesma*) or absent (*Mytilimeria* and adult *Entodesma*). Complex PALLIAL EYES are embedded in the excurrent siphon of some species of *Lyonsia*. HYPOBRANCHIAL GLANDS have not been reported. The FOOT is small (*Entodesma*) to large (*Lyonsia*), laterally compressed, unheeled, and has a BYSSAL GROOVE; the adult is byssate in some species.

The LABIAL PALPS are narrow and elongated. The CTENIDIA are EULAMELLIBRANCH (SYNAPTORHABDIC) and HETERORHABDIC; the outer demibranchs consist only of upturned descending lamellae. The gills are united posteriorly to the SIPHONAL SEPTUM, separating INFRA- and SUPRABRANCHIAL CHAMBERS. The gills are inserted into and fused with the distal oral groove of the palps (CATEGORY II association). Incurrent and excurrent water flows are posterior. The STOMACH is TYPE IV; the MIDGUT is coiled. The HINDGUT passes through the ventricle of the heart and leads to a freely hanging rectum. Lyonsiids are SIMULTANEOUS HERMAPHRODITES and produce large, yolky eggs that hatch LECITHOTROPHIC VELIGER larvae. The nervous system has not been well studied, but examination of available data suggests it is not concentrated. STATOCYSTS (TYPE B1, with single large STATOLITHS) are known in adult *Lyonsia*. ABDOMINAL SENSE ORGANS have not been reported.

Lyonsiids are marine or estuarine SUSPENSION FEEDERS, in most cases shallowly and vertically INFAUNAL in fine sand (*Lyonsia*), or less often as nestlers (*Entodesma*) in rock crevices, algal holdfasts, or holes in sponges or tunicates. The Bladderclam (*Mytilimeria nuttalli* Conrad, 1837), from the northeastern Pacific, is totally sessile, living embedded in compound ascidian colonies.

The family Lyonsiidae is known since the Paleocene and is represented by ca. 9 living genera and ca. 45 species, distributed worldwide.

---

### *Lyonsia floridana* Conrad, 1849 – **Florida Lyonsia**

Elongated oval, posteriorly narrowly rostrate, anteriorly inflated, thin-walled, translucent pearly, with numerous raised radial striae; interior margin smooth. Florida, Gulf of Mexico, Caribbean Central America, South America (Colombia). Length 7 mm (to 16 mm).

**The periostracum** of *Lyonsia floridana*, shown here in a specimen from Apalachee Bay, Florida, is often adorned with sand particles.

## References

Ansell, A. D. 1967. Burrowing in *Lyonsia norvegica* (Gmelin) (Bivalvia: Lyonsiidae). *Proceedings of the Malacological Society of London*, 37(6): 387–393.

Campos M., B., and L. Ramorino M. 1981. Huevos, larvas e postlarva de *Entodesma cuneata* (Gray, 1828) (Bivalia: Pandoracea: Lyonsiidae). *Revista de Biologia Marina*, 17(2): 229–251.

Chanley, P. E., and M. Castagna. 1966. Larval development of the pelecypod *Lyonsia hyalina*. *The Nautilus*, 79(4): 123–128.

Morgan, R. E., and J. A. Allen. 1976. On the functional morphology and adaptations of *Entodesma saxicola* (Bivalvia: Anomalodesmacea). *Malacologia*, 15(2): 233–240.

Morton, B. 1987. The mantle margin and radial mantle glands of *Entodesma saxicola* and *E. inflata* (Bivalvia: Anomalodesmata: Lyonsiidae). *Journal of Molluscan Studies*, 53(2): 139–151.

Narchi, W. 1968. The functional morphology of *Lyonsia californica* Conrad, 1837 (Bivalvia). *The Veliger*, 10(4): 305–313.

Prezant, R. S. 1981a. The arenophilic radial mantle glands of the Lyonsiidae (Bivalvia: Anomalodesmata) with notes on lyonsiid evolution. *Malacologia*, 20(2): 267–289.

Prezant, R. S. 1981b. Comparative shell ultrastructure of lyonsiid bivalves. *The Veliger*, 23(4): 289–299, 12 pls.

Prezant, R. S. 1981c. Taxonomic re-evaluation of the bivalve family Lyonsiidae. *The Nautilus*, 95(2): 58–72.

Thomas, K. A. 1993. The functional morphology of the digestive system of *Lyonsia hyalina* Conrad, 1831 (Bivalvia: Anomalodesmata: Pandoroidea). *Journal of Molluscan Studies*, 59(2): 175–186.

Yonge, C. M. 1952. Structure and adaptation in *Entodesma saxicola* (Baird) and *Mytilimeria nuttallii* Conrad with a discussion on evolution within the family Lyonsiidae (Eulamellibranchia). *University of California Publications in Zoölogy*, 55(10): 439–450.

Yonge C. M. 1976. Primary and secondary ligaments with the lithodesma in the Lyonsiidae (Bivalvia: Pandoracea). *Journal of Molluscan Studies*, 42(3): 395–408.

# Family Periplomatidae – Spoon Clams

**Classification**
AUTOLAMELLIBRANCHIATA Grobben, 1894
HETEROCONCHIA Hertwig, 1895
Heterodonta Neumayr, 1883
Anomalodesmata Dall, 1889
Periplomatidae Dall, 1895

**Featured species**
*Periploma margaritaceum* (Lamarck, 1801) – **Unequal Spoon Clam**

pallial line

Oblong oval, posteriorly truncate, moderately inflated (right valve more so), thin-walled, white, slightly iridescent, smooth, with oblique low keel (bounded posteriorly by groove) from umbo to anterior ventral margin; interior white, with radial rib from under hinge to posterior margin, margin smooth. South Carolina to Florida, West Indies, Gulf of Mexico, Caribbean Central America, South America (Venezuela, Brazil). Length 13 mm (to 14 mm). Syn. *anguliferum* Philippi, 1847.

**A living *Periploma margaritaceum*** (with its siphons fully retracted), from the Indian River Lagoon, eastern Florida, shows pigmentation of its body through the thin shell. The black splotch is the digestive gland surrounding the stomach; the pinkish coloration most likely represents the gills.

# Family description

The periplomatid shell is small to medium-sized (to 90 mm), thin-walled to solid, and irregularly oval to elongated or rhomboidal (the ventral margin has strong interdigitating projections in *Albimanus*). The anterior end is rounded, the posterior end truncate or ROSTRATE with the tip directed slightly dorsally. It is usually INEQUIVALVE (right valve larger

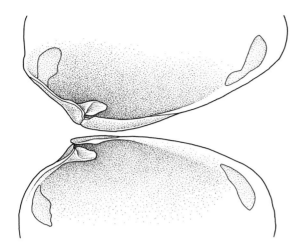

and more convex, overlapping the left valve at the umbo), moderately inflated, and either not gaping or gaping anteriorly and posteriorly. The shell is usually INEQUILATERAL (umbones posterior), with PROSO- or OPISTHOGYRATE UMBONES. A dorsoventral UMBONAL CRACK or fissure is present in each valve, equal in length to the chondrophore, and covered by periostracum. Shell microstructure is ARAGONITIC "PRISMATONACREOUS" and three-layered, with a simple PRISMATIC outer layer, a lenticular NACREOUS middle layer, and a sheet-nacreous inner layer. TUBULES are apparently absent. Exteriorly periplomatids are white, iridescent where worn, covered by a thin adherent PERIOSTRACUM and sand grains (in some taxa, e.g., *Cochlodesma*; apparently not incorporated into the periostracum but simply adhering to the rough surface). Sculpture is smooth or pustulate with weak commarginal striae, rarely with radial ribs. LUNULE and ESCUTCHEON are absent. Interiorly the shell is thinly nacreous to chalky. The PALLIAL LINE has a (usually deep) SINUS. The inner shell margins are smooth or rarely broadly denticulate. The HINGE PLATE is

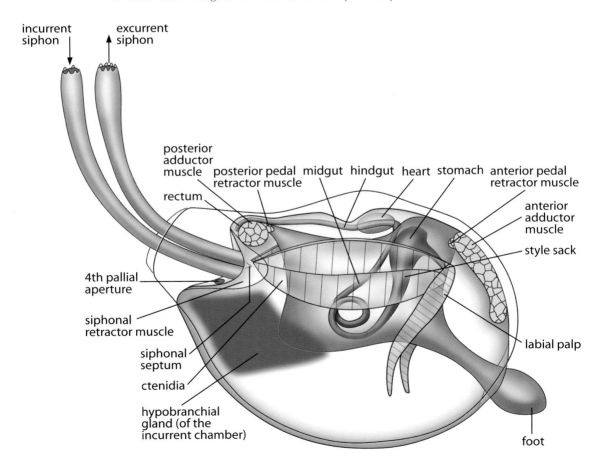

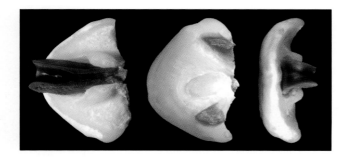

**The free lithodesma** of *Periploma planiusculum* (G. B. Sowerby I, 1834) from California, in ventral, dorsal, and anterior views. The dark brown material in dorsal view is the ligament, which has become detached from the chondrophores; the ventral view shows tissue remnants from within the umbones.

EDENTATE in adults. The LIGAMENT is internal (RESILIUM) and OPISTHODETIC, set on prominent CHONDROPHORES that are supported behind by shelly buttresses (or "clavicles"), usually extending posteriorly toward the muscle scar. A strong U-shaped free LITH-ODESMA bridges the valves anterior and dorsal to the chondrophores in most taxa; *Offadesma* instead has a proteinaceous pad; *Takashia* has neither structure. A secondary external ligament of fused periostracum unites the valves along the dorsal hinge margin.

The animal is HETEROMYARIAN (posterior ADDUCTOR MUSCLE smaller), with the anterior adductor muscle elongated arcuate and the posterior adductor muscle more rounded

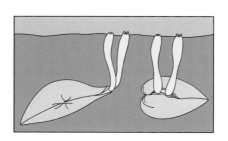

in cross section; pedal retractor muscles are small and in some cases do not attach to the shell. Pedal elevator and protractor muscles have not been reported. The MANTLE margins are extensively fused ventrally, with a small anteroventral pedal gape. Marginal adductor muscles (ORBITAL MUSCLES) span the fused inner mantle folds acting as accessory adductors. Posterior EXCURRENT and IN-CURRENT SIPHONS are long and separate, with a strong siphonal retractor muscle. A FOURTH PALLIAL APERTURE is present in some, ventral to the base of the incurrent siphon. Arenophilic radial mantle glands (involved in adhering sand grains to the external shell surface) are present in some taxa (e.g., *Offadesma*). Presumed HYPOBRANCHIAL GLANDS have been reported in both the infra- and suprabranchial chambers of *Offadesma angasi* (Crosse & Fischer, 1864). The FOOT is small, heeled, and in some has a BYSSAL GROOVE; the adult is nonbyssate.

The LABIAL PALPS are small to large, narrow, and elongated. The CTENIDIA are EU-LAMELLIBRANCH (SYNAPTORHABDIC), HETERORHABDIC, and relatively small; the outer demibranchs consist only of the reflected descending lamellae. The gills are not inserted into (or fused with) the distal oral groove of the palps (CATEGORY III association). Distally the tips of the ctenidia are attached to the SIPHONAL SEPTUM, separating INFRA- and SUPRABRANCHIAL CHAMBERS. Incurrent and excurrent water flows are posterior. The STOM-ACH is TYPE IV; the MIDGUT is coiled. The HINDGUT passes through the ventricle of the heart and leads to a sessile rectum. Unusually for bivalves, the heart is located anteriorly to the umbones. Periplomatids are SIMULTANEOUS HERMAPHRODITES and produce large, yolky eggs that hatch LECITHOTROPHIC VELIGERS or DIRECT-DEVELOPING larvae. The nervous system is slightly concentrated, with the visceral ganglia displaced anteriorly from the posterior adductor. STATOCYSTS (TYPE B2, with single large STATOLITHS plus STATOCONIA) are known in adult *Cochlodesma*. ABDOMINAL SENSE ORGANS have not been reported.

Periplomatids are marine SUSPENSION FEEDERS, moderately deeply INFAUNAL in soft sediment. Certain morphological adaptations (e.g., elongated palps, hooded mouth, and hypobranchial glands) are associated with life in turbid waters. They lie horizontally with the more convex right valve uppermost, and with the siphons in separate burrows (in *Cochlodesma*, only the incurrent siphon reaches the surface, with the excurrent discharging into the subsurface sand); the interiors of the siphonal canals are hardened by mucus produced by the siphons. Peristaltic movements reported in periplomatid siphons seem to be associated with tube formation and/or PSEUDOFECES removal. Some species are passive burrowers, i.e., they cannot rebury once dislodged from the sediment; however, *Periploma margaritaceum* repeatedly reburies itself under experimental laboratory conditions. Predators include boring gastropods.

The family Periplomatidae is known since the Cretaceous and is represented by ca. 6 living genera and ca. 35 species, distributed worldwide in temperate and tropical seas (a few are boreal), from intertidal to abyssal zones.

---

*Periploma tenerum* (P. Fischer, 1882) – **Delicate Spoon Clam**

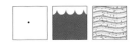

Oval, posteriorly rostrate, moderately inflated (right valve more so), thin-walled, white, slightly iridescent, smooth with commarginal growth lines only; interior white, with radial rib from under hinge to posterior margin. North Carolina to Florida. Length 14 mm. Note: The valves photographed are of possible SYNTYPE specimens of *Periploma tenerum*.

---

### References

Allen, J. A. 1958. Observations on *Cochlodesma praetenue* (Pulteney) [Eulamellibranchia]. *Journal of the Marine Biological Association of the United Kingdom*, 37: 97–112.

Bernard, F. R. 1989. Living Periplomatidae of the Pacific and Indo-Pacific regions. *Venus*, 48(1): 1–11.

Morton, B. 1981. The biology and functional morphology of *Periploma* (*Offadesma*) *angasai* [sic] (Bivalvia: Anomalodesmata: Periplomatidae). *Journal of Zoology*, 193: 39–70.

Narchi, W., and O. Domaneschi. 1995. Morphology of *Periploma ovata* d'Orbigny, 1846 (Bivalvia: Periplomatidae) and the adaptive radiation in the Anomalodesmata. *Abstracts, Twelvth International Malacological Congress, Unitas Malacologica*, pp. 223–224.

Rosewater, J. 1968. Notes on Periplomatidae (Pelecypoda: Anomalodesmata), with a geographical checklist. *The American Malacological Union, Annual Reports for 1968*, pp. 37–39.

Rosewater, J. 1980. Predator boreholes in *Periploma margaritaceum*, with a brief survey of other Periplomatidae. *The Veliger*, 22(3): 248–251, 1 pl.

Rosewater, J. 1984. Burrowing activities of *Periploma margaritaceum* (Lamarck, 1801) (Bivalvia: Anomalodesmata: Periplomatidae). *American Malacological Bulletin*, 2: 35–40.

# Family Spheniopsidae – Gripp's Clams

**Classification**
AUTOLAMELLIBRANCHIATA Grobben, 1894
HETEROCONCHIA Hertwig, 1895
Heterodonta Neumayr, 1883
Anomalodesmata Dall, 1889
? Spheniopsidae Gardner, 1928

**Featured species**
*Grippina* sp. A – Gripp's Clam

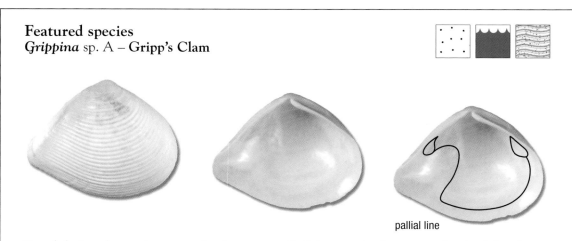

pallial line

Rounded trigonal, posterior narrowed and truncate, with slight ventral indentation, inflated, solid, with coarse, rounded commarginal ridges and 2 radial ridges from umbo to rostrum, white exteriorly and interiorly; interior glossy. Florida Keys endemic. Length 2 mm.

**Scanning electron micrographs** of the shells (external left valve, internal right valve) and hinges (right and left) of *Grippina* sp. A emphasize greater surface detail than do light photographs.

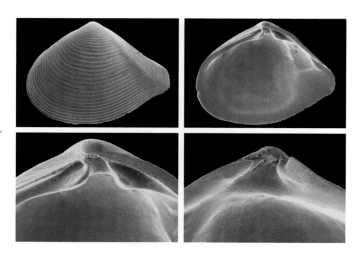

# Family description

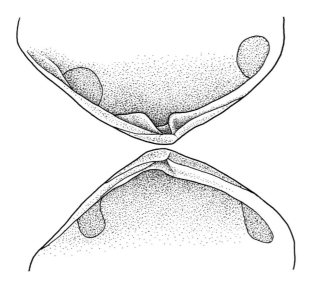

The spheniopsid shell is small (to ca. 5 mm), solid, and oval to elongate to trigonal; the anterior end is rounded, and the posterior end pointed and truncate. It is EQUIVALVE or slightly INEQUIVALVE (left valve slightly smaller), compressed, and not gaping. The shell is EQUILATERAL or INEQUILATERAL (umbones slightly posterior), with ORTHOGYRATE UMBONES. Shell microstructure is ARAGONITIC and two-layered, with a CROSSED LAMELLAR outer layer and a COMPLEX CROSSED LAMELLAR inner layer. TUBULES have not been reported. Exteriorly spheniopsids are white or translucent with white patches, covered by inconspicuous PERIOSTRACUM that in some species is set with minute calcareous granules. Sculpture is commarginal, in some species with widely spaced, rounded commarginal ribs; the surface can also be granular or punctate. LUNULE and ESCUTCHEON are present and bound by radial ridges. Interiorly the shell is non-NACREOUS. The PALLIAL LINE has a short rounded SINUS. The inner shell margins are smooth. The HINGE PLATE is weak and EDENTATE in adults,

---

*Grippina sp.* B – Gripp's Clam

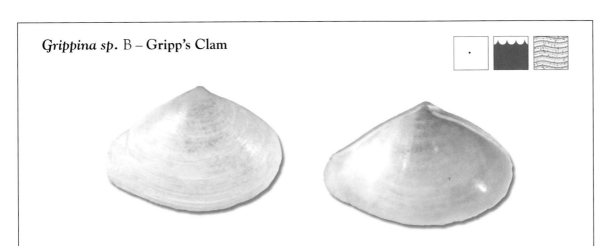

Rounded trigonal, posterior narrowed, compressed, thin-walled, smooth with low, rounded commarginal ridges, white exteriorly and interiorly; interior glossy with radial groove/ridge just posterior of midvalve. Florida Keys. Length 3 mm. Note: Redfern (2001) described and figured three species of *Grippina* from Abaco, Bahamas, as G. sp. A, sp. B, and sp. C. Our G. sp. A is likewise a new unnamed species; our sp. B might be the same as Redfern's sp. A. The right valve of the specimen used for scanning electron microscopy (p. 202) has been drilled by a predatory gastropod.

but the dorsal margins of the right valve are flattened and extended as oblique "teeth" flanking the resilifer and fitting under the dorsal margin of the left valve. The LIGAMENT is internal (RESILIUM), set in a subumbonal sunken triangular RESILIFER, and is supported by a stout LITHODESMA.

Almost nothing is known of the anatomy of spheniopsids. The few available data are from impressions left on the shells and the dried soft parts of one eastern Pacific species, *Grippina californica* Dall, 1912. The animal is ISOMYARIAN. Pedal retractor muscles are present dorsomedially to each adductor muscle (based on muscle scars). The MANTLE margins are fused ventrally, with a small anteroventral pedal gape. Posterior EXCURRENT and INCURRENT SIPHONS are short and united. Larvae are brooded in the adult and larval development is LECITHOTROPHIC or DIRECT (based on prodissoconch size). Spheniopsids are marine and most likely INFAUNAL in soft sediments. Nothing else is known of the morphology or ecology of this family, and their recent transfer to the Anomalodesmata is controversial.

The family Spheniopsidae is known since the Oligocene and is represented by 2 living genera and ca. 17 species, so far reported in the Recent fauna only from shores of the Americas and New Zealand.

**Scanning electron micrographs** of the shell (external left valve) and right hinge (right) of *Grippina* sp. B.

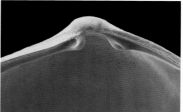

## References

Coan, E. V. 1990. The eastern Pacific species of the bivalve family Spheniopsidae. *The Veliger*, 33(4): 394–401.

Marshall, B. A. 2002. Some Recent Thraciidae, Periplomatidae, Myochamidae, Cuspidariidae and Spheniopsidae (Anomalodesmata) from the New Zealand region and referral of *Thracia reinga* Crozier, 1966 and *Scintillona benthicola* Dell, 1956 to *Tellimya* Brown, 1827 (Montacutidae) (Mollusca: Bivalvia). *Molluscan Research*, 22: 221–288.

Redfern, C. 2001. *Bahamian Seashells: A Thousand Species from Abaco, Bahamas*. Bahamianseashells.com, Inc., Boca Raton, Florida, ix + 280 pp., 124 pls.

# Family Thraciidae — Thracia Clams

**Classification**
AUTOLAMELLIBRANCHIATA Grobben, 1894
Heteroconchia Hertwig, 1895
Heterodonta Neumayr, 1883
Anomalodesmata Dall, 1889
Thraciidae Stoliczka, 1870 [1830]

**Featured species**
*Thracia morrisoni* Petit, 1964 – **Morrison's Wrinkled Thracia**

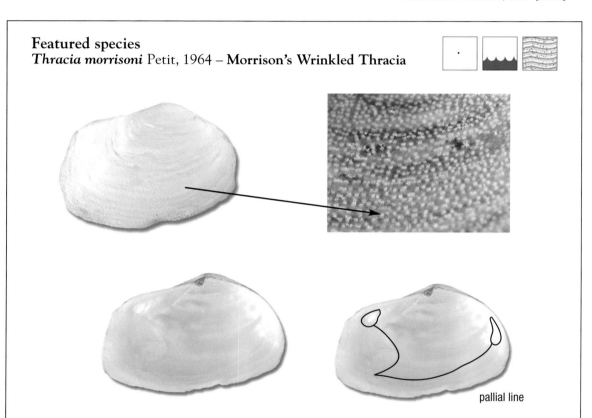

pallial line

Irregularly quadrangular to oval, with margins often distorted, right valve more inflated than left, umbones slightly anterior, close together, surface of right umbo often eroded or punctured by abrasion against left umbo, thin-walled, with commarginal growth lines, wrinkles, and granules (see inset), and a poorly defined radial ridge separating posterior quarter of valve, chalky white; interior white. North Carolina to Florida, Bahamas, West Indies. Length 15 mm (to 18 mm). Note: Frequently misidentified as *Thracia corbuloides* Blainville, 1824, a European species.

# Family description

The thraciid shell is usually small to medium-sized (to 90 mm), thin-walled (usually fragile) to solid, and irregularly quadrangular to rounded (often distorted in nestling *Iaxartia*); the posterior end is truncate or bluntly ROSTRATE. It is EQUIVALVE or INEQUIVALVE (right

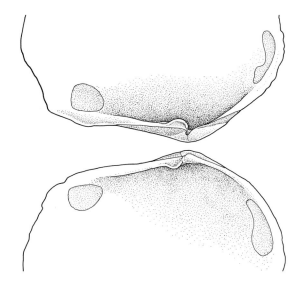

valve larger and more convex, overlapping the left valve), compressed to inflated, and in some species with the posterior end gaping. The shell is EQUILATERAL OR INEQUILATERAL (umbones often posterior, in some anterior), with ORTHOGYRATE and often eroded UMBONES; an apical perforation is occasionally present in the right valve, caused by abrasion by the left apex. Shell microstructure is ARAGONITIC and two-layered, with finely HOMOGENOUS outer and inner layers (unusually non-NACREOUS for an anomalodesmatan, although Mesozoic thraciids were nacreous). TUBULES are apparently absent. Exteriorly thraciids are white, covered by a thin to thick PERIOSTRACUM that can be collarlike at the base of the siphons and covered with sand grains and detritus. Sculpture is smooth or (usually) finely granulate to coarsely pustulate, occasionally also with commarginal ribs or striae, oblique undulations (*Cyathodonta*), or (rarely) radial ribs. LUNULE and ESCUTCHEON are usually absent; an escutcheon is present (larger in one valve) in, e.g., *Skoglundia* and *Pseudocyathodonta*. Interiorly the shell is non-nacreous. The PALLIAL LINE has a shallow to deep SINUS. The inner shell margins are smooth. The HINGE PLATE is EDENTATE in adults and is equipped with an external or internal hinge (both in *Bushia* and *Lampeia*) plus a free LITHODESMA. The external LIGAMENT when present (e.g., in *Bushia* and *Cyathodonta*) is OPISTHODETIC, PARIVINCULAR, and set on NYMPHS. The internal portion (RESILIUM) when present (e.g., *Asthenothaerus* and *Thracia*) sets in posteriorly oblique CHONDROPHORES, which differ in relative size among species. A small to large semicircular or butterfly-shaped lithodesma bridges the valves anterior to the chondrophores; the lithodesma is proportionally larger

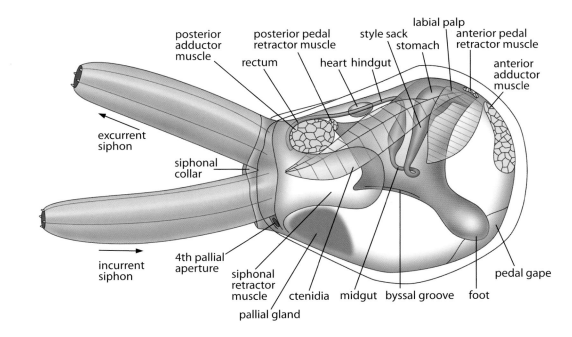

in juveniles of some species and in some cases is absent or inconspicuous in the adult. A secondary external ligament of fused periostracum unites the valves along the dorsal hinge margin.

The animal is slightly HETEROMYARIAN (anterior ADDUCTOR MUSCLE narrower), with the anterior adductor muscle elongated arcuate in cross section and the posterior adductor muscle rounded; pedal retractor muscles are small. Pedal elevator and protractor have not been reported. The MANTLE margins are extensively fused ventrally, with a small to minute anteroventral pedal gape. Posterior EXCURRENT and INCURRENT SIPHONS are long (with the incurrent siphon longer) or short (in *Ixartia*), separate, and highly extensible, sheathed basally by periostracum. The siphonal retractor muscle is large and strong. A FOURTH PALLIAL APERTURE is present ventral to the base of the incurrent siphon in most species for which anatomy is known. Large pallial (HYPOBRANCHIAL?) GLANDS are reported lining the inner surface of the posterior INFRABRANCHIAL CHAMBER in some *Trigonothracia* and *Thracia*; these are asymmetrical (larger on right) in *Thracia meridionalis* E. A. Smith, 1885. The FOOT is small, unheeled, and has a BYSSAL GROOVE; the adult is nonbyssate.

The LABIAL PALPS are large, and trigonal or elongated. The CTENIDIA are EULAMELLIBRANCH (SYNAPTORHABDIC), HETERORHABDIC, and relatively small; the outer demibranchs consist only of reflected descending lamellae. The gills are not inserted into (or fused with) the distal oral groove of the palps (CATEGORY III association). Incurrent and excurrent water flows are posterior. INFRA- and SUPRABRANCHIAL CHAMBERS are in direct communication (i.e., the gills are not united posteriorly with the SIPHONAL SEPTUM). The STOMACH is TYPE IV; the MIDGUT is weakly coiled and expanded in diameter posteriorly (known in *T. meridionalis* only). The HINDGUT passes through the ventricle of the heart and leads to a sessile or freely hanging rectum. Thraciids are SIMULTANEOUS HERMAPHRODITES and produce large, yolky eggs that hatch LECITHOTROPHIC (perhaps DIRECT-DEVELOPING) VELIGER larvae; brooding occurs in some in the suprabranchial chamber. The nervous system has not been studied. STATOCYSTS (TYPE B1, with single large STATOLITHS) are present in the adults of several species. ABDOMINAL SENSE ORGANS have not been reported.

Thraciids are marine SUSPENSION or DEPOSIT FEEDERS, INFAUNAL in fine or coarse sand or mud, or less typically nestling in rock crevices (*Iaxartia*). They are slow burrowers,

*Thracia stimpsoni* Dall, 1886 – **Stimpson's Thracia**

Rounded trigonal, posterior rostrate, right valve more inflated than left, umbones slightly anterior, close, right punctured by left, thin-walled, with commarginal growth lines and two strong radial ridges (one marginal) from the umbo to the rostrum, white exteriorly and interiorly. Florida Keys, Gulf of Mexico. Length 65 mm.

capable of reburying. Two species of *Thracia* are known to lie horizontally deep in muddy sediment with the right valve uppermost, and with the siphons in separate burrows to the surface; the siphonal canals are hardened by mucus produced by the siphons, allowing the siphonal tips to remain below the sediment surface, away from predators (e.g., fish and seabirds). The siphons are known to undergo peristaltic contractions within the canals.

The family Thraciidae is known since the Jurassic and is represented by 18 living genera and ca. 30 species, distributed worldwide mainly in temperate to cold waters, from the shallow subtidal to moderately deep zones. They are generally rare.

Note: Additional Thraciidae species appear after the Verticordiidae, p. 212.

### References

Allen, J. A. 1961. The British species of *Thracia* (Eulamellibranchia). *Journal of the Marine Biological Association of the United Kingdom*, 41(3): 723–735.

Coan E. V. 1990. The Recent eastern Pacific species of the bivalve family Thraciidae. *The Veliger*, 33(1): 20–55.

Kamanev, G. M. 2002. Genus *Parvithracia* (Bivalvia: Thraciidae) with descriptions of a new subgenus and two new species from the northwestern Pacific. *Malacologia*, 44: 107–134.

Morse, E. S. 1913. Notes on *Thracia conradi*. *The Nautilus*, 27(7): 73–77.

Morton, B. 1995. The ecology and functional morphology of *Trigonothracia jinxingae* (Bivalvia: Anomalodesmata: Thracioidea) from Xiamen, China. *Journal of Zoology*, 237(3): 445–468.

Sartori, A. F., and O. Domaneschi. 2005. The functional morphology of the Antarctic bivalve *Thracia meridionalis* Smith, 1885 (Anomalodesmata: Thraciidae). *Journal of Molluscan Studies*, 71: 199–210.

Thomas, M. L. H. 1967. *Thracia conradi* in Malpeque Bay, Prince Edward Island. *The Nautilus*, 80(3): 84–87.

# Family Verticordiidae – Verticordia Clams

**Classification**
AUTOLAMELLIBRANCHIATA Grobben, 1894
HETEROCONCHIA Hertwig, 1895
Heterodonta Neumayr, 1883
Anomalodesmata Dall, 1889
Septibranchia Pelseneer, 1888
Verticordiidae Stoliczka, 1870

**Featured species**
*Spinosipella acuticostata* (Philippi, 1844) – **Sharp-ribbed Verticord**

Quadrangular, inflated, solid, surface granular with strong, sharp radial ribs, white, margin smooth but plicated by exterior ribs. North Carolina, Florida Keys, West Indies, Gulf of Mexico, South America (Colombia, Brazil). Length 13 mm. Formerly in *Verticordia*.

## Family description

The verticordiid shell is small to medium-sized (to 100 mm), thin-walled to solid, and dorsoventrally oval to roundly trapezoidal; the anterodorsal margin is often concave. It is EQUIVALVE or somewhat INEQUIVALVE (right valve usually slightly larger and overlapping the left; left overlapping right in some), usually inflated, and either not gaping or gaping posteriorly. The shell is INEQUILATERAL (umbones anterior), with PROSOGYRATE UMBONES. Shell microstructure is ARAGONITIC and three-layered, with a simple PRISMATIC outer layer, a lenticular NACREOUS middle layer, and a sheet-nacreous inner layer. TUBULES are apparently absent. Exteriorly verticordiids are whitish, covered by thin, yellow to pale olive to brownish PERIOSTRACUM that can be thickened posteriorly or infolded around the valve margins. Sculpture is smooth or granulose, usually with radial ribs or striae and aligned beads or minute spines, some with broad undulations producing a deeply plicate margin; some have sand grains adhering to the rough granular surface. The LUNULE is usually pres-

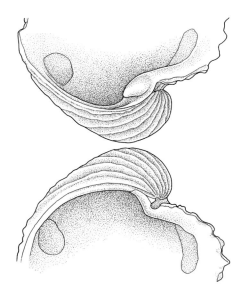

ent, impressed, and well marked (shallow to weak in some genera, e.g., *Halirus* and *Lyonsiella*); the ESCUTCHEON is weakly developed. Interiorly the shell is nacreous. The PAL-LIAL LINE is usually ENTIRE or with a shallow SINUS (*Bentholyonsia*). The inner shell margins are smooth but in some are plicated by the external ribs. The HINGE PLATE is weak, in some cases thickened anteriorly, and EDENTATE in adults (*Lyonsiella* and *Bentholyonsia*) or HETERODONT with one or two peglike CARDINAL TEETH in the right valve (and corresponding sockets in the left valve) plus lateral lamellae in some (*Trigonulina* and *Halirus*). The LIGAMENT is internal (RESILIUM) and OPISTHODETIC, set in a RESILIFER that is not projecting in most species, and supported by a LITHODESMA that in some taxa is posteriorly forked; a small portion of the ligament extends exteriorly in *Lyonsiella*. A secondary external ligament of fused periostracum unites the valves along the dorsal hinge margin.

The animal is ISOMYARIAN or HETEROMYARIAN (anterior or posterior ADDUCTOR MUSCLE smaller); pedal retractor muscles are reduced with the anterior one absent in some species. Pedal elevator and protractor are absent; however, *Bentholyonsia* has unique suspensory muscles that originate in the mantle lobes and insert anterodorsally near the umbones. The MANTLE margins are fused midventrally, with a small to medium-sized anteroventral pedal gape and large EXCURRENT and (still larger, in some species three to four times larger) INCURRENT APER-

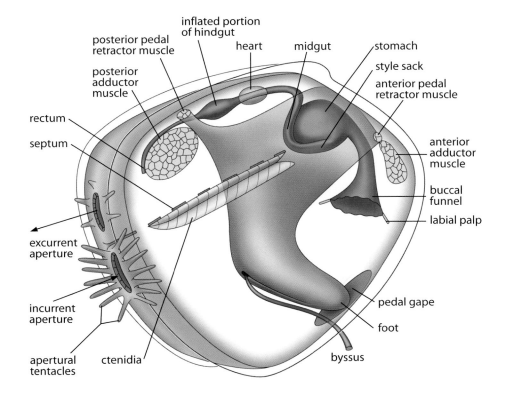

TURES each usually surrounded by a ring of numerous, large (occasionally aborescent), muscular, sensory tentacles. Some taxa (e.g., *Lyonsiella*, *Halicardia*, and *Laevicordia*) also have a membranous valve or "siphon" in one or both apertures. Siphonal retractor muscles are usually absent (except in *Bentholyonsia*). Pallial retractor muscles are particularly well developed posteroventrally, and TAENIOID MUSCLES in the fused ventral mantle margin of some species serve to retract the incurrent apparatus (i.e., tentacles, valve, or siphon); in *Lyonsiella* and *Bentholyonsia* the taenioid muscles leave a distinct muscle scar interior to the anterior adductor. A FOURTH PALLIAL APERTURE is present posteroventrally in some taxa. The mantle edge is glandular in some species, with arenophilic radial glands along the margins (involved in adhering sand grains to the external shell surface). HYPO-BRANCHIAL GLANDS have been reported in the suprabranchial chamber of *Bentholyonsia*. The FOOT is usually small and digitiform (large in *Lyonsiella*), usually unheeled, and laterally compressed; the foot of *Halicardia* is extended posteriorly by a keel-like heel (OPISTHOPODIUM). The adult is usually weakly byssate (except in *Euciroa*).

The LABIAL PALPS are small and triangular (although completely nonplicate) in *Bentholyonsia*, but in all other genera are minute and intimately associated with the lips (see following); none of these palps are capable of particle sorting. The CTENIDIA are usually EULAMELLIBRANCH (SYNAPTORHABDIC), HOMORHABDIC, relatively to extremely small, horizontally oriented (rather than vertically as in most bivalves), and (except in *Bentholyonsia*) attached to the ventral surface of a membranous perforated membrane (SEPTUM) with open OSTIA that separates INFRA- and SUPRASEPTAL CHAMBERS; some authors call these gills "SEPTIBRANCH" because of their association with a septum although the gill filaments are not confined to the septal ostia. Gills are absent in some genera (e.g., *Dallicordia*). The gill filaments are short and few in number and probably function solely in respiration. The outer demibranchs vary from complete (some *Policordia*), to consisting solely of reflected descending lamellae, to absent (in *Halicardia*, *Laevicordia*, some *Lyonsiella*, and some *Policordia*). The gills are not inserted into (or fused with) the distal oral groove of the palps (CATEGORY III association). Incurrent and excurrent water flows are posterior. The alimentary canal is modified for a carnivorous diet. The lips (except in *Bentholyonsia*) are usually expanded to form a ciliated, posteriorly facing, trumpet-shaped BUCCAL FUNNEL, with the palps firmly attached at the margins (or to the mantle or septum), opening near the anterior end of the gills; in *Euciroa*, the lips are produced into LATERAL BULBS. The strongly muscular esophagus is expandable. The STOMACH is TYPE II, and has a dorsal hood; the MIDGUT is usually not coiled (loosely coiled in *Bentholyonsia*). The HINDGUT is short, passes through or ventral to (*Bentholyonsia*) the ventricle of the heart (which is relatively small), usually includes an inflated portion (of uncertain function; absent in *Bentholyonsia*) posterior to the heart, and leads to a sessile rectum. The kidneys are accompanied laterally by an extensive system of LACUNAE (possibly serving as a blood reservoir to effect siphonal expansion during prey capture, see below), extending into the ventral and anterior parts of the body, often as far as the level of the mouth, and communicating with the kidney via ducts. Verticordiids are usually SIMULTANEOUS HERMAPHRODITES (*Bentholyonsia* is GONOCHORISTIC, based on a single specimen) and produce few large, yolky eggs that hatch LECITHOTROPHIC VELIGER larvae. The nervous system is not concentrated; in some taxa, the visceral and/or pedal ganglia are enlarged (approaching the size of the posterior adductor muscle; presumably in part associated with the predatory function of the posterior siphons). STATOCYSTS (TYPE B1, B2, or B3) are known in adult *Lyonsiella* and *Bentholyonsia*. ABDOMINAL SENSE ORGANS have not been reported.

Verticordiids are marine carnivores on benthic prey (including foraminiferans, polychaete worms, copepods, and other small crustaceans), and are INFAUNAL in deepwater muddy substrata. Prey capture has never been directly observed, but morphology suggests that the tentacles act as mechanical sensors, detecting nearby prey that is then snared by the tentacles themselves or an everted incurrent valve or siphon (as in Poromyidae). Once secured, the taenioid muscles retract to bring the prey into the MANTLE CAVITY. The prey could be transferred to the gills, which convey it to the buccal funnel for ingestion, or the siphon could extend forward to place prey in the funnel directly. The foot also has been hypothesized to assist in transferring captured prey to the mouth.

The family Verticordiidae is known since the Cretaceous and is represented by 18 living genera and ca. 50 species, distributed worldwide in deep to abyssal waters. The group generally is rare. Composition of this family and the relationships of its constituent genera are in need of revision. Euciroidae (with *Euciroa* and *Acereuciroa*) and Lyonsiellidae (with *Lyonsiella*, *Policordia*, *Bentholyonsia*, and others) are often treated as independent families.

### References

Allen, J. A., and J. F. Turner. 1974. On the functional morphology of the family Verticordiidae (Bivalvia) with descriptions of new species from the abyssal Atlantic. *Philosophical Transactions of the Royal Society of London, Series B, Biological Sciences*, 268(894): 401–532.

Bernard, F. R. 1974. Septibranchs of the eastern Pacific (Bivalvia Anomalodesmata). *Allan Hancock Monographs in Marine Biology*, 8: 1–279.

Dreyer, H., G. Steiner, and E. M. Harper. 2003. Molecular phylogeny of Anomalodesmata (Mollusca: Bivalvia) inferred from 18S rRNA sequences. *Zoological Journal of the Linnean Society*, 139: 229–246.

Harper, E. M., H. Dreyer, and G. Steiner. 2006. Reconstructing the Anomalodesmata (Mollusca: Bivalvia): morphology and molecules. *Zoological Journal of the Linnean Society*, 148: 395–420.

Morton, B. 1985 ("1984"). Prey capture in *Lyonsiella formosa* (Bivalvia: Anomalodesmata: Verticordiacea). *Pacific Science*, 38(4): 283–297.

Morton, B. 2003. The functional morphology of *Bentholyonsia teramachii* (Bivalvia: Lyonsiellidae): clues to the origin of predation in the deep water Anomalodesmata. *Journal of Zoology*, 261: 363–380.

Nakazima, M. 1967. Some observations on the soft part [*sic*] of *Halicardia nipponensis* Okutani. *Venus*, 25(3–4): 147–158, pls. 6–9.

Poutiers, J.-M., and F. R. Bernard. 1995. Carnivorous bivalve molluscs (Anomalodesmata) from the tropical western Pacific Ocean, with a proposed classification and a catalogue of Recent species. Pages 107–188, in: P. Bouchet, ed., *Résultats des Campagnes MUSORSTOM, 14. Mémoires du Muséum National d'Histoire Naturelle*, 167.

Yonge, C. M., and B. Morton. 1980. Ligament and lithodesma in the Pandoracea and the Poromyacea with a discussion on evolutionary history in the Anomalodesmata (Mollusca: Bivalvia). *Journal of Zoology*, 191: 263–292.

## THRACIIDAE

### *Asthenothaerus hemphilli* Dall, 1886 – **Hemphill's Square Thraciid**

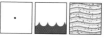

Quadrangular to elongated oval, posterior truncate and slightly gaping, left valve slightly smaller than right valve, thin-walled, surface granular with commarginal ridges, translucent white with thin, yellowish periostracum, pallial sinus deep, margin smooth. North Carolina, Florida Keys, Gulf of Mexico. Length 11 mm. Syn. *balesi* Rehder, 1943.

### *Bushia elegans* (Dall, 1886) – **Elegant Bush Clam**

Rounded trigonal, posterior sharply truncated and not gaping, left valve slightly smaller than right valve, thin-walled, surface with even commarginal ridges, white with thin, yellowish periostracum, pallial sinus deep, margin smooth. North Carolina to Florida, West Indies. Length 11 mm (to 13 mm). Note: The valves photographed are of a SYNTYPE specimen of *Bushia elegans*.

### *Thracia* cf. *phaseolina* Lamarck, 1822 – **Kidney-Bean Thracia**

Elongated quadrangular, posterior rostrate, sharply truncated and gaping, left valve slightly smaller than right valve, umbones close together, surface of right umbo often eroded or punctured by abrasion against left umbo, thin-walled, surface very finely granular with even commarginal ridges, chalky white, pallial sinus deep, margin smooth. Florida Keys, Gulf of Mexico, Caribbean Central America. Length 20 mm. Note: True *T. phaseolina* is European, and specimens reported under this name from the western Atlantic belong to an undescribed species (Coan, 1990: 36, 39); no specimens from the western Atlantic have been seen by the authors; the description and figures are based on true specimens from England.

## VERTICORDIIDAE

### *Euciroa elegantissima* (Dall, 1881) – **Elegant Verticord**

Oval, inflated, solid, with scaly radial ribs (distinctly smaller posteriorly), dull white with thin, pale periostracum, interior radially wrinkled, margin denticulate (by radial wrinkles). Florida, West Indies, South America (Colombia). Length 34 mm.

### *Haliris fischeriana* (Dall, 1881) – **Fischer's Beaded Verticord**

Quadrangular, inflated, lunule shallow, solid, with finely beaded or serrated radial ribs, white, margin smooth. North Carolina to Florida, West Indies, Gulf of Mexico, South America (Brazil). Length 8 mm (to 9 mm).

### *Trigonulina ornata* d'Orbigny, 1853 – **Ornate Verticord**

Quadrangular, compressed, left valve slightly smaller than right valve, lunule impressed, solid, surface radially granular with ca. 12 strong, sharp radial ribs on anterior half to three-fourths, dull white, margin smooth but plicated by exterior ribs. Massachusetts to Florida, Bermuda, Bahamas, West Indies, Gulf of Mexico, South America (Suriname, Brazil). Length 3 mm (to 5 mm). Formerly in *Verticordia*.

*Asthenothaerus hemphilli*

*Bushia elegans*

*Thracia* cf. *phaseolina*

*Euciroa elegantissima*

*Haliris fischeriana*

*Trigonulina ornata*

# Family Poromyidae – Poromya Clams

**Classification**

AUTOLAMELLIBRANCHIATA Grobben, 1894
HETEROCONCHIA Hertwig, 1895
Heterodonta Neumayr, 1883
Anomalodesmata Dall, 1889
Septibranchia Pelseneer, 1888
Poromyidae Dall, 1886

**Featured species**
*Poromya granulata* (Nyst & Westendorp, 1839) –
**Granular Poromya**

Rounded trigonal, inflated, fragile, with radial lines of fine granules, white. North Atlantic, North Carolina, Florida, West Indies, South America (Colombia, Brazil), also western Europe, Arctic seas. Length 5 mm.

# Family description

The poromyid shell is small to medium-sized (usually < 30 mm), thin-walled, and orbicular to elongated oval to quadrangular; the anterior end is rounded, and the posterior end truncate or with a short ROSTRUM. It is (usually) EQUIVALVE or INEQUIVALVE (right valve overlapping left), (usually) inflated or compressed, and slightly gaping posteriorly at the siphonal opening. The shell is EQUILATERAL to slightly INEQUILATERAL (umbones anterior), with PROSOGYRATE UMBONES. Shell microstructure is ARAGONITIC and two- or three-layered, with a PRISMATIC or HOMOGENOUS outer layer, a lenticular NACREOUS middle layer, and a sheet-nacreous inner layer; some species have only homogenous inner and outer layers. TUBULES are apparently absent. Exteriorly poromyids are white to pale gray to yellow, covered by a thin, dehiscent or adherent, dark brown to yellow PERIOSTRACUM. Sculpture is smooth to radially striate, often with fine radial rows of small granules or

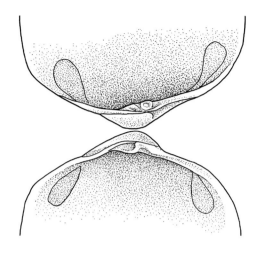

spinules. LUNULE and ESCUTCHEON are absent or weak. Interiorly the shell is nacreous. The PALLIAL LINE is deeply impressed with a truncate or slightly sinuous posterior slope (called a small PALLIAL SINUS by some authors). The inner shell margins are smooth. The HINGE PLATE is EDENTATE in adults or with a (secondary) cardinal tubercle in the right valve anterior to the resilium and a corresponding socket, or in some cases a small tooth, in the left valve. The external LIGAMENT is OPISTHODETIC, PLANIVINCULAR, set on PSEUDONYMPHS, and in some species split posteriorly into two longitudinal halves; an internal portion (RESILIUM) is set in an oblique RESILIFER, and lacks a LITHODESMA. A secondary external ligament of fused periostracum unites the valves along the dorsal hinge margin both anteriorly and posteriorly.

The animal is ISOMYARIAN or HETEROMYARIAN (posterior ADDUCTOR MUSCLE smaller in *Cetoconcha*); pedal retractor muscles are thin; the posterior pedal retractor is single (rather than left and right muscles as in most bivalves) but bifurcates close to its insertion on the shell. Pedal elevator and protractor muscles have not been reported. The MANTLE margins are generally thin, thickened marginally, and largely not fused ventrally, with a large pedal gape. Posteriorly a small, short EXCURRENT SIPHON and a much larger INCUR-

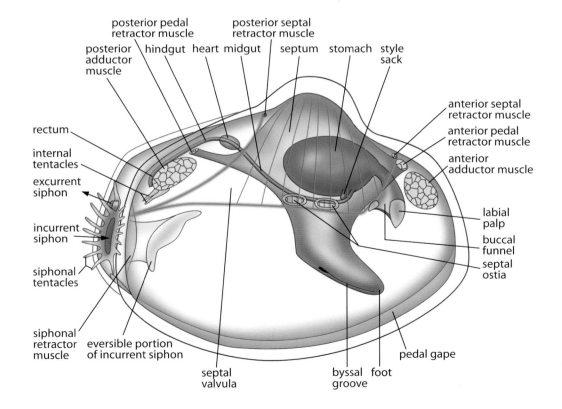

RENT SIPHON are together surrounded by a species-specific number of large, brightly colored, sensory (probably also secretory) tentacles; intertentacular knobs or papillae also are present in some taxa. The incurrent siphon has a large, muscular, internal portion that is eversible into a large raptorial hood or cowl. Short siphonal retractor muscles are present. A pair of internal tentacles are suspended from the posterior adductor muscle in some species; these are of unconfirmed function but could serve to cleanse the supraseptal cavity. The mantle edges can be thickened and can have embedded spicules. A FOURTH PALLIAL APERTURE and arenophilic glands are absent. HYPOBRANCHIAL GLANDS have not been reported. The FOOT is long and digitiform, has a BYSSAL GROOVE, and projects through a tight gap in the septum; the adult is nonbyssate.

The LABIAL PALPS are large, ventrally directed, cup-shaped, with ciliated inner surfaces, and function in the transfer of food items from the incurrent siphon. The CTENIDIA are SEPTIBRANCH (SYNAPTORHABDIC) with filaments occupying one or two pairs of large OSTIA (or two or three groups of small pores) equipped with gill filaments with or (secondarily?) without interfilamentar junctions (also called BRANCHIAL SIEVES) and with thickened edges (that act as closing valves) set in a muscular SEPTUM suspended by small septal retractor muscles and larger lateral septal muscles, and separating INFRA- and SUPRASEPTAL CHAMBERS. Contraction and relaxation of the septum changes hydrostatic pressure in the MANTLE CAVITY, assisting prey capture. Posterior to the foot, the septum continues as two thick, dish- or crescent-shaped pads (SEPTAL VALVULAE), continuous with the SIPHONAL SEPTUM, and connecting medially between the siphons; these are hemocoels that store blood used to expand and contract the incurrent siphonal hood during prey capture. The gills are not inserted into (or fused with) the distal oral groove of the palps (CATEGORY III association). Incurrent and excurrent water flows are posterior. The alimentary canal is modified for a carnivorous diet. The lips are enlarged into a BUCCAL FUNNEL, with the palps at anterior and posterior corners (anterior pair larger); the mouth is broad and the esophagus is short and muscular. The STOMACH is TYPE II; food is carried from the mouth to the stomach by muscular peristalsis, rather than by cilia as in other bivalves. The MIDGUT is not coiled. The HINDGUT passes through the ventricle of the heart, and leads to a sessile rectum. The kidneys are accompanied laterally by a system of LACUNAE (as in Verticordiidae, although smaller), extending into the posterior parts of the body near the supraseptal chamber and base of the siphons; cells similar to those in the lacunae are found in the septal valvulae in some *Cetoconcha*. Poromyids are SIMULTANEOUS HERMAPHRODITES and produce few large, yolky eggs that probably hatch LECITHOTROPHIC VELIGER larvae; the larvae are possibly brooded in the supraseptal chamber. The nervous system is not concentrated. STATOCYSTS (TYPE B1, with single large STATOLITHS) are known in adult *Poromya*. ABDOMINAL SENSE ORGANS have not been reported.

Poromyids are marine carnivores, shallowly INFAUNAL in deepwater muddy substrata. The foot is used in active burrowing. Recorded prey organisms include foraminiferans, sponges, chaetognaths, and (most importantly) polychaete worms and small crustaceans. Prey capture (suggested by morphology) involves (1) detection by the sensory tentacles; (2) relaxation of the septal retractor muscles and simultaneous eversion of the incurrent siphonal hood (by filling with blood from the valvulae) to capture prey; (3) contraction of the septal and siphonal retractor muscles, which draws the hood with the prey into the INFRASEPTAL CHAMBER; and (4) expansion and posterior extension of the labial palps to transfer the prey to the mouth. The foot might also assist in pushing captured prey into the mouth.

The family Poromyidae is known since the Cretaceous and is represented by ca. 6 liv-

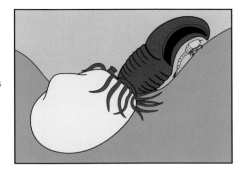

**Carnivorous poromyids** feed on small annelids or crustaceans using the enlarged incurrent siphon. Prey is brought into the infraseptal chamber by contraction of the siphonal retractor muscles, which draws in the incurrent siphonal hood. The labial palps then extend backward to transfer the prey to the mouth.

ing genera and ca. 50 species, distributed worldwide in deep and abyssal waters. Some authors place *Cetoconcha* in its own family Cetoconchidae, based on septal characters.

### References

Allen, J. A., and R. E. Morgan. 1981. The functional morphology of Atlantic deep water species of the families Cuspidariidae and Poromyidae (Bivalvia): an analysis of the evolution of the septi-branch condition. *Philosophical Transactions of the Royal Society of London, Series B, Biological Sciences*, 294(1073): 413–546.

Bernard, F. R. 1974. Septibranchs of the eastern Pacific (Bivalvia Anomalodesmata). *Allan Hancock Monographs in Marine Biology*, 8: 1–279.

Krylova, E. M. 1997. New taxa and the system of Recent representatives of the family Poromyidae (Bivalvia, Septibranchia, Poromyidae). *Ruthenica*, 7(2): 141–147.

Krylova, E. M. 2001. Septibranchiate molluscs of the family Poromyidae (Bivalvia: Poromyoidea) from the tropical western Pacific Ocean. Pages 165–200, in: P. Bouchet, and B. A. Marshall, eds., *Tropical Deep-Sea Benthos, volume 22. Mémoires du Muséum National d'Histoire Naturelle*, 185.

Morton, B. 1981. Prey capture in the carnivorous septibranch *Poromya granulata* (Bivalvia: Anomalodesmata: Poromyacea). *Sarsia*, 66: 241–256.

Odhner, N. H. 1960. Mollusca. *Reports of the Swedish Deep-Sea Expedition, 2, Zoology*, 22: 365–400.

Poutiers, J.-M., and F. R. Bernard. 1995. Carnivorous bivalve molluscs (Anomalodesmata) from the tropical western Pacific Ocean, with a proposed classification and a catalogue of Recent species. Pages 107–188, in: P. Bouchet, ed., *Résultats des Campagnes MUSORSTOM, 14. Mémoires du Muséum National d'Histoire Naturelle*, 167.

Yonge, C. M. 1928. Structure and function of the organs of feeding and digestion in the septibranchs, *Cuspidaria* and *Poromya*. *Philosophical Transactions of the Royal Society of London, Series B, Containing Papers of a Biological Character*, 216(1928): 221–263.

Yonge, C. M., and B. Morton. 1980. Ligament and lithodesma in the Pandoracea and the Poromyacea with a discussion on evolutionary history in the Anomalodesmata (Mollusca: Bivalvia). *Journal of Zoology*, 191: 263–292.

***Poromya albida*** Dall, 1886 – **White Poromya**

Rounded trigonal, inflated, fragile, with finely granular surface, white. North Carolina, Florida Keys. Length 21 mm. Note: The valve photographed is of the HOLOTYPE specimen of *Poromya albida* (right valve only, anterior end damaged).

***Poromya rostrata*** Rehder, 1943 – **Rostrate Poromya**

Rounded trigonal, posterior rostrate and slightly indented ventrally, inflated, fragile, with coarsely granular surface, white. North Carolina to Florida, West Indies, South America (Colombia, Uruguay). Length 7 mm. Compare *Plectodon granulatus* in cuspidariidae (p. 226), which is less fragile, more rostrate, and has a small lithodesma.

**Deepwater muddy sediments** are the typical habitat for carnivorous poromyids, where they prey upon other small organisms. A stalked glass sponge, *Hyalonema* sp., shares the environment (here at 806 m off the Bahamas).

*Poromya albida*

RV

*Poromya rostrata*

# Family Cuspidariidae – Dipper Clams

## Classification
AUTOLAMELLIBRANCHIATA Grobben, 1894
HETEROCONCHIA Hertwig, 1895
Heterodonta Neumayr, 1883
Anomalodesmata Dall, 1889
Septibranchia Pelseneer, 1888
Cuspidariidae Dall, 1886

**Featured species**
*Cuspidaria rostrata* (Spengler, 1793) – **Rostrate Dipper Clam**

Oval with long, keeled, tubelike posterior rostrum pointing slightly ventrally, thin-walled, smooth with coarse commarginal growth lines, white with yellowish periostracum; interior white, margin smooth, resilifer posteriorly inclined. North Atlantic to Florida, West Indies, South America (Brazil). Length 20 mm.

## Family description

The cuspidariid shell is small to medium-sized (to ca. 50 mm), thin-walled, and oval to trigonal or pyriform; the anterior end is rounded, and the posterior end truncate or (in many) with a long spoutlike ROSTRUM. It is EQUIVALVE to slightly INEQUIVALVE (right or left valve slightly smaller or less convex), inflated (usually) or compressed, and gaping posteriorly only at the tip of the rostrum. The shell is slightly to strongly INEQUILATERAL (umbones anterior), with PROSO- or ORTHOGYRATE UMBONES. The posterodorsal margin (along the rostrum) is usually concave. Shell microstructure is ARAGONITIC and two-layered, with finely HOMOGENOUS outer and inner layers. TUBULES are apparently absent. Exteriorly cuspidariids are white, chalky in some species, covered by a thin or coarse, yellow to brown, adherent or dehiscent PERIOSTRACUM, with attached sand particles in some species. Sculpture is smooth, strongly radial, commarginally ribbed, or oblique (*Soyomya*), often with surface granules or small pits. A weak rostral ridge extends from the umbone to the posterior margin of each valve in some species. A LUNULE is absent; the ESCUTCHEON is absent or weak. Interiorly the shell is non-NACREOUS and can be weakly radially striate

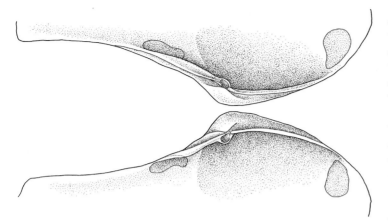

or strongly grooved (as a reflection of the external ribs). *Halonympha* has a distinct buttress or raised ridge on the interior dorsal edge; in some species this extends onto the posterior adductor muscle scar. The PALLIAL LINE is ENTIRE or has a small SINUS. The inner shell margins are smooth but can be scalloped by the external ribs. The hinge and ligament vary widely in the family. The HINGE PLATE is either EDENTATE in adults or (secondarily) HETERODONT with single cardinal tubercles and/or lateral ridges. The LIGAMENT is internal (RESILIUM), and AMPHI- or (more often) OPISTHODETIC, set in a small oblique or triangular RESILIFER with a small LITHODESMA. A secondary external ligament of fused periostracum unites the valves along the dorsal hinge margin.

The animal is slightly HETEROMYARIAN (posterior ADDUCTOR MUSCLE slightly smaller; anterior adductor irregular in cross section); the pedal retractor muscles are thin. Pedal elevator and protractor muscles have not been reported. The MANTLE margins are generally thin, thickened marginally, and fused ventrally, with a small to medium-sized anteroventral pedal gape. Posterior EXCURRENT and INCURRENT SIPHONS are short to long, asymmetrical, united, and surrounded by a siphonal sheath (a SYNAPOMORPHY of the family) that secretes the rostrum; the incurrent siphon is larger in diameter and slightly longer than the excurrent siphon, with an extendable hood or cowl at the tip that can be directed toward potential prey. An incurrent siphonal valve or sphincter, pierced by a simple closeable slit, seals the incurrent siphon internally at the base. The siphons are them-

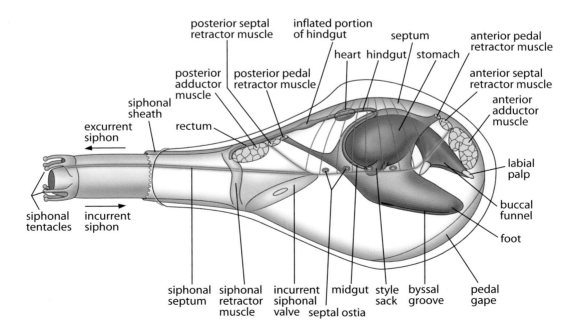

selves highly muscular and also can be equipped with a thin band of siphonal retractor muscles; both siphons can be retracted and the small excurrent siphon also can be inverted. Prominent club-shaped or frilled sensory tentacles (consistently 3 excurrent and 4 incurrent) with mechanoreceptors tip the siphons. The mantle edges can be thickened. Posterodorsal mantle sinuses store blood that is used in expanding the incurrent siphon for prey capture (see following). A FOURTH PALLIAL APERTURE and arenophilic glands are absent. HYPOBRANCHIAL GLANDS have not been reported. The FOOT is small to long and digitiform, has a BYSSAL GROOVE, and projects through a tight gap in the septum; the adult is nonbyssate.

The LABIAL PALPS are usually small (large in, e.g., *Protocuspidaria* and *Halonympha*), horn- or cup-shaped, ciliated but unridged, intimately associated with the buccal funnel (see below), and play a limited role in particle sorting. The CTENIDIA are SEPTIBRANCH (SYNAPTORHABDIC) and usually occupy OSTIA in a variably muscular SEPTUM suspended by paired septal retractor muscles (absent in *Protocuspidaria* and a few others) and in some also lateral septal muscle bundles, and separating (together with the SIPHONAL SEPTUM) INFRA- and SUPRASEPTAL CHAMBERS. In most species the ostia are simple ciliated pores (these cilia believed homologous with those on typical gill filaments) with thickened edges that open and close the pores and allow water flow only in the infra- to supraseptal direction; in some taxa the openings also have tissuelike valves. The pores are set in a single row on either side of the septum, usually 4 or 5 pores (up to 20 in, i.e., *Octoporia*) on each side, but are not necessarily aligned in pairs. The septa of *Protocuspidaria*, *Bidentaria*, and *Edentaria* are less well developed, without septal retractor muscles (each is supported by lateral muscle fibers inserting in the dorsal mantle) and with recognizably EULAMELLIBRANCH gill filaments aligned within a longitudinal "window" on each side. Contraction and relaxation of the septum changes hydrostatic pressure in the MANTLE CAVITY, assisting in prey capture. The gills are not inserted into (or fused with) the distal oral groove of the palps (CATEGORY III association). Incurrent and excurrent water flows are posterior. The alimentary canal is modified for a carnivorous diet. The lips are enlarged into a BUCCAL FUNNEL, with the palps at anterior and posterior corners (posterior pair usually larger); the mouth has a large lumen and the esophagus is short and muscular, with convoluted walls. The STOMACH is TYPE II; the MIDGUT is not coiled. Proteolytic enzymes have been reported in the digestive tract. The HINDGUT passes through the ventricle of the heart, includes in some species an inflated portion (of uncertain function, as in Verticordiidae) posterior to the heart, and leads to a sessile rectum. The kidneys are accompanied laterally by a system of LACUNAE (as in Verticordiidae, although smaller), extending into the posterior parts of the body near the supraseptal chamber, visceral ganglia, and base of the siphons. Parts of the gonad and digestive gland are arborescent on the external surface of the visceral mass in some *Myonera*. Most cuspidariids are GONOCHORISTIC; some have been suggested to be PROTANDRIC HERMAPHRODITES but this reproductive strategy has not been confirmed. Larval development is either DIRECT or LECITHOTROPHIC. The nervous system is not concentrated; an unpaired siphonal ganglion has been reported in some species but questioned by other authors. STATOCYSTS (TYPE C, with single, large, immobile STATOLITHS) are known in adult *Cuspidaria*. ABDOMINAL SENSE ORGANS have not been reported.

Cuspidariids are marine carnivores, INFAUNAL in soft sediments, with the tips of the siphons at the surface where the sensory tentacles can detect prey (e.g., polychaete worms, chaetognaths, small crustaceans, and foraminiferans). The process of prey capture (documented in the laboratory) is similar to that described for Poromyidae, but evidence points

**Living *Cardiomya gemma*** (see shell, p. 225) reveals its long, tentaculate, united siphons, as is typical for the family. This specimen is from the Indian River Lagoon, eastern Florida.

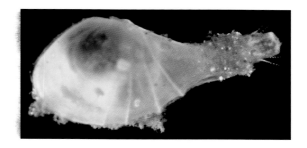

more toward swimming than benthic prey. The stomach also usually contains sand grains and small pebbles, probably brought in with prey, but possibly also aiding in trituration.

The family Cuspidariidae is known since the Jurassic and is represented by ca. 20 living genera and ca. 200 species, distributed worldwide in deep and abyssal waters.

### References

Allen, J. A., and R. E. Morgan. 1981. The functional morphology of Atlantic deep water species of the families Cuspidariidae and Poromyidae (Bivalvia): an analysis of the evolution of the septibranch condition. *Philosophical Transactions of the Royal Society of London, Series B, Biological Sciences*, 294(1073): 413–546.

Bernard, F. R. 1974. Septibranchs of the eastern Pacific (Bivalvia Anomalodesmata). *Allan Hancock Monographs in Marine Biology*, 8: 1–279.

Grobben, C. 1893. Beiträge zur Kenntniss des Baues von Cuspidaria (Neaera) cuspidata Olivi, nebst Betrachtungen über das System der Lamellibranchiaten. *Arbeiten aus dem Zoologischen Institute der Universität Wien und der Zoologischen Station in Triest*, 10(2): 1–46, pls. 1–4.

Marshall, B. A. 2002. Some Recent Thraciidae, Periplomatidae, Myochamidae, Cuspidariidae and Spheniopsidae (Anomalodesmata) from the New Zealand region and referral of *Thracia reinga* Crozier, 1966 and *Scintillona benthicola* Dell, 1956 to *Tellimya* Brown, 1827 (Montacutidae) (Mollusca: Bivalvia). *Molluscan Research*, 22: 221–288.

Poutiers, J.-M., and F. R. Bernard. 1995. Carnivorous bivalve molluscs (Anomalodesmata) from the tropical western Pacific Ocean, with a proposed classification and a catalogue of Recent species. Pages 107-188, in: P. Bouchet, ed., *Résultats des Campagnes MUSORSTOM, 14. Mémoires du Muséum National d'Histoire Naturelle*, 167.

Reid, R. G. B., and S. P. Crosby. 1980. The raptorial siphonal apparatus of the carnivorous septibranch *Cardiomya planetica* Dall (Mollusca: Bivalvia), with notes on feeding and digestion. *Canadian Journal of Zoology*, 58: 670–679.

Reid, R. G. B., and A. M. Reid. 1974. The carnivorous habit of members of the septibranch genus *Cuspidaria* (Mollusca: Bivalvia). *Sarsia*, 56: 47–56.

Yonge, C. M. 1928. Structure and function of the organs of feeding and digestion in the septibranchs, *Cuspidaria* and *Poromya*. *Philosophical Transactions of the Royal Society of London, Series B, Containing Papers of a Biological Character*, 216(1928): 221–263.

Yonge, C. M., and B. Morton. 1980. Ligament and lithodesma in the Pandoracea and the Poromyacea with a discussion on evolutionary history in the Anomalodesmata (Mollusca: Bivalvia). *Journal of Zoology*, 191: 263–292.

*Cardiomya alternata* (d'Orbigny, 1853) – **Alternate Cardiomya**

Oval with short, keeled posterior rostrum, thin-walled, with strong radial ribs on main body, crossed by distinct commarginals, white; interior grooved by external ribs, resilifer vertical. Florida, West Indies, South America (Colombia). Length 4 mm. Formerly in *Cuspidaria*. Note: The sepia-toned drawings are the original illustrations of the species.

*Cardiomya costellata* (Deshayes, 1833) – **Little-ribbed Cardiomya**

Oval with keeled posterior rostrum, thin-walled, with rounded radial ribs on main body, in some cases of alternating size, close together anteriorly but weaker and spaced farther apart posteriorly, white exteriorly and interiorly; interior glossy and grooved by external ribs, resilifer vertical. North Carolina to Florida, Bahamas, West Indies, Caribbean Central America. Length 11 mm.

*Cardiomya gemma* Verrill & Bush, 1898 – **Precious Cardiomya**

Oval with short, broad posterior rostrum, fragile, with few strong radial ribs on posterior half, translucent white with light yellow periostracum; interior glossy and weakly grooved by external ribs, resilifer vertical. North Carolina to Florida. Length 5 mm (to 6 mm). Formerly considered a synonym or form of *C. costellata* (see previous). Note: See p. 223 for photograph of living specimen.

*Cardiomya perrostrata* (Dall, 1881) – **Rostrate Cardiomya**

Oval with posterior rostrum with three longitudinal keels, fragile, with numerous strong radial ribs on main body, close together anteriorly but spaced farther apart posteriorly, interspaces with commargial striae, white; interior glossy and grooved by external ribs, resilifer vertical. Massachusetts to Florida, West Indies, Gulf of Mexico, Caribbean Central America, South America (Brazil). Length 5 mm (to 10 mm).

*Cardiomya ornatissima* (d'Orbigny, 1853) – **Ornate Cardiomya**

Oval with keeled posterior rostrum, thin-walled, with few, widely spaced, strong radial ribs on main body, interspaces concave and commarginally ridged, white exteriorly and interiorly; interior grooved by external ribs, resilifer vertical. North Carolina to Florida, West Indies, Gulf of Mexico, Caribbean Central America, South America (Colombia, Brazil). Length 13 mm. Syn. *glypta* Bush, 1898. Note: The sepia-toned drawings at right are the original illustrations of the species.

*Cardiomya alternata*

dorsal

*Cardiomya costellata*

*Cardiomya gemma*

*Cardiomya perrostrata*

*Cardiomya ornatissima*

dorsal

### *Cardiomya striata* (Jeffreys, 1876) – **Striate Cardiomya**

Oval with short, upturned posterior rostrum, thin-walled, with numerous fine, sharp radial ribs on main body, white exteriorly and interiorly; interior glossy and grooved by external ribs, resilifer vertical. Northern Atlantic to Florida, Bermuda, Gulf of Mexico, South America (Brazil). Length 13 mm (to 19 mm).

### *Cuspidaria obesa* (Lovén, 1846) – **Obese Dipper Clam**

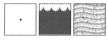

Oval with straight posterior rostrum, fragile, smooth with commarginal growth lines only, translucent white exteriorly and interiorly; interior glossy, margin smooth, resilifer posteriorly inclined. North Atlantic to Florida, West Indies, also western Europe. Length 7 mm.

### *Myonera gigantea* (Verrill, 1884) – **Giant Myonera**

Oval with very short, broad posterior rostrum, smooth with commarginal growth lines only, white exteriorly and interiorly, margin smooth, resilifer vertical or posteriorly directed. Maryland, Florida Keys. Length 20 mm.

### *Myonera lamellifera* (Dall, 1881) – **Shingled Myonera**

Oval with broad posterior rostrum, fragile, with sharp commarginal ridges with fine commarginal growth lines in the interspaces, white exteriorly and interiorly; interior glossy, margin smooth, resilifer vertical or posteriorly directed. Florida Keys, West Indies, Gulf of Mexico. Length 6 mm (to 12 mm).

### *Myonera paucistriata* Dall, 1886 – **Few-ribbed Myonera**

Oval with very short, broad posterior rostrum, fragile, with low commarginal ridges and two radial ridges at junction of main body and rostrum, white exteriorly and interiorly; interior glossy and grooved by external ridges, margin smooth, resilifer vertical or posteriorly directed. North Carolina to Florida, West Indies, South America (Brazil), also eastern Pacific, Hawaiian Islands, and western Europe. Length 10 mm (to 13 mm). Note: The photographed specimen is a left valve only, with minor damage to the early part of the shell.

### *Plectodon granulatus* (Dall, 1881) – **Grainy Plectodon**

Oval with broad, straight posterior rostrum, solid, surface granular, white exteriorly and interiorly; interior glossy, margin smooth, resilifer posteriorly directed. North Carolina, Florida Keys, West Indies, Gulf of Mexico, South America (to Venezuela). Length 12 mm. Compare *Poromya rostrata* in Poromyidae (p. 218), which is more fragile, less rostrate, and lacks a lithodesma.

*Cardiomya striata*

*Cuspidaria obesa*

*Myonera gigantea*

*Myonera lamellifera*

LV

*Myonera paucistriata*

*Plectodon granulatus*

# Family Lucinidae – Lucine Clams

### Classification
AUTOLAMELLIBRANCHIATA Grobben, 1894
HETEROCONCHIA Hertwig, 1895
Heterodonta Neumayr, 1883
Veneroida H. Adams & A. Adams, 1856
Lucinoidea Fleming, 1828
Lucinidae Fleming, 1828

## Featured species
### *Codakia orbicularis* (Linnaeus, 1758) – **Tiger Lucine**

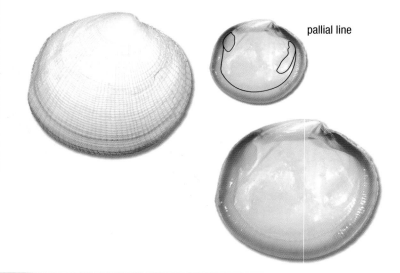

pallial line

Circular to oval, compressed, lunule small and deeply impressed (pitlike), solid, with numerous close-set radial ribs crossed by narrow commarginal ridges (forming beads) and periodic coarser growth lines, white; interior flushed with yellow or pink, margin very finely denticulate. North Carolina to Florida, Bermuda, Bahamas, West Indies, Gulf of Mexico, Caribbean Central America, South America (to Brazil). Length 55 mm (to 80 mm).

*Codakia orbicularis* **lives** in coralline sand in the Florida Keys. This group of juveniles is from Newfound Harbor.

## Family description

The lucinid shell is small to large (to 150 mm), solid, and circular to oval or trapezoidal. It is EQUIVALVE, compressed to inflated, and not gaping. The shell is EQUILATERAL, with PROSO- or OPISTHOGYRATE UMBONES. Shell microstructure is ARAGONITIC and three-layered, with a composite or spherulitic PRISMATIC outer layer, a CROSSED LAMELLAR middle

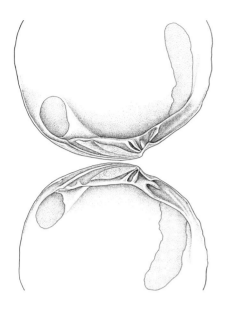

layer, and a COMPLEX CROSSED LAMELLAR inner layer. TUBULES are present, penetrating only the inner shell layer. Exteriorly lucinids are usually whitish, covered by inconspicuous, or in some species scaly and dehiscent, PERIOSTRACUM, which can be partially calcified; *Rastafaria* builds long periostracal "dreadlock"-like tubes that control water flow in and out of the mantle cavity. Sculpture is smooth or with commarginal ridges and (usually weaker) radial, sometimes obliquely oriented, ribs that are rarely beset with small spines. A radial indentation (sulcus) is often present posteriorly, and in some species also anteriorly. The LUNULE is shallow to deep, often small, and asymmetrical; an ESCUTCHEON is absent. Interiorly the shell is non-NACREOUS and sometimes strikingly colored pink, orange, or yellow. The PALLIAL LINE is ENTIRE; the pallial area is often granular or punctate. The inner shell margins are smooth to weakly denticulate. A deep impression on the internal shell surface running diagonally from the heart to the ventral tip of the anterior adductor muscle is the result of a large pallial blood vessel at that site. The HINGE PLATE is distinct but weak, and EDENTATE in adults or HETERODONT, typically with two CARDINAL TEETH per valve (fewer in some species; the posterior tooth often larger and bifid), plus elongated, strong to weak (or absent) anterior and posterior LATERAL TEETH. The LIGAMENT is PARIVINCULAR, OPISTHODETIC, and supported by NYMPHS. A secondary external ligament of fused periostracum unites the valves along the dorsal hinge margin.

The animal is HETEROMYARIAN (posterior ADDUCTOR MUSCLE rounded in cross section; the often much longer anterior adductor narrowly elongated); the ventral tip of the

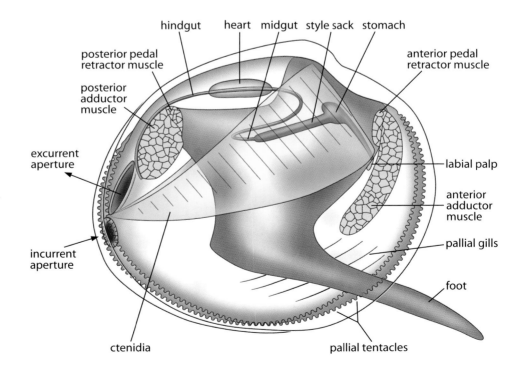

anterior adductor lies within the pallial line. The anterior surface of the anterior adductor muscle is ciliated; this might serve as a supplementary food particle collector. Pedal retractor muscles are small. Pedal elevators and protractors are not reported. The MANTLE margins are not fused ventrally, but are fused posteriorly to form EXCURRENT and (usually) INCURRENT APERTURES (the latter not fused in some); the excurrent aperture is equipped with an eversible tube or "SIPHON" (without siphonal retractor muscle) of variable length and extensibility. The inner mantle surface near the anterior adductor muscle can be thickened by blood spaces or in some species bears a series of ridges, complex folds, or pectinate structures, sometimes set on septa. These PALLIAL GILLS serve as accessory respiratory organs and aid in functionally separating respiratory surfaces from the location of the endosymbionts. HYPOBRANCHIAL GLANDS have not been reported. The FOOT is long, cylindrical, heeled, and highly extensible (to more than 10 times the shell length); a BYSSAL GROOVE and BYSSUS are absent in the adult. The foot forms an incurrent tube in the sediment; the differentiated tip includes a mucus gland that secretes a lining for the tube.

The LABIAL PALPS are small folds at the edge of the lips. The CTENIDIA are large, EU-LAMELLIBRANCH (SYNAPTORHABDIC), HOMORHABDIC, and unusually thickened (by fusion of the ascending and descending lamellae), harboring endosymbiotic sulfide-oxidizing bacteria within bacteriocytes in the filamental tissues in all species so far examined; the outer demibranchs are absent. The gills are not inserted into (or fused with) the distal oral groove of the palps (CATEGORY III association). They are united posteriorly to each other and to the mantle edge between the incurrent and excurrent apertures, separating INFRA- and SUPRABRANCHIAL CHAMBERS. The tissues of living individuals can be highly colored due to intracellular hemoglobin present in the blood, which is likely associated with the sulfide-oxidizing symbiosis. Incurrent and excurrent water flows are posterior; incurrent flow is also anterior. The STOMACH is a highly simplified TYPE IV; the MIDGUT is not or only weakly coiled. Lateral extensions of the visceral mass, containing part of the reproductive organs, can form simple domelike external pouches (VISCERAL LOBES). The HINDGUT is extremely narrow in cross section, passes through the ventricle of the heart, and leads to a sessile rectum. Lucinids are GONOCHORISTIC and produce planktonic VELIGER larvae. *Codakia orbicularis* is known to be a facultative planktotroph with an unusual two-step metamorphosis including long planktonic and benthic stages. The nervous system is not concentrated. STATOCYSTS (with STATOLITHS) have been reported in adult *Loripes*. ABDOMINAL SENSE ORGANS are absent.

Lucinids are marine or estuarine SUSPENSION FEEDERS that have recently been shown to live in a likely obligate chemosymbiosis with sulfide-oxidizing (and some perhaps with methane-oxidizing) endosymbiotic bacteria in the gills. They are typical inhabitants of reduced sediments where other bivalves are scarce, and live deeply INFAUNAL in fine to coarse soft sediments with high sulfide content (e.g., anoxic sediments, mangrove swamps, and sewage outfalls) to support the symbionts (which, where known, are environmentally rather than parentally transmitted). Many are common and numerically dominant bivalves in shallow seagrass beds, in unvegetated sands of reef habitats, or in organic-rich sediments in mangrove areas. Other species are associated with cold seeps and hydrothermal vents—again, environments with simultaneous access to both oxygen and hydrogen sulfide. The mucus-lined incurrent tube allows continuous suspension feeding in the absence of long siphons while remaining deeply buried.

The family Lucinidae is known since the Silurian and (by a recent estimate including undescribed taxa) is represented by 90 living genera and up to 500 species, distributed worldwide in depths from the intertidal zone to over 7,200 m.

**Bacteriocytes** packed with bacteria (in green) fill the gill tissues of the Corrugate Lucine (*Austriella corrugata* (Deshayes, 1843)), from Dampier, Western Australia.

### References

Allen, J. A. 1958. On the basic form and adaptations to habitat in the Lucinacea (Eulamellibranchia). *Philosophical Transactions of the Royal Society of London, Series B, Biological Sciences*, 241(684): 421–484.

Allen, J. A. 1960. The ligament of the Lucinacea (Eulamellibranchia). *Quarterly Journal of Microscopical Science*, 101(1): 25–36.

Boss, K. J. 1969. Lucinacea and their heterodont affinities (Bivalvia). *The Nautilus*, 82(4): 128–131.

Bouchet, P., and R. von Cosel. 2004. The world's largest lucinid is an undescribed species from Taiwan (Mollusca: Bivalvia). *Zoological Studies*, 43(4): 704–711.

Bretsky, S. S. 1976. Evolution and classification of the Lucinidae (Mollusca; Bivalvia). *Palaeontographica Americana*, 8(50): 219–337.

Dall, W. H. 1901. Synopsis of the Lucinacea and of the American species. *Proceedings of the United States National Museum*, 23(1237): 779–833.

Distel, D. L., and H. Felbeck. 1987. Endosymbiosis in the lucinid clams *Lucinoma aequizonata*, *Lucinoma annulata* and *Lucina floridana*: a reexamination of the functional morphology of the gills as bacteria-bearing organs. *Marine Biology*, 96: 79–86.

Fisher, M. R., and S. C. Hand. 1984. Chemoautotrophic symbionts in the bivalve *Lucina floridana* from seagrass beds. *Biological Bulletin*, 167(2): 445–459.

Frenkiel, L., and M. Mouëza. 1995. Gill ultrastructure and symbiotic bacteria in *Codakia orbicularis* (Bivalvia, Lucinidae). *Zoomorphology*, 115: 51–61.

Giere, O. 1985. Structure and position of bacterial endosymbionts in the gill filaments of Lucinidae from Bermuda (Mollusca, Bivalvia). *Zoomorphology*, 105(5): 296–301.

Gros, O., L. Frenkiel, and M. Mouëza. 1997. Embryonic, larval, and post-larval development in the symbiotic clam *Codakia orbicularis* (Bivalvia: Lucinidae). *Invertebrate Biology*, 116(2): 86–101.

Gros, O., M. Liberge, and H. Felbeck. 2003. Interspecific infection of aposymbiotic juveniles of *Codakia orbicularis* by various tropical lucinid gill-endosymbionts. *Marine Biology*, 142(1): 57–66.

Hickman, C. S. 1994. The genus *Parvilucina* in the eastern Pacific: making evolutionary sense of a chemosymbiotic species complex. *The Veliger*, 37(1): 43–61.

Narchi, W., and R. C. Farani Assis. 1980. Anatomia funcional de *Lucina pectinata* (Gmelin, 1791) Lucinidae—Bivalvia. *Boletim de Zoologia, Universidade de São Paulo*, 5: 79–110.

Petit, R. E. 2001. A note on *Lucina multilineata* "Tuomey & Holmes" (Bivalvia: Lucinidae). *The Nautilus*, 115(1): 35–36.

Reid, R. G. B. 1990. Evolutionary implications of sulphide-oxidizing symbioses in bivalves. Pages 127–140, in: B. Morton, ed., *The Bivalvia, Proceedings of a Memorial Symposium in Honour of Sir Charles Maurice Yonge (1899–1986), Edinburgh, 1986*. Hong Kong University Press, Hong Kong.

Reid, R. G. B., and D. G. Brand. 1986. Sulfide-oxidizing symbiosis in lucinaceans: implications for bivalve evolution. *The Veliger*, 29(1): 3–24.

Taylor, J. D., and E. A. Glover. 1997. The lucinid bivalve genus *Cardiolucina* (Mollusca, Bivalvia, Lucinidae): systematics, anatomy and relationships. *Bulletin of The Natural History Museum, Zoology Series*, 63(2): 93–122.

Taylor, J. D., and E. A. Glover. 2000. Functional anatomy, chemosymbiosis and evolution of the Lucinidae. Pages 207–225, in: E. M. Harper, J. D. Taylor, and J. A. Crame, eds., *The Evolutionary Biology of the Bivalvia, Geological Society of London, Special Publication*, 177.

Taylor, J. D., and E. A. Glover. 2005. Cryptic diversity of chemosymbiotic bivalves: a systematic revision of worldwide *Anodontia* (Mollusca: Bivalvia: Lucinidae). *Systematics and Biodiversity*, 3(3): 281–338.

Taylor, J. D., and E. A. Glover. 2006. Lucinidae (Bivalvia): the most diverse group of chemosymbiotic molluscs. *Zoological Journal of the Linnean Society*, 148: 421–438.

Williams, S. T., J. D. Taylor, and E. A. Glover. 2004. Molecular phylogeny of the Lucinoidea (Bivalvia): non-monophyly and separate acquisition of bacterial chemosymbiosis. *Journal of Molluscan Studies*, 70(2): 187–202.

### *Anodontia alba* Link, 1807 – **Buttercup Lucine**

Circular, inflated, with shallow radial furrow from umbones to anterior margin, lunule absent, thin-walled, smooth with commarginal growth lines increasingly roughened ventrally, crossed by microscopic radial threads, dull white; interior yellow to orange, margin white with slight radial folds, hinge edentate. North Carolina to Florida, Bermuda, Bahamas, West Indies, Gulf of Mexico, Caribbean Central America, South America (to Venezuela). Length 46 mm (to 57 mm).

### *Anodontia schrammi* (Crosse, 1876) – **Chalky Buttercup Lucine**

Circular, inflated, lunule absent, solid, smooth with coarse commarginal growth lines, chalky white with dark brown periostracum; interior chalky white and pustulose, margin with weak radial folds, hinge edentate. North Carolina to Florida, Bermuda, West Indies, Gulf of Mexico. Length 80 mm (to 104 mm). Formerly known as *philippiana* Reeve, 1850 (Indo-Pacific).

### *Lucina pensylvanica* (Linnaeus, 1758) – **Pennsylvania Lucine**

Circular, inflated, umbones sharply hooked anteriorly, lunule large, impressed and raised at center into thin blade, posterior quarter with strong radial groove that notches posteroventral margin, solid, with delicate commarginal ridges and periodic coarser growth lines, white with yellowish periostracum forming brittle, toothed, recurved commarginal ridges; interior white, margin smooth. Maryland to Florida, Bermuda, Bahamas, West Indies, Gulf of Mexico, Caribbean Central America, South America (Colombia). Length 34 mm (to 51 mm). Formerly in *Linga*.

### *Phacoides pectinata* (Gmelin, 1791) – **Thick Lucine**

Circular, compressed, lunule large, impressed and raised at center into thin blade, posterior quarter with strong radial groove that notches posteroventral margin, solid, with sharp, unequally spaced commarginal ridges and periodic coarser growth lines, white or light orange exteriorly and interiorly; interior margin often apricot and smooth or with weak radial folds. North Carolina to Florida, West Indies, Gulf of Mexico, Caribbean Central America, South America (to Brazil). Length 44 mm (to 58 mm). Formerly in *Lucina*.

### *Stewartia floridana* (Conrad, 1833) – **Florida Lucine**

Circular, compressed, lunule elongated and deeply impressed (pitlike), solid, with rough commarginal growth lines, white exteriorly and interiorly; interior margin smooth or with weak radial folds. Florida Keys, Gulf of Mexico. Length 35 mm (to 36 mm). Formerly in *Pseudomiltha*.

### *Lucinoma filosa* (Stimpson, 1851) – **Northeastern Lucine**

Circular, compressed, lunule elongated and impressed, solid, with sharp, raised, evenly spaced commarginal ridges, white with thin yellowish periostracum; interior white, margin smooth. Eastern Canada to Florida, Gulf of Mexico. Length 25 mm (to 60 mm).

*Anodontia alba*

*Anodontia schrammi*

*Lucina pensylvanica*

*Phacoides pectinata*

*Stewartia floridana*

*Lucinoma filosa*

### *Ctena orbiculata* (Montagu, 1808) – **Dwarf Tiger Lucine**

Obliquely oval, lunule large, elongated and only slightly impressed, solid, cancellate with divaricating radials and periodic coarser growth lines, white to yellowish exteriorly and interiorly (never pink); interior margin weakly denticulate. North Carolina to Florida, Bermuda, Bahamas, West Indies, Gulf of Mexico, Caribbean Central America, South America (to Brazil). Length 11 mm (to 25 mm).

### *Ctena pectinella* (C. B. Adams, 1852) – **Little-comb Lucine**

Circular, slightly produced ventrally, lunule small and impressed, solid, cancellate with divaricating radial ribs crossed by regular raised commarginal threads, white exteriorly and interiorly; interior margin coarsely denticulate. Florida, Bermuda, West Indies, Caribbean Central America, South America (to Uruguay). Length 4 mm (to 9 mm).

### *"Parvilucina" costata* (d'Orbigny, 1845) – **Costate Lucine**

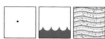

Obliquely oval, inflated, lunule small and impressed, solid, with strong, paired radial ribs crossed by narrow commarginal ridges forming scales, occasionally with periodic coarser growth lines, radial ribs becoming obsolete ventrally, white to yellowish exteriorly and interiorly; interior margin denticulate. North Carolina to Florida, Bermuda, Bahamas, West Indies, Caribbean Central America, South America (to Brazil). Length 9 mm (to 13 mm). Formerly in *Codakia*. Note: A new genus is needed for this species (J. Taylor & E. Glover, pers. comm., August 2005, referring to phylogenetic studies underway at the time of writing).

### *Callucina keenae* Chavan, 1971 – **Dosinia-like Lucine**

Circular, lunule small and impressed, solid, smooth with faint radial threads, white tinted with yellow; interior white with glossy radial rays, margin denticulate. North Carolina to Florida, Bermuda, Bahamas, West Indies, Caribbean Central America, South America (to Venezuela). Length 17 mm. Syn. *radians* Conrad, 1841, non Bory de St. Vincent, 1824.

### *Parvilucina crenella* (Dall, 1901) – **Many-lined Lucine**

Circular to slightly oblique oval, inflated, lunule elongated and impressed, solid, with even network of fine commarginal ridges and radial ribs, white to yellowish exteriorly and interiorly; interior margin finely denticulate. North Carolina to Florida, Bermuda, West Indies, Gulf of Mexico, Caribbean Central America, South America (Colombia, Brazil). Length 5 mm (to 10 mm). Formerly known as "*multilineata* Tuomey Holmes, 1857," an invalid name. Note: The digitiform foot and thick opaque gills characteristic of lucinids are evident in this laboratory photograph of living *Parvilucina crenella* from soft anoxic mud off Big Coppitt Key.

*Ctena orbiculata*

*Ctena pectinella*

*"Parvilucina" costata*

*Parvilucina crenella*

*Callucina keenae*

### *Cavilinga blanda* (Dall, 1901) – **Three-ridged Lucine**

Obliquely oval, inflated, umbones sharply turned anteriorly, lunule small and deeply impressed (pitlike), solid, with fine commarginal ribs and three or four deep commarginal furrows, white, occasionally yellow or salmon; interior chalky white, margin finely denticulate. North Carolina to Florida, West Indies, Gulf of Mexico, Caribbean Central America, South America (Colombia, Brazil). Length 4 mm (to 7 mm). Formerly known as *trisulcata* Conrad, 1841 (Miocene–Pliocene fossil, eastern United States); formerly in *Parvilucina*.

### *Pleurolucina leucocyma* (Dall, 1886) – **Four-ribbed Lucine**

Oval to rounded trigonal, lunule large, broad and only slightly impressed, solid, with four large rounded radial waves crossed by fine commarginal ridges, white exteriorly and interiorly; interior margin finely denticulate. North Carolina to Florida, Bahamas, West Indies, Gulf of Mexico, Caribbean Central America. Length 5 mm (to 7 mm). Formerly in *Linga*.

### *Pleurolucina sombrerensis* (Dall, 1886) – **Sombrero Lucine**

Circular, inflated, lunule small and impressed, solid, with numerous sharp commarginals, occasionally wavy, chalky white exteriorly and interiorly; interior margin denticulate. Florida Keys, West Indies, Gulf of Mexico, Caribbean Central America, South America (Brazil). Length 6 mm. Formerly in *Linga*.

### *Radiolucina amianta* (Dall, 1901) – **Miniature Lucine**

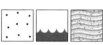

Circular, inflated, lunule small and impressed, solid, with eight or nine large radials crossed by thick, narrow, rounded commarginal ridges forming pits at intersections, often with periodic coarser growth lines, white; interior glossy white, margin denticulate. North Carolina to Florida, West Indies, Gulf of Mexico, Caribbean Central America, South America (to Uruguay). Length 4 mm (to 9 mm). Formerly in *Linga*.

### *Myrtea sagrinata* (Dall, 1886) – **Arrow Lucine**

Circular, compressed, lunule elongated and impressed, thin-walled, with commarginal ridges underlain by radial ribs, white exteriorly and interiorly; interior margin smooth. Florida Keys, West Indies, Gulf of Mexico. Length 6 mm. Note: The valves photographed are of a SYNTYPE specimen of *Myrtea sagrinata*.

### *Myrteopsis lens* (Verrill & Smith, 1880) – **Lens Lucine**

Circular, somewhat truncated posteriorly, compressed, lunule elongated and impressed, thin-walled, smooth centrally with commarginal ridges most evident posteriorly, white exteriorly and interiorly; interior margin smooth. Massachusetts, North Carolina, Florida Keys, South America (Brazil). Length 13 mm (to 18 mm). Formerly in *Myrtea*.

*Cavilinga blanda*

*Pleurolucina leucocyma*

*Pleurolucina sombrerensis*

*Radiolucina amianta*

*Myrtea sagrinata*

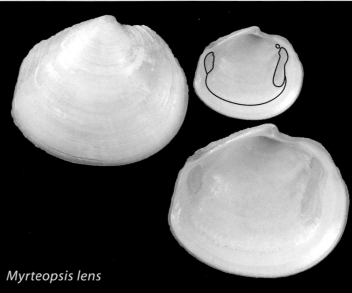

*Myrteopsis lens*

### *Lucinisca nassula* (Conrad, 1846) – **Woven Lucine**

Circular, compressed, lunule large, impressed and raised at center into thin blade, strongly cancellate with numerous equal radial ribs (more or less equal in strength on posterior slope as on main body of shell) adorned with small, broad, flat scales, white exteriorly and interiorly; interior margin strongly denticulate. North Carolina to Florida, Bahamas, Gulf of Mexico, Caribbean Central America, South America (Colombia). Length 13 mm.

### *Lucinisca muricata* (Spengler, 1798) – **Spinose Lucine**

Circular, compressed, lunule small and impressed, solid, with numerous unequal (sometimes alternating large and small) radial ribs (fewer and coarser in strength on posterior slope than on main body of shell) adorned with small, broad, arched (underside hollow) scales, white exteriorly and interiorly; interior margin smooth to weakly radially folds to denticulate. Florida Keys, West Indies, Caribbean Central America, South America (to Brazil). Length 15 mm (to 19 mm).

### *Divalinga quadrisulcata* (d'Orbigny, 1845) – **Cross-hatched Lucine**

Circular, inflated, lunule small and deeply impressed (pitlike), thin-shelled, with incised oblique grooves forming a radial row of chevrons on anterior third, with periodic coarser growth lines, glossy white exteriorly and interiorly; interior margin minutely denticulate, anterior muscle scar football-shaped. Massachusetts to Florida, Bermuda, Bahamas, West Indies, Gulf of Mexico, Caribbean Central America, South America (to Brazil). Length 18 mm (to 24 mm). Syn. *americana* C. B. Adams, 1852. Compare *Divaricella dentata*, which is generally larger, with a straighter and dentate dorsal hinge line.

### *Divaricella dentata* (Wood, 1815) – **Dentate Lucine**

Circular, inflated, lunule small and deeply impressed (pitlike), thin-shelled, with incised oblique grooves forming a radial row of chevrons on anterior third, with periodic coarser growth lines, ventral margin and dorsal hinge line dentate, white, often with yellowish tint, exteriorly and interiorly; interior margin smooth, anterior muscle scar cucumber-shaped. North Carolina to Florida, Bermuda, Bahamas, West Indies, Caribbean Central America. Length 30 mm (to 37 mm). Compare *Divalinga quadrisulcata*, which is generally smaller, with a more rounded and untoothed dorsal hinge line.

**Mangrove thickets**, such as this one in Lake Surprise (a landlocked tidal basin on Key Largo), serve as important high-nutrient nursery environments for a wide variety of marine organisms. The sulfide-rich anoxic sediment among the prop roots is a typical habitat for lucinid clams.

*Lucinisca nassula*

*Lucinisca muricata*

*Divalinga quadrisulcata*

*Divaricella dentata*

# Family Ungulinidae – Diplodon Clams

**Classification**
AUTOLAMELLIBRANCHIATA Grobben, 1894
HETEROCONCHIA Hertwig, 1895
Heterodonta Neumayr, 1883
Veneroida H. Adams & A. Adams, 1856
Incertae sedis
Ungulinidae H. Adams & A. Adams, 1857

**Featured species**
*Diplodonta punctata* (Say, 1822) – **Atlantic Diplodon**

Orbicular to slightly obliquely oval, smooth near umbones, elsewhere with numerous fine, incised commarginal lines and microscopic radial rows of pustules, white exteriorly and interiorly. North Carolina to Florida, Bermuda, Bahamas, West Indies, Gulf of Mexico, Caribbean Central America, South America (Colombia, Venezuela, Brazil, also off Chile). Length 15 mm (to 19 mm). Note: The right valve of this specimen has been drilled by a predatory gastropod.

# Family description

The ungulinid shell is small to medium-sized (to at least 30 mm), thin-walled, often fragile, and circular to oval, frequently oblique, or oblong trigonal; nestling species can be irregular in outline. It is EQUIVALVE, usually inflated, and not gaping. The shell is EQUILATERAL or slightly INEQUILATERAL (umbones anterior), with OPISTHOGYRATE UMBONES. Shell microstructure is ARAGONITIC and three-layered, with a composite PRISMATIC outer layer, a CROSSED LAMELLAR middle layer, and a COMPLEX CROSSED LAMELLAR inner layer. TUBULES have not been reported. Exteriorly ungulinids are usually whitish, covered by a thin, inconspicuous, dehiscent PERIOSTRACUM. Sculpture is smooth or with low commarginal striae, in some species minutely punctate or granular. LUNULE and ESCUTCHEON are ab-

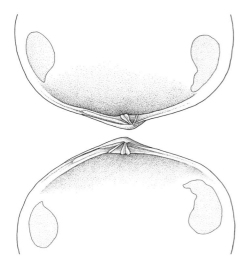

sent. Interiorly the shell is non-NACREOUS. The PALLIAL LINE is ENTIRE. The inner shell margins are smooth. The HINGE PLATE is weak and HETERODONT, with two CARDINAL TEETH per valve (the left anterior and right posterior ones bifid = "diplodont"), plus weak or absent anterior and posterior LATERAL TEETH. The LIGAMENT is PARIVINCULAR and OPISTHODETIC, supported by sunken NYMPHS. A secondary external ligament of fused periostracum unites the valves along the posterodorsal hinge margin only.

The animal is HETEROMYARIAN (posterior ADDUCTOR MUSCLE large and rounded in cross section; anterior adductor narrowly elongated); the ventral tip of the anterior adductor is coextensive with the pallial line. Pedal retractor muscles are present, but pedal elevator and protractor muscles have not been reported. The MANTLE margins are not fused ventrally, but are fused posteriorly to form a posterodorsal EXCURRENT and (smaller) posteroventral aperture; SIPHONS are absent. HYPOBRANCHIAL GLANDS have not been reported. The FOOT is elongated, somewhat laterally compressed, highly extensible (to three times the shell length), unheeled, and with a terminal bulb; a BYSSUS and BYSSAL GROOVE are absent. The terminal bulb of the foot forms an incurrent tube in the sediment; the tip includes a mucus gland that secretes a lining for the incurrent tube and in some species builds nests of agglutinated sand and detritus.

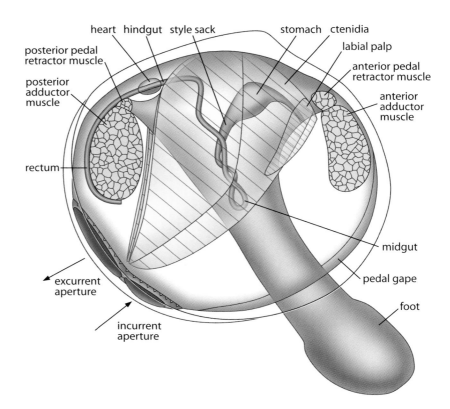

The LABIAL PALPS are moderate to large and trigonal. The CTENIDIA are EULAMELLI-BRANCH (SYNAPTORHABDIC) and HOMORHABDIC. The demibranchs are thin (in contrast to Lucinidae and Thyasiridae); the outer demibranchs are smaller than the inner demibranchs with a characteristic angular margin. The gills are not inserted into (or fused with) the distal oral groove of the palps (CATEGORY III association). Incurrent and excurrent water flows are posterior; incurrent flow is also anterior. The STOMACH is TYPE V; the MIDGUT is weakly coiled. Lateral extensions of the visceral mass, containing part of the reproductive organs, can form simple domelike pouches. The HINDGUT passes through the ventricle of the heart, and leads to a sessile rectum. Ungulinids are GONOCHORISTIC and produce planktonic VELIGER larvae. The nervous system is not concentrated. STATOCYSTS and ABDOMINAL SENSE ORGANS have not been reported.

Ungulinids are marine or estuarine SUSPENSION FEEDERS that are shallowly to deeply INFAUNAL in soft sediments or gravel; none are known to harbor chemosymbiotic bacteria. Some species form nests of agglutinated sand and mucus around themselves.

The family Ungulinidae is known since the Cretaceous and is represented by 12 living genera and ca. 50 species, distributed worldwide in temperate to tropical zones (rarely in cooler waters).

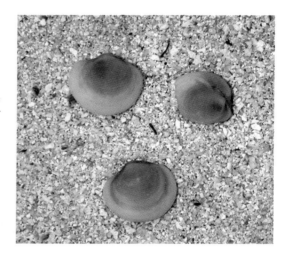

*Phlyctiderma semiaspera* lives in clean coralline sand. These specimens are from seagrass beds off Newfound Harbor, Lower Florida Keys.

**References**

Allen, J. A. 1958. On the basic form and adaptations to habitat in the Lucinacea (Eulamellibranchia). *Philosophical Transactions of the Royal Society of London, Series B, Biological Sciences,* 241(684): 421–484.

Allen, J. A. 1960. The ligament of the Lucinacea (Eulamellibranchia). *Quarterly Journal of Microscopical Science,* 101(1): 25–36.

Dall, W. H. 1899. Synopsis of the American species of the family Diplodontidae. *Journal of Conchology,* 9(8): 244–246.

Domaneschi, O. 1982 ("1979"). Aspectos da biologia de *Diplodonta punctata* (Say, 1822) (Bivalvia—Lucinacea—Ungulinidae). *Revista Nordestina de Biologia,* 2(1–2): 21–25.

Duvernoy, G. L. 1842. Mémoire sur l'animal de l'onguline couleur de laque (*Ungulina rubra* Daud.) et sur les rapports de ce Mollusque acéphale. *Annales des Sciences Naturelles, Partie Zoologique, Série 2,* 18: 110–122, pl. 5B.

Kilburn R. N. 1996. The family Ungulinidae in southern Africa and Mozambique (Mollusca: Bivalvia: Lucinoidea). *Annals of the Natal Museum,* 37: 267–286.

Mittre, H. 1850. Notice sur les genres *Diplodonta* et *Scacchia. Journal de Conchyliologie,* 1: 238–246, pl. 12.

Williams, S. T., J. D. Taylor, and E. A. Glover. 2004. Molecular phylogeny of the Lucinoidea (Bivalvia): non-monophyly and separate acquisition of bacterial chemosymbiosis. *Journal of Molluscan Studies,* 70(2): 187–202.

### *Diplodonta notata* Dall & Simpson, 1901 – **Pitted Diplodon**

Circular to oval, with minute pustules and low commarginal growth lines becoming more crowded ventrally, translucent white exteriorly and interiorly. Florida Keys, Bahamas, West Indies, Gulf of Mexico, Caribbean Central America, South America (Colombia). Length 8 mm (to 11 mm). Note: Also known as Marked Diplodon.

### *Diplodonta nucleiformis* (Wagner, 1840) – **Nut-shaped Diplodon**

Orbicular, smooth and glossy, white exteriorly and interiorly. North Carolina to Florida, Bahamas, West Indies, South America (Colombia, Brazil). Length 8 mm (to 12 mm).

### *Phlyctiderma semiaspera* (Philippi, 1836) – **Pimpled Diplodon**

Orbicular, with microscopically pitted surface, some of which are concentrically arranged, translucent to chalky white exteriorly to interiorly. North Carolina to Florida, Bahamas, West Indies, Gulf of Mexico, Caribbean Central America, South America (to Argentina). Length 9 mm (to 13 mm).

### *Phlyctiderma soror* (C. B. Adams, 1852) – **Sister Diplodon**

Circular to oval, with microscopically pitted surface, white exteriorly and interiorly. North Carolina to Florida, West Indies, Gulf of Mexico, Caribbean Central America, South America (Colombia, Venezuela). Length 12 mm (to 20 mm). Note: The valves photographed are of a SYN-TYPE specimen of *Phlyctiderma soror*.

**Scanning electron micrographs** of the surface sculpture of *Diplodonta punctata* (left), *D. notata* (center), and *Phlyctiderma soror* (right), each at 100 (top) and 500 (bottom) times actual magnification, show the microscopic pustules characteristic of *Diplodonta* and the microscopic pits characteristic of *Phlyctiderma*.

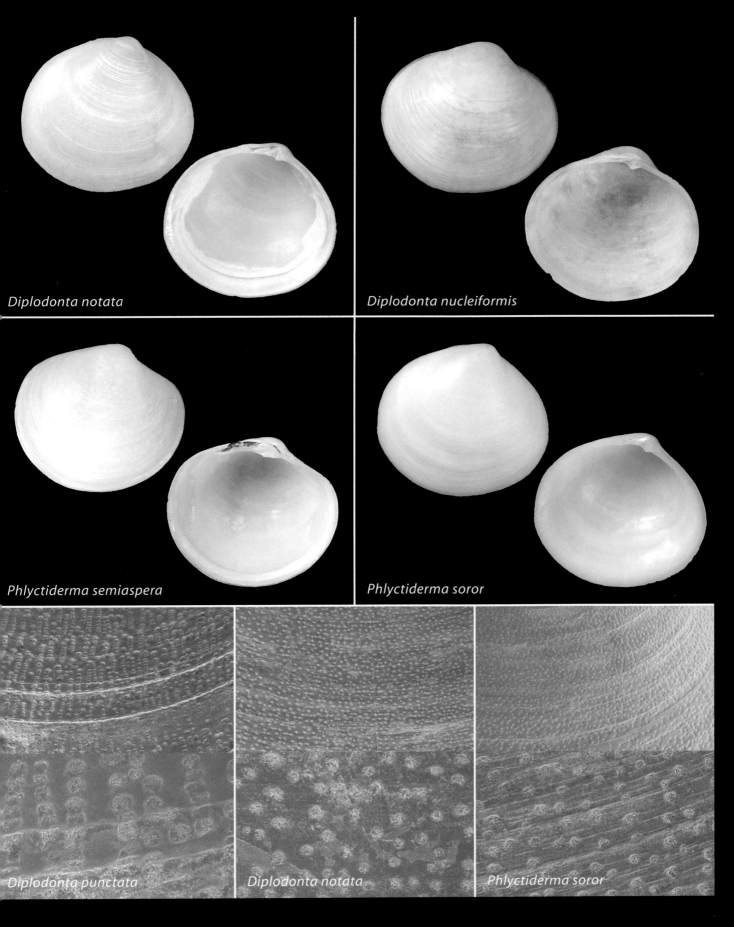

*Diplodonta notata*

*Diplodonta nucleiformis*

*Phlyctiderma semiaspera*

*Phlyctiderma soror*

*Diplodonta punctata*

*Diplodonta notata*

*Phlyctiderma soror*

# Family Thyasiridae – Cleft or Hatchet Clams

**Classification**
AUTOLAMELLIBRANCHIATA Grobben, 1894
HETEROCONCHIA Hertwig, 1895
Heterodonta Neumayr, 1883
Veneroida H. Adams & A. Adams, 1856
Incertae sedis
Thyasiridae Dall, 1900 [1895]

**Featured species**
*Thyasira trisinuata* (d'Orbigny, 1853) – **Atlantic Cleft Clam**

Rounded trigonal, posterior slope with two strong radial waves, with fine commarginal grooves and periodic coarser growth lines, translucent white exteriorly and interiorly; interior with fine radial grooves. Eastern Canada to Florida, West Indies, Gulf of Mexico, South America (Colombia, Brazil). Length 7 mm (to 18 mm).

# Family description

The thyasirid shell is minute to large (to >100 mm; however, *Mendicula verrilli* (Payne & Allen, 1991) matures at 1 mm), thin-walled, often fragile, and oval to trigonal or quadrangular, frequently oblique. It is usually EQUIVALVE, often inflated, and not gaping. The shell is EQUILATERAL, with PROSOGYRATE UMBONES. Shell microstructure is ARAGONITIC and three-layered, with a composite PRISMATIC outer layer, a CROSSED LAMELLAR middle layer, and a COMPLEX CROSSED LAMELLAR inner layer. TUBULES have not been reported. Exteriorly thyasirids are white, sometimes covered by a usually thin, yellowish, inconspicuous PERIOSTRACUM. Most species have ferruginous deposits around the incurrent and excurrent apertures or coating larger parts of the shell. Sculpture is smooth or with commarginal lirae; some are granulose. One or more posterior radial indentations (sulci) are characteristic of larger members of the family, giving rise to the name "cleft clams." The LUNULE is well defined or obscure; the ESCUTCHEON is variably expressed, from absent to

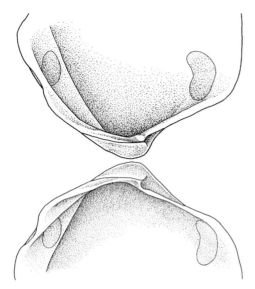

deep. Interiorly the shell is non-NACREOUS. The PALLIAL LINE is ENTIRE; the pallial area is often punctate. A pallial blood vessel scar, extending from the lunule margin to the posterior ventral margin, often is visible. The inner shell margins are smooth. The HINGE PLATE is weak, EDENTATE in adults or with poorly defined subumbonal tubercules. The LIGAMENT is PARIVINCULAR and OPISTHODETIC, set in deep NYMPHS. A secondary external ligament of fused periostracum unites the valves along the dorsal hinge margin.

The animal is HETEROMYARIAN (posterior ADDUCTOR MUSCLE rounded or elongated in cross section; the larger anterior adductor often narrowly elongated); the ventral tip of the anterior adductor is confluent with the pallial line. Pedal retractor muscles are small. Pedal elevator and protractor muscles have not been reported. The MANTLE margins are not fused ventrally, but are fused posteriorly to form an EXCURRENT APERTURE. The anteroventral mantle margins can be thickened and glandular. HYPOBRANCHIAL GLANDS have not been reported. The FOOT is vermiform, extremely extensible (to 15–30 times the shell length), with small or no heel, and with a terminal bulb; a BYSSUS and BYSSAL GROOVE are absent. The terminal bulb of the foot forms an incurrent tube in the sediment; the tip includes a mucus gland that secretes a lining for the incurrent tube.

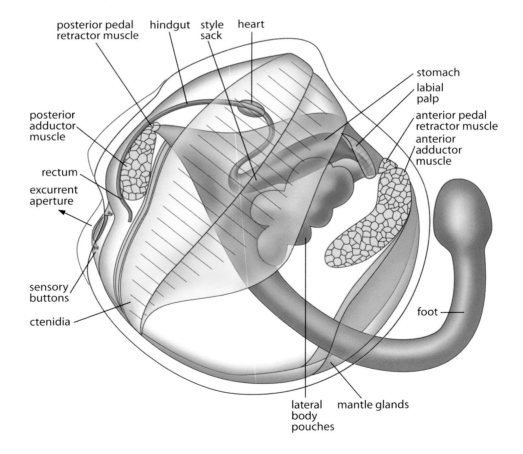

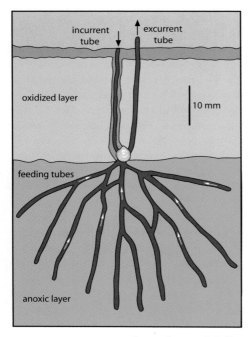

incurrent tube

excurrent tube

oxidized layer

10 mm

feeding tubes

anoxic layer

**A typical thyasirid builds** a three-dimensional burrow in mud or muddy sand with its extensible foot that can reach up to 30 times the length of the shell. The shell lies with the umbones uppermost inside the small living chamber, with incurrent and often also excurrent tubes to the surface and ventral feeding tubes into the sulfide-rich sediments. Water entering from the ventral tubes brings necessary nutrients to bacteria in the gills.

The LABIAL PALPS are small. The CTENIDIA are PSEUDO-LAMELLIBRANCH or EULAMELLIBRANCH (SYNAPTORHABDIC), HOMORHABDIC, and thickened. Both demibranchs are usually present, although the outer one can be much reduced in size or absent in small-bodied forms through PEDOMORPHOSIS. Symbiotic sulfur- or methane-oxidizing bacteria within the filamental tissues have been reported for some, but not all, species of this family. This association is different in thyasirids than for other bivalves in several respects, including: (1) the bacteria are packed together within large bacteriocytes (not in separate vesicles as is typical for Lucinidae); (2) most are extracellular, although two endosymbiotic cases (the condition in other bivalves) are known; and (3) the bacteria are members of several different phylogenetic groups. The gills are not inserted into (or fused with) the distal oral groove of the palps (CATEGORY III association). Incurrent and excurrent water flows are posterior; incurrent flow is also anterior. The anterior incurrent aperture is formed by apposition of the foot and anterior mantle margins. The STOMACH is a simplified TYPE IV; the digestive diverticula and the greater part of the gonad protrude outside the visceral mass as grapelike lateral body pouches; the MIDGUT is not coiled. The HINDGUT passes through the ventricle of the heart, and leads to a freely hanging rectum. Most thyasirids are GONOCHORISTIC and produce planktonic VELIGER larvae; in *Thyasira gouldii* (Philippi, 1845), fertilization is internal, and large eggs are expelled onto the surface of the mud bottom where they DEVELOP DIRECTLY into crawl-away juveniles. The nervous system is not concentrated. STATOCYSTS and ABDOMINAL SENSE ORGANS have not

---

### *Axinus grandis* (Verrill & Smith, 1885) – **Great Cleft Clam**

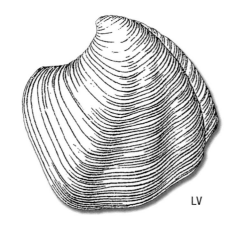

LV

Rounded trigonal, umbones strongly hooked anteriorly, anterodorsal margin upturned, posterior dorsal margin with strong groove from umbo to ventral margin, with fine, closely spaced commarginal grooves forming irregular, raised, wavelike ridges, chalky white exteriorly; interior glossy white with faint radial striae. Virginia, North Carolina New Jersey to Florida Keys, West Indies, Gulf of Mexico, Caribbean Central America, also North Atlantic, Azores, Bay of Biscay, Cape Verde Islands. Length 21 mm.

been reported; however, two pairs of low knobs (sensory buttons) of uncertain homology with the latter lie near the excurrent aperture.

Thyasirids are marine SUSPENSION FEEDERS, deeply INFAUNAL in sandy and muddy sediment. Symbiotic sulfide- or methane-oxidizing chemosymbiotic bacteria in the gills allow some species to thrive in sulfide-rich sediments (even diesel-contaminated habitats) as well as in methane-seep and hydrothermal sites. The digestive system is simplified, perhaps in response to partial dependence on the products of bacterial symbiosis. Diatoms have been found in the digestive systems of some species.

The family Thyasiridae is known since the Cretaceous and is represented by 12 living genera and ca. 150 species, distributed mainly in boreal and antiboreal, colder waters, and ranging from sublittoral to hadal depths (with some exceeding depths of 10,000 m in the Kermadec and Tonga trenches), with greatest diversity at slope depths.

## References

Allen, J. A. 1958. On the basic form and adaptations to habitat in the Lucinacea (Eulamellibranchia). *Philosophical Transactions of the Royal Society of London, Series B, Biological Sciences,* 241(684): 421–484.

Allen, J. A. 1960. The ligament of the Lucinacea (Eulamellibranchia). *Quarterly Journal of Microscopical Science,* 101(1): 25–36.

Bernard, F. R. 1972. The genus *Thyasira* in western Canada (Bivalvia: Lucinacea). *Malacologia,* 11(2): 365–389.

Blacknell, W. N., and A. D. Ansell. 1974. The direct development of bivalve *Thyasira gouldi* (Philippi). *Thalassia Jugoslavica,* 10(1–2): 23–43.

Dando, P. R., and A. J. Southward. 1986. Chemautotrophy in bivalve molluscs of the genus *Thyasira. Journal of the Marine Biological Association of the United Kingdom,* 66(4): 915–929.

Dufour, S. C. 2005. Gill anatomy and the evolution of symbiosis in the bivalve family Thyasiridae. *The Biological Bulletin,* 208(3): 200–212.

Dufour, S. C., and H. Felbeck. 2003. Sulphide mining by the superextensile foot of symbiotic thyasirid bivalves. *Nature,* 426(6): 65–67.

Jones, G. F., and B. E. Thompson. 1986. The ecology of *Adontorhina cyclia* Berry (1947) (Bivalvia: Thyasiridae) on the southern California borderland. *Internationale Revue der Gesamten Hydrobiologie,* 71: 687–700.

Kamenev, G. M., V. A. Nadtochy, and A. P. Kuznetsov. 2001. *Conchocele bisecta* (Conrad, 1849) (Bivalvia: Thyasiridae) from cold-water methane-rich areas of the Sea of Okhotsk. *The Veliger,* 44(1): 84–94.

Miloslavskaya, N. M. 1977. Mollusks of the family Thyasiridae (Bivalvia, Lucinoidea) of the Arctic seas of the USSR. *Issledovaniia Fauny Morei, Akademiia Nauk SSSR, Zoologicheskii Institut,* 14(22): 391–417 [In Russian with English abstract.]

Nakazima, M. 1958. Notes on gross anatomy of *Cochocele disjuncta. Venus,* 20(2): 186–197.

Oliver, P. G., and I. J. Killeen. 2002. The Thyasiridae (Mollusca, Bivalvia) of the British continental shelf and North Sea oilfields. An identification manual. *Studies in Marine Biodiversity and Systematics from the National Museum of Wales. BIOMÔR Reports,* 3: vi + 73 pp.

Oliver, P. G., and J. Sellanes. 2005. New species of Thyasiridae from a methane seepage area off Concepción, Chile. *Zootaxa,* 1092: 1–20.

Payne, C. M., and J. A. Allen. 1991. The morphology of deep-sea Thyasiridae (Mollusca: Bivalvia) from the Atlantic Ocean. *Philosophical Transactions of the Royal Society of London, Series B, Biological Sciences,* 334(1272): 481–562.

Slack-Smith, S. M. 1994. *Axinopsida serricata* (Carpenter, 1864), its burrowing behavior and the functional anatomy of its pallial organs (Mollusca: Thyasiridae). *The Veliger,* 37(1): 30–35.

Southward, E. C. 1986. Gill symbionts in thyasirids and other bivalve molluscs. *Journal of the Marine Biological Association of the United Kingdom,* 66(4): 889–914.

Williams, S. T., J. D. Taylor, and E. A. Glover. 2004. Molecular phylogeny of the Lucinoidea (Bivalvia): non-monophyly and separate acquisition of bacterial chemosymbiosis. *Journal of Molluscan Studies,* 70(2): 187–202.

# Family Chamidae – Jewelboxes or Rock Oysters

Classification
AUTOLAMELLIBRANCHIATA Grobben, 1894
HETEROCONCHIA Hertwig, 1895
Heterodonta Neumayr, 1883
Veneroida H. Adams & A. Adams, 1856
Chamoidea Lamarck, 1809
Chamidae Lamarck, 1809

## Featured species
***Chama macerophylla*** Gmelin, 1791 – **Leafy Jewelbox**

Circular to irregularly oval, with frilly commarginal ridges and spines that are radially striated, white to yellow to purple to reddish or in combination, umbones occasionally differing in color; interior white occasionally suffused with brownish purple, margin denticulate. North Carolina to Florida, Bermuda, Bahamas, West Indies, Gulf of Mexico, Caribbean Central America, South America (to Brazil); also introduced to Guam and the Hawaiian Islands. Length 60 mm (to 90 mm).

***Chama macerophylla* lives** in gregarious clumps, like these specimens from subtidal rocks on the bayside of West Summerland Key.

# Family description

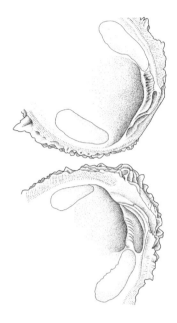

The chamid shell is small to medium-sized (to 115 mm), solid, irregularly rounded, and cemented by the left or right valve by a large attachment area (which determines final shape) or secondarily detached during ontogeny (in *Arcinella*). Recent studies indicate that some species can cement by either valve at an early stage. The shell is strongly IN-EQUIVALVE (lower attached valve deeply cupped with a deep UMBONAL CAVITY; free valve smaller and flatter), inflated, and not gaping. It is EQUILATERAL (in *Arcinella*) or more often INEQUILATERAL, with PROSO-GYRATE UMBONES separated by a wide CARDINAL AREA, and that spirally wind away from the margin with growth and ultimately become submerged in the shell. Shell microstructure is primarily ARAGONITIC and two- or three-layered, with a CROSSED LAMELLAR outer layer and a COM-PLEX CROSSED LAMELLAR inner layer, occasionally with an additional outer CALCITIC layer or patches. TUBULES are present, usually penetrating only the inner shell layer. Exteriorly (and often interiorly) chamids can be highly colored in shades of red, pink, purple, yellow, brown, or green, covered by an inconspicuous PERIOSTRACUM. Sculpture is prominent (although frequently eroded in shallow-water species or due to encrusting organisms), usually commarginal, radial, or both, with elon-

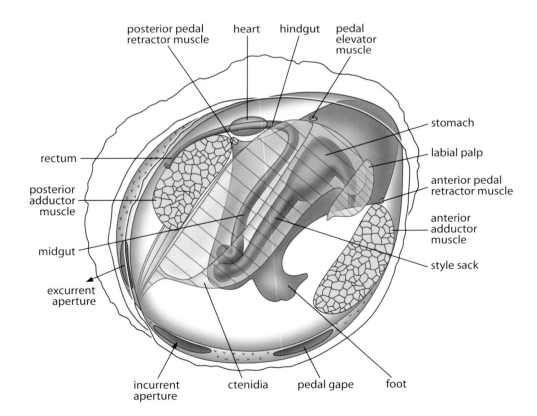

gated (sometimes HYOTE) spines or frills, often different on each valve. A LUNULE is present in free-living *Arcinella*; the ESCUTCHEON is poorly defined. Interiorly the shell is non-NACREOUS. A defined shell margin (COMMISSURAL SHELF) is flat and either smooth, denticulate, or pustulose. The PALLIAL LINE is ENTIRE. The HINGE PLATE is strong and arched, HETERODONT in juveniles, and degenerating with growth into one or two large, curved, grooved "cardinal" teeth in each valve and corresponding sockets in the opposite valve; poorly defined "LATERAL" TEETH are also present in some species. The LIGAMENT is PARIVINCULAR, OPISTHODETIC, set in heavy NYMPHS, and split (thus losing function) anteriorly due to growth that progressively separates the two umbones.

The animal is ISOMYARIAN or slightly HETEROMYARIAN (posterior ADDUCTOR MUSCLE smaller), with adductors that are large and elongated or kidney-shaped in cross section; each adductor muscle scar is elevated off the inner shell surface. Pedal retractor and elevator muscles are small. Pedal protractor muscles have not been reported. The MANTLE margins are extensively fused ventrally, with a small pedal gape and posteroventral EXCURRENT and INCURRENT APERTURES that can bear short invaginable siphons (the latter distinctly muscular only in *Arcinella*); the inner mantle fold is glandular around the pedal aperture in some species. HYPOBRANCHIAL GLANDS have not been reported. The FOOT is small, laterally compressed, and very extensible, apparently functioning in cleaning the gill surfaces and other organs of the infrabranchial chamber; a BYSSUS and BYSSAL GROOVE are absent.

The LABIAL PALPS are small to medium-sized (large in *Arcinella*) and, like the gills and pedal retractors, bilaterally asymmetrical in strongly inequivalve *Chama* species. The CTENIDIA are EULAMELLIBRANCH (SYNAPTORHABDIC) and HETERORHABDIC; the outer demibranchs are smaller than the inner demibranchs. The gills are united posteriorly to form a diaphragm separating INFRA- and SUPRABRANCHIAL CHAMBERS. They are inserted into and fused with the distal oral groove of the palps (CATEGORY II association). Incurrent and excurrent water flows are posterior. The STOMACH is TYPE IV or V; the MIDGUT is weakly coiled. The HINDGUT passes through the ventricle of the heart, and leads to a sessile rectum. Chamids are GONOCHORISTIC or PROTANDRIC HERMAPHRODITES and produce planktonic VELIGER larvae. The nervous system is not concentrated. STATOCYSTS are present in adults. ABDOMINAL SENSE ORGANS have not been reported.

Chamids are marine SUSPENSION FEEDERS, EPIFAUNAL on hard substrata (e.g., rocks, coral, and shells). Most species are stenohaline and intolerant of turbid conditions. They often are heavily encrusted with a diverse assemblage of fouling organisms (e.g., sponges, polychaetes, and tunicates) and/or penetrated by borers.

The family Chamidae is known since the Cretaceous and is represented by 6 living genera and ca. 70 species, distributed in shallow temperate or tropical seas. The common genus *Pseudochama* has recently been synonymized with *Chama* on the basis that cementation by the right or left valve (and hence apparent valve coiling) can vary, even within species (Campbell et al., 2004). Chamids are most abundant in the sublittoral zone of open coasts and on rock and coral reefs (see also p. 8). In the Florida Keys, chamids are a major component of the encrusting fauna on artificial reefs (shipwrecks). Together with other sessile invertebrate species they are increasingly targeted by "live rock" collectors for the aquarium trade (harvesting of live rock in Florida's federal waters has been illegal since 1997; some supplies are now coming from aquaculture ventures).

## References

Allen, J. A. 1977. On the biology and functional morphology of *Chama gryphoides* Linné (Bivalvia: Chamidae). *Vie et Milieu, Série A, Biologie Marine*, 26(2A): 243–260.

Bayer, F. M. 1943. The Florida species of the family Chamidae. *The Nautilus*, 56(4): 116–124, pls. 12–15.

Bernard, F. R. 1976. Living Chamidae of the eastern Pacific (Bivalvia: Heterodonta). *Natural History Museum of Los Angeles County Contributions in Science*, 278: 1–43.

Campbell, M. R., G. Steiner, L. D. Campbell, and H. Dreyer. 2004. Recent Chamidae (Bivalvia) from the western Atlantic Ocean. In: R. Bieler and P. M. Mikkelsen, eds., *Bivalve Studies in the Florida Keys*, Proceedings of the International Marine Bivalve Workshop, Long Key, Florida, July 2002. *Malacologia*, 46(2): 381–415.

Ferreira, S. C., and S. Z. Xavier. 1981. Notas sobre a ontogenia da família Chamidae (Mollusca—Bivalvia). *Boletim do Museo Nacional Nova Série Geologia*, 38: 1–6, 1 pl.

Grieser, E. 1913. Über die Anatomie von *Chama pellucida* Broderip. *Zoologische Jahrbücher, Supplement 13, Fauna Chilensis*, 4(2): 207–280, pl. 18.

Harper, E. M. 1998. Calcite in chamid bivalves. *Journal of Molluscan Studies*, 64(3): 391–399.

Healy, J. M., K. L. Lamprell, and J. Stanisic. 1993. Description of a new species of *Chama* from the Gulf of Carpentaria with comments on *Pseudochama* Odhner (Mollusca: Bivalvia: Chamidae). *Memoirs of the Queensland Museum*, 33(1): 211–216.

Kennedy, W. J., N. J. Morris, and J. D. Taylor. 1970. The shell structure, mineralogy and relationships of the Chamacea (Bivalvia). *Palaeontology*, 13(3): 379–413, pls. 70–77.

LaBarbera, M., and P. E. Chanley. 1971. Larval and postlarval development of the corrugated jewel box clam *Chama congregata* Conrad (Bivalvia: Chamidae). *Bulletin of Marine Science*, 21(3): 733–744.

Matsukuma, A. 1996. A new genus and four new species of Chamidae (Mollusca, Bivalvia) from the Indo-West Pacific with reference to transposed shells. *Bulletin du Muséum National d'Histoire Naturelle, Section A, Zoologie, Biologie et Écologie Animales, Série 4*, 18(1–2): 23–53.

Odhner, N. H. 1919. Studies on the morphology, the taxonomy and the relations of Recent Chamidae. *Kungliga Svenska Vetenskapsakademiens Handlingar*, 59(3): 1–102, pls. 1–8.

Pilsbry, H. A., and T. L. McGinty. 1938. Review of Florida Chamidae. *The Nautilus*, 51(3): 73–79, pl. 7.

Vance, R. R. 1978. A mutualistic interaction between a sessile marine clam and its epibionts. *Ecology*, 59(4): 679–685.

Yonge, C. M. 1967. Form, habit and evolution in the Chamidae (Bivalvia) with reference to conditions in the rudists (Hippuritacea). *Philosophical Transactions of the Royal Society of London, Series B, Biological Sciences*, 252(775): 49–105.

***Chama congregata*** Conrad, 1833 – **Little Corrugate Jewelbox**

Circular to irregularly oval, free valve with undulating surface, both valves with short, flat, smooth scales in wavy radial rows, whitish stained with reddish purple to brown specklings; interior of free valve stained with purple-red, margin denticulate. North Carolina to Florida, Bermuda, Bahamas, West Indies, Gulf of Mexico, Caribbean Central America, South America (to Brazil). Length 24 mm (to 35 mm).

***Chama florida*** Lamarck, 1819 – **Florida Jewelbox**

Circular to oval, with concentric frills and fluted spines, white with pink umbo, color spiraling toward margin as radial bands; interior white, occasionally stained with pink, margin denticulate. Florida, Bahamas, West Indies, Caribbean Central America, South America (to Brazil). Length 20 mm.

***Chama inezae*** (F. M. Bayer, 1943) – **Alabaster Jewelbox**

Circular to irregularly oval, thin-walled, with very thin, widely spaced, commarginal, erect, ruffled but unridged lamellae on both valves, alabaster white; interior margin smooth. Florida, West Indies, Caribbean Central America. Length 20 mm. Formerly in *Pseudochama*.

***Chama radians*** Lamarck, 1819 – **Atlantic Jewelbox**

Circular to irregularly oval, with low commarginal ridges and weak radial folds, white with brown to orange markings; interior white stained with purple-brown, margin denticulate. North Carolina to Florida, Bermuda, Bahamas, West Indies, Gulf of Mexico, Caribbean Central America, South America (to Brazil), also western Europe. Length 40 mm (to 47 mm). Formerly in *Pseudochama*.

***Chama sarda*** Reeve, 1847 – **Cherry Jewelbox**

Irregularly oval, with commarginal foliations and short spines, white and bright red exteriorly and interiorly, margin denticulate. Florida, Bermuda, Bahamas, West Indies, Gulf of Mexico, Caribbean Central America, South America (Colombia, Brazil). Length 20 mm (to 38 mm).

***Chama sinuosa*** Broderip, 1835 – **Smooth-edged Jewelbox**

Circular to irregularly oval, with large, fluted, smooth (rarely striated) fronds, free valve with central radial depressed band bordered by deep narrow grooves, white; interior white flushed with dull green, margin smooth, pallial line runs into anterior muscle scar (rather than past it, as in most species). Florida, Bermuda, West Indies, Gulf of Mexico, Caribbean Central America, South America (Colombia, Brazil). Length 40 mm (to 75 mm).

*Chama congregata*

*Chama florida*

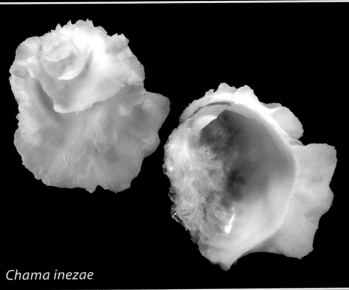

*Chama inezae*

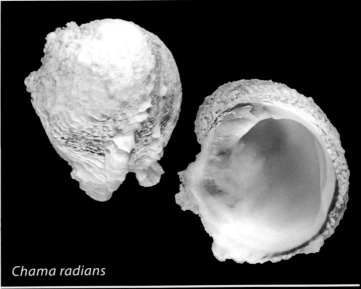

*Chama radians*

*Chama sarda*

*Chama sinuosa*

***Chama lactuca*** Dall, 1886 – **Milky Jewelbox**

Circular to irregularly oval, attached valve with commarginal lamellae, free valve with short, radial and commarginal spinelike fronds, processes of both valves radially grooved, white, occasionally with brown bands; interior white. North Carolina to Florida, West Indies, Gulf of Mexico, South America (Venezuela). Length 20 mm.

***Arcinella cornuta*** Conrad, 1866 – **Florida Spiny Jewelbox**

Quadrangular to obliquely trigonal, equivalve, attached only during early growth stage, thereafter free-living, lunule prominent and bounded by groove, with seven to nine radial rows of pleated radial ribs covered by large spines, with coarse pitting between ribs, creamy white; interior white flushed with pink and/or yellow. North Carolina to Florida, Gulf of Mexico, Caribbean Central America, South America (Venezuela). Length 30 mm (to 40 mm). Note: Inset shows the distinctive juvenile shell of this species.

**Coral reefs** (here at Conch Reef, with adult Yellow Damselfish) are the typical habitat of *Chama* species.

*Chama lactuca*

*Arcinella cornuta*

# Family Lasaeidae – Lasaeid Clams

Classification

AUTOLAMELLIBRANCHIATA Grobben, 1894
HETEROCONCHIA Hertwig, 1895
Heterodonta Neumayr, 1883
Veneroida H. Adams & A. Adams, 1856
Galeommatoidea J. E. Gray, 1840
Lasaeidae J. E. Gray, 1842

---

**Featured species**
*Lasaea adansoni* (Gmelin, 1791) – **Adanson's Lepton**

Oval, inflated, umbones subcentral, smooth with fine radial scratches, and periodic coarser growth lines, dark gray to white or purple-rose with yellow-brown periostracum; interior similarly colored; living nestled among barnacle/turf communities. Florida, Bermuda, Bahamas, Caribbean Central America, South America (to Argentina), also eastern Pacific and western Europe. Length 3 mm. Syn. *rubra* Montagu, 1803. Note: Also known as Reddish Lepton.

---

# Family description

The lasaeid shell is minute to small (to 30 mm), thin-walled, often fragile, and usually rounded to elongated oval. It is EQUIVALVE, compressed to inflated, and not gaping. The shell is EQUILATERAL OR INEQUILATERAL (umbones posterior), with PROSOGYRATE UMBONES. Shell microstructure is ARAGONITIC; the layers of Lasaeidae have not been specifically examined, but those of the closely related Galeommatidae are two-layered, with a CROSSED LAMELLAR outer layer and a COMPLEX CROSSED LAMELLAR inner layer. TUBULES are apparently not present. Exteriorly lasaeids are translucent, covered by a thin to thick PERIOSTRACUM. Sculpture is smooth, rarely with radial riblets, commarginal lirae, or both. LUNULE and ESCUTCHEON are usually absent; however, a small, depressed lunule has been described in some *Montacuta*. Interiorly the shell is non-NACREOUS. The PALLIAL LINE is ENTIRE. The inner shell margins are smooth. The HINGE PLATE is strong and variable, often characteristically indented in an inverted "V" under the umbones, and HETERODONT, with up to two small cardinal tubercles in each valve plus laminar LATERAL TEETH. The

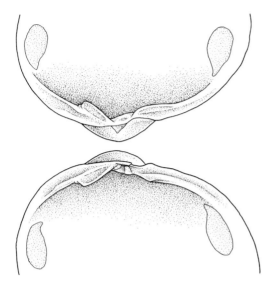

LIGAMENT is PARIVINCULAR, OPISTHODETIC, and set in small NYMPHS; an internal portion (RESILIUM) is also developed in some species, set in a weak RESILIFER.

The animal is ISOMYARIAN or slightly HETEROM-YARIAN (anterior or posterior ADDUCTOR MUSCLE slightly larger); pedal retractor and (occasionally) protractor muscles are present. Pedal elevator muscles are absent. The MANTLE margins are not extensively fused ventrally, and form an extensive pedal gape, an anterior INCURRENT APERTURE (that in some is formed only by temporary appression of the mantle lobes), and a posterior EXCURRENT APERTURE. The inner mantle fold is hypertrophied, and often covers much of the shell exterior (but can usually be completely withdrawn). Sensory tentacles (paired or unpaired) are often present, especially anteriorly or posteriorly. HYPOBRANCHIAL GLANDS have not been reported. The FOOT is large, heeled, laterally compressed, and with a ventral planar sole; a BYSSAL GROOVE and well-developed byssal gland are usually present, although the adults of some species do not produce a BYSSUS.

The LABIAL PALPS are small to medium-sized. The CTENIDIA are EULAMELLIBRANCH (SYNAPTORHABDIC), usually HOMORHABDIC, and generally with few filaments; the outer

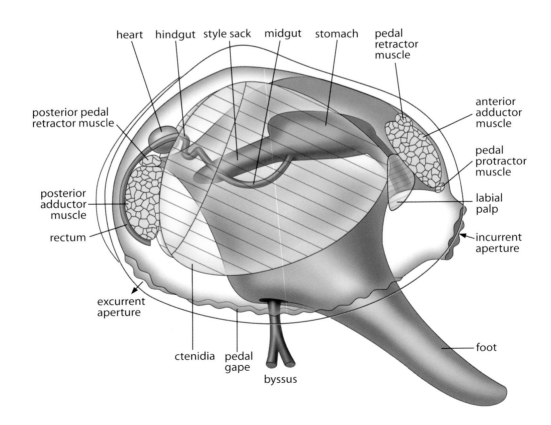

demibranchs can be smaller than the inner demibranchs (e.g., in *Kellia*), can consist only of reflected descending lamellae (e.g., in *Lasaea*), or can be absent (e.g., in *Montacuta*). The gills are not inserted into (or fused with) the distal oral groove of the palps (CATE-GORY III association). Posteriorly the tips of the ctenidia are attached one another and to the mantle, separating INFRA- and SUPRABRANCHIAL CHAMBERS; water flow is anteroposterior. The STOMACH is TYPE IV; the MIDGUT is not coiled. The HINDGUT passes through the ventricle of the heart, and leads to a sessile rectum. Lasaeids are PROTOGYNOUS or SI-MULTANEOUS HERMAPHRODITES; some species produce spermatophores that are taken up by the partner and stored in the suprabranchial chamber of the gills. Most species brood larvae in the suprabranchial chamber, either for a short while (releasing planktonic VELIGERS) or until the crawl-away juvenile stage (DIRECT DEVELOPMENT). Molecular and karyological studies have shown that *Lasaea* also self-fertilizes and reproduces asexually, producing polyploid clonal populations. The nervous system is not concentrated, but has relatively large ganglia. STATOCYSTS (with single STATOLITHS) are present in adults. AB-DOMINAL SENSE ORGANS have not been reported.

Lasaeids are marine SUSPENSION FEEDERS, either free-living in protected environ-ments (e.g., kelp holdfasts, barnacles, algal turf, and rock crevices), or commensal with various INFAUNAL or EPIFAUNAL invertebrates (e.g., crustaceans, chitons, sea urchins, polychaetes), byssally attaching to them and feeding on their mucous secretions or on particles from water currents created by them, while also deriving shelter and protection. Many species actively crawl on the foot-sole, resembling small gastropods.

The family Lasaeidae is known since the Cretaceous and is represented by ca. 70 liv-ing genera and an undetermined number of species, distributed worldwide. Certain fea-tures (e.g., reduction of the outer demibranchs and number of gill filaments, brooding, and creeping foot) have been referred to secondary simplification associated with PEDO-MORPHOSIS (NEOTENY). The distinction and relationships of this family to other proposed galeommatoidean families (e.g., Galeommatidae, Kelliidae, Erycinidae, and Montacuti-dae) are controversial. All recognized Florida Keys species are here treated as belonging to Lasaeidae. Common names within this group are largely vernacularized taxonomic names, often referring to prior generic or familial placement.

**Living lasaeids,** like this *Mysella planulata* from Indian Key Fill, have a mobile probing foot on which most can actively crawl.

# References

Boyko, C. B., and P. M. Mikkelsen. 2002. Anatomy and biology of *Mysella pedroana* (Mollusca: Bivalvia: Galeommatoidea), and its commensal relationship with *Blepharipoda occidentalis* (Crustacea: Anomura: Albuneidae). *Zoologischer Anzeiger*, 241: 149–160.

Narchi, W. 1969. On *Pseudopythina rugifera* (Carpenter, 1864) (Bivalvia). *The Veliger*, 12(1): 43–52.

Ó Foighil, D. 1985. Sperm transfer and storage in the brooding bivalve *Mysella tumida*. *Biological Bulletin*, 169: 602–614.

Ó Foighil, D. 1986. Prodissoconch morphology is environmentally modified in the brooding bivalve *Lasaea subviridis*. *Marine Biology*, 92(4): 517–524.

Ó Foighil, D. 1987. Cytological evidence for self-fertilization in *Lasaea subviridis* (Galeommatacea: Bivalvia). *International Journal of Invertebrate Reproduction and Development*, 12: 83–90.

Ó Foighil, D., and M. J. Smith. 1995. Evolution of asexuality in the cosmopolitan marine clam *Lasaea*. *Evolution*, 49(1): 140–150.

Ó Foighil, D., and M. J. Smith. 1996. Phylogeography of an asexual marine clam complex, *Lasaea*, in the northeastern Pacific based on cytochrome oxidase III sequence variation. *Molecular Phylogenetics and Evolution*, 6(1): 134–142.

Oldfield, E. 1955. Observations on the anatomy and mode of life of *Lasaea rubra* (Montagu) and *Turtonia minuta* (Fabricius). *Proceedings of the Malacological Society of London*, 31: 226–249.

Oldfield, E. 1961. The functional morphology of *Kellia suborbicularis* (Montagu), *Montacuta ferruginosa* (Montagu) and *M. substriata* (Montagu), (Mollusca, Lamellibranchiata). *Proceedings of the Malacological Society of London*, 34(5): 255–295.

Ponder, W. F. 1971. Some New Zealand and subantarctic bivalves of the Cyamiacea and Leptonacea with descriptions of new taxa. *Records of the Dominion Museum*, 7(13): 119–141.

Popham, M. L. 1940. The mantle cavity of some of the Erycinidae, Montacutidae and Galeommatidae with special reference to the ciliary mechanisms. *Journal of the Marine Biological Association of the United Kingdom*, 24(2): 549–587.

### *Erycina periscopiana* Dall, 1899 – **Periscope Erycina**

Elongated oval, compressed, umbones near posterior end, dorsal and ventral margins near-parallel, smooth with commarginal growth lines only, white exteriorly and interiorly. North Carolina, Florida Keys, Bahamas, West Indies, Caribbean Central America, South America (Colombia). Length 4 mm (to 5 mm).

### *Kellia suborbicularis* (Montagu, 1803) – **Suborbicular Kellyclam**

Oval, inflated, umbones subcentral, smooth with prominent, periodic coarse growth lines, translucent white. North Atlantic to North Carolina, Florida Keys, Gulf of Mexico, South America (to Argentina), also eastern Pacific and western Europe. Length 7 mm (to 9 mm).

### *Mysella planulata* (Stimpson, 1851) – **Plate Mysella**

Oval, compressed, umbones closer to posterior end, dorsal margin recessed anterior and posterior of umbones, smooth, translucent white exteriorly and interiorly; living commensally in crustacean and polychaete burrows. North Atlantic to North Carolina, Florida, Gulf of Mexico. Length 2 mm. Note: Placed in family Montacutidae by some authors. See also living specimen on p. 260.

### *Orobitella floridana* (Dall, 1899) – **Florida Lepton**

Oval, compressed, umbones near posterior end, with weak radial striae and commarginal growth lines that strengthen ventrally into well-defined ridges, white exteriorly and interiorly; interior glossy; living commensally in *Onuphis* polychaete tubes. North Carolina, Florida, Bahamas, Gulf of Mexico, Caribbean Central America. Length 10 mm (to 16 mm). Formerly in *Neaeromya*. Note: Also known as Giant Montacutid.

**Members of the allied family Galeommatidae** have yet to be reported from the Florida Keys, but occur elsewhere in South Florida. Yoyo Clams from the Indian River Lagoon, eastern Florida (left to right, *Divariscintilla yoyo* Mikkelsen & Bieler, 1989, *D. luteocrinita* Mikkelsen & Bieler, 1992, and two *D. octotentaculata* Mikkelsen & Bieler, 1992) live commensally in burrows of the stomatopod *Lysiosquilla scabricauda* (Lamarck, 1818). These greatly modified "snail-like" clams have internal shells, sensory tentacles, and a highly extensible crawling foot.

*Erycina periscopiana*

*Kellia suborbicularis*

*Mysella planulata*

*Orobitella floridana*

# Family Hiatellidae – Hiatella Clams and Geoducks

**Classification**
AUTOLAMELLIBRANCHIATA Grobben, 1894
HETEROCONCHIA Hertwig, 1895
Heterodonta Neumayr, 1883
Veneroida H. Adams & A. Adams, 1856
Hiatelloidea J. E. Gray, 1824
Hiatellidae J. E. Gray, 1824

**Featured species**
*Hiatella arctica* (Linnaeus, 1767) – **Arctic Hiatella**

pallial line

LV articulated

Quadrangular but highly irregular, dorsal and ventral margins parallel, umbones anterior, with coarse irregular growth lines and occasionally one or two, occasionally scaled radial ribs on posterior slope, chalky white; interior glossy white. North Atlantic to Argentina, also eastern Pacific, western Europe, Antarctic islands. Length 12 mm (to 45 mm). Note: Recent studies on Brazilian specimens under this name show strong evidence of more than one species, including different spawning times, egg color, postlarval form, radial ribs, presence of adult byssus, position of adductor muscles, and siphonal tip color.

*Hiatella arctica* **maintains** its position in a crevice or its own borehole by expansion of the postvalvular extension of the mantle that forces the valves apart to help grip the burrow walls.

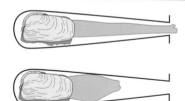

# Family description

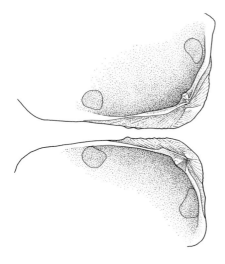

The hiatellid shell is small to large (to ca. 300 mm; i.e., Pacific *Panopea abrupta* (Conrad, 1849), the largest burrowing bivalve, with siphons two to three times its shell length, to nearly 1 m), solid, longer than high, oval to elongated quadrangular to trapezoidal, and frequently irregular and distorted by the constraints of its habitat. Radial rows of spines are present in some individuals. It is roughly EQUIVALVE, inflated, and usually gaping both anteriorly and posteriorly (anteriorly only in *Saxicavella*; posteriorly only in *Cyrtodaria*). The shell is EQUILATERAL or INEQUILATERAL (umbones anterior), with PROSO- or ORTHOGYRATE UMBONES that are frequently touching and eroded. Shell microstructure is ARAGONITIC and two- or three-layered, with a simple PRISMATIC or HOMOGENOUS outer layer (in *Panopea* only), a homogenous middle layer, and a COMPLEX CROSSED LAMELLAR or homogenous inner layer. TUBULES are apparently absent. Exteriorly hiatellids are chalky whitish gray, covered by a thick, yellow to brown to black, dehiscent PERIOSTRACUM, that is coated with adherent sand grains in *Panomya*. The periostracum often spans the mantle edge both dorsally and ventrally and is posteriorly prolonged into a siphonal sheath. Sculpture is smooth or weakly and irregularly commarginal. LUNULE and ESCUTCHEON are absent. Interiorly the shell is white and non-NACREOUS. The PALLIAL LINE is deeply impressed, often discontinuous, usually with a distinct SINUS (absent in *Saxicavella* and *Cyrtodaria*). The inner shell margins are smooth. The HINGE PLATE is strong, irregular, and weakly HETERODONT, with one or two weak denticles and sockets that are usually absent in larger adults; LATERAL TEETH are absent. The LIGAMENT is massive, PARIVINCULAR, OPISTHODETIC, set in well-developed NYMPHS, covered and extended posteriorly by a secondary external portion of thick, fused periostracum; this ligament type permits wide separation of the valves.

The animal is ISOMYARIAN or slightly HETEROMYARIAN (anterior ADDUCTOR MUSCLE smaller in *Hiatella*; posterior adductor smaller in *Cyrtodaria*), with the anterior adductor mus-

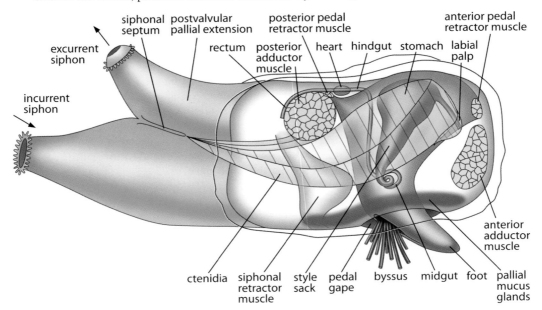

cle placed more ventrally than the posterior adductor; the two adductor muscles can contract simultaneously or alternately. Pedal retractor muscles are present; pedal elevator and protractor muscles are apparently absent. In *Panopea*, *Panomya*, and *Cyrtodaria*, the pallial muscles (ORBITAL MUSCLES) are massive, united ventrally across the valves, and create a muscular floor of the MANTLE CAVITY that acts as a secondary adductor muscle. The MANTLE margins are extensively fused ventrally, with a small anteroventral pedal gape. PALLIAL MUCUS GLANDS (absent in *Saxicavella*) line both sides of the pedal gape. Posteriorly the mantle cavity usually extends past the ends of the shell as a POSTVALVULAR PALLIAL EXTENSION. Posterior EXCURRENT and INCURRENT SIPHONS are usually large, muscular, partially or completely united, and sheathed by periostracum; the shorter excurrent siphon is directed dorsally or laterally and bears a terminal cone (VALVULAR MEMBRANE). The postvalvular extension and siphons can together extend 2–10 times the shell length and are not fully retractile into the shell. In *Hiatella*, they can be brightly colored orange-yellow with white or red tips. *Saxicavella* lacks siphons, siphonal retractor muscles, and the postvalvular pallial extension, and instead has widely separated excurrent and (larger) incurrent apertures with a common ring of tentacles. HYPOBRANCHIAL GLANDS have not been reported. The FOOT is usually small (large in *Cyrtodaria*) and digitiform; it has a BYSSAL GROOVE and the adults of most species are byssate.

The LABIAL PALPS are small to large and elongated. The CTENIDIA are EULAMELLIBRANCH (SYNAPTORHABDIC) and HOMORHABDIC (although plicate in some); the outer demibranchs are smaller than the inner demibranchs. The gills usually extend into the postvalvular extension, and often further into the excurrent siphon, uniting with the SIPHONAL SEPTUM to separate INFRA- and SUPRABRANCHIAL CHAMBERS. The association between the gill and the distal oral groove of the palps has not been recorded. Incurrent and excurrent water flows are posterior. The STOMACH is TYPE IV; the MIDGUT is coiled (not coiled in *Saxicavella*). The HINDGUT passes through the ventricle of the heart, and leads to a sessile or freely hanging rectum. Hiatellids are GONOCHORISTIC and produce planktonic VELIGER larvae. The nervous system is not concentrated, and also has a pair of siphonal ganglia. STATOCYSTS in the adult are present but small (reported for *Panopea*). ABDOMINAL SENSE ORGANS have not been reported.

Hiatellids are marine SUSPENSION FEEDERS; they can be shallow to deep burrowers, nestlers in crevices, or weak borers in soft rock by mechanical abrasion; some species (e.g., British *Hiatella gallicana* (Lamarck, 1818)) either nestle or bore, depending on the type of substratum onto which the larvae settle. Shell shape and foot morphology suggest that *Cyrtodaria* is capable of moving horizontally through the sediment (the only member of the superfamily to do so). They are frequent members of fouling communities and are also considered minor bioeroders in some localities. *Hiatella* maintains its position in a crevice or its own burrow by expanding the postvalvular extension of the body using water pressure in the mantle cavity (rather than hydrostatic pressure in the blood) to force the shell valves apart. Off New England, *Cyrtodaria* is often the most abundant prey item of fish such as cod and haddock; other typical predators on hiatellids include carnivorous snails, sea stars, and crabs. Pea crabs (Pinnotheridae) and flatworms have been found in the mantle cavity of geoducks, presumably in commensal relationship with the clam.

The family Hiatellidae is known since the Permian and is represented by 5 living genera and ca. 25 species, distributed worldwide, including cold and subantarctic waters to deeper than 900 m. *Hiatella* is a genus of worldwide distribution and extreme interspecific variability because of its boring/nestling habit; the group is in need of revision. The best-known member of the family, for its enormous size (more than 9 kg, reported maximum life span of 146 years) and edibility, is the Pacific Geoduck (also known as King Clam or Elephant Trunk Clam), *Panopea abrupta*. Annual commercial and recreational harvest of this species from Washington State and western Canada once reached >5,000 tons in the late 1980s but later declined due to overharvesting, pollution, and development of coastal areas; annual catch

**The Pacific Geoduck** (*Panopea abrupta* (Conrad, 1849)), the largest burrowing clam in the world (up to 30 cm in shell length), is a valuable species in commercial and sport fisheries as well as aquaculture ventures of the Pacific Northwest. Most are harvested individually by divers using a directed water jet. Much of the catch is sold live to Asian markets (as King Clam). Its enormous edible siphons can reach three times the shell length.

---

*Hiatella azaria* (Dall, 1881) – **Dirt Hiatella**

Rounded trigonal but irregular, umbones subcentral, with coarse irregular growth lines, light colored with dark periostracum; interior white. Florida, Gulf of Mexico. Length 25 mm.

---

limits and mariculture projects throughout the region now seek to replenish natural populations as well as sustain the fishery.

### References

Beu, A. G. 1971. New light on the variation and taxonomy of the bivalve *Hiatella*. *New Zealand Journal of Geology and Geophysics*, 14(1): 64–66.

Bower, S. M., and J. Blackbourn. 2003. *Geoduck Clam* (Panopea abrupta): *Anatomy, Histology, Development, Pathology, Parasites and Symbionts*. http://www-sci.pac.dfo-mpo.gc.ca/geoduck/title_e.htm, last accessed 05 January 2006.

Goodwin, C. L., and B. Pease. 1989. *Species Profiles: Life Histories and Environmental Requirements of Coastal Fishes and Invertebrates (Pacific Northwest)—Pacific Geoduck Clam*. U.S. Fish and Wildlife Service, Biological Report 82(11.120). U.S. Army Corps of Engineers, TR El-82-4, 14 pp. Online version http://www.nwrc.usgs.gov/wdb/pub/0125.pdf; last accessed 03 December 2006.

Hunter, W. R. 1949. The structure and behaviour of *Hiatella gallicana* (Lamarck) and *Hiatella arctica* (L.) with special reference to the boring habit. *Proceedings of the Royal Society of Edinburgh, Section B, Biology*, 63: 271–289.

Lindsay, C. E. 1967. The geoduck. *The American Malacological Union, Annual Reports for 1966*, pp. 67–68.

Narchi, W. 1973. On the functional morphology of *Hiatella solida* (Hiatellidae: Bivalvia). *Marine Biology*, 19: 332–337.

Tiba, R. 1988. Revision of the genus *Panomya* in the north-western Pacific with a description of a new species (Bivalvia Hiatellidae). *Bulletin of the Institute of Malacology Tokyo*, 2(6): 93–95, pls. 35–38.

Yonge, C. M. 1971. On functional morphology and adaptive radiation in the bivalve superfamily Saxicavacea (*Hiatella* (= *Saxicava*), *Saxicavella*, *Panomya*, *Panope*, *Cyrtodaria*). *Malacologia*, 11(1): 1–44.

# Family Gastrochaenidae – Chimney or Flask Clams

**Classification**
AUTOLAMELLIBRANCHIATA Grobben, 1894
HETEROCONCHIA Hertwig, 1895
Heterodonta Neumayr, 1883
Veneroida H. Adams & A. Adams, 1856
Gastrochaenoidea J. E. Gray, 1840
Gastrochaenidae J. E. Gray, 1840

## Featured species
***Lamychaena hians*** (Gmelin, 1791) – **Atlantic Chimney Clam**

  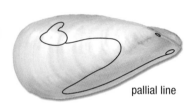

pallial line

Oval to spatulate (anterior narrowed), posterior bluntly rounded, umbones at anterior end, smooth with fine commarginal ridges, whitish to yellowish exteriorly and interiorly. North Carolina to Florida, Bermuda, Bahamas, West Indies, Gulf of Mexico, Caribbean Central America, South America (to Uruguay). Length 25 mm. Formerly in *Gastrochaena*. Compare *Gastrochaena* cf. *ovata*, which is more pointed posteriorly and has its anterior end extending forward of the umbones.

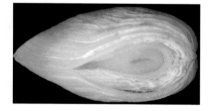

**A ventral view** of a preserved *Lamychaena hians* shows the wide pedal gape.

**The extreme length** of the gastrochaenid siphons is shown in this extracted specimen of *Lamychaena hians* from Molasses Reef. The siphons are sensitive to changes in light intensity, and are fully retractable into the shell.

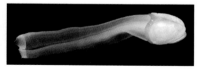

## Family description

The gastrochaenid shell is small to medium-sized (to 50 mm), thin-walled, longer than high, oval to elongated quadrangular with a broadly curving anterior margin that is thickened in some taxa. In some (e.g., *Spengleria*), a strong radial groove divides the shell into anteroventral and posterodorsal sections that can be differently sculptured. The shell is EQUIVALVE, inflated below the umbones, compressed posteriorly, and broadly gaping an-

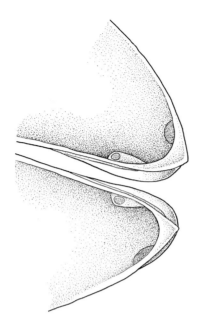

teroventrally or along the entire ventral margin. It is INEQUILATERAL (umbones anterior), with PROSOGYRATE UMBONES. Shell microstructure is ARAGONITIC and three-layered, with a composite PRISMATIC outer layer, a CROSSED LAMELLAR middle layer, and a COMPLEX CROSSED LAMELLAR or HOMOGENOUS inner layer. TUBULES are apparently absent. The mantle-secreted burrow lining is also aragonitic. Exteriorly gastrochaenids are whitish and covered by a thin, adherent PERIOSTRACUM that bears aragonitic spines in *Spengleria* (most evident posteriorly). Sculpture is smooth with commarginal growth lines or reflected ridges, which are stronger anteriorly. LUNULE and ESCUTCHEON are absent. Interiorly the shell is non-NACREOUS and chalky. The PALLIAL LINE is slightly impressed and discontinuous, with a shallow to deep SINUS. The inner shell margins are smooth. The HINGE PLATE is weak and EDENTATE (DESMODONT) in adults, in some with a weak "cardinal" ridge and/or an elongated trigonal MYOPHORE or an irregular tubercle under the hinge plate. The LIGAMENT is PARIVINCULAR, OPISTHODETIC, and set in slightly raised NYMPHS. A secondary external ligament of fused periostracum unites the valves dorsally.

The animal is HETEROMYARIAN (anterior ADDUCTOR MUSCLE smaller) or rarely ISOMYARIAN (*Eufistulana*). In some species, some of the well-defined pallial muscle bundles form distinct accessory adductors. Pedal retractor muscles are present, the anterior of which can be larger than the anterior adductor and insert on the myophore (if present). Pedal protractors are present (bifurcate in *Spengleria*). Pedal elevator muscles are present in *Spengleria*. The siphonal

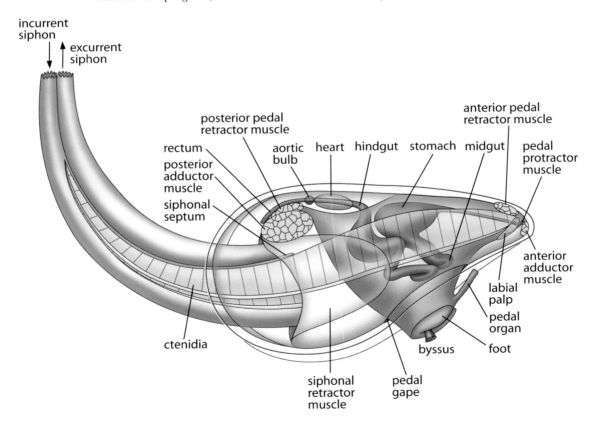

retractor muscles are very long. The MANTLE margins are extensively fused and muscular ventrally, with a small anteroventral pedal gape; the inner and middle mantle folds are reflected over much of the shell valves in *Eufistulina* and *Cucurbitula*. The inner and middle folds are glandular, serving in various aspects of boring, shell formation, and secretion of the calcareous burrow lining. PALLIAL MUCUS GLANDS are absent. Reports of bioluminescence in Japanese *Gastrochaena cuneiformis* Spengler, 1783, from white glandular areas near the periostracal groove of the mantle (also present in other species), are probably in error. Posterior EXCURRENT and INCURRENT SIPHONS are greatly elongated, naked, and united (in *Gastrochaena* and *Eufistulina*) or separate (in *Spengleria*). The incurrent siphon is equipped with a basal PROXIMAL VALVE in some species. Full retraction of the extensive mantle and siphons into the shell is not possible in some taxa. HYPOBRANCHIAL GLANDS have not been reported. The FOOT is small, disk- or cylinder-shaped, with a BYSSAL GROOVE, and usually with an anterior digitiform or papillate PEDAL ORGAN that assists in boring (absent in *Spengleria*); the foot is suctorial, used as an anchor within the burrow, supplementing the minute BYSSUS in the adult (absent in some species).

The LABIAL PALPS are small and trigonal. The CTENIDIA are EULAMELLIBRANCH (SYNAPTORHABDIC) and HOMORHABDIC or rarely HETERORHABDIC (in *Spengleria*); the outer demibranch is usually smaller than the inner demibranch, and both demibranchs can be greatly prolonged into the excurrent siphon. The gills are inserted and fused with the distal oral groove of the palps (CATEGORY II association). Incurrent and excurrent water flows are posterior. The STOMACH is TYPE IV; the MIDGUT is coiled. The HINDGUT passes through the ventricle of the heart, and leads to a sessile rectum. An AORTIC BULB is present. Gastrochaenids are GONOCHORISTIC and produce planktonic VELIGER larvae. The nervous system is not concentrated. STATOCYSTS (with STATOLITHS) are present in adults (reported in *Eufistulana*). ABDOMINAL SENSE ORGANS have not been reported. Gastrochaenids are responsive to light changes but EYES are not present.

Gastrochaenids are marine SUSPENSION FEEDERS. Most species are endolithic, burrowing into and constructing calcareous tubes in living and dead coral, shells, and soft rock. Some taxa build smooth, tapering, ADVENTITIOUS, free tubes oriented vertically in soft sediment (*Eufistulina*) or superficial flask-shaped IGLOOS, composed of multiple segments (CUPULES) and cemented and shallowly bored into the surfaces of other shells (*Cucurbitula*). Boring is achieved mechanically (at least in part) through abrasion by the anteriormost shell ridge (which lacks conspicuous rasping denticles except in *Spengleria*), by repeated opening and closing of the shell valves, effected by the action of the posterior adductor and anterior pedal retractor muscles. Rotation within the burrow in at least some species creates a circular cross section. Burrows can be straight, curved, internally ornamented (e.g., concentric ridges in *Lamychaena hians*, minute pustules in *Spengleria rostrata*), and/or several times longer than the shell. The aragonitic lining exists in two parts (shell and siphonal chambers, often separated by pointed processes, a ridge, or SEPTUM [DIAPHRAGM]) and can include a PEDAL SCAR indicating the position of the pedal disk and/or PROBING TUBULES (in, e.g., *L. hians*) at the anterior end of the shell chamber. It can be secreted without direct contact with the substratum, as a self-sustaining mucous membrane that later becomes calcified. Chemical boring is probably also used by some gastrochaenids, through secretions produced by the pedal organ and mantle, to create the siphonal tubes, the probing tubules, and to widen the burrow. Delicate-shelled *Eufistulana*, which builds free tubes in sediment, is probably exclusively a chemical borer (to penetrate obstacles it encounters during burrowing). Extensive repair of the tube is possible, and under experimental laboratory conditions a complete igloo can be produced, independent of calcareous substratum, within a few days; this ability allows gastrochaenids, unlike other boring bivalves, to inhabit relatively thin and rapidly eroding substrata. The

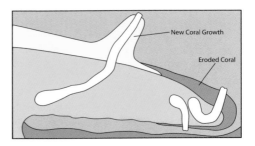

**Gastrochaenid burrows**, shown by latex casts of three *Lamychaena hians* in a *Diploria* coral head, can be several times longer than the shell, and are continuously extended to keep up with their growing coral substratum. The diet of these clams consists primarily of re-suspended matter from the substratum rather than from plankton.

tube opening is narrowed against predators and often extends above the substratum to avoid overgrowth by encrusting organisms. This CHIMNEY (hence the common name) usually ends in a distinctive figure-eight- or dumbbell-shaped aperture.

The family Gastrochaenidae is known since the Triassic and is represented by 5 living genera and ca. 15 species, distributed worldwide, in shallow tropical, subtropical, and warm-temperate seas. Boring species contribute to the breakdown (bioerosion) of coralline limestone.

### References

Bales, B. R. 1940. The rock dwellers of the Florida Keys. *The Nautilus*, 54(2): 39–42.

Boss, K. J. 1967. On the evolution of *Spengleria* (Gastrochaenidae: Bivalvia). *The American Malacological Union, Annual Reports for 1967*, pp. 15–17.

Bromley, R. G. 1978. Bioerosion of Bermuda reefs. *Palaeogeography, Palaeoclimatology, Palaeoecology*, 23(3/4): 169–197.

Carter, J. G. 1978. Ecology and evolution of the Gastrochaenacea (Mollusca, Bivalvia) with notes on the evolution of the endolithic habitat. *Bulletin of the Peabody Museum of Natural History, Yale University*, 41: 92 pp.

Gohar, H. A. F., and G. N. Soliman. 1963. On the rock-boring lamellibranch *Rocellaria rüpelli* (Deshayes). *Publications of the Marine Biological Station, Ghardaqa (Red Sea)*, 12: 145–157, pl. 1.

Morton, B. 1982. Pallial specializations in *Gastrochaena (Cucurbitula) cymbium* Spengler 1783 (Bivalvia: Gastrochaenacea). Pages 859–872, in: B. Morton and C. K. Tseng, eds., *Proceedings of the First International Marine Biological Workshop: The Marine Flora and Fauna of Hong Kong and Southern China, Hong Kong, 1980, volume 2, Ecology, Morphology, Behaviour and Physiology*. Hong Kong University Press, Hong Kong.

Morton, B. 1983a. The biology and functional morphology of *Eufistulana mumia* (Bivalvia: Gastrochaenacea). *Journal of Zoology*, 200: 381–404.

Morton, B. 1983b. Evolution and adaptive radiation in the Gastrochaenacea (Bivalvia). In: A. Bebbington, ed., *Proceedings of the Second Franco–British Symposium on Molluscs. Journal of Molluscan Studies, Supplement*, 12A: 117–121.

Morton, B. 1985. Tube formation in the Bivalvia—csövek kialakulása kagylónál. *Soosiana*, 13: 11–26.

Purchon, R. D. 1954. A note on the biology of the lamellibranch *Rocellaria (Gastrochaena) cuneiformis* Spengler. *Proceedings of the Zoological Society of London*, 124(1): 17–33.

Savazzi, E. 1982. Adaptations to tube dwelling in the Bivalvia. *Lethaia*, 15(3): 275–297.

Soliman, G. N. 1973. On the structure and behaviour of the rock-boring bivalve *Rocellaria retzii* (Deshayes) from the Red Sea. *Proceedings of the Malacological Society of London*, 40(4): 313–318, pl. 1.

Valentich-Scott, P., and G. E. Dinesen. 2004. Rock and coral boring Bivalvia (Mollusca) of the middle Florida Keys, U.S.A. In R. Bieler and P. M. Mikkelsen, eds., *Bivalve Studies in the Florida Keys*, Proceedings of the International Marine Bivalve Workshop, Long Key, Florida, July 2002. *Malacologia*, 46(2): 339–354.

### *Gastrochaena* cf. *ovata* G. B. Sowerby I, 1834 – **Winged Chimney Clam**

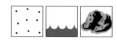

Oval to spatulate (anterior narrowed), posterior roundly pointed, anterior end protruding as a trigonal, winglike extension in front of umbo, with sharp commarginal ridges, white or yellowish exteriorly and interiorly. North Carolina to Florida, Bermuda, Bahamas, West Indies, Gulf of Mexico, South America (Brazil). Length 13 mm (to 22 mm). Formerly in *Rocellaria*. Compare *Lamychaena hians*, which is rounded posteriorly and has its umbones at the extreme anterior end. Note: The articulated view is a preserved specimen, showing the retracted soft parts. Also known as Ovate Chimney Clam. *Gastrochaena ovata* was originally described from the Gulf of Panama; the western Atlantic species is believed to be distinct (J. Carter, pers. comm., February 2005).

### *Spengleria rostrata* (Spengler, 1783) – **Rostrate Chimney Clam**

Elongated quadrangular, posterior truncate, smooth anteriorly, with strong transverse lamellations on elevated trigonal area radiating from umbones to posterior end, whitish, with light yellow-brown periostracum; interior white. North Carolina to Florida, Bermuda, West Indies, Gulf of Mexico, Caribbean Central America, South America (Colombia, Brazil). Length 30 mm. Note: The articulated view is a preserved specimen, showing the retracted soft parts.

**Gastrochaenid burrows** are signaled by a distinctive figure-eight-shaped chimney on the surface of soft coral rocks. Although the larvae are believed to settle only on dead coral, living coral tissue often surrounds the gastrochaenid opening (left).

Gastrochaena cf. ovata

LV

Spengleria rostrata

dorsal

# Family Trapezidae – Coral Clams

**Classification**
AUTOLAMELLIBRANCHIATA Grobben, 1894
HETEROCONCHIA Hertwig, 1895
Heterodonta Neumayr, 1883
Veneroida H. Adams & A. Adams, 1856
Arcticoidea R. B. Newton, 1891
Trapezidae Lamy, 1920 [1895]

## Featured species
### *Coralliophaga coralliophaga* (Gmelin, 1791) – **Coral Clam**

pallial line

Elongated oval, cylindrical, somewhat variable due to nestling habit, smooth with variable radial threads, elevated commarginal lamellations most prominent posteriorly, translucent white with two minute streaks of brown radiating from each umbo, prodissoconch light brown; living in crevices of soft rock or dead coral. North Carolina to Florida, Bermuda, Bahamas, West Indies, Gulf of Mexico, Caribbean Central America, and South America (Colombia, Brazil), also Indo-West Pacific. Length 40 mm (to 63 mm). Compare *Lithophaga* species, which are thinner shelled and edentate, with dark periostracum (as adults).

**The compressed foot** and opaque retracted siphons show prominently in this living juvenile *Coralliophaga coralliophaga* from the Florida Keys.

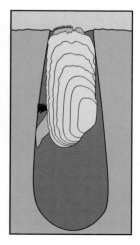

*Coralliophaga coralliophaga* **lives** in the abandoned burrows of rock-boring mollusks, secured by its byssus and frilly commarginal shell lamellae.

# Family description

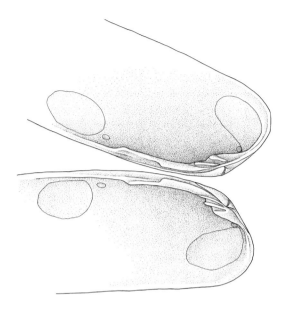

The trapezid shell is small to medium-sized (to 70 mm), thin-walled to solid, and elongated cylindrical to trapezoidal, distorted in nestlers by the shape of the habitat. Some (e.g., *Glossocardia*) have prominent shell keels from the umbones to the posteroventral corner. The shell is EQUIVALVE OR INEQUIVALVE (right valve slightly larger in *Fluviolanatus*), compressed to inflated, and not or only slightly gaping. The shell is INEQUILATERAL (umbones anterior), with PROSOGY-RATE UMBONES. Shell microstructure is ARAGONITIC and two-layered, with a CROSSED LAMELLAR outer layer and a COMPLEX CROSSED LAMELLAR inner layer, the latter confined within the pallial line. TUBULES are apparently absent. Exteriorly trapezids are usually chalky and covered by a thin, dehiscent, brownish PERIOS-TRACUM that is two-layered and in *Trapezium* extends beyond the ventral shell margins as a strong membrane. Sculpture is smooth or with commarginal ridges and rarely radial lirae or microscopic scales. LUNULE and ESCUTCHEON are absent. Interiorly the shell is non-NACREOUS. The PALLIAL LINE is ENTIRE or shallowly sinuate posteriorly. The inner shell margins are smooth. The HINGE PLATE is narrow, straight, and HETERODONT, with two CARDINAL TEETH in each valve; posterior LATERAL TEETH are usually present, whereas the anterior laterals are absent in some

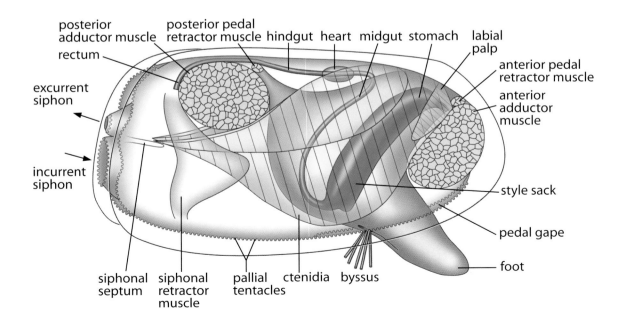

taxa. All teeth are reduced in *Coralliophaga* and absent in *Fluviolanatus*. The LIGAMENT is PARIVINCULAR, OPISTHODETIC, and set in weak NYMPHS. A secondary external ligament of fused periostracum unites the valves along the antero-, and in some also postero-, dorsal hinge margin.

The animal is weakly to markedly HETEROMYARIAN (anterior ADDUCTOR MUSCLE smaller); pedal retractor muscles are present. Pedal protractors and elevators have not been reported. The MANTLE margins are fused ventrally, with a moderate to large anteroventral pedal gape. Marginal adductor muscles (ORBITAL MUSCLES) span the fused mantle folds in *Trapezium*, acting as accessory adductors. Posterior EXCURRENT and INCURRENT SIPHONS are short; the excurrent siphon has a terminal cone (VALVULAR MEMBRANE). A pair of long, branched, sensory tentacles originate between the two siphons in *Trapezium*. HYPOBRANCHIAL GLANDS have not been reported. The FOOT is small, laterally compressed, and has a BYSSAL GROOVE; most species are byssate as adults. Large hemocoels in the mantle folds and foot of *Fluviolanatus* house algal cells that appear to be symbiotic ZOOXANTHELLAE.

The LABIAL PALPS are small to moderately large, and triangular. The CTENIDIA are EULAMELLIBRANCH (SYNAPTORHABDIC) and usually HETERORHABDIC (HOMORHABDIC in *Fluviolanatus*). The outer demibranchs are smaller than the inner demibranchs. The gills are not inserted into (or fused with) the distal oral groove of the palps (CATEGORY III association). Posteriorly the gills are united with the SIPHONAL SEPTUM, separating INFRA- and SUPRABRANCHIAL CHAMBERS; both demibranchs expand posteroventrally and project into the incurrent siphon in some species. Incurrent and excurrent water flows are posterior. The mouth is close to the posterodorsal surface of the anterior adductor muscle. The STOMACH is TYPE V; the MIDGUT can be coiled or not coiled. The HINDGUT passes through the ventricle of the heart, and leads to a sessile or freely hanging rectum. Trapezids are GONOCHORISTIC, but no data are available on larval development. The nervous system is not concentrated. STATOCYSTS are present in adult *Coralliophaga* and *Fluviolanatus*, each containing a single large STATOLITH. ABDOMINAL SENSE ORGANS have not been reported.

Trapezids are marine or estuarine SUSPENSION FEEDERS, predominantly in shallow waters, INFAUNAL in sand or nestling in crevices or boreholes of other organisms in coral rock; Australian *Fluviolanatus subtortus* (Dunker, 1857) attaches to algae, wood, or reeds. Although the name *Coralliophaga* literally means "coral eater," morphological evidence does not support the notion that the animals can create their own holes in coral rock. Predators include fish and octopuses.

The family Trapezidae is known since the Cretaceous and is represented by ca. 5 living genera and ca. 20 species, distributed mainly in shallow tropical and subtropical seas.

## References

Matsukuma, A., and T. Habe. 1995. Systematic revision of living species of *Myocardia*, Glossidae and *Glossocardia*, Trapezidae (Bivalvia). Pages 75–106, in: P. Bouchet, ed., *Résultats des Campagnes MUSORSTOM, 14. Memoirs du Muséum National d'Histoire Naturelle*, 167.

Morton, B. 1979. Some aspects of the biology and functional morphology of *Trapezium* (*Neotrapezium*) *sublaevigatum* (Lamarck) (Bivalvia: Arcticacea). *Pacific Science*, 33(2): 177–194.

Morton, B. 1980. Some aspects of the biology and functional morphology of *Coralliophaga* (*Coralliophaga*) *coralliophaga* (Gmelin 1791) (Bivalvia: Arcticacea): a coral associated nestler in Hong Kong. Pages 311–330, in B. Morton, ed., *The Malacofauna of Hong Kong and Southern China*. Hong Kong University Press, Hong Kong.

Morton, B. 1982. The biology, functional morphology and taxonomic status of *Fluviolanatus subtorta* (Bivalvia: Trapeziidae), a heteromyarian bivalve possessing "zooxanthellae." *Journal of the Malacological Society of Australia*, 5(3–4): 113–140.

Solem, A. 1954. Living species of the pelecypod family Trapeziidae. *Proceedings of the Malacological Society of London*, 31(2): 64–84, pls. 5–7.

# Family Sportellidae – Sportellid Clams

Classification
AUTOLAMELLIBRANCHIATA Grobben, 1894
HETEROCONCHIA Hertwig, 1895
Heterodonta Neumayr, 1883
Veneroida H. Adams & A. Adams, 1856
Cyamioidea G. O. Sars, 1878
? Sportellidae Dall, 1899

**Featured species**
*Basterotia elliptica* (Récluz, 1850) – **Elliptical Sportellid**

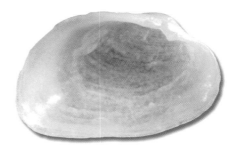

Rounded quadrangular, inflated, posterior sometimes weakly carinate, solid, granular with coarse commarginal growth lines, chalky white; interior glossy white. North Carolina to Florida, Bermuda, Bahamas, West Indies, Gulf of Mexico, Caribbean Central America, South America (Colombia). Length 6 mm (to 13 mm).

# Family description

The sportellid shell is small to medium-sized (to ca. 20 mm), thin-walled to solid, and transversely oval to quadrangular or elongated. It is EQUIVALVE, compressed or inflated, and occasionally somewhat gaping at the anterior and ventral margins. The shell is slightly to strongly INEQUILATERAL (umbones anterior), with PROSOGYRATE UMBONES; the posterior slope is often angled, occasionally carinate. Shell microstructure is ARAGONITIC, with finely granular HOMOGENOUS outer and inner layers. TUBULES have not been reported. Exteriorly sportellids are usually chalky white and covered by a thin PERIOSTRACUM. Sculpture is smooth or irregularly commarginal, frequently with pustules or granulations. A LUNULE is absent; an ESCUTCHEON is often present, but usually more developed on the left valve. Interiorly the shell is non-NACREOUS and chalky. The PALLIAL LINE is usually ENTIRE or shallowly sinuate posteriorly. The inner shell margins are smooth or denticulate. The HINGE PLATE usually is broad and HETERODONT, with (often projecting) anterior CARDINALS in each valve (in some cases with extensions resembling LATERAL TEETH); the left valve often has a conspicuous gap to accept the cardinal of the right valve. A second (central) cardinal tooth can be developed in either valve. The LIGAMENT

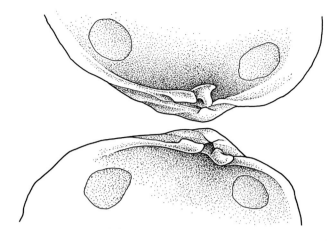

is strong to weak (or rarely absent, e.g., in *Ensitellops*), PARIVINCULAR, OPISTHODETIC, and usually set on NYMPHS; an internal portion (RESILIUM) usually sits on a small to wide triangular RESILIFER-like area, often extending to a pit under the umbones.

Composition and position of this questionably MONOPHYLETIC family (now usually placed in Cyamioidea) is uncertain. Very little is known about the anatomy of sportellids; our knowledge is based mostly on *Basterotia quadrata* (Hanley, 1843) by Fischer (1860, 1886) and *Anisodonta alata* (Powell, 1952) by Ponder (1971). The animal is ISOMYARIAN or weakly HETEROMYARIAN (anterior ADDUCTOR MUSCLE somewhat smaller) with the adductor muscle scars in some cases strengthened by a radial rib on each inner border. Pedal retractor muscles are present. Pedal protractors are absent. Pedal elevator muscles have not been reported. The MANTLE margins are fused ventrally, with a small to medium-sized anteroventral pedal gape. Posterior EXCURRENT and INCURRENT APERTURES are surrounded by stout tentacles; one or both of them can be equipped with a very short siphonlike extension (without siphonal retractor muscles). HYPOBRANCHIAL GLANDS have not been reported. The FOOT is small to medium-sized and heeled; a BYSSAL GROOVE is present in some species (e.g., in *Basterotia*) but a BYSSUS is absent in the adult.

The LABIAL PALPS are small. The CTENIDIA are EULAMELLIBRANCH (SYNAPTORHABDIC) and HETERORHABDIC; the descending lamellae of the smaller outer demibranchs are reflected above the ctenidial axis. The association of the gills and palps has not been

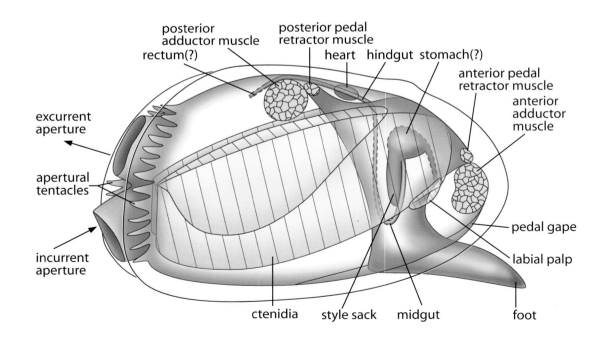

reported. Incurrent and excurrent water flows are assumed to be posterior. The STOMACH type has not been determined; the MIDGUT is not coiled. The HINDGUT passes through the ventricle of the heart, and leads to the rectum. Reproduction in sportellids is poorly known but they apparently brood VELIGER larvae inside the mantle, along the ventral margin. Shelled larvae have been found in *Anisodonta alata*, byssally attached to the periostracum at the posteroventral end of the pedal opening. The nervous system has not been studied.

Sportellids are marine SUSPENSION FEEDERS that have been reported as commensal with larger invertebrates such as tube-dwelling polychaetes. They are rarely encountered.

The family Sportellidae is known since the Jurassic and is represented by ca. 8 living genera and ca. 50 species, distributed worldwide in temperate and tropical seas.

### References

Coan, E. V. 1999. The eastern Pacific Sportellidae (Bivalvia). *The Veliger*, 42(2): 132–151.

Fischer, P. 1860. De genre *Eucharis*. *Journal de Conchyliologie*, 8(1): 23-26.

Fischer, P. 1886. Nouvelles observations sur le genre *Eucharis*, Recluz. *Journal de Conchyliologie*, 34(3): 193–203, pl. 11.

Ponder, W. F. 1971. Some New Zealand and subantarctic bivalves of the Cyamiacea and Leptonacea with descriptions of new taxa. *Records of the Dominion Museum*, 7(13): 119–141.

### *Basterotia quadrata* (Hinds, 1843) – **Square Sportellid**

Quadrangular to rounded trigonal, somewhat irregular, inflated, solid, granular with cordlike ridge from umbo to posteroventral margin, chalky white; interior glossy white. North Carolina to Florida, Bermuda, Bahamas, West Indies, Gulf of Mexico, Caribbean Central America, South America (Colombia, Brazil). Length 12 mm (to 14 mm).

### *Ensitellops protexta* (Conrad, 1841) – **Textured Sportellid**

Elongated oval, compressed, with scattered sharp pustules and divaricating microscopic striae, translucent white exteriorly and interiorly; living as nestlers or borers in soft material such as sponges. North Carolina, Florida Keys, Gulf of Mexico. Length 7 mm (to 10 mm).

**Fossilized dugong rib bones** were found at the bottom of a sinkhole at 521 m on the Pourtales Terrace off Marathon. The sediment here was coarse, dark brown sand (unusual for the Florida Keys) and contained echinoid spines, pteropod and other mollusk shells, foraminiferan tests, and stylasterid coral fragments.

*Basterotia quadrata*

*Ensitellops protexta*

# Family Corbiculidae – Marsh Clams and Asian Clams

**Classification**
AUTOLAMELLIBRANCHIATA Grobben, 1894
HETEROCONCHIA Hertwig, 1895
Heterodonta Neumayr, 1883
Veneroida H. Adams & A. Adams, 1856
Sphaerioidea Deshayes, 1854 [1820]
Corbiculidae J. E. Gray, 1847 [1840]

**Featured species**
*Polymesoda floridana* (Conrad, 1846) – **Southern Marsh Clam**

pallial line

Roundly oval to elongated trigonal, narrowed posteriorly (occasionally rostrate), smooth with irregular commarginal growth lines, white or yellowish to deep purple or pink covered with sparse, yellow-brown periostracum; interior white or purple with commarginal bands of deep purple; living in brackish environments. Florida, Bahamas, West Indies, Gulf of Mexico, Caribbean Central America, South America (Uruguay). Length 24 mm. Syn. *maritima* d'Orbigny, 1853. Note: Also known as Florida Marsh Clam.

**The habitat of *Polymesoda floridana* (left)** is typically brackish tidal flats in protected creek areas. On Ohio Key, this mangrove-fringed pond frequently dries to cracking mud and isolated pools of high-salinity water.

***Polymesoda* juveniles (center)** concentrate in moist cracks in the drying mud on Ohio Key.

**The shells of *Polymesoda floridana* (right)** are highly polychromic, showing a wide range of colors interiorly and exteriorly.

# Family description

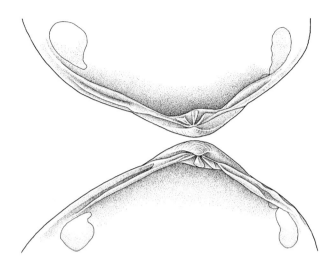

The corbiculid shell is small to large (to 150 mm), thin-walled to solid, and rounded trigonal to oval, in some species slightly posteriorly ROSTRATE. It is nearly EQUIVALVE, moderately inflated, and not gaping or slightly gaping anteriorly or posteriorly. The shell is EQUILATERAL, with PROSOGYRATE UMBONES. Shell microstructure is ARAGONITIC and two-layered, with a CROSSED LAMELLAR outer layer and a COMPLEX CROSSED LAMELLAR inner layer, the latter confined within the pallial line. TUBULES are present in *Corbicula fluminea* (Müller, 1774) (absent in *Polymesoda*). Exteriorly corbiculids are brown or olive green to brightly colored in shades of pink, yellow, and purple, covered by a thick, fibrous, adherent PERIOSTRACUM that is two-layered in *Geloina*. Sculpture is commarginal, ranging from growth lines only to strong ridges. LUNULE and ESCUTCHEON are absent. Interiorly the shell is non-NACREOUS and in some species brightly colored. The PALLIAL LINE is ENTIRE or has a small SINUS. The inner shell margins are smooth. The HINGE PLATE is narrow and HETERODONT, with usually three (often bifid) CARDINAL TEETH in each valve; anterior and posterior LATERAL TEETH are strong, long, and smooth or serrated. The LIGAMENT is strong, PARIVINCULAR, and OPISTHODETIC.

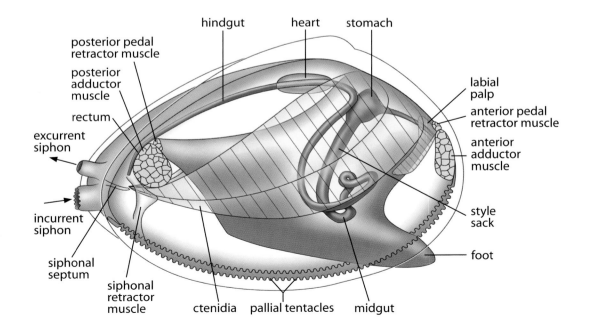

The animal is ISOMYARIAN or weakly HETEROMYARIAN (anterior ADDUCTOR MUSCLE smaller); anterior and posterior pedal retractor muscles are present. Pedal protractors are present in *Corbicula,* absent in others. Pedal elevators have not been reported. A small siphonal retractor muscle is present in *Polymesoda.* The MANTLE margins are not fused ventrally. Posterior EXCURRENT and INCURRENT SIPHONS are short and separate. HYPO-BRANCHIAL GLANDS are present in *Corbicula fluminea,* lining the interlamellar and interfilamental junctions of both inner demibranchs. The FOOT is strong and trigonal; a BYSSAL GROOVE and BYSSUS are absent in the adult.

The LABIAL PALPS are small to large and trigonal. The CTENIDIA are EULAMELLI-BRANCH (SYNAPTORHABDIC) and HOMORHABDIC or HETERORHABDIC; the demibranchs are united to each other and to the SIPHONAL SEPTUM, separating INFRA- and SUPRABRANCH-IAL CHAMBERS. The gills are not inserted into (or fused with) the distal oral groove of the palps (CATEGORY III association). Incurrent and excurrent water flows are posterior. The STOMACH is TYPE V; the MIDGUT is coiled. The HINDGUT passes through the ventricle of the heart, and leads to a freely hanging rectum. Most corbiculids are GONOCHORISTIC and produce planktonic VELIGER larvae, with freshwater species often using complex and variable reproductive strategies including different types of hermaphroditism. Fertilized eggs are usually brooded in the gills to the PEDIVELIGER stage, in the modified inner demibranchs (MARSUPIA). Reproductive strategy can vary with environmental fluctuations. The nervous system is not concentrated. STATOCYSTS and ABDOMINAL SENSE ORGANS have not been reported.

Corbiculids are estuarine or freshwater, INFAUNAL SUSPENSION FEEDERS, capable of burrowing or ploughing through the substratum. Tolerance to desiccation and salinity fluctuations is high in many species, which often can live under suboptimal circumstances, aided by the ability of taking up food through the pedal gape without siphonal access to the sediment surface. Members of the family are collected for food. especially in Asia and South America; some species of *Corbicula* have been implicated as vectors of schistosomiasis in humans.

The family Corbiculidae is known since the Triassic and is represented by ca. 8 living genera and ca. 100 species, distributed worldwide mainly in tropical and subtropical regions. The Asian Clam (*Corbicula fluminea*) is a well-known and widespread freshwater pest species in North America, introduced from Asia (probably for food) in the 1930s.

**The pest species *Corbicula fluminea*** (Müller, 1774), the Asian Clam, introduced to North American freshwaters from Asia, is a member of the family Corbiculidae.

## References

Britton, J. C., and B. Morton 1982. A dissection guide, field and laboratory manual for the introduced bivalve *Corbicula fluminea*. *Malacological Review, Supplement* 3: vi + 82 pp.

Fox, R. 2004. *Invertebrate Anatomy Online*, Corbicula fluminea, *Asian Clam*. http://www.lander.edu/rsfox/310CorbiculaLab.html, last accessed 30 November 2005.

Morton, B. S. 1976. The biology and functional morphology of the Southeast Asian mangrove bivalve, *Polymesoda* (*Geloina*) *erosa* (Solander, 1786) (Bivalvia: Corbiculidae). *Canadian Journal of Zoology*, 54(4): 482–500.

Morton, B. 1985. The reproductive strategy of the mangrove bivalve *Polymesoda* (*Geloina*) *erosa* (Bivalvia: Corbiculoidea) in Hong Kong. *Malacological Review*, 18: 83–89.

Morton, B. 1989. The functional morphology of the organs of the mantle cavity of *Batissa violacea* (Lamarck, 1797) (Bivalvia: Corbiculacea). *American Malacological Bulletin*, 7: 73–79.

Prime, T. 1865. Monograph of American Corbiculadae [*sic*] (Recent and fossil). *Smithsonian Miscellaneous Collections*, 145: 1–119.

Tan Tiu, A., and R. S. Prezant. 1989. Shell tubules in *Corbicula fluminea* (Bivalvia: Heterodonta): functional morphology and microstructure. *The Nautilus*, 103(1): 36–39.

van der Schalie, H. 1933. Notes on the brackish water bivalve, *Polymesoda caroliniana* (Bosc). *Occasional Papers of the Museum of Zoology, University of Michigan*, 11(258): 1–8, pl. 1.

Woodward, F. R. 1964. Studies on *Polymesoda expansa* (Mousson) (Corbiculidae, Bivalvia) from the Bismark Archipelago. *Videnskabelige Meddelelser fra Dansk Naturhistorisk Forening i København*, 127: 149–158, pl. 19.

# Family Cardiidae – Heart Cockles and Giant Clams

**Classification**
AUTOLAMELLIBRANCHIATA Grobben, 1894
Heteroconchia Hertwig, 1895
Heterodonta Neumayr, 1883
Veneroida H. Adams & A. Adams, 1856
Cardioidea Lamarck, 1809
Cardiidae Lamarck, 1809

**Featured species**
*Laevicardium serratum* (Linnaeus, 1758) – **Egg Cockle**

Oval to obliquely oval, thin-walled, smooth and glossy with microscopic radial ribs, juveniles with double ridge on posterior slope, white flushed with orange, yellow, or rose, occasionally with small patches of purple or brown; interior flushed with purple, deep pink, or yellow, margins finely denticulate. North Carolina to Florida, Bermuda, Bahamas, West Indies, Gulf of Mexico, Caribbean Central America, South America (to Suriname). Length 30 mm (to 65 mm). Compare *Laevicardium mortoni*, which is smaller and yellowish with zigzag brown lines, and generally inhabits more estuarine locations. Note: Formerly known as *laevigatum*, which has an Indo-Pacific LECTOTYPE (Vidal, 1999).

**The muscular foot** of cardiids, such as this juvenile *Laevicardium serratum* from Sister Creek, allows the animal to move by short leaps or to swim for short distances, as an antipredator tactic.

# Family description

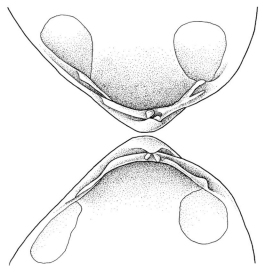

The cardiid shell is small to massive (to 1.5 m; *Tridacna gigas* reaches wet weights of >330 kg), solid or thin-walled (occasionally translucent), ovoid to rounded, trigonal, or quadrangular, and frequently higher than long; tridacnines are longer than high and have extensively fluted margins. The shell is EQUIVALVE, inflated, and usually not gaping (or with large byssal gape near the hinge in Tridacninae). It is usually EQUILATERAL (but see *Papyridea*), with PROSOGYRATE, nearly touching, UMBONES. Shell microstructure is ARAGONITIC and two-layered, with a CROSSED LAMELLAR outer layer and a COMPLEX CROSSED LAMELLAR inner layer, the latter confined within the pallial line. TUBULES are apparently absent. Exteriorly cardiids are often brightly colored in yellow, pink, or orange, covered by a thin (rarely thick), adherent, smooth or HIRSUTE, frequently abraded PERIOSTRACUM. Sculpture is smooth in some species, but most bear radial ribs that are often strong and spinose, nodulose, or cup-shaped; secondary sculpture can occur in the intercostal spaces. Sculpture on the posterior area can be different, often more pronounced, than that on the main body of the shell. LUNULE and ESCUTCHEON are absent. Interiorly the shell is non-NACREOUS. The PALLIAL LINE is weak and ENTIRE. The inner shell margins are usually denticulate (as a reflection of the external ribs) and interlocking. The HINGE is HETERODONT/CYCLODONT, with two (or one in tridacnines) conical,

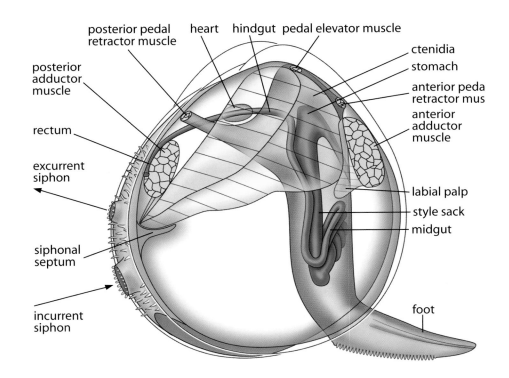

hooked CARDINAL TEETH in each valve. Anterior and posterior LATERAL TEETH are short and strong; only the posterior laterals are present in tridacnines. The LIGAMENT is short, PARIVINCULAR, OPISTHODETIC, and set on strong NYMPHS; a secondary external ligament of fused periostracum unites the valves anteriorly and posteriorly.

The animal is usually weakly HETEROMYARIAN (anterior ADDUCTOR MUSCLE slightly larger or smaller); anterior and posterior pedal retractor muscles are present, as are pedal elevator muscles. Pedal protractors have not been reported. Tridacnine anatomy is "rotated" through differential ventral growth, resulting in a ventral hinge and dorsal siphons. Adult *Tridacna* are MONOMYARIAN (anterior adductor muscle absent), with the posterior adductor and the enlarged posterior pedal retractor muscles centralized; the small anterior pedal retractor muscle inserts on the shell margin posterior to the umbo, ventral to the ligament. The MANTLE margins are usually largely unfused ventrally. Well-developed EXCURRENT and INCURRENT APERTURES are present posteriorly, extended tubelike in some species (but with siphonal retractor muscles only in Tridacninae), and often surrounded by sensory tentacles. In tridacnines the mantle lobes are extensively fused, with very large apertures, the excurrent being tubular and the incurrent marginally papillate. Symbiotic ZOOXANTHELLAE (*Symbiodinium*) live in a branched tubular system extending from the digestive system into the greatly enlarged siphonal tissues, the color of which (ranging from grayish white to greenish to dark blue) is determined by the symbionts; thousands of PALLIAL EYES (HYALINE ORGANS, equipped with lenses in *Tridacna*) on the expanded tissues transmit light to the symbionts. *Tridacna* also exhibits greatly enlarged kidneys, related to this symbiosis. Such symbionts also are present in Indo-Pacific *Corculum cardissa* (Linnaeus, 1758) and *Fragum* spp., which have translucent valve areas to transmit light. HYPOBRANCHIAL GLANDS have not been reported in any cardiid. The FOOT is strong, muscular, digitiform, heeled, sickle-shaped, and bears lateral and ventral ridges (possible SYNAPOMORPHIES for the family); a rudimentary BYSSAL GLAND is present, although a BYSSUS is rarely present in the adult. Tridacnines are again very different; the foot is small but strongly byssate in *Tridacna* (grooved but nonbyssate in *Hippopus*).

The LABIAL PALPS are small and narrow to large and trigonal. The CTENIDIA are EULAMELLIBRANCH (SYNAPTORHABDIC) and HETERORHABDIC; the outer demibranchs are usually smaller than the inner demibranchs. The demibranchs are united to each other and to the SIPHONAL SEPTUM, separating INFRA- and SUPRABRANCHIAL CHAMBERS. The gills are usually inserted into and fused with the distal oral groove of the palps (CATEGORY II association). Incurrent and excurrent water flows are posterior. The gills of *Tridacna* are S- or J-shaped, with the outer demibranchs dorsally reflected in some species. The cardiid STOMACH is TYPE V; the MIDGUT is coiled. The HINDGUT passes through the ventricle of the heart, and leads to a sessile rectum (that is terminally bulbous in some species). An AORTIC BULB has been reported in some species (including *Tridacna*). Most cardiids appear to be PROTANDRIC or SIMULTANEOUS HERMOPHRODITES and produce planktonic VELIGER larvae. The nervous system is not concentrated. STATOCYSTS (with STATOLITHS) have been reported in adult *Cardium*. ABDOMINAL SENSE ORGANS have not been reported. Pallial ("siphonal") eyes are present in some species of *Cardium*, associated with the sensory tentacles surrounding the excurrent and incurrent apertures; the extent of their presence throughout the family is not known.

Cardiids are marine, estuarine, or occasionally freshwater SUSPENSION FEEDERS that are mainly shallowly INFAUNAL or EPIFAUNAL in soft sand or mud. The muscular foot is capable of rapid burrowing, and of leaping and swimming several centimeters at a time to evade predators (e.g., sea stars, rays, seabirds, and carnivorous gastropods). While buried, the foot projects downward to form an anchor, occasionally leaving tracks in the substratum during brief ploughing events. Tridacnines are sessile, embedded hinge downward in coral reefs; burrowing is achieved mechanically through the action of the shell valves against the substratum. The valves gape to expose the mantle to sunlight; the symbiotic zooxanthellae provide the majority of the adult's nutrition. Tridacnines use a jet of water through the excurrent aperture to deter grazing fish. Heart cockles are fished in many parts of the world for human consumption and often are of substantial economic importance. Mechanized cockle fisheries have led to regional depletion of the beds, thus threatening the food basis of local bird populations (because of this, the Dutch Wadden Sea was closed to such commercial fishing in 2005). The large adductor muscle of tridacnines was once a prized delicacy in Southeast Asia and these long-lived clams (some *Tridacna* are believed to live >100 years) have been extensively harvested for their meat and to supply the aquarium and curio trade. Overfishing in the 1960s and 1970s led to the inclusion of all *Tridacna* species as "vulnerable" on the IUCN *Red List of Threatened Species* and of all tridacnines on Appendix II of CITES (Convention on International Trade in Endangered Species of Wild Flora and Fauna). Various small-bodied species are now supplied to the aquarium trade by mariculture ventures.

The family Cardiidae is known since the Triassic and is represented by 27 living genera and ca. 250 species, distributed worldwide mainly in temperate and tropical/subtropical seas, predominantly in shallow coastal environments. The common name "heart cockle" refers to the shell outline in anterior/posterior view. A recent phylogenetic analysis has shown the giant clams (Tridacninae) to be a subfamily of Cardiidae.

**Dinocardium robustum** is favorite food for Ring-billed Gulls (*Larus delawarensis*) in eastern Florida, which carry shells to a height of 6–9 m and then drop them onto the hard sand to break the shell and gain access to the flesh. According to R. Tucker Abbott, President Franklin Roosevelt was presented with a silver platter set with a large gold-plated *Dinocardium* to mark the start of the Florida Sea-Level Canal Project in 1935 (abandoned in 1943).

**At Shell Beach, Western Australia,** the endemic cardiid *Microfragum erugatum* (Tate, 1889) forms dense populations in shallow carbonate sands of hyper- or mesosaline bays. Empty shells are washed ashore by the millions, become consolidated over time into solidified deposits, known locally as Hamelin Coquina, and are harvested in blocks for decorative building material.

## References

Baker, E. B., and A. S. Merrill. 1965. An observation of *Laevicardium mortoni* actually swimming. *The Nautilus*, 78(3): 104.

Berry, P. F., and P. E. Playford. 1997. Biology of modern *Fragum erugatum* (Mollusca, Bivalvia, Cardiidae) in relation to deposition of the Hamelin Coquina, Shark Bay, Western Australia. *Marine and Freshwater Research*, 48: 415–420.

Clench, W. J., and L. C. Smith. 1944. The family Cardiidae in the western Atlantic. *Johnsonia*, 1(13): 32 pp.

Drost, K. 1886. Über das Nervensystem und die Sinnesepithelien der Herzmuschel (*Cardium edule* L.) nebst einigen Mittheilungen über den histologischen Bau ihres Mantels und ihrer Siphonen. *Morphologisches Jahrbuch*, 12(2): 163–201, pl. 10.

Farmer, M. A., W. K. Fitt, and R. K. Trench. 2001. Morphology of the symbiosis between *Corculum cardissa* (Mollusca: Bivalvia) and *Symbiodinium corculorum* (Dinophyceae). *The Biological Bulletin*, 200(3): 336–343.

Grobben, C. 1898. Expedition S. M. Schiff "Pola" in das Rothe Meer. Nördliche Hälfte (October 1895—Mai 1896). VIII. Zoologische Ergebnisse. Beiträge zur Morphologie und Anatomie der Tridacniden. *Denkschriften der Kaiserlichen Akademie der Wissenschaften, Mathematisch-Naturwissenschaftliche Classe*, 65: 433–444, pls. 1–3.

Hylleberg, J. 2004. *Lexical Approach to Cardiacea: Fossil and Living Cockles Mainly in the Families Cardiidae and Lymnocardiidae, 3 vols.* Phuket Marine Biological Center Special Publication, 29–30. Phuket Marine Biological Center, Phuket, Thailand, 939 pp., 241 pls.

Jacksonville Shell Club. 2004. The Atlantic Giant Cockle. *Shell-O-Gram*, 45(2): 3–4.

Keen, A. M. 1980. The pelecypod family Cardiidae: a taxonomic summary. *Tulane Studies in Geology and Paleontology*, 16(1): 1–40.

McLean, R. A. 1939. The Cardiidae of the western Atlantic. *Memorias Sociedad Cubana de Historia Naturale*, 13(3): 157–173, pls. 23–26.

Norton, J. H., M. A. Shepherd, H. M. Long, and W. K. Fitt. 1992. The zooxanthellal tubular system in the giant clam. *The Biological Bulletin*, 183(3): 503–506.

Penchaszadeh, P. E., and J. J. Salaya. 1983. Reproduction and gonadal changes in *Laevicardium laevigatum* (Mollusca: Bivalvia: Cardiidae) of Golfo Triste, Venezuela. *The Veliger*, 25(4): 343–346.

Rosewater, J. 1965. The family Tridacnidae in the Indo-Pacific. *Indo-Pacific Mollusca*, 1(6): 347-394.

Schneider, J. A. 1992. Preliminary cladistic analysis of the bivalve family Cardiidae. *American Malacological Bulletin*, 9(2): 145–155.

Schneider, J. A. 1994. On the anatomy of the alimentary tracts of the bivalves *Nemocardium* (*Keenaea*) *centifilosum* (Carpenter, 1864) and *Clinocardium nuttallii* (Conrad, 1837) (Cardiidae). *The Veliger*, 37(1): 36–42.

Schneider, J. A. 1995. Phylogeny of the Cardiidae (Mollusca: Bivalvia): Protocardiinae, Laevicardiinae, Lahilliinae, Tulongoncardiinae subfam. n. and Pleuriocardiinae subfam. n. *Zoologica Scripta*, 24(4): 321–346.

Schneider, J. A. 1998. Phylogeny of the Cardiidae (Bivalvia): phylogenetic relationships and morphological evolution within the subfamilies Clinocardiinae, Lymnocardiinae, Fragiinae and Tridacninae. *Malacologia*, 40(1–2): 321–373.

Stasek, C. R. 1962. The form, growth, and evolution of the Tridacnidae (giant clams). *Archives de Zoologie Expérimentale et Générale*, 101(1): 1–40.

Stasek, C. R. 1963. Orientation and form in the bivalved Mollusca. *Journal of Morphology*, 112(3): 195–214.

Vidal, J. 1999. Taxonomic review of the elongated cockles: genera *Trachycardium*, *Vasticardium* and *Acrosterigma* (Mollusca, Cardiidae). *Zoosystema*, 21(2): 259–335.

Voskuil, R. P. A., and W. J. H. Onverwagt. 1991a. Studies on Cardiidae. 3. The Recent species of *Maoricardium* Marwick, 1944 (Mollusca: Bivalvia), with description of a new species. *Basteria*, 55(1–3): 25–33.

Voskuil, R. P. A., and W. J. H. Onverwagt. 1991b. Studies on Cardiidae. 4. The taxonomy of the genus *Trachycardium* (part 1) with descriptions of three new species. *Vita Marina*, 41(2): 56–72.

Watters, G. T. 2002. The status and identity of *Papyridea soleniformis* (Bruguière, 1789) (Bivalvia: Cardiidae). *The Nautilus*, 116(4): 118–128.

Wilson, B. R., and S. E. Stevenson. 1977. Cardiidae (Mollusca, Bivalvia) of Western Australia. *Western Australian Museum Special Publication*, 9: 114 pp.

Yonge, C. M. 1981. Functional morphology and evolution in the Tridacnidae (Mollusca: Bivalvia: Cardiacea). *Records of the Australian Museum*, 33(17): 735–777.

Zugmayer, E. 1904. Über Sinnesorgane an den Tentakeln des Genus *Cardium*. *Zeitschrift für Wissenschaftliche Zoologie*, 76(3): 478–508, pl. 29.

### *Laevicardium mortoni* (Conrad, 1831) – **Yellow Egg Cockle**

Obliquely oval to rounded trigonal, with straight posterior slope and somewhat pointed posteroventral corner, thin-walled, glossy smooth, juveniles without double ridge on posterior slope; white or yellow crossed by wavy or zigzag brown lines and dark brown ray on posterior slope; interior glossy yellow with zigzag brown lines and dark posterior band, margins smooth or finely denticulate. Eastern Canada to Florida, Bermuda, Bahamas, Gulf of Mexico, Caribbean Central America. Length 25 mm. Compare *Laevicardium serratum*, which is larger (as an adult) without zigzag brown lines, and generally inhabits more oceanic locations. Note: Also known as Morton's Egg Cockle.

### *Laevicardium pictum* (Ravenel, 1861) – **Painted Egg Cockle**

Obliquely oval to trigonal, thin-walled, smooth with faint radial and commarginal lines, glossy white or cream with rose or brown mottlings or zigzag lines, umbones often pink; interior glossy cream or brownish, margins denticulate. Virginia to Florida, Bermuda, West Indies, Gulf of Mexico, Caribbean Central America, South America (to Brazil). Length 18 mm.

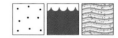

### *Microcardium peramabile* (Dall, 1881) – **Eastern Microcockle**

Oval to quadrangular, thin-walled, with ca. 90 radial ribs that are spinose on posterior slope, crossed by minute commarginal ridges, division between fine and coarse sculpture marked by single crested spinose radial rib, white, occasionally mottled with tan on posterior slope; interior white, margin denticulate. Rhode Island to Florida, West Indies, Gulf of Mexico, Caribbean Central America, South America (Brazil). Length 7 mm (to 18 mm). Formerly in *Nemocardium*. Compare *Microcardium tinctum*, which is pinkish in color and has more numerous radial ribs.

### *Microcardium tinctum* (Dall, 1881) – **Dyed Microcockle**

Oval to quadrangular, thin-walled to fragile, with ca. 150 minute radial ribs that are spinose on posterior slope, crossed by minute commarginal ridges, white with rose red stain at umbones; interior white with rose flush posteriorly, margin denticulate. Florida, West Indies, Gulf of Mexico, Caribbean Central America, South America (Suriname, Brazil). Length 10 mm (to 19 mm). Formerly in *Nemocardium*. Compare *Microcardium peramabile*, which is white and has fewer radial ribs.

**Tridacna gigas** Röding, 1798, is the largest bivalve that has ever lived, to 1.5-m shell length. A recent phylogenetic analysis, based on anatomy and paleontological evidence, has shown that the giant clams (family Tridacnidae) of the Indo-Pacific are actually members of Cardiidae. Cardiids (including tridacnines) are the best-known living bivalves harboring symbiotic zooxanthellae.

*Laevicardium mortoni*

*Laevicardium pictum*

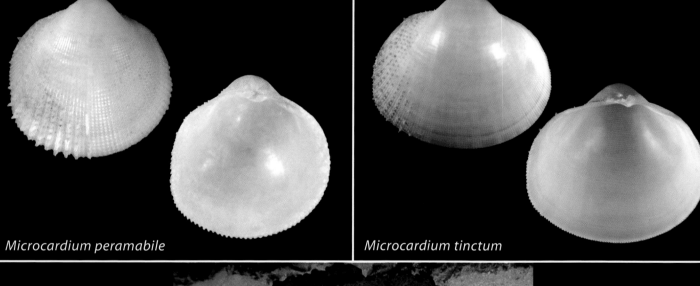

*Microcardium peramabile*

*Microcardium tinctum*

### *Trachycardium egmontianum* (Shuttleworth, 1856) – **Florida Prickly Cockle**

Elongated oval (juveniles circular), solid, with coarse, squarish radial ribs, ca. 15 crossing umbo and 27–31 on main body, with commarginal striae between radials (especially in juveniles), ribs with sharp horseshoe-shaped (hollow) prickles (see inset) over most of each rib (even in juveniles), creamy white with small patches of purple-brown; interior glossy, flushed with salmon and purple, occasionally pure white, with grooves corresponding to external ribs, margin strongly scalloped. North Carolina to Florida, Bahamas, West Indies, Gulf of Mexico. Length 40 mm (to 84 mm). Compare *Dallocardia muricata*, which is more evenly round in outline with solid (not hollow) spines, is yellowish (lacking pink or purplish coloration), and is smooth between the radial ribs in the juvenile stage; also compare *Acrosterigma magum*, in which most radial ribs are almost smooth.

### *Dallocardia muricata* (Linnaeus, 1758) – **Yellow Prickly Cockle**

Oval, solid, with coarse radial ribs, ca. 18 across umbo and 30–40 on main body, anterior and posterior ribs with flat or pointed (not hollow) prickles (see inset), central area with mostly smooth ribs (especially in juveniles), smooth between radials (especially in juveniles), cream to yellow with irregular patches of brown-red or purple; interior glossy white flushed with yellow, often with radial purple streaks beneath umbones, with grooves corresponding to external ribs, margin sharply scalloped. North Carolina to Florida, Bahamas, West Indies, Gulf of Mexico, Caribbean Central America, South America (to Argentina). Length 40 mm (to 70 mm). Formerly in *Trachycardium*. Compare *Trachycardium egmontianum*, which is more elongated oval in outline with hollow spines, usually has pink or purplish coloration, and has commarginal striae between the radial ribs in the juvenile stage.

### *Acrosterigma magnum* (Linnaeus, 1758) – **Magnum Prickly Cockle**

Elongated oval, solid, with 30–35 near-smooth radial ribs, anterior ribs with low transverse bars, posterior ribs with small toothlike scales, whitish at umbones, elsewhere yellow to purple-pink, with patches of reddish brown; interior glossy white tinted with yellow posteriorly, orange at deepest part, and pale purple at margin, with grooves corresponding to external ribs, margin sharply scalloped. Florida, Bermuda, Bahamas, West Indies, Gulf of Mexico, Caribbean Central America, South America (to Uruguay). Length 60 mm (to 62 mm). Formerly in *Trachycardium*. Compare *Trachycardium egmontianum*, which has strong, hollow spines over most radial ribs.

### *Dinocardium robustum* (Lightfoot, 1786) – **Giant Atlantic Cockle**

Rounded trigonal, thin-walled, with 32–36 rounded, low, smooth radial ribs, straw yellow with brown mottlings, posterior slope mahogany red to purple; interior rose-brown with white anterior margin, with grooves corresponding to external ribs, margin strongly scalloped. Maryland to Florida, West Indies, Gulf of Mexico, Caribbean Central America. Length 60 mm (to 119 mm). Note: Subspecies *vanhyningi* Clench & L. C. Smith, 1944, common in western Florida, is larger with external patches of mahogany red and a more oblique posterior slope.

*Trachycardium egmontianum*

*Dallocardia muricata*

*Acrosterigma magnum*

*Dinocardium robustum*

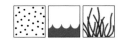

***Ctenocardia media*** (Linnaeus, 1758) – **Atlantic Strawberry Cockle**

Quadrangular to obliquely trigonal, solid, with posterior third strongly truncate and slightly concave, with 33–36 flat-topped radial ribs with low, crescent-shaped scales (often eroded), white with reddish brown, mottlings; interior glossy white, often flushed with yellow, rose-brown or purple, margin strongly scalloped. North Carolina to Florida, Bermuda, Bahamas, West Indies, Gulf of Mexico, Caribbean Central America, South America (to Brazil). Length 30 mm (to 35 mm). Formerly in *Americardia*.

***Ctenocardia guppyi*** (Thiele, 1910) – **Guppy's Strawberry Cockle**

Quadrangular, thin-walled, with posterior and ventral margins meeting at right angle, with 26–28 beaded flattened radial ribs with low, crescent-shaped scales, with commarginal threads between ribs, white to yellowish, spotted with orange-brown or purplish; interior white, sometimes with brownish flecks or streaks, margin strongly scalloped. Florida Keys, Bahamas, West Indies, Caribbean Central America. Length 4 mm (to 15 mm). Formerly in *Trigoniocardia* or *Americardia*.

***Papyridea lata*** (Born, 1778) – **Broad Paper Cockle**

Oval, thin-walled to fragile, posterior margin dentate, posterior of umbones evenly rounded (lacking ridge), gaping anteriorly and posteriorly, with 43–59 primary trigonal radial ribs that are wider and flattened on posterior half, each with scales in a narrow medial band, white with large patches of yellow, orange, rose, or purple, occasionally forming commarginal bands; interior with exterior colors visible, often with two conspicuous, broad radial rays from umbo, margin strongly scalloped. South Carolina to Florida, Bermuda, Bahamas, West Indies, Gulf of Mexico, Caribbean Central America, South America (Brazil). Length 30 mm (to 44 mm). Compare *Papyridea soleniformis*, which is less colorful, and has more-anterior umbones and radial ribs that are scaled posterodorsally.

***Papyridea soleniformis*** (Bruguière, 1789) – **Spiny Paper Cockle**

Elongated oval, thin-walled to fragile, umbones anterior, posterior of umbones angled with ridge, posterodorsal margin dentate, gaping anteriorly and posteriorly, with 40–48 primary radial ribs becoming wider posteriorly and that are trigonal in cross section, each with scales on the posterodorsal slope, white or yellowish with irregular rose or purple mottlings, occasionally forming commarginal bands; interior white with exterior colors visible, margin strongly scalloped. Florida, Bermuda, Bahamas, West Indies, Gulf of Mexico, Caribbean Central America, South America (Venezuela, Brazil), also eastern Atlantic from Angola, Ascension Island, St. Helena, St. Vincent, São Tomé and Príncipe, and Cape Verde Islands. Length 50 mm (to 60 mm). Compare *Papyridea lata*, which is more colorful (orange, pink, or purple), and has more-central umbones and radial ribs that are scaled medially.

***Papyridea semisulcata*** (Gray, 1825) – **Frilled Paper Cockle**

Elongated oval, fragile, umbones anterior, anteriorly gaping, with narrow radial ribs, becoming broader and flatter posteriorly where they are edged by (occasionally spiny) radial ridges, forming a rasplike surface anteriorly, ribs extending beyond margin forming 8–12 posterodorsal projections that interdigitate when closed, white (occasionally orange or yellow) with orange-brown spots on posterior side of umbones, occasionally stained brown or purple near margin; interior white, margin strongly scalloped posteriorly. Florida, Bermuda, Bahamas, West Indies, Gulf of Mexico, Caribbean Central America, South America (Colombia, Brazil). Length 7 mm (to 14 mm).

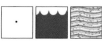

***Trigoniocardia antillarum*** (d'Orbigny, 1853) – **Antillean White Cockle**

Trigonal to quadrangular, solid, with 16–18 coarse, sharp radial ribs crossed by strong, smooth commarginal ridges, 5 or 6 center ribs enlarged and beaded; interior white, margin strongly scalloped. Florida Keys, West Indies, Caribbean Central America, South America (to Brazil). Length 8 mm (to 10 mm).

*Ctenocardia media*

*Ctenocardia guppyi*

*Papyridea lata*

*Papyridea soleniformis*

*Papyridea semisulcata*

*Trigoniocardia antillarum*

# Family Veneridae — Venus Clams

**Classification**
AUTOLAMELLIBRANCHIATA Grobben, 1894
HETEROCONCHIA Hertwig, 1895
Heterodonta Neumayr, 1883
Veneroida H. Adams & A. Adams, 1856
Veneroidea Rafinesque, 1815
Veneridae Rafinesque, 1815

## Featured species
*Periglypta listeri* (Gray, 1838) – **Caribbean Reef Clam**

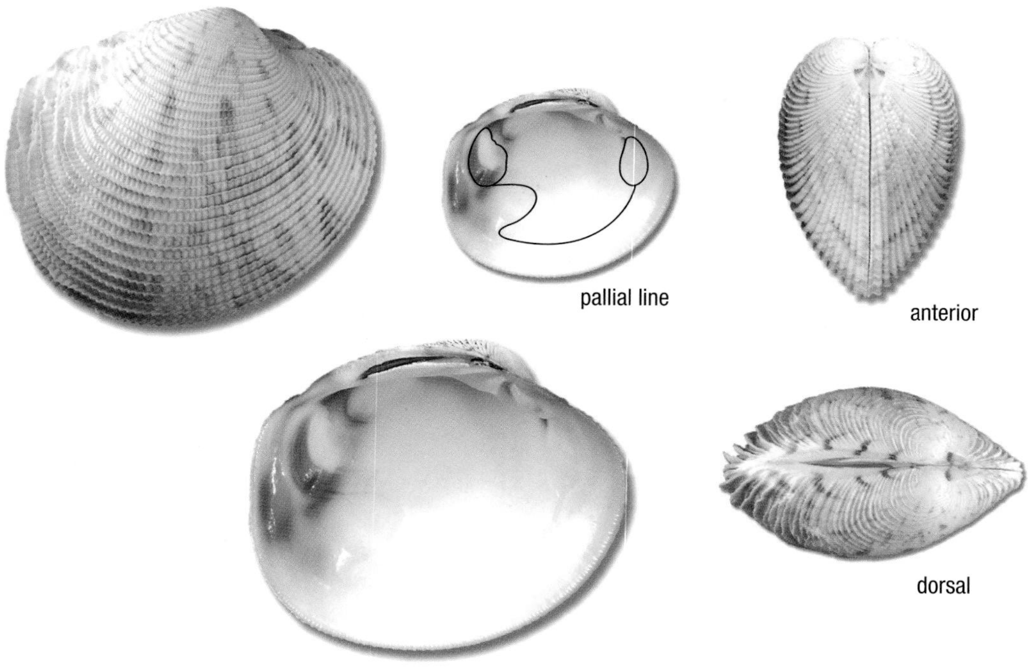

pallial line

anterior

dorsal

Elongated oval, with bluntly truncated posterior margin, inflated, with prominent erect commarginal ridges underlain by dense, flattened radial ribs, cream with brown speckles and flames, periostracum inconspicuous, lunule flush and bounded by groove, escutcheon bounded by groove; interior yellowish white with purple-brown stain posteriorly, pallial sinus wide and rounded, margin denticulate, hinge with cardinal teeth 1, 2b, and 3b bifid, 3a and 1 diverging, anterior lateral tooth minute and often purple-brown stained. North Carolina, Florida, Bahamas, West Indies, Gulf of Mexico, Caribbean Central America, South America (to French Guiana). Length 40 mm (to 75 mm). Note: Also known as Princess Venus.

**The zipperlike mantle edge** of *Periglypta listeri* forms a tightly sealed cavity, entered and exited only by the extended siphons.

**The expanded siphons** of *Periglypta listeri* lie flush with the surface vegetation in shallow water at Spanish Harbor Key. The incurrent siphon is larger in diameter than the excurrent one.

# Family description

The venerid shell is minute to large (to 170 mm in *Mercenaria campechiensis*; *Turtonia minuta* (Fabricius, 1780) is sexually mature at 1.5 mm), thin-walled to solid, oval to orbicular to chordate or rounded-trigonal, or distorted in nestlers and borers due to irregularities of the substratum. It is EQUIVALVE or rarely and secondarily INEQUIVALVE (right valve overlapping the left, in, e.g., mature *Claudiconcha* and gerontic *Choristodon*), moderately compressed to inflated, and usually not gaping (gaping posteriorly in Petricolinae). The shell is EQUILATERAL or INEQUILATERAL (umbones anterior), with PROSOGYRATE UMBONES. Shell microstructure is ARAGONITIC and two- or three-layered, with a composite PRIS-

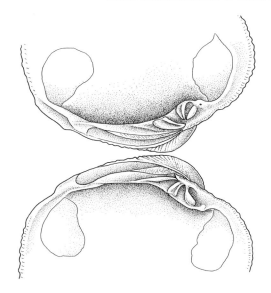

MATIC outer layer (absent in some species), a CROSSED LAMELLAR or HOMOGENOUS middle layer, and a COMPLEX CROSSED LAMELLAR or homogenous inner layer. TUBULES are apparently absent. Exteriorly venerids are covered by an inconspicuous to glossy or thick PERIOSTRACUM, which is two-layered in some (Petricolinae and *Callocardia*) and harbors calcareous needles in others (especially Pitarinae). The shell surface is heavily encrusted with "cemented" sand in Samarangiinae (a SYNAPOMORPHY for the subfamily) and with mud and detritus in *Callocardia*, each affixed by mucus during shell formation. Sculpture is predominantly commarginal but exceedingly variable, from smooth (often highly polished) to strongly radial to cancellate, and occasionally spinose, nodulose, or lamellose especially on the posterior slope (or especially anteriorly in the boring petricolines). Strong commarginal sculpture has been shown to act as a defensive mechanism against boring and other predators in some

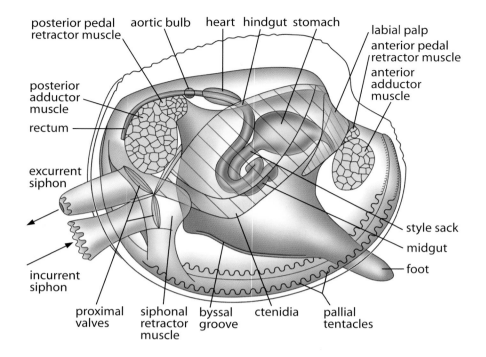

posterior pedal retractor muscle  aortic bulb  heart  hindgut  stomach

labial palp
anterior pedal retractor muscle
anterior adductor muscle

posterior adductor muscle

rectum

excurrent siphon

style sack
midgut
foot

incurrent siphon

proximal valves  siphonal retractor muscle  byssal groove  ctenidia  pallial tentacles

species. LUNULE and ESCUTCHEON are usually slightly to deeply impressed, bordered by a shallow to deep groove, and often asymmetric (larger and/or overlapping in one valve) and/or differently colored or sculptured than the main body of the shell; one or both are poorly defined or absent in many taxa, e.g., *Clementia*, *Cyclina*, *Gemma*, *Irus*, *Nutricola*, *Petricola*, and *Turtonia*; a deeply excavated escutcheon is a synapomorphy for Sunettiinae. Interiorly the shell is non-NACREOUS, and white or with purple, orange, or pink staining, especially posteriorly. The PALLIAL LINE has a shallow to deep, rounded or tapering SINUS, that is especially deep and dorsally directed in Dosiniinae, very shallow in Samarangiinae and Gouldiinae, and completely absent in *Turtonia*. The inner shell margins are smooth, denticulate (in some as a result of external or underlying radial structure), or with oblique (in *Transennella*) or commarginal grooves (in *Lioconcha*). The HINGE PLATE is narrow to wide and HETERODONT, with two or more (typically three; often bifid or variously thickened; peglike in *Turtonia*) CARDINAL TEETH in each valve; the anterior and middle cardinals of the right valve are diagnostically parallel (instead of diverging) in Pitarinae and two other subfamilies (but this is also true in some members of other subfamilies). The hinge line can be excavated, wherein the cardinal teeth overhang the margin. Posterior LATERAL TEETH are usually absent (present in *Irus* and *Turtonia*), whereas anterior laterals are present in some taxa and (by ontogenetic evidence) probably represent at least two separate origins (termed lateral and pseudolateral by some authors). The LIGAMENT is PARIVINCULAR, usually OPISTHODETIC (rarely AMPHIDETIC), and set on strong NYMPHS that can be rugose, striated, or strongly subdivided into two portions called "pseudocardinals" (in *Tivela*). A secondary ligament of fused periostracum unites the valves over a small part of the dorsal margin in a few species.

Venerids are well known for their lack of special adaptations, and considerable anatomical uniformity is present throughout the family; few synapomorphies are confirmed for the family or its subfamilial components. The animal is usually ISOMYARIAN or weakly HETEROMYARIAN (anterior ADDUCTOR MUSCLE slightly smaller). Anterior and posterior pedal retractor muscles are present; pedal protractors and elevators are absent (although visceral retractor muscles that insert in the UMBONAL CAVITY have been described in *Periglypta* and Petricolinae). *Petricolaria* has an accessory posterior adductor muscle anterior to the posterior adductor and pedal retractors, dorsal to the rectum. The MANTLE margins are usually not largely fused ventrally, with a small to (usually) large pedal gape. The wavy mantle edges and associated marginal tentacles of *Periglypta listeri* can closely interact to close the pedal gape like a zipper. Large pallial glands near the pedal gape in *Petricola* and *Cooperella* have been postulated to secrete chemicals or mucus involved in rock boring or sand agglutination, respectively. Posterior EXCURRENT and INCURRENT SIPHONS are short to long, separated to fully united, and can be variously pigmented (sometimes differently in closely related species of the same genus). The excurrent siphon is usually slightly smaller in diameter and usually has a terminal cone (VALVULAR MEMBRANE); the incurrent siphon is ventrally open ("absent") in *Turtonia*. Each siphon possesses a basal siphonal membrane (PROXIMAL VALVE) consisting of a thin tissue flap narrowing the lumen; *Tivela* has a second siphonal membrane inside the incurrent siphon. A FOURTH PALLIAL APERTURE is absent (although reported in error in *Gemma* by one author). HYPOBRANCHIAL GLANDS have not been reported. The FOOT is small to large, usually wedge-shaped (or lunate in Dosiniinae, a synapomorphy of that subfamily), heeled, and laterally compressed; a BYSSAL GROOVE is present or absent and a BYSSUS is usually absent in the adult. When present, the adult byssus (in *Irus*, *Nutricola*, *Turtonia*, and a few others) is usually single-stranded and delicate, interpreted as the NEOTENOUS retention of a postlarval feature; however, that of *Venerupis galactites* (Lamarck, 1818) is an elaborate, comblike, multibranched structure that is clearly derived in many aspects.

The LABIAL PALPS are small to medium-sized and trigonal. The CTENIDIA are EULAMELLIBRANCH (SYNAPTORHABDIC) and usually HETERORHABDIC (HOMORHABDIC in *Gemma*, *Turtonia*, and a few others); the outer demibranchs are usually smaller than the inner demibranchs (comprised of the ascending lamella only and reflected in *Gemma*; entire outer demibranch absent in *Turtonia*). The demibranchs are united to each other posteriorly, separating INFRA- and SUPRABRANCHIAL CHAMBERS. The gills are usually inserted into and fused with the distal oral groove of the palps (CATEGORY II association), or are not inserted (or fused) (CATEGORY III) in, e.g., *Gemma*, *Nutricola*, and petricolines. Incurrent and excurrent water flows are posterior. The STOMACH is usually TYPE V (simplified in *Gemma* and *Nutricola*; TYPE IV in *Turtonia*). The MIDGUT is loosely to highly coiled; it is greatly expanded in diameter in *Cooperella*. The HINDGUT passes through the ventricle of the heart, and leads to a sessile or rarely freely hanging rectum. An AORTIC BULB is usually present and conspicuous (absent in a few species, e.g., *Gemma* and *Turtonia*). Oxygen-transporting myoglobin is present in the muscle tissues of some species, imparting a few species (e.g., the Beef Steak Clam [*Saxidomus nuttalli* Conrad, 1837]) with brightly colored tissues. Venerids are usually GONOCHORISTIC and produce planktonic VELIGER larvae; some species are PROTANDRIC HERMAPHRODITES (e.g., *Mercenaria* and *Nutricola*), whereas others brood larvae in the gills (e.g., *Gemma*) and/or have DIRECT DEVELOPMENT (e.g., *Liocyma* and *Nutricola*). The females of *Turtonia minuta* produce one to four gelatinous egg capsules (secreted by the two innermost mantle folds) attached to their own byssus threads, and which hatch direct-developing juveniles (up to 16 per capsule). Many venerids can be sexually mature within 1 year, and some can live for more than 40 years. The nervous system is not concentrated; small siphonal ganglia are present in some species.

Statocysts (with single statoliths) have been reported in adults of many species. Abdominal sense organs have not been reported.

Venerids are marine or estuarine suspension feeders (or possibly rarely deposit feeders, i.e., *Cooperella*), usually shallowly infaunal in soft sediments including sand, mud, rubble, and seagrass beds; epibionts (e.g., polydorid polychaetes, algae, and small corals) and erosion on the posterior margins indicate that some species live partially exposed with the posterior end upward. *Dosinia* lies buried vertically with the dorsal margin upward. Most venerids are capable of movement both horizontally and vertically through the sediment; their burrowing sequence has been well studied. A few species (especially petricolines) are byssally attached nestlers, or mechanical tunnelers (capable of only enlarging the crevice they inhabit) or borers (by simple opening–closing and twisting the roughened shell valves; however, *Petricola* has mantle glands that suggest chemical boring) in soft rock, clay, seagrass, or dead coral. *Petricola* has separated siphons in a common, wide tube with a weakly dumbbell-shaped opening; *Cooperella* builds an infaunal "cocoon" of sand particles held together by mucus secreted by glands surrounding the pedal gape. Some species (e.g., *Gemma gemma* and *Turtonia minuta* (Fabricius, 1780)) can occur in extremely dense populations. Commensal crabs (Pinnotheridae) have been reported in the mantle cavities of some species. Predators include boring and otherwise carnivorous gastropods, sea stars, octopuses, crustaceans, fish (who nip at the siphon tips), and water birds. Venerids include many economically important species in the world's fisheries, and are widely commercially harvested or raised in mariculture. The Northern Quahog or

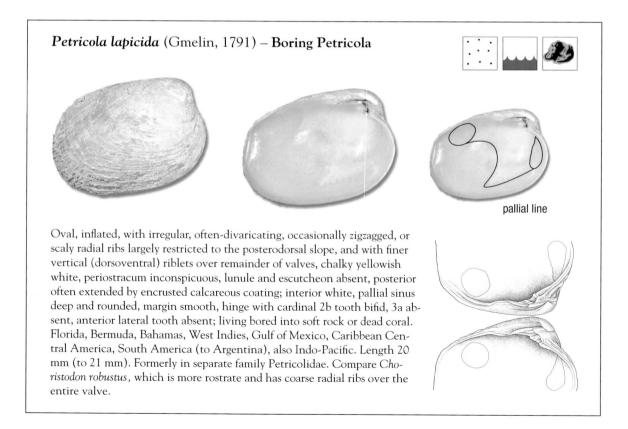

*Petricola lapicida* (Gmelin, 1791) – **Boring Petricola**

pallial line

Oval, inflated, with irregular, often-divaricating, occasionally zigzagged, or scaly radial ribs largely restricted to the posterodorsal slope, and with finer vertical (dorsoventral) riblets over remainder of valves, chalky yellowish white, periostracum inconspicuous, lunule and escutcheon absent, posterior often extended by encrusted calcareous coating; interior white, pallial sinus deep and rounded, margin smooth, hinge with cardinal 2b tooth bifid, 3a absent, anterior lateral tooth absent; living bored into soft rock or dead coral. Florida, Bermuda, Bahamas, West Indies, Gulf of Mexico, Caribbean Central America, South America (to Argentina), also Indo-Pacific. Length 20 mm (to 21 mm). Formerly in separate family Petricolidae. Compare *Choristodon robustus*, which is more rostrate and has coarse radial ribs over the entire valve.

Hard-shelled Clam (*Mercenaria mercenaria*) was exploited for food by pre-Columbian peoples along the eastern U.S. coastline; its purple-and-white shells were carved into beads called WAMPUM that were used in belts commemorating significant events and as currency after European contact. The same species now accounts for a large percentage of the total aquaculture production in Florida, ranking third in dollar value behind tropical fish and aquatic plants. Some other well-known and economically important species include the Eastern Pacific Pismo Clam (*Tivela stultorum* Mawe, 1854), and the Manila Clam (*Ruditapes philippinarum* (A. Adams & Reeve, 1850), originally from the Indo-Pacific and now widely introduced to many regions of the world). *Gemma gemma* forms an important portion of the diet of economically valuable game ducks in New Jersey.

The family Veneridae is known since the Cretaceous and is represented by ca. 50 living genera and more than 800 species (in 12 commonly recognized subfamilies), making this family the most diverse of extant Bivalvia. Venerids are distributed worldwide mainly in shallow temperate to tropical seas. The common name "venus clam" refers to the heart-shaped shell outline in anterior–posterior view. A recent phylogenetic analysis, including both morphological and molecular characters, found the family to be MONOPHYLETIC; Turtoniidae, Petricolidae, and Cooperellidae were concluded to be subfamilies within Veneridae, with morphological differences of turtoniines attributed to neoteny, and those of petricolines to the boring habit.

**The dark brown siphons** of the otherwise white-bodied *Petricola lapicida* protrude from a coral rock in the shallow subtidal zone of Spanish Harbor Key. See another image of living *Petricola lapicida* in its rock burrow on p. 91.

**Beached shells** of *Petricolaria pholadiformis* at Bayville, New York, suggest high population densities of these boring bivalves offshore.

**Living animals** of *Petricola* sp. (left), *Petricolaria pholadiformis* (center), and *Choristodon robustus* (right) show some of the variation in the length and appearance of the siphons in the Petricolinae.

**Venerids include** some of the most edible mollusks. *Mercenaria mercenaria*, here as a plate of "steamers" at a Florida Keys restaurant, has been harvested for food for about 300 years in its native range and is now increasingly raised in aquaculture.

### References

Ansell, A. D. 1961. The functional morphology of the British species of Veneracea (Eulamellibranchia). *Journal of the Marine Biological Association of the United Kingdom*, 41: 489–515.

Ansell, A. D. 1962. Observations on burrowing in the Veneridae (Eulamellibranchia). *Biological Bulletin*, 123(3): 521–530.

Ansell, A. D. 1970. Boring and burrowing mechanisms in *Petricola pholadiformis* Lamarck. *Journal of Experimental Marine Biology and Ecology*, 4(3): 211–220.

Bieler, R., I. Kappner, and P. M. Mikkelsen. 2004. *Periglypta listeri* (Gray, 1838) (Bivalvia: Veneridae) in the western Atlantic: taxonomy, anatomy, life habits, and distribution. In: R. Bieler and P. M. Mikkelsen, eds., *Bivalve Studies in the Florida Keys*, Proceedings of the International Marine Bivalve Workshop, Long Key, Florida, July 2002. *Malacologia*, 46(2): 427–458.

Bieler, R., P. M. Mikkelsen, and R. S. Prezant. 2005. Byssus-attachment by infaunal clams: seagrass-nestling *Venerupis* in Esperance Bay, Western Australia (Bivalvia: Veneridae). Pages 177–197, in: F. E. Wells, D. I. Walker, and G. A. Kendrick, eds., *The Marine Flora and Fauna of Esperance, Western Australia*. Western Australian Museum, Perth.

Carriker, M. R. 1961. Interrelation of functional morphology, behavior, and autecology in early stages of the bivalve *Mercenaria mercenaria*. *Journal of the Elisha Mitchell Scientific Society*, 77(2): 168–241.

Clench, W. J. 1942. The genera *Dosinia*, *Macrocallista* and *Amiantis* in the western Atlantic. *Johnsonia*, 1(3): 1–8.

Coan, E. V. 1997. Recent species of the genus *Petricola* in the eastern Pacific (Bivalvia: Veneroidea). *The Veliger*, 40(4): 298–340.

Dall, W. H. 1902. Synopsis of the family Veneridae and of the North American Recent species. *Proceedings of the United States National Museum*, 26(1312): 335–412.

Duval, D. M. 1963. The biology of *Petricola pholadiformis* Lamark [*sic*] (Lamellibranchiata Petricolidae). *Proceedings of the Malacological Society of London*, 35(223): 89–100.

Fox, R. 2004. *Invertebrate Anatomy Online*, Mercenaria mercenaria, *Quahog, with Notes on* Tapes japonicus. http://www.lander.edu/rsfox/310mercenariaLab.html, last accessed 05 January 2006.

Gray, S. 1982. Morphology and taxonomy of two species of the genus *Transennella* (Bivalvia: Veneridae) from western North America and a description of *T. confusa* sp. nov. *Malacological Review*, 15: 107–117.

Guéron, C. O. C., and A. C. S. Coelho. 1989. Considerações taxonômicas e morfologia de *Dosinia* (*Dosinia*) *concentrica* (Born, 1778) (Mollusca, Bivalvia, Veneridae). *Boletim do Museu Nacional, Nová Série, Zoologia*, 334: 1–19.

Guéron, C. O. C., and W. Narchi. 2000. Anatomia functional de *Protothaca* (*Leukoma*) *pectorina* (Lamarck) (Bivalvia, Veneridae). *Revista Brasileira de Zoologia*, 17(4): 1007–1039.

Hansen, B. 1953. Brood protection and sex ratio of *Transennella tantilla* (Gould), a Pacific bivalve. *Videnskabelige Meddelelser fra Dansk Naturhistorisk Forening*, 115: 313–324.

Jones, C. C. 1979. Anatomy of *Chione cancellata* and some other chionines (Bivalvia: Veneridae). *Malacologia*, 19(1): 157–199.

Kraeuter, J. N., and M. Castagna, eds. 2001. *Biology of the Hard Clam*. Elsevier Science, Amsterdam, The Netherlands, xix + 751 pp.

Lindberg, D. R. 1990. *Transennella* Dall versus *Nutricola* Bernard (Bivalvia: Veneridae): an argument for evolutionary systematics. *Journal of Molluscan Studies*, 56(1): 129–132.

Mikkelsen, P. M., R. Bieler, T. A. Rawlings, and I. Kappner. 2006. The phylogeny of Veneroidea (Mollusca: Bivalvia: Heterodonta) based on morphology and molecules. *Zoological Journal of the Linnean Society*, 148: 439–521.

Morton, B. 1985. Aspects of the biology and functional morphology of *Irus irus* (Bivalvia: Veneridae: Tapetinae) with a comparison of *Bassina calophylla* (Chioninae). Pages 321–336, in: B. Morton and D. Dudgeon, eds., *Proceedings of the Second International Workshop on the Malacofauna of Hong Kong, and Southern China*. Hong Kong University Press, Hong Kong.

Morton, B. 1995. The biology and functional morphology of *Cooperella subdiaphana* (Carpenter) (Bivalvia: Petricolidae). *The Veliger*, 38(2): 162–170.

Morton, B. 2000. The anatomy of *Callocardia hungerfordi* (Bivalvia: Veneridae) and the origin of its shell camouflage. *Journal of Molluscan Studies*, 66: 21–30.

Narchi, W. 1972a. Structure and adaptation in *Transennella tantilla* (Gould) and *Gemma gemma* (Totten) (Bivalvia: Veneridae). *Bulletin of Marine Science*, 21(4): 866–885.

Narchi, W. 1972b. Comparative study of the functional morphology of *Anomalocardia brasiliana* (Gmelin, 1791) and *Tivela mactroides* (Born, 1778) (Bivalvia, Veneridae). *Bulletin of Marine Science*, 22(3): 643–670.

Narchi, W. 1975. Functional morphology of a new *Petricola* (Mollusca Bivalvia) from the littoral of São Paulo, Brazil. *Proceedings of the Malacological Society of London*, 41: 451–465.

Narchi, W., and F. di Dario. 2002. The anatomy and functional morphology of *Tivela ventricosa* (Gray, 1838) (Bivalvia: Veneridae). *The Nautilus*, 116(1): 13–24.

Ockelmann, K. W. 1964. *Turtonia minuta* (Fabricius), a neotenous veneracean bivalve. *Ophelia*, 1(1): 121–146.

Ohno, T. 1996. Intra-periostracal calcified needles of the bivalve family Veneridae. *Bulletin de l'Institut Océanographique, Monaco*, spec. no. 14, 4: 305–314.

Oldfield, E. 1955. Observations on the anatomy and mode of life of *Lasaea rubra* (Montagu) and *Turtonia minuta* (Fabricius). *Proceedings of the Malacological Society of London*, 31(5/6): 226–249.

Palmer, K. van Winkle. 1927–1929. The Veneridae of eastern America, Cenozoic and Recent. *Palaeontographica Americana*, 1(5): 209–522. [March 1927; pls. 32–76, February 1929.]

Purchon, R. D. 1955. The functional morphology of the rock-boring lamellibranch *Petricola pholadiformis*. *Journal of the Marine Biological Association of the United Kingdom*, 34: 257–278.

Roopnarine, P. D., and G. J. Vermeij. 2000. One species becomes two: the case of *Chione cancellata*, the resurrected *C. elevata*, and a phylogenetic analysis of *Chione*. *Journal of Molluscan Studies*, 66: 517–534.

Sellmer, G. P. 1967. Functional morphology and ecological life history of the gem clam, *Gemma gemma* (Eulamellibranchia: Veneridae). *Malacologia*, 5(2): 137–223.

Taylor, J. D., E. A. Glover, and C. J. R. Braithwaite. 1999. Bivalves with 'concrete overcoats': *Granicorium* and *Samarangia*. *Acta Zoologica (Stockholm)*, 80: 285–300.

Valentich-Scott, P., and G. E. Dinesen. 2004. Rock and coral boring Bivalvia (Mollusca) of the middle Florida Keys, U.S.A. In: R. Bieler and P. M. Mikkelsen, eds., *Bivalve Studies in the Florida Keys*, Proceedings of the International Marine Bivalve Workshop, Long Key, Florida, July 2002. *Malacologia*, 46(2): 339–354.

Yonge, C. M. 1958. Observations on *Petricola carditoides* (Conrad). *Proceedings of the Malacological Society of London*, 33(1): 25–31, pl. 4.

### Macrocallista maculata (Linnaeus, 1758) – Calico Clam

Oval, compressed, smooth and polished, glossy cream with small checkerboard patches of brownish red, periostracum varnishlike, lunule flush and bounded by groove, escutcheon demarcated by color only and not bounded by groove; interior white with yellow and/or purple blush, pallial sinus tapering, reaching angular limit near center-valve, margin smooth, hinge with cardinal tooth 3b bifid, 3a and 1 parallel, anterior lateral tooth elongated. North Carolina to Florida, Bermuda, Bahamas, West Indies, Gulf of Mexico, Caribbean Central America, South America (to Brazil). Length 55 mm (to 77 mm).

### Macrocallista nimbosa (Lightfoot, 1786) – Sunray Venus

Elongated oval, compressed, smooth and glossy, dull salmon or beige to mauve with broken radial bands of darker hue, periostracum varnishlike, lunule flush and bounded by groove, escutcheon demarcated by color only and not bounded by groove; interior dull white with reddish blush, pallial sinus tapered, margin smooth, hinge with cardinal tooth 3b bifid, 3a and 1 nearly parallel, anterior lateral tooth elongated. North Carolina to Florida, Gulf of Mexico. More typical of western Florida coasts. Length 85 mm (to 150 mm).

### Callista eucymata (Dall, 1890) – Glory-of-the-Seas Venus

Rounded trigonal, with ca. 50 flattened commarginal ridges with short dorsal and long ventral slopes, glossy white to pale brown with blotches and zigzags of reddish brown, periostracum varnishlike, lunule flush and bounded by groove, escutcheon absent; interior with pink and/or yellow blush, pallial sinus wide and rounded, margin smooth, hinge with cardinal tooth 3b bifid, 3a and 1 parallel, anterior lateral tooth elongated. New Jersey to Florida, West Indies, Gulf of Mexico, South America (to Brazil). Length 25 mm (to 27 mm).

### Mercenaria mercenaria (Linnaeus, 1758) – Northern Quahog

Rounded trigonal, with coarse commarginal ridges, smooth at center valve, chalky whitish gray to yellowish gray, *notata* form with brown zigzag markings, periostracum inconspicuous, lunule flush or impressed and bounded by groove, escutcheon not bounded by groove; interior white with purple stain especially posteriorly, pallial sinus tapered, margin denticulate, hinge with cardinal teeth 1, 2a, 2b, and 3b bifid, 3a and 1 diverging, anterior lateral tooth absent. Eastern Canada to Florida, Bermuda, Gulf of Mexico, also introduced to eastern Pacific, Puerto Rico, and western Europe. Length 70 mm (to 150 mm). Compare *Mercenaria campechiensis*, which reaches a much larger maximum size, and is commarginally ridged at center valve and internally white. Note: Inset shows a free pearl formed by M. *mercenaria*. See the anatomy of M. *mercenaria* on pp. 11-15.

### Mercenaria campechiensis (Gmelin, 1791) – Southern Quahog

Rounded trigonal to oval, inflated, with coarse commarginal ridges over entire valve, chalky whitish gray, periostracum inconspicuous, lunule flush or impressed and bounded by groove, escutcheon not bounded by groove; interior entirely white, pallial sinus tapered, margin denticulate, hinge with cardinal teeth 1, 2b, and 3b bifid, 3a and 1 diverging, anterior lateral tooth absent. New Jersey to Florida, West Indies, Gulf of Mexico, Caribbean Central America. Length 98 mm (to 110 mm). Compare *Mercenaria mercenaria*, which is smooth at center valve (as an adult) and is purple-stained internally.

*Macrocallista maculata*

*Macrocallista nimbosa*

*Callista eucymata*

*Mercenaria mercenaria*

*Mercenaria campechiensis*

### *Dosinia discus* (Reeve, 1850) – **Disk Dosinia**

Circular to oval, compressed, with fine commarginal grooves (ca. 20/cm), glossy beige, periostracum varnishlike, lunule impressed and bounded by groove, escutcheon absent; interior white, pallial sinus deeply tapered, margin smooth, hinge with cardinal tooth 3b bifid, 3a and 1 nearly parallel, anterior lateral tooth minute. Virginia to Florida, Bahamas, Gulf of Mexico. Length 60 mm (to 74 mm). Compare *Dosinia elegans*, which has fewer commarginal grooves.

### *Dosinia elegans* (Conrad, 1846) – **Elegant Dosinia**

Circular, compressed, with coarse commarginal grooves (8–10/cm), glossy beige, periostracum varnishlike, lunule impressed and bounded by groove, escutcheon absent; interior white, pallial sinus deeply tapered, margin smooth, hinge with cardinal tooth 3b bifid, 3a and 1 nearly parallel, anterior lateral tooth minute. North Carolina to Florida, Gulf of Mexico, Caribbean Central America. Length 60 mm (to 92 mm). Compare *Dosinia discus*, which has more numerous commarginal grooves.

### *Globivenus rugatina* (Heilprin, 1887) – **Queen Venus**

Circular, inflated, with prominent anterior projection corresponding to lower lunular boundary, with erect commarginal ribs and five to eight fine commarginals in the interspaces, becoming rounded and crossed by fine radials ventrally, cream to whitish with light mauve mottlings, periostracum inconspicuous, lunule impressed and bounded by groove, escutcheon not bounded by groove; interior white, pallial sinus tapered, margin denticulate, hinge with cardinal teeth 1 2b, and 3b bifid, 3a and 1 diverging, anterior lateral tooth minute. North Carolina to Florida, Bermuda, West Indies. Length 20 mm. Formerly in *Ventricolaria*. Compare *Globivenus rigida*, which has erect, dorsally recurved commarginal ridges and fewer fine commarginals in the interspaces.

### *Globivenus rigida* (Dillwyn, 1817) – **Rigid Venus**

Circular, inflated, with erect, dorsally recurved commarginal ridges and one to three finer commarginal striae in the interspaces, brownish white with dark brown mottlings, periostracum inconspicuous, lunule impressed and bounded by groove, escutcheon confined largely to left valve, not bounded by groove and with reddish brown bars on left valve; interior white or cream, pallial sinus small and tapered, margin denticulate, hinge with cardinal teeth 1, 2b, and 3b bifid, 3a and 1 diverging, anterior lateral tooth minute. North Carolina to Florida, Bahamas, West Indies, Gulf of Mexico, Caribbean Central America, South America (to Brazil). Length 70 mm. Formerly in *Ventricolaria*. Compare *Globivenus rugatina*, which has erect commarginal ridges (that become rounded ventrally), with more numerous fine commarginals in the interspaces.

### *Circomphalus strigillinus* (Dall, 1902) – **Empress Venus**

Rounded trigonal, inflated, with distinct raised commarginal ridges separated by fine commarginal threads, whitish, periostracum inconspicuous, lunule impressed and bounded by groove, escutcheon not bounded by groove; interior white, pallial sinus small and tapered, margin thickened and denticulate, hinge with cardinal teeth 1 and 3b bifid, 3a and 1 diverging, anterior lateral tooth minute. North Carolina to Florida, Bermuda, West Indies, Gulf of Mexico, South America (to Uruguay). Length 40 mm (to 45 mm). Compare to *Mercenaria* species, which are generally larger and each have a small anterior lateral tooth.

### *Cyclinella tenuis* (Récluz, 1852) – **Atlantic Cyclinella**

Circular, moderately compressed, thin-walled, smooth with irregular growth lines, dull white, periostracum varnishlike, lunule flush and bounded by groove, escutcheon absent; interior white, pallial sinus deeply tapered, margin smooth, hinge with cardinal teeth 2a and 3b bifid, 3a and 1 nearly parallel, anterior lateral tooth absent. Virginia to Florida, West Indies, Gulf of Mexico, Caribbean Central America, South America (to Brazil). Length 10 mm (to 36 mm).

*Dosinia discus*

*Dosinia elegans*

*Globivenus rugatina*

*Globivenus rigida*

*Circomphalus strigillinus*

*Cyclinella tenuis*

### Chione elevata Say, 1822 – Florida Cross-barred Venus

Rounded trigonal, more pointed posteriorly, with coarse, rounded radial ribs and thin, widely spaced, erect (not recurved) commarginal ridges, creamy white often with fine pale orange-brown zigzags, occasionally with brown rays, periostracum inconspicuous, lunule flush and bounded by groove, with both radial and commarginal threads, escutcheon not bounded by groove and crossed by purple-brown bars; interior dark purple to reddish brown, pallial sinus small and tapered, margin denticulate, hinge with no cardinal teeth bifid, 3a and 1 diverging, anterior lateral tooth absent. North Carolina to Florida, Bahamas, West Indies, Gulf of Mexico, Caribbean Central America. Length 25 mm. Formerly known as *C. cancellata* Linnaeus, 1767 (of the Caribbean, see following).

### Chione mazyckii Dall, 1902 – Mazyck's Venus

Rounded trigonal, more pointed posteriorly, with fine radial ribs and thin, closely spaced, erect commarginal ridges, whitish, periostracum inconspicuous, lunule large, flush and bounded by groove, escutcheon not bounded by groove and crossed by purple-brown bars; interior white to rose, pallial sinus small and tapered, margin denticulate, hinge with no cardinal teeth bifid, 3a and 1 diverging, anterior lateral tooth absent. North Carolina to Florida, South America (Colombia). Length 7 mm. Formerly considered a form of *C. cancellata*.

### Chione cancellata (Linnaeus, 1767) – Caribbean Cross-barred Venus

This species is the western Caribbean sister-species to *Chione elevata* of the Florida Keys, southeastern United States, and eastern Caribbean. It differs from the latter in having more closely spaced and recurved commarginal ridges, and the lunule with radial ribs only.

### Lirophora clenchi (Pulley, 1952) – Clench's Thick-ringed Venus

Rounded trigonal, solid, with coarse, thick, rounded commarginal ridges that are not reflected dorsally or thinned into elevated ridges, glossy cream with rose and brown mottlings, periostracum varnishlike but thin, lunule impressed and bounded by groove, escutcheon not bounded by groove; interior white sometimes flushed posteriorly with light orange or purple, pallial sinus small and tapered, margin denticulate, hinge with no cardinal teeth bifid, 3a and 1 diverging, anterior lateral tooth absent. Florida Keys, Gulf of Mexico. Length 25 mm.

### Lirophora cf. latilirata (Conrad, 1841) – Imperial Venus

Rounded trigonal, solid, with five to nine coarse, thick, rounded commarginal ridges that are reflected dorsally but not thinned into elevated ridges, glossy cream with rose and brown mottlings, periostracum varnishlike, lunule impressed and bounded by groove, escutcheon not bounded by groove; interior white flushed with orange, pallial sinus small and tapered, margin denticulate, hinge with no cardinal teeth bifid, 3a and 1 diverging, anterior lateral tooth absent. North Carolina to Florida, Bermuda, West Indies, Gulf of Mexico, Caribbean Central America, South America (to Brazil). Length 30 mm. Note: *Lirophora latilirata* is a Miocene fossil; the living species that has been referred by this name remains unnamed (P. Roopnarine, pers. comm., August 2005).

### Lirophora paphia (Linnaeus, 1767) – King Venus

Rounded trigonal, with concave margin at lunule, solid, with 10–12 coarse, thick, rounded commarginal ridges that are crimped ventrally by radial grooves, reflected dorsally, and thinned into elevated ridges posteriorly, glossy white with numerous brown triangles and broken radial rays, periostracum varnishlike, lunule impressed and bounded by groove, escutcheon not bounded by groove; interior white often flushed with purple or brown, pallial sinus small and tapered, margin denticulate, hinge with no cardinal teeth bifid, 3a and 1 diverging, anterior lateral tooth absent. Florida Keys, Bahamas, West Indies, Gulf of Mexico, Caribbean Central America, South America (to Brazil). Length 45 mm. Formerly in *Chione*.

*Chione elevata*

*Chione mazyckii*

*Chione cancellata*

*Lirophora clenchi*

*Lirophora* cf. *latilirata*

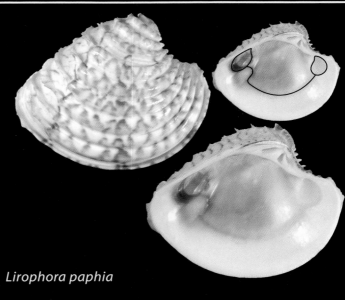

*Lirophora paphia*

***Chionopsis intapurpurea*** (Conrad, 1849) – **Lady-in-Waiting Venus**

Rounded trigonal, solid, with dense, smooth, low commarginal ridges (sharp and flaring anteriorly and posteriorly) and close radial ribs in the interspaces, glossy white to cream with light brown chevrons, periostracum inconspicuous, lunule brownish and bounded by groove, escutcheon not bounded by groove; interior glossy white with purple radial band or blush posteriorly, pallial sinus tapered, margin denticulate, hinge with cardinal teeth 1 and 2b bifid, 3a and 1 diverging, anterior lateral tooth absent. North Carolina to Florida, Bermuda, West Indies, Gulf of Mexico, Caribbean Central America, South America (to Brazil). Length 34 mm (to 43 mm). Formerly in *Chione* or *Puberella*. Compare *Chionopsis pubera*, which is white interiorly.

***Chionopsis pubera*** (Bory Saint-Vincent, 1827) – **Downy Venus**

Rounded trigonal, solid, with dense, fine radial ribs crossed by dense, smooth, low commarginal ridges (sharp and higher anteriorly and posteriorly) that are crenulated ventrally by the radials, glossy white to cream with brown chevrons or spots, periostracum inconspicuous, lunule impressed, brownish and bounded by groove, escutcheon flush and bounded by groove; interior glossy white, pallial sinus tapered, margin denticulate, hinge with cardinal teeth 1 and 2b bifid, 3a and 1 diverging, anterior lateral tooth absent. Florida Keys, West Indies, Gulf of Mexico, South America (to Brazil). Length 70 mm (to 75 mm). Formerly in *Chione* or *Puberella*. Compare *Chionopsis intapurpurea*, which has a purplish radial band or blush interiorly.

***Timoclea pygmaea*** (Lamarck, 1818) – **White Pygmy Venus**

Elongated quadrangular, with fine, often paired, radial ribs crossed by commarginal ridges becoming larger and frilly posteriorly, creamy white with orange-brown flecks, umbones often pink, periostracum inconspicuous, lunule white and bounded by groove, escutcheon crossed by brown bars and not bounded by groove; interior white, posterior end of hinge plate usually purple, pallial sinus small and tapered, margin weakly denticulate, hinge with cardinal teeth 1, 2b, and 3b bifid, 3a and 1 diverging, anterior lateral tooth absent. North Carolina to Florida, Bermuda, Bahamas, West Indies, Caribbean Central America. Length 7 mm (to 16 mm). Formerly in *Chione*. Compare *Timoclea grus*, which has a brown lunule and broad brown band posterodorsally.

***Timoclea grus*** (Holmes, 1858) – **Gray Pygmy Venus**

Elongated quadrangular, with unpaired radial ribs crossed by finer commarginal ridges, posterior 12 or so ribs cut by radial groove, dull gray to white, occasionally pink or orange, periostracum inconspicuous, lunule brown and bounded by groove, escutcheon bounded by groove; interior white with a strong purple-brown ray posteriorly, both ends of hinge plate purple-brown, pallial sinus small and tapered, margin denticulate, hinge with cardinal teeth 1, 2b, and 3b bifid, 3a and 1 diverging, anterior lateral tooth absent. North Carolina to Florida, West Indies, Gulf of Mexico, Caribbean Central America. Length 7 mm (to 11 mm). Compare *Timoclea pygmaea*, which has a white lunule and scattered orange-brown flecks exteriorly.

***Gouldia cerina*** (C. B. Adams, 1845) – **Atlantic Gould Clam**

Oval to roundly trigonal, equilateral, compressed, with flat, close-set commarginal ridges crossed by narrow radial ribs anteriorly and posteriorly forming weakly cancellate surface, creamy white to yellow or orange-brown, occasionally rayed with darker hue, periostracum inconspicuous, lunule long, flush and bounded by groove, escutcheon not bounded by groove; interior white, some with pink or peach blush, pallial sinus very shallow, margin smooth, hinge with no cardinal teeth bifid, 3a and 1 diverging, anterior lateral tooth elongated. Maryland to Florida, Bermuda, Bahamas, West Indies, Gulf of Mexico, Caribbean Central America, South America (to Brazil). Length 5 mm (to 13 mm).

***Anomalocardia cuneimeris*** (Conrad, 1846) – **Pointed Venus**

Trigonal, rostrate posteriorly, with rounded commarginal ridges, more prominent near umbones, glossy white to yellow, orange, or purple, with orange-brown or purple specks and narrow radial streaks, periostracum varnishlike, lunule impressed and bounded by groove, escutcheon not bounded by groove; interior color as exterior or often purple, pallial sinus small and tapered, margin denticulate, hinge with no cardinal teeth bifid, 3a and 1 diverging, anterior lateral teeth absent. North Carolina to Florida, Bahamas, West Indies, Gulf of Mexico, Caribbean Central America. Length 20 mm. Syn. *auberiana* d'Orbigny, 1853.

*Chionopsis intapurpurea*

*Chionopsis pubera*

*Timoclea pygmaea*

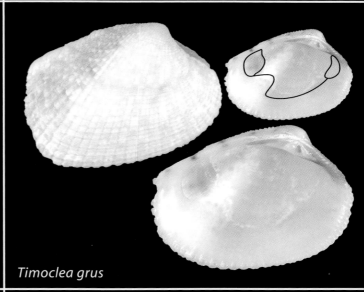

*Timoclea grus*

*Gouldia cerina*

*Anomalocardia cuneimeris*

### *Pitarenus cordatus* (Schwengel, 1951) – **Corded Pitar**

Rounded trigonal, inflated, solid, with close commarginal threads, white, periostracum with calcareous needles, lunule flush and bounded by groove, escutcheon absent; interior white with pinkish blush, pallial sinus tapered, margin denticulate, hinge with cardinal tooth 3b bifid, 3a and 1 parallel, anterior lateral tooth minute. Florida, Gulf of Mexico, South America (Brazil). Length 40 mm. Formerly in *Pitar*. Note: The valves photographed are of a PARATYPE specimen of *Pitarenus cordatus*.

### *Pitar albidus* (Gmelin, 1791) – **White Pitar**

Rounded trigonal to quadrangular, smooth, white, periostracum with calcareous needles, lunule large, flush and bounded by groove, escutcheon absent; interior white, pallial sinus rounded, margin smooth, hinge with cardinal tooth 3b bifid, 3a and 1 parallel, anterior lateral tooth elongated. Florida Keys, West Indies, Caribbean Central America, South America (to Brazil). Length 25 mm.

### *Pitar circinatus* (Born, 1778) – **Purple Pitar**

Rounded trigonal, solid, with regular thin to lamellose commarginal ribs and microscopic commarginal striae in the interspaces, white sometimes with reddish or pinkish radial bands, periostracum thin and inconspicuous, lunule flush and bounded by groove, escutcheon very narrow and not bounded by groove, both lunule and escutcheon deep purple; interior white, pallial sinus rounded and displaced ventrally, margin smooth, hinge with cardinal tooth 3b bifid, 3a and 1 parallel, anterior lateral tooth elongate. Florida Keys, West Indies, Caribbean Central America, South America (to Brazil). Length 30 mm (to 41 mm).

### *Pitar dione* (Linnaeus, 1758) – **Royal Comb Pitar**

Rounded trigonal, with wide, erect commarginal ribs, wider anteriorly and ending in two radial rows of long spines posteriorly, creamy white to pale pink with violet or magenta markings, periostracum inconspicuous, lunule flush, unridged and bounded by groove, escutcheon not bounded by a groove; interior white, pallial sinus tapered and displaced ventrally, margin smooth, hinge with cardinal tooth 3b bifid, 3a and 1 parallel, anterior lateral tooth elongated. Florida Keys, West Indies, Gulf of Mexico, Caribbean Central America, South America (to Venezuela). Length 45 mm.

### *Pitar fulminatus* (Menke, 1828) – **Lightning Pitar**

Rounded trigonal, smooth, white with orange-brown zigzags and chevrons, periostracum with calcareous needles, lunule very large, flush and bounded by groove, escutcheon absent; interior white, pallial sinus tapered and displaced ventrally, margin smooth, hinge with cardinal tooth 3b bifid, 3a and 1 parallel, anterior lateral tooth elongated. North Carolina to Florida, Bermuda, Bahamas, West Indies, Gulf of Mexico, Caribbean Central America, South America (to Brazil). Length 40 mm (to 48 mm). Compare *Pitar simpsoni*, which is smaller (as an adult), sometimes has orange-brown chevrons externally, and is white or purple interiorly.

### *Pitar simpsoni* (Dall, 1895) – **Simpson's Pitar**

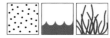

Rounded trigonal, smooth, creamy white or purplish with orange-brown zigzag lines and purple at umbones, periostracum with calcareous needles (see inset), lunule flush and bounded by groove, escutcheon absent; interior white or purple, pallial sinus tapered, margin smooth, hinge with cardinal tooth 3b bifid, 3a and 1 parallel, anterior lateral tooth elongated. Florida, Bahamas, West Indies, Gulf of Mexico, Caribbean Central America. Length 10 mm (to 14 mm). Compare *Pitar fulminatus*, which is larger (as an adult), has orange-brown zigzags and chevrons externally, and is always white internally. Note: The right valve of this specimen has been drilled by a predatory gastropod.

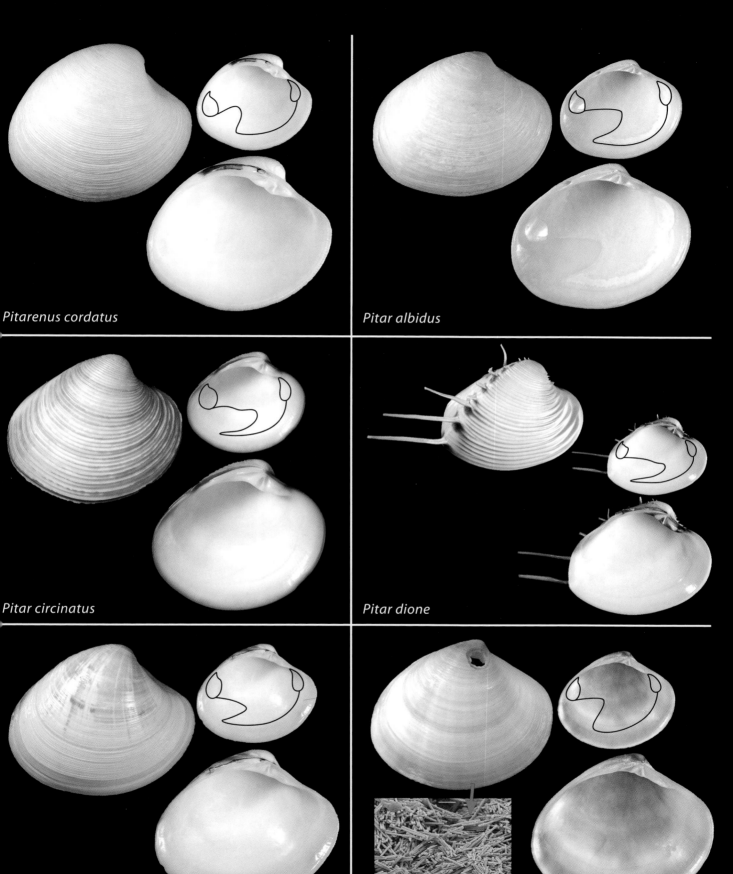

*Pitarenus cordatus*

*Pitar albidus*

*Pitar circinatus*

*Pitar dione*

*Pitar fulminatus*

*Pitar simpsoni*

***Transennella conradina*** (Dall, 1884) – **Conrad's Transennella**

Rounded trigonal, rostrate posteriorly, with fine, raised commarginal ridges, smooth centrally, cream to brown, in some with brown zigzags, prodissoconch purple-rose, periostracum varnishlike, lunule raised in center and bounded by groove, escutcheon not bounded by groove; interior white with posterior purple ray, pallial sinus tapered, margin with oblique grooves, hinge with cardinal teeth 2b and 3b bifid, 3a and 1 parallel, anterior lateral tooth elongated. North Carolina to Florida, Bahamas, Gulf of Mexico. Length 5 mm (to 17 mm). Compare other *Transennella*, which all have rounded pallial sinuses; *T. simpsoni* is completely smooth externally, *T. cubaniana* and *T. culebrana* are more evenly trigonal (nonrostrate), and *T. cubaniana* has dense commarginal ridges throughout.

***Transennella cubaniana*** (d'Orbigny, 1853) – **Cuban Transennella**

Rounded trigonal, not rostrate posteriorly, with fine, dense commarginal ridges over entire surface, white, rarely flecked with brown, occasionally with purple-brown umbones, periostracum varnishlike, lunule raised in center and bounded by groove, escutcheon not bounded by groove; interior white, pallial sinus rounded, margin with oblique grooves (see inset, area highlighted in green), hinge with cardinal teeth 2b and 3b bifid, 3a and 1 parallel, anterior lateral tooth elongated. Florida Keys, Bermuda, Bahamas, West Indies, Caribbean Central America, South America (to Brazil). Length 5 mm (to 11 mm). Compare other *Transennella*, which are all smooth-shelled at least centrally and all weakly to strongly rostrate.

***Transennella culebrana*** (Dall & Simpson, 1901) – **Puerto Rico Transennella**

Rounded trigonal (more evenly trigonal than *T. cubaniana*, or shortly rostrate), with commarginal ridges becoming smooth centrally, whitish to brownish, periostracum yellow-brown, lunule raised in center and bounded by groove, escutcheon not bounded by groove; interior white, pallial sinus rounded, margin with oblique grooves, hinge with cardinal teeth 2b and 3b bifid, 3a and 1 parallel, anterior lateral tooth elongated. Florida Keys, Bahamas, West Indies, South America (Colombia, Uruguay). Length 7 mm. Compare other *Transennella*; *T. simpsoni* and *T. conradina* are more strongly rostrate, whereas *T. cubaniana* is less so, *T. conradina* has a tapered pallial sinus, *T. simpsoni* is completely smooth externally, and *T. cubaniana* has dense commarginal ridges throughout. Note: The valves photographed are of the HOLOTYPE specimen of *Transennella culebrana*.

***Transennella stimpsoni*** Dall, 1902 – **Banded Transennella**

Rounded trigonal, rostrate posteriorly, smooth and glossy, whitish, occasionally with brown or violet chevrons, occasionally with purplish flush and brown rays at umbones, periostracum varnishlike, lunule raised in center and bounded by groove, escutcheon not bounded by groove; interior white flushed with purple, pallial sinus rounded, margin with oblique grooves, hinge with cardinal teeth 1, 2b, and 3b bifid, 3a and 1 parallel, anterior lateral tooth elongated. North Carolina to Florida, Bahamas, Gulf of Mexico, Caribbean Central America, South America (Colombia, Brazil). Length 14 mm (to 15 mm). Compare other *Transennella*; *T. conradina* and *T. culebrana* have raised commarginal ridges (except at center valve), *T. conradina* has a tapered pallial sinus, *T. cubaniana* and *T. culebrana* are more evenly trigonal (nonrostrate), and *T. cubaniana* has dense commarginal ridges throughout. Note: The valve photographed is of the HOLOTYPE specimen of *Transennella stimpsoni*. Also known as Stimpson's Transennella.

***Gemma gemma*** (Totten, 1834) – **Amethyst Gem Clam**

Rounded trigonal, smooth with fine commarginal ridges, glossy whitish flushed with purple-brown at umbones and posteriorly, periostracum varnishlike, lunule and escutcheon absent; interior white with purple-brown hinge, pallial sinus small and tapered, margin denticulate, hinge with cardinal tooth 3b bifid, 3a and 1 diverging, anterior lateral tooth absent. Eastern Canada to Florida, Bahamas, West Indies, Gulf of Mexico, also introduced to eastern Pacific. More typical of northern regions. Length 4 mm.

***Parastarte triquetra*** (Conrad, 1846) – **Brown Gem Clam**

Trigonal, equilateral, smooth and glossy, tan to brown flushed with pink, periostracum varnishlike, lunule and escutcheon absent; interior purplish, pallial sinus very shallow, margin denticulate, hinge with cardinal tooth 2b bifid, 3a and 1 diverging, anterior lateral tooth absent. Florida, Bahamas, West Indies, Gulf of Mexico, Caribbean Central America. Length < 3 mm (to 4 mm).

*Transennella conradina*

*Transennella cubaniana*

*Transennella culebrana*

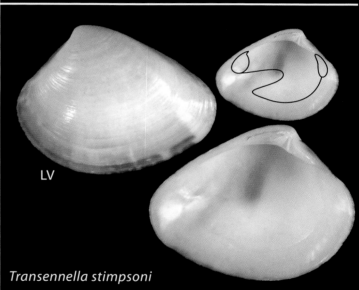

LV

*Transennella stimpsoni*

*Gemma gemma*

*Parastarte triquetra*

### *Tivela abaconis* Dall, 1902 – **Abaco Tivela**

Rounded trigonal, equilateral, smooth, white with deep rose at umbones, central flush of purplish pink or orange, periostracum varnishlike with calcareous needles, lunule large, flush and bounded by groove, escutcheon absent; interior whitish with pinkish blush, pallial sinus rounded, margin smooth, hinge with cardinal tooth 2b bifid, 4b split forming pseudocardinal tooth in left valve, 3a and 1 diverging, anterior lateral tooth elongated. Florida Keys, Bahamas, West Indies, Caribbean Central America, South America (Colombia). Length 6 mm (to 11 mm).

### *Tivela floridana* Rehder, 1939 – **Florida Tivela**

Rounded trigonal, equilateral, smooth and polished with microscopic growth lines near margins, glossy tan or purplish, periostracum varnishlike, lunule large, flush and bounded by groove, escutcheon absent; interior tan to purplish with purple-brown radial ray, pallial sinus rounded, margin smooth, hinge with cardinal tooth 2b bifid, 4b split forming pseudocardinal tooth in left valve, 3a and 1 diverging, anterior lateral tooth elongated. Florida, Caribbean Central America. Length 7 mm (to 11 mm).

### *Tivela mactroides* (Born, 1778) – **Trigonal Tivela**

Rounded trigonal, posterior pointed in large specimens, equilateral, inflated, smooth and polished with fine growth lines, whitish or tan with rays and mottled with brown, anterior sometimes violet or purplish, periostracum varnish-like, lunule flush and bounded by groove, escutcheon absent; interior purple-brown dorsally, pallial sinus rounded, margin smooth, hinge with cardinal tooth 2b bifid, 4b split forming pseudocardinal tooth in left valve, 3a and 1 diverging, anterior lateral tooth elongated. Florida Keys, West Indies, Gulf of Mexico, Caribbean Central America, South America (to Brazil). Length 40 mm.

### *Cooperella atlantica* Rehder, 1943 – **Atlantic Cooper's Clam**

Oval, equilateral, fragile, smooth and glossy, translucent white, periostracum inconspicuous, lunule and escutcheon absent; interior white, pallial sinus wide and rounded, margin smooth, hinge with cardinal tooth 2b bifid, 3a absent, anterior lateral tooth absent. Florida, Bahamas, West Indies, South America (Colombia, Brazil). Length 4 mm (to 16 mm).

### *Choristodon robustus* (G. B. Sowerby I, 1834) – **Boring Choristodon**

Oval, posterior bluntly elongated to rostrate, anteriorly and posteriorly gaping, solid, moderately inflated to compressed, with numerous coarse, irregular radial ribs crossed by heavy commarginal growth lines forming irregular scaling (strongest posteriorly), chalky grayish white, periostracum inconspicuous, lunule and escutcheon absent; interior white with light brown staining, pallial sinus wide and rounded, margin smooth, hinge with cardinal teeth 2b and 3b bifid, 3a absent, anterior lateral tooth absent; living bored into soft rock or dead coral. North Carolina to Florida, Bermuda, Bahamas, West Indies, Gulf of Mexico, Caribbean Central America, South America (Colombia, Brazil), also eastern Pacific. Length 25 mm. Syn. *typica* Jonas, 1844. Formerly in *Rupellaria*, formerly Petricolidae. Compare *Petricola lapicida* (p. 304), which is less rostrate and has coarse radial ribs posterodorsally and finer dorsoventral riblets at midvalve.

### *Petricolaria pholadiformis* (Lamarck, 1818) – **False Angelwing**

Elongated oval, fragile, with ca. 60 radial ribs, anteriormost 10 larger and with prominent scales, finer posteriorly, chalky white or yellowish white, periostracum thin and most evident at margins, lunule and escutcheon absent; interior white, pallial sinus rounded, margin smooth, hinge line excavated with cardinal teeth 2b and 3b bifid, 3a absent, anterior lateral tooth absent; living bored into soft rock, clay, or consolidated mud. Eastern Canada to Florida, West Indies, Gulf of Mexico, also recorded as introduced to eastern Pacific and western Europe to the Congo coast. Length 30 mm (to 80 mm). Formerly in separate family Petricolidae.

*Tivela abaconis*

*Tivela floridana*

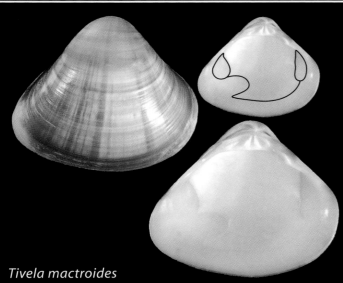

*Tivela mactroides*

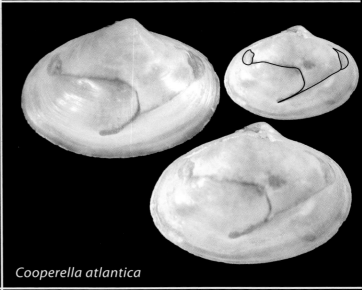

*Cooperella atlantica*

*Choristodon robustus*

*Petricolaria pholadiformis*

# Family Tellinidae – Tellin Clams

**Classification**
AUTOLAMELLIBRANCHIATA Grobben, 1894
HETEROCONCHIA Hertwig, 1895
Heterodonta Neumayr, 1883
Veneroida H. Adams & A. Adams, 1856
Tellinoidea Blainville, 1814
Tellinidae Blainville, 1814

**Featured species**
*Scissula similis* (J. Sowerby, 1806) – **Candystick Tellin**

pallial line

Elongated oval, posterior end blunt, thin-walled, smooth and polished with incised oblique striae, white or yellow commonly with 6–12 pink radial rays; interior glossy yellowish with red rays or solid pink or yellow, pallial sinus deep and close to anterior muscle scar, dropping vertically to join ventral pallial line, hinge commonly with red splotch anterior to cardinal teeth. Florida, Bermuda, Bahamas, West Indies, Gulf of Mexico, Caribbean Central America. Length 16 mm (to 28 mm). Compare *Angulus probrinus* (p.333), which is glossy smooth, without oblique grooves (scissulations).

**A living *Scissula similis*** from Newfound Harbor displays its partially extended siphons and foot.

# Family description

The tellinid shell is small to medium-sized (to 125 mm), usually thin-walled, elongated oval to quadrangular, usually with a rounded anterior end and an elongated or blunt posterior end, and usually with slight posterior flexure toward the right side. It is EQUIVALVE to INEQUIVALVE (with either right or left valve flatter than the other), compressed, and

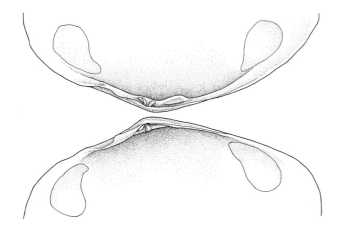

usually not gaping. The shell is usually INEQUI-LATERAL (umbones posterior), with PROSO-, OPISTHO-, or ORTHOGYRATE UMBONES. Shell microstructure is ARAGONITIC and three-layered, with a composite PRISMATIC outer layer (absent in some species), a CROSSED LAMELLAR middle layer, and a COMPLEX CROSSED LAMELLAR or HOMOGENOUS inner layer, the last confined within the pallial line. TUBULES are absent. Exteriorly tellinids are often brightly colored, frequently with color arranged in radial rays, and are covered by thin, varnishlike, inconspicuous PERIOSTRACUM. Sculpture is smooth to weakly commarginal, in some cases with SCISSULATE grooves; a few species have prominent radial ribs or filelike scales on one valve only. LUNULE and ESCUTCHEON are absent. Interiorly the shell is non-NACREOUS. The PALLIAL LINE is strong, with a deep SINUS, frequently differing in the two valves; in some species with an INTERLINEAR SCAR. The inner shell margins are smooth. The HINGE PLATE is narrow and HETERODONT, with two (bifid in some species) CARDINAL TEETH in each valve; anterior and posterior LATERAL TEETH are present (in Tellininae) or absent (in Macominae). The LIGAMENT is PARIVINCULAR, OPISTHODETIC, and usually set on weak, elongated NYMPHS.

The animal is ISOMYARIAN or weakly HETEROMYARIAN (posterior ADDUCTOR MUSCLE slightly smaller); anterior and posterior pedal retractor muscles are present. Pedal protractors are present; pedal elevators can be present or absent. The MANTLE margins are usually largely unfused ventrally. Posterior EXCURRENT and INCURRENT SIPHONS are long, separate, and independently mobile, with a CRUCIFORM MUSCLE at their base. A specialized, open, sensory organ accompanies the cruciform muscle. HYPOBRANCHIAL GLANDS have not been reported. The FOOT is small to large, wedge-shaped, heeled, laterally com-

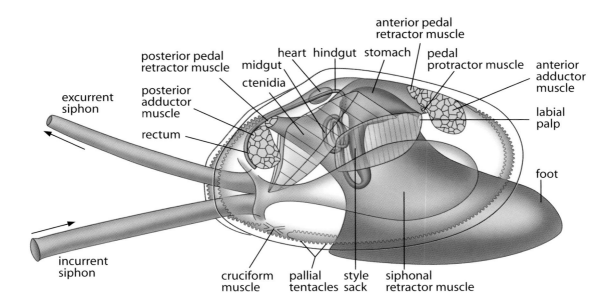

pressed, and in some species is capable of flattening ventrally into a sole; a BYSSAL GROOVE and BYSSUS are absent in the adult.

The LABIAL PALPS are very large and quadrangular. The CTENIDIA are EULAMELLI-BRANCH (SYNAPTORHABDIC) and HOMORHABDIC; the outer demibranchs are smaller than the inner demibranchs and are reflected dorsally. The gills are not inserted into (or fused with) the distal oral groove of the palps (CATEGORY III association). Incurrent and excurrent water flows are posterior. The STOMACH is TYPE V, with a posterodorsal, in some cases lobed, gastric appendix; the MIDGUT is coiled. The HINDGUT passes through the ventricle of the heart, and leads to a sessile rectum. An AORTIC BULB has been reported for some species. Tellinids are GONOCHORISTIC or PROTANDRIC HERMAPH-RODITES and produce planktonic VELIGER larvae; late-stage larvae have a distinctively trigonal shell with an extended anterior end. The nervous system is not concentrated. STATOCYSTS (with STA-TOLITHS) have been reported in adult *Macoma* and *Tellina*. AB-DOMINAL SENSE ORGANS have not been reported.

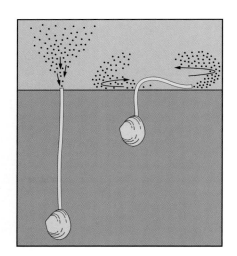

Tellinids are marine or estuarine SUSPENSION FEEDERS, deeply INFAUNAL in soft sediments. Most species are also capable of opportunistic DEPOSIT FEEDING, using the long extensible in-current siphon to sweep the superficial detritus; the excurrent siphon extends more or less horizontally below the surface. Tellinids are capable of rapid reburying if dislodged. In northern Europe's Wadden Sea, juvenile *Macoma balthica* (Lin-naeus, 1758) migrate several kilometers during their ontogeny, each secreting a long byssal thread that drags the animal along in the water column. Many tellins are important ecologically because of their often-great abundance, and are staple food items for bottom-feeding commercial fish (e.g., flounder).

The family Tellinidae is known since the Cretaceous and is represented by ca. 90 living genera and ca. 350 species, distributed worldwide but more species-rich in tropical seas.

## References

Arruda, E. P., and O. Domaneschi. 2005. New species of *Macoma* (Bivalvia: Tellinoidea: Tellinidae) from southeastern Brazil, and with description of its gross anatomy. *Zootaxa*, 1012: 13–22.

Barón, P. J., and N. F. Ciocco. 1997. Anatomía de la almeja *Tellina petitiana* d'Orbigny, 1846. I. Organización general, valvas, manto, sifones, pie y branquias (Bivalvia, Tellinidae). *Revista de Biología Marina y Oceanografía,* 32(2): 95–110.

Barón, P. J., and N. F. Ciocco. 1998. Anatomía de la almeja *Tellina petitiana* d'Orbigny, 1846. III. Sistema nervioso y gónada (Bivalvia, Tellinidae). *Revista de Biología Marina y Oceanografía,* 33(1): 139–154.

Boss, K. J. 1966. The subfamily Tellininae in the western Atlantic. The genus *Tellina. Johnsonia,* 4(45): 217–272.

Boss, K. J. 1968. The subfamily Tellininae in the western Atlantic. The genera *Tellina* (part II) and *Tellidora. Johnsonia,* 4(46):273–344.

Boss, K. J. 1969. The subfamily Tellininae in the western Atlantic. The genus *Strigilla. Johnsonia,* 4(47): 345–366.

Ciocco, N. F., and P. J. Barón. 1998. Anatomía de la almeja *Tellina petitiana* d'Orbigny, 1846. II. Sistema digestivo, corazón, riñones, cavidad y glándulas pericárdicas (Bivalvia, Tellinidae). *Revista de Biología Marina y Oceanografía,* 33(1): 73–87.

Dall, W. H. 1900. Synopsis of the family Tellinidae and of the North American species. *Proceedings of the United States National Museum,* 23(1210): 285–326, pls. 2–4.

Gilbert, M. A. 1977. The behaviour and functional morphology of deposit feeding in *Macoma balthica* (Linne, 1758), in New England. *Journal of Molluscan Studies,* 43(1): 18–27.

Gilbert, M. A. 1978. Aspects of feeding in three species of *Tellina* (Tellinidae: Bivalvia) from Tuckertown Bay, Bermuda. *Bulletin of the American Malacological Union for 1977,* pp. 56–60.

Goeij, P. de, and P. Luttikhuizen. 1998. Deep-burying reduces growth in intertidal bivalves: field and mesocosm experiments with *Macoma balthica. Journal of Experimental Marine Biology and Ecology,* 228: 327–337.

Hiddink, J. G. 2002. *The Adaptive Value of Migrations for the Bivalve* Macoma balthica. Ph.D. Dissertation, Rijksuniversiteit Groningen, The Netherlands, 172 pp. Online version http://irs.ub.rug.nl/ppn/240848039, last accessed 03 December 2006.

Reid, R. G. B., and A. Reid. 1969. Feeding processes of members of the genus *Macoma* (Mollusca: Bivalvia). *Canadian Journal of Zoology,* 47(4): 649–657.

Trueman, E. R. 1949. The ligament of *Tellina tenuis. Proceedings of the Zoological Society of London,* 119(3): 717–742.

Wilson, J. G. 1990. Gill and palp morphology of *Tellina tenuis* and *T. fabula* in relation to feeding. Pages 141–150, in B. Morton, ed., *The Bivalvia, Proceedings of a Memorial Symposium in Honour of Sir Charles Maurice Yonge (1899–1986), Edinburgh, 1986.* Hong Kong University Press, Hong Kong.

Yonge, C. M. 1949. On the structure and adaptations of the Tellinacea, deposit-feeding Eulamellibranchia. *Philosophical Transactions of the Royal Society of London, Series B, Biological Sciences,* 234(609): 29–76.

### *Scissula iris* (Say, 1822) – **Iris Tellin**

Elongated trigonal, posterior end pointed, thin-walled, smooth and polished with incised oblique striae, white exteriorly and interiorly; interior with wavy oblique lines and two radial thickenings or weak, white radial ribs posteriorly, pallial sinus deep and close to anterior muscle scar, dropping almost vertically to join ventral pallial line. North Carolina to Florida, Bermuda, Gulf of Mexico, Caribbean Central America. Length 10 mm (to 15 mm). Note: Also known as Rainbow Tellin.

### *Scissula candeana* (d'Orbigny, 1853) – **Candé's Wedge Tellin**

Elongated trigonal, posterior end pointed, smooth and polished with incised oblique striae anteriorly, white flushed with pale yellow or pink (especially at umbones); interior glossy white flushed with yellow, pallial sinus deep and close to anterior muscle scar, dropping almost vertically to join ventral pallial line. Florida, Bermuda, Bahamas, West Indies, Caribbean Central America, South America (Uruguay). Length 10 mm (to 16 mm).

### *Scissula consobrina* (d'Orbigny, 1853) – **Sibling Tellin**

Elongated oval, posterior end blunt, fragile, smooth and polished with incised oblique striae, white with red or pink rays exteriorly and interiorly, pallial sinus deep and close to anterior muscle scar, dropping vertically to join ventral pallial line. North Carolina to Florida, Bermuda, West Indies, Gulf of Mexico, South America (Colombia). Length 12 mm (to 14 mm).

### *Strigilla carnaria* (Linnaeus, 1758) – **Large Strigilla**

Oval, approximately equilateral, thin-walled, smooth with incised oblique striae forming a single row of chevrons on posterior slope, white or pink with white border, pallial sinus deep and close to anterior muscle scar, dropping vertically to join ventral pallial line. North Carolina to Florida, West Indies, Gulf of Mexico, Caribbean Central America, South America (Brazil to Argentina). Length 13 mm (to 28 mm).

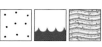

### *Strigilla mirabilis* (Philippi, 1841) – **White Strigilla**

Oval, umbones slightly anterior, thin-walled, smooth with incised oblique striae forming four to six rows of chevrons on posterior slope, white tinted with yellow exteriorly and interiorly, pallial sinus deep and close to anterior muscle scar, dropping vertically to join ventral pallial line. North Carolina to Florida, Bermuda, Bahamas, West Indies, Gulf of Mexico, Caribbean Central America, South America (to Brazil). Length 10 mm (to 15 mm).

### *Strigilla pisiformis* (Linnaeus, 1758) – **Pea Strigilla**

Oval, umbones slightly anterior, thin-walled, smooth with incised oblique striae forming two sets of chevrons on posterior slope and a third on anterior slope, white; interior white with pink in deepest part, pallial sinus deep and close to anterior muscle scar, dropping vertically to join ventral pallial line. Florida, Bahamas, West Indies, Caribbean Central America, South America (to Brazil). Length 10 mm (to 15 mm).

*Scissula iris*

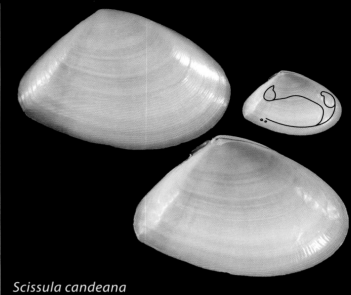

*Scissula candeana*

*Scissula consobrina*

*Strigilla carnaria*

*Strigilla mirabilis*

*Strigilla pisiformis*

*Arcopagia fausta* (Pulteney, 1799) – **Lucky Tellin**

Rounded triangular to oval, posterior truncated, solid, smooth with microscopic radial threads, creamy white flushed with yellow at umbones, with flaky brown periostracum at margins; interior glossy white flushed with yellow, pallial sinus reaching just past midpoint, turning posteriorly before joining ventral pallial line, interlinear scar present. North Carolina to Florida, Bermuda, Bahamas, West Indies, Caribbean Central America, South America (to Venezuela). Length 85 mm (to 88 mm). Compare *Laciolina laevigata*, which has peach-pink radial rays, inconspicuous periostracum, and a less truncate posterior end; its pallial sinus drops vertically to join ventral pallial line. Note: Also known as Favored or Faust Tellin.

*Leporimetis intastriata* (Say, 1826) – **Twisted Duck Clam**

Oval, posterior keeled on right valve (furrowed on left valve) and strongly twisted, thin-walled, smooth, white occasionally tinted with yellow exteriorly and interiorly, adductor muscle scars unequal (anterior one long and narrow), pallial sinus very large, turning posteriorly before joining ventral pallial line. South Carolina to Florida, Bermuda, Bahamas, West Indies, Gulf of Mexico, Caribbean Central America, South America (Colombia). Length 45 mm (to 66 mm). Formerly in *Psammotreta* (Psammobiidae). Note: Also known as Atlantic Fat-Tellin or Atlantic Grooved Macoma.

*Laciolina laevigata* (Linnaeus, 1758) – **Smooth Tellin**

Oval, posterior bluntly pointed with moderate twist, solid, smooth with raised commarginal threads and faint radial scratches, left valve smoother and glossier than right, white flushed with pale yellow, with peach-pink radial rays occasionally confined to margin; interior glossy white flushed with yellow, pallial sinus reaching midvalve, dropping vertically to join ventral pallial line. North Carolina, Florida, Bermuda, Bahamas, West Indies, Gulf of Mexico, Caribbean Central America, South America (Colombia). Length 60 mm. Compare *Arcopagia fausta*, which is evenly white to yellow, and has flaky brown periostracum and a more strongly truncate posterior end; its pallial sinus turns posteriorly before joining ventral pallial line.

*Laciolina magna* (Spengler, 1798) – **Great Tellin**

Elongated oval, posterior blunt with weak radial ridge posterodorsally, solid, glossy smooth with commarginal growth lines (stronger on right valve) strongest on posterior slope where they are crossed by microscopic diagonal threads, left valve glossy white (rarely faintly yellow), right valve glossy orange to pinkish with pink umbones fading to yellow-orange, occasionally with short, pink radial rays at umbones; interior colors reflecting exterior, pallial sinus reaching midvalve, turning posteriorly before joining ventral pallial line. North Carolina to Florida, Bermuda, Bahamas, West Indies, Gulf of Mexico, Caribbean Central America, South America (Colombia). Length 95 mm (to 122 mm).

*Tellinella listeri* (Röding, 1798) – **Speckled Tellin**

Elongated oval, posterior narrowed, strongly twisted and slightly hooked ventrally, right posterior with two radial ridges, solid, with erect commarginal ridges and dense radial threads, creamy white to pink, yellowish at umbones, with dashes and zigzag markings (occasionally forming radial rays) of light purplish brown; interior glossy white flushed with yellow, pallial sinus deep, turning posteriorly before joining ventral pallial line, interlinear scar present. North Carolina to Florida, Bermuda, Bahamas, West Indies, Gulf of Mexico, Caribbean Central America, South America (to Brazil). Length 50 mm (to 100 mm).

*Tellidora cristata* (Récluz, 1842) – **White-crested Tellin**

Rounded trigonal, posterior pointed, approximately equilateral, very compressed, solid, with coarse commarginal ridges extended dorsally both anteriorly and posteriorly into "sawtooth" margins, white exteriorly and interiorly, pallial sinus deep and close to anterior muscle scar, turning posteriorly before joining ventral pallial line. North Carolina to Florida, Gulf of Mexico, Caribbean Central America. Length 30 mm (to 37 mm).

*Arcopagia fausta*

*Leporimetis intastriata*

*Laciolina laevigata*

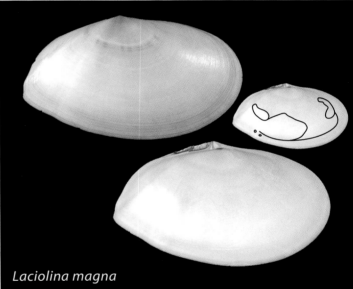

*Laciolina magna*

*Tellinella listeri*

*Tellidora cristata*

### *Tellina radiata* Linnaeus, 1758 – **Sunrise Tellin**

Elongated oval, posterior narrowed, solid, smooth and highly polished, white or cream flushed with yellow, usually with pink or rose radial rays, umbones usually bright red; interior flushed with yellow or orange, pallial sinus deep and close to anterior muscle scar, turning posteriorly before joining ventral pallial line, short interlinear scar present. South Carolina to Florida, Bermuda, Bahamas, West Indies, Gulf of Mexico, Caribbean Central America, South America (to Brazil). Length 90 mm (to 105 mm).

### *Phyllodina squamifera* (Deshayes, 1855) – **Eastern Crenulate Tellin**

Elongated oval, equilateral, posterior narrowed and slightly hooked ventrally, with fine commarginal ridges extended into strong crenulations at dorsal margins, white with yellow or orange flush exteriorly and interiorly, pallial sinus deep and ascending, turning posteriorly before joining ventral pallial line, interlinear scar present. North Carolina to Florida, Gulf of Mexico, South America (Brazil). Length 20 mm (to 27 mm).

### *Angulus merus* (Say, 1834) – **Pure Tellin**

Oval, fragile, smooth with fine commarginal ridges, white exteriorly and interiorly, pallial sinus very deep and close to anterior muscle scar, turning posteriorly before joining ventral pallial line. Florida, Bermuda, Bahamas, West Indies, Gulf of Mexico, Caribbean Central America, South America (Colombia, Brazil). Length 20 mm (to 26 mm). Compare *Angulus paramerus* and *A. tampaensis*, which have vertically descending pallial sinuses; *A. paramerus* has a larger anterior lateral tooth, and *A. tampaensis* is pinkish or peach-colored.

### *Angulus tampaensis* (Conrad, 1866) – **Tampa Tellin**

Oval, posterior pointed, fragile, smooth with very fine commarginal threads, white with faint pink or peach blush exteriorly and interiorly, pallial sinus very deep and close to anterior muscle scar, dropping vertically to join ventral pallial line. Florida, West Indies, Gulf of Mexico, Caribbean Central America. Length 20 mm. Compare *Angulus merus* and *A. paramerus*, which are white externally; *A. paramerus* has a larger anterior lateral tooth, and *A merus* has a pallial sinus that turns posteriorly before joining the ventral pallial line.

### *Angulus paramerus* (Boss, 1964) – **Perfect Tellin**

Oval, posterior blunt, fragile, with fine, dense commarginal ridges and weak radial threads, dull white exteriorly and interiorly, umbones mottled with semitranslucent patches; interior glossy white, occasionally with radial squiggles corresponding to exterior threads, pallial sinus very deep and close to anterior muscle scar (occasionally touching it), dropping vertically to join ventral pallial line, anterior lateral tooth large. Florida, Bermuda, Bahamas, West Indies. Length 13 mm (to 15 mm). Compare *Angulus merus* and *A. tampaensis*, which have less prominent anterior lateral teeth; *A. tampaensis* is pinkish or peach-colored, and *A. merus* has a pallial sinus that turns posteriorly before joining the ventral pallial line.

### *Angulus agilis* (Stimpson, 1857) – **Northern Dwarf Tellin**

Rounded triangular, fragile, smooth with very fine commarginal striae, glossy white with pinkish opalescent sheen; interior white, pallial sinus deep and close to anterior muscle scar, dropping almost vertically to join ventral pallial line. Eastern Canada to Florida, Gulf of Mexico. Length 13 mm (to 16 mm).

*Tellina radiata*

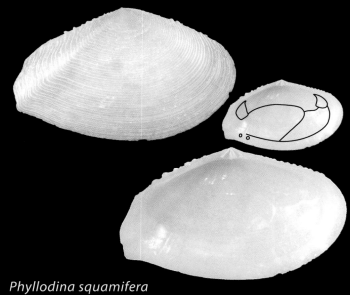

*Phyllodina squamifera*

*Angulus merus*

*Angulus tampaensis*

RV

*Angulus paramerus*

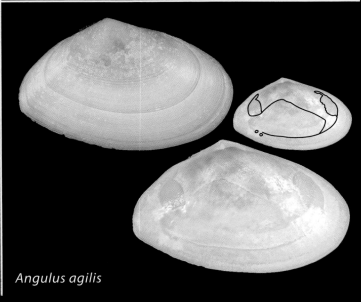

*Angulus agilis*

### *Angulus texanus* (Dall, 1900) – **Texas Tellin**

Elongated trigonal, posterior pointed and sharply twised to right, fragile, smooth with very fine commarginal striae, umbones with oblique creases, white occasionally suffused with yellow and with faint opalescent sheen; interior glossy white or yellowish, pallial sinus deep and close to anterior muscle scar, dropping vertically to join ventral pallial line. North Carolina to Florida, Bermuda, Bahamas, Gulf of Mexico, Caribbean Central America. Length 13 mm (to 14 mm).

### *Angulus probrinus* (Boss, 1964) – **Slandered Tellin**

Elongated oval to quadrangular, posterior blunt, fragile, glossy smooth with very fine commarginal striae, pink or yellow with pinkish opalescent sheen; interior colors reflecting exterior, pallial sinus deep and close to anterior muscle scar, dropping vertically to join ventral pallial line. North Carolina, Florida, West Indies, Gulf of Mexico, South America (Colombia, Brazil). Length 10 mm (to 25 mm). Compare *Scissula similis*, which has oblique grooves (scissulations) externally.

### *Angulus sybariticus* (Dall, 1881) – **Dall's Dwarf Tellin**

Elongated oval, posterior blunt, fragile, glossy smooth with numerous raised commarginal threads, more crowded anteriorly, white flushed with pink and yellow, with orange-brown or pink radial ray on posterior slope; interior flushed with pink and yellow, pallial sinus deep and close to anterior muscle scar, dropping nearly vertically to join ventral pallial line. North Carolina to Florida, Bermuda, Bahamas, West Indies, Gulf of Mexico, Caribbean Central America, South America (Colombia, Brazil). Length 5 mm (to 13 mm).

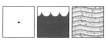

### *Elliptotellina americana* (Dall, 1900) – **American Dwarf Tellin**

Oval, solid, smooth with raised rounded commarginal ridges and radial threads on posterior third, white or yellowish with red or brown spot on dorsal margin near each end; interior white, pallial sinus reaching midvalve, turning posteriorly before joining ventral pallial line. North Carolina, Florida, West Indies, Gulf of Mexico, South America (Brazil). Length 8 mm.

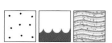

### *Angulus versicolor* (DeKay, 1843) – **Many-colored Dwarf Tellin**

Elongated oval to quadrangular, posterior blunt, fragile, smooth with very fine commarginal grooves, white, pink or rayed with pinkish opalescent sheen; interior colors reflecting exterior, pallial sinus deep and close to anterior muscle scar, dropping directly posteriorly to join ventral pallial line. Massachusetts to Florida, Bermuda, West Indies, Gulf of Mexico, Caribbean Central America, South America (Colombia, Brazil). Length 9 mm (to 17 mm). Note: The living *Angulus versicolor* were photographed from bayside of Big Coppitt Key.

*Angulus texanus*

*Angulus probrinus*

*Angulus sybariticus*

*Angulus versicolor*

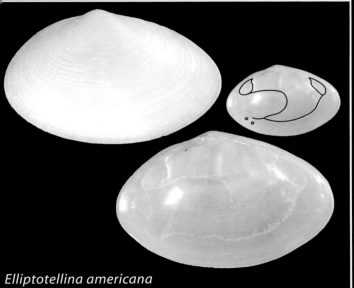

*Elliptotellina americana*

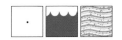

### *Merisca aequistriata* (Say, 1824) – **Striate Tellin**

Rounded trigonal, posterior pointed with distinct posterior twist, solid, with crowded, sharp commarginal ridges strongest posteriorly, smooth centrally, posterior dorsal slope with two radial ridges in right valve and one in left valve, white exteriorly and interiorly, pallial sinus deep and close to anterior muscle scar, dropping vertically to join ventral pallial line. Maryland to Florida, Bahamas, West Indies, Gulf of Mexico, Caribbean Central America, South America (Colombia, Brazil). Length 11 mm (to 25 mm). Compare *Merisca martinicensis*, which has raised commarginal ridges throughout, a pallial sinus that turns posteriorly before joining the ventral pallial line, and lateral teeth that are especially strong in the right valve.

### *Merisca cristallina* (Spengler, 1798) – **Crystal Tellin**

Rounded trigonal, posterior narrowed and rostrate, fragile, with strong, regular, widely spaced commarginal ridges, translucent white exteriorly and interiorly; interior glossy, pallial line deep and close to anterior muscle scar, dropping directly to join ventral pallial line. South Carolina to Florida, West Indies, Gulf of Mexico, Caribbean Central America, South America (Colombia). Length 15 mm (to 24 mm).

### *Merisca martinicensis* (d'Orbigny, 1853) – **Martinique Tellin**

Rounded trigonal, posterior pointed and bluntly rostrate, slightly concave posteroventrally, equilateral, with raised commarginal ridges, white exteriorly and interiorly, pallial sinus deep and close to anterior muscle scar, turning posteriorly before joining ventral pallial line, lateral teeth especially strong in right valve. Florida, West Indies, Gulf of Mexico, Caribbean Central America, South America (Colombia, Brazil). Length 7 mm (to 13 mm). Compare *Merisca aequistriata*, which is smoother exteriorly, and has a pallial sinus that drops vertically to join the ventral pallial line.

### *Eurytellina alternata* (Say, 1822) – **Alternate Tellin**

Elongated oval, posterior narrowed, solid, with regular, dense commarginal ridges (more dense on right valve), smooth near umbones, glossy creamy white or yellowish (rarely pink); interior glossy white or yellow, pallial sinus deep and close to anterior muscle scar, dropping vertically to join ventral pallial line, in some with short interlinear scar, with weak radial rib from umbo to just inside anterior muscle scar. North Carolina to Florida, Bermuda, Bahamas, West Indies, Gulf of Mexico, Caribbean Central America, South America (Brazil). Length 55 mm (to 72 mm). Compare *Eurytellina angulosa*, which is more trigonal with regular, fine commarginal ridges that are equally dense on both valves.

### *Eurytellina lineata* (Turton, 1819) – **Rose-petal Tellin**

Elongated trigonal, posterior narrowed, solid, smooth with fine commarginal grooves, glossy white or watermelon red with slight opalescent sheen; interior colors reflecting exterior, pallial sinus deep and just touching anterior muscle scar, dropping vertically to join ventral pallial line, short interlinear scar present, with weak radial rib from umbo to just inside anterior muscle scar. North Carolina to Florida, Bermuda, West Indies, Gulf of Mexico, Caribbean Central America, South America (to Brazil). Length 30 mm (to 38 mm).

### *Eurytellina angulosa* (Gmelin, 1791) – **Angulate Tellin**

Rounded trigonal, posterior pointed, solid, with smooth, regular, fine commarginal ridges equally dense on both valves, glossy white or pinkish with greenish yellow periostracum, umbones with three orange-red or yellow rays; interior white, pallial sinus deep and close to (but not touching) anterior muscle scar, dropping vertically to join ventral pallial line, in some with short interlinear scar, with weak radial rib from umbo to just inside anterior muscle scar. Florida, West Indies, Gulf of Mexico, Caribbean Central America, South America (to Brazil). Length 60 mm. Compare *Eurytellina alternata*, which is more elongated oval with regular commarginal ridges that are more dense on the right valve.

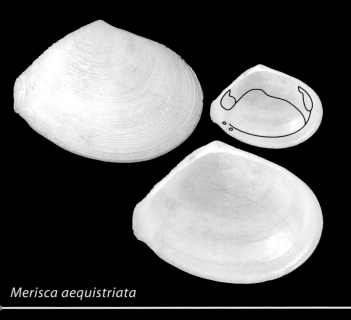

*Merisca aequistriata*

*Merisca cristallina*

*Merisca martinicensis*

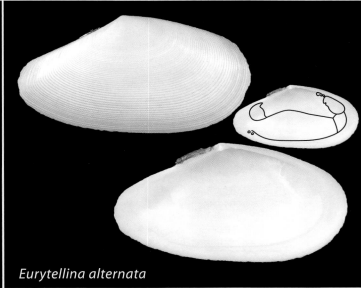

*Eurytellina alternata*

*Eurytellina lineata*

*Eurytellina angulosa*

### *Eurytellina punicea* (Born, 1778) – **Watermelon Tellin**

Rounded trigonal, posterior pointed, solid, with weak commarginal grooves separated by wide bands, bright watermelon pink or purplish red exteriorly and interiorly, pallial sinus deep and just touching anterior muscle scar, dropping vertically to join ventral pallial line, with weak radial rib from umbo to just inside anterior muscle scar. North Carolina, Florida Keys, West Indies, Caribbean Central America, South America (to Brazil). Length 30 mm (to 45 mm).

### *Eurytellina nitens* (C. B. Adams, 1845) – **Shiny Dwarf Tellin**

Elongated oval, posterior narrowed and blunt, thin-walled, with fine commarginal grooves, posterior third of right valve with sharply cut radial striae, pink or apricot with commarginal bands of white; interior colors reflecting exterior, pallial sinus deep and close to anterior muscle scar, dropping vertically to join ventral pallial line, with weak radial rib from umbo to just inside anterior muscle scar (stronger in left valve). North Carolina to Florida, West Indies, Gulf of Mexico, Caribbean Central America, South America (Colombia, Brazil). Length 30 mm (to 39 mm).

### *Macoma cerina* (C. B. Adams, 1845) – **Waxy Macoma**

Oval, posterior pointed and slightly hooked ventrally, fragile, smooth, white with apricot flush exteriorly and interiorly, pallial sinus deep and near anterior muscle scar, turning posteriorly before joining ventral pallial line. Florida, West Indies, Gulf of Mexico. Length 8 mm. Note: The commonly accepted taxonomic concept of this species is used here; however, Clench & Turner (1950: 250, 265, pl. 45, figs. 3, 4) have shown that the HOLOTYPE of this species is a specimen of the widespread tropical western Atlantic psammobiid *Heterodonax bimaculatus* (Linnaeus, 1758). However, aspects of the original written description do not match the holotype specimen. This taxonomic incompatibility has not been resolved.

### *Macoma brevifrons* (Say, 1834) – **Short Macoma**

Elongated oval to quadrangular, posterior blunt, thin-walled, with dense commarginal growth lines, umbones smooth, glossy white usually with blush of iridescent orange centrally and umbonally, with light brown periostracum; interior colors reflecting exterior, pallial sinus deep and close to anterior muscle scar, turning posteriorly before joining ventral pallial line. North Carolina to Florida, Gulf of Mexico, Caribbean Central America, South America (to Argentina). Length 12 mm (to 36 mm).

### *Macoma constricta* (Bruguière, 1792) – **Constricted Macoma**

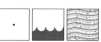

Rounded triangular, posterior pointed and slightly hooked, smooth with commarginal growth lines, white with gray periostracum; interior white, pallial sinus deep and close to anterior muscle scar, turning posteriorly before joining ventral pallial line. North Carolina to Florida, West Indies, Gulf of Mexico, Caribbean Central America, South America (to Brazil). Length 30 mm (to 50 mm).

### *Macoma extenuata* Dall, 1900 – **Slender Macoma**

Elongated oval, posteriorly narrowed, thin-walled, smooth, dull white exteriorly, glossy white interiorly, pallial sinus reaching just past midvalve, turning posteriorly before joining ventral pallial line. Florida Keys, Gulf of Mexico. Length 12 mm (to 14 mm).

*Eurytellina punicea*

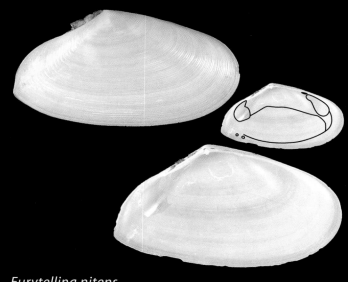

*Eurytellina nitens*

*Macoma cerina*

*Macoma brevifrons*

*Macoma constricta*

*Macoma extenuata*

### *Macoma limula* Dall, 1895 – **Little-file Macoma**

Elongated trigonal, posterior pointed and blunt, with radial ridge on posterior slope, thin-walled, smooth with finely granular surface (especially at ventral edge and posterior point), white exteriorly and interiorly, pallial sinus reaching past midvalve, turning posteriorly before joining ventral pallial line. North Carolina to Florida, West Indies, Gulf of Mexico. Length 16 mm. Compare *Macoma tenta*, which has an opalescent sheen exteriorly.

### *Macoma pseudomera* Dall & Simpson, 1901 – **Mera-like Macoma**

Oval, thin-walled, smooth, white exteriorly and interiorly; interior surface frequently with microscopic punctations at center, pallial sinus reaching just past midvalve, turning posteriorly before joining ventral pallial line. Florida Keys, Bermuda, West Indies, Gulf of Mexico, South America (Colombia, Brazil). Length 20 mm.

### *Macoma tageliformis* Dall, 1900 – **Tagelus-like Macoma**

Elongated trigonal, posterior narrowed and blunt, solid, smooth except for growth lines, white exteriorly and interiorly, pallial sinus reaching midvalve, turning posteriorly before joining ventral pallial line. Florida, West Indies, Gulf of Mexico, South America (to Brazil). Length 60 mm.

### *Macoma tenta* (Say, 1834) – **Elongated Macoma**

Elongated trigonal, posterior narrowed, thin-walled, smooth, white with opalescent sheen; interior glossy white, pallial sinus reaching past midvalve, turning posteriorly before joining ventral pallial line. Eastern Canada to Florida, Bermuda, West Indies, Gulf of Mexico, South America (Colombia, Brazil). Length 30 mm. Compare *Macoma limula*, which has a finely granular surface. Note: The right valve of this specimen has been drilled by a predatory gastropod.

### *Acorylus gouldii* (Hanley, 1846) – **Gould's Wedge Tellin**

Obliquely oval, solid, smooth centrally, with incised commarginal ridges anteriorly and posteriorly, glossy white exteriorly and interiorly, pallial sinus deep and joining anterior muscle scar and ventral pallial line at the same point. North Carolina to Florida, Bermuda, Bahamas, West Indies, Gulf of Mexico, Caribbean Central America, South America (Venezuela, Brazil). Length 8 mm (to 11 mm). Note: Also called Cuneate Dwarf Tellin.

### *Cymatoica hendersoni* Rehder, 1939 – **Henderson's Tellin**

Elongated trigonal, posterior pointed and blunt, fragile, with low, undulating commarginal ridges exteriorly and interiorly, translucent white exteriorly and interiorly, pallial sinus reaching past midvalve, turning posteriorly before joining ventral pallial line. Florida and Cuba. Length 7 mm (to 10 mm). Formerly considered a subspecies of *C. orientalis* (Dall, 1890), which it overlaps in distribution.

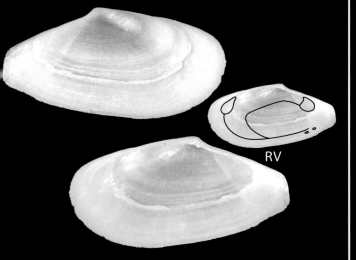

*Macoma limula*

RV

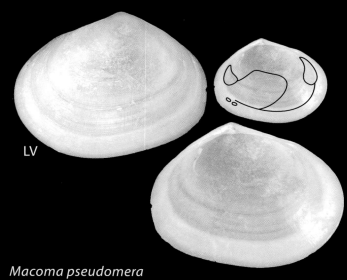

LV

*Macoma pseudomera*

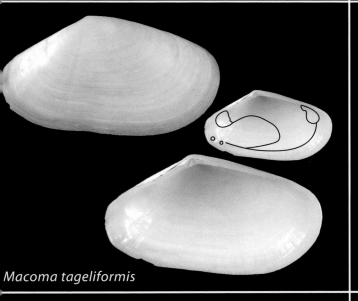

*Macoma tageliformis*

*Macoma tenta*

*Acorylus gouldii*

*Cymatoica hendersoni*

# Family Donacidae – Wedge Clams and Coquinas

**Classification**
AUTOLAMELLIBRANCHIATA Grobben, 1894
HETEROCONCHIA Hertwig, 1895
Heterodonta Neumayr, 1883
Veneroida H. Adams & A. Adams, 1856
Tellinoidea Blainville, 1814
Donacidae J. Fleming, 1828

## Featured species
### *Donax variabilis* Say, 1822 – **Variable Coquina**

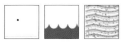

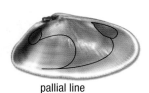

pallial line

Elongated trigonal, solid, smooth with radial scratches (strongest posteriorly), color highly variable, commonly bright white, yellow, pink, purple, bluish, occasionally with darker rays; interior similarly variable, often white or dark purple, pallial sinus reaching midvalve, turning posteriorly before joining ventral pallial line. New Jersey to Florida, Gulf of Mexico, Caribbean Central America. Length 25 mm.

**Donax variabilis is found** in great numbers within the surf zone of sandy beaches. Although these clams spend most of their time buried in the sand, they emerge several times per tidal cycle to migrate with the waves. This process is not merely the passive erosion of clams from the sand; rather the clams actively jump out of the sand, using sound to identify large waves. Activity is greatest during the rising tide, when the clams emerge to ride only the largest (loudest) 20% of waves, which move them furthest on the beach.

**Donax variabilis is renowned** for its polychromism, believed to act as an antipredator device, preventing shorebirds from forming a single search image of the clam.

# Family description

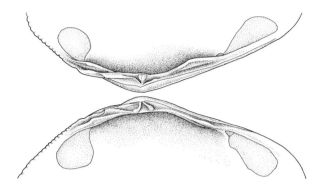

The donacid shell is small to medium-sized (to 150 mm), solid, and elongated to dorsoventrally tall-trigonal, often wedge-shaped (hence the common name), with the anterior end narrower and rounded, the posterior end obliquely truncate; the last can be set off by a radial keel or sulcus. It is EQUIVALVE, compressed, and not or only slightly gaping. The shell is INEQUILATERAL (umbones slightly to strongly posterior), with OPISTHOGYRATE UMBONES. Shell microstructure is ARAGONITIC and three-layered, with a composite PRISMATIC outer layer, a CROSSED LAMELLAR middle layer, and a COMPLEX CROSSED LAMELLAR or HOMOGENOUS inner layer, the last confined within the pallial line. TUBULES are absent. Exteriorly donacids are usually pale or muted in color, but highly polychromic within species, and are covered by usually thin and dehiscent PERIOSTRACUM (adherent in freshwater species). Sculpture is smooth to radial, with the posterodorsal margin often differently or more strongly sculptured. LUNULE and ESCUTCHEON are absent. Interiorly the shell is non-NACREOUS. The PALLIAL LINE strong, with a distinct SINUS. The inner shell margins are smooth or denticulate. The HINGE PLATE is strong and HETERODONT, with two (bifid in some species) CARDINAL TEETH in each valve; anterior and posterior LATERAL TEETH are present or weak. The LIGAMENT is PARIVINCULAR, OPISTHODETIC, and usually set on NYMPHS.

The animal is ISOMYARIAN; anterior and posterior pedal retractor muscles are present. Pedal protractors and elevators are present. The MANTLE margins are usually largely unfused ventrally. Posterior EXCURRENT and INCURRENT SIPHONS are short (with the excur-

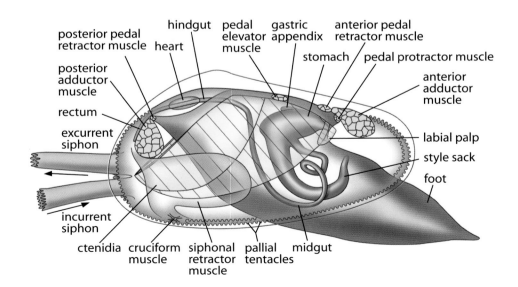

rent siphon often distinctly longer than the incurrent) and separate, with a CRUCIFORM MUSCLE at their base. A specialized, closed, sensory organ accompanies the cruciform muscle. HYPOBRANCHIAL GLANDS have not been reported. The FOOT is very large, muscular, highly laterally compressed (bladelike), unheeled, and pointed anteriorly; a BYSSAL GROOVE and BYSSUS are absent in the adult.

The LABIAL PALPS are small. The CTENIDIA are EULAMELLIBRANCH (SYNAPTORHABDIC) and HOMORHABDIC or HETERORHABDIC; the outer demibranchs are smaller than the inner demibranchs. The gills are not inserted into (or fused with) the distal oral groove of the palps (CATEGORY III association). The gills are united posteriorly with the SIPHONAL SEPTUM, separating INFRA- and SUPRABRANCHIAL CHAMBERS. Incurrent and excurrent water flows are posterior. The STOMACH is TYPE IV or V, with a posterodorsal, in some cases lobed, gastric appendix; the MIDGUT is coiled. The HINDGUT passes through the ventricle of the heart, and leads to a sessile rectum. Donacids are GONOCHORISTIC and produce planktonic VELIGER larvae; late-stage larvae have a distinctively trigonal shell with an extended anterior end. The nervous system has not been well studied, but examination of available data suggests it is not concentrated. STATOCYSTS and ABDOMINAL SENSE ORGANS have not been reported. Donacids are said to respond to light changes but EYES are not present.

Donacids are marine SUSPENSION FEEDERS, and INFAUNAL, with *Donax* spp. characteristically in the surf zone of exposed beaches, often in extreme abundance. They are capable of dislodging themselves, followed by rapid reburying between wave surges. Others (*Iphigenia* and *Galatea*) live in estuarine and freshwaters of South America and Africa. Various species of donacids are harvested worldwide for food or bait. Predators include crabs, boring gastropods, fish, water birds, and pigs (which feed on *Donax denticulatus* (Linnaeus, 1758) on West Indies' beaches).

The family Donacidae is known since the Cretaceous and is represented by ca. 5 living genera and ca. 60 species, distributed worldwide.

---

### *Iphigenia brasiliana* (Lamarck, 1818) – **Giant False Coquina**

Rounded trigonal, solid, smooth, cream with purple-stained umbones and thin, glossy brown periostracum; interior white, pallial sinus reaching midvalve, turning posteriorly before joining ventral pallial line. Florida, West Indies, Gulf of Mexico, Caribbean Central America, South America (Suriname, Brazil). Length 40 mm (to 48 mm). Note: *Iphigenia brasiliana* inhabits estuarine areas, and thus is rare in the Florida Keys.

# References

Adamkewicz, S. L., and M. G. Harasewych. 1996. Systematics and biogeography of the genus *Donax* (Bivalvia: Donacidae) in eastern North America. *American Malacological Bulletin*, 13(1/2): 97–103.

Ansell, A. D. 1981. Functional morphology and feeding of *Donax serra* Röding and *Donax sordidus* Hanley (Bivalvia: Donacidae). *Journal of Molluscan Studies*, 47(1): 59–72.

Ansell, A. D. 1983. The biology of the genus *Donax*. Pages 607–635, in: A. McLachlin and T. Erasmus, eds., *Sandy Beaches as Ecosystems, Based on the Proceedings of the First International Symposium on Sandy Beaches, Held in Port Elizabeth, South Africa, 17–21 January 1983*. Dr. W. Junk Publishers, The Hague, The Netherlands.

Ellers, O. 1995a. Behavioral control of swash-riding in the clam *Donax variabilis*. *The Biological Bulletin*, 189(2): 120–127.

Ellers, O. 1995b. Discrimination among wave-generated sounds by a swash-riding clam. *The Biological Bulletin*, 189(2): 128–137.

Mikkelsen, P. S. 1981. A comparison of two Florida populations of the Coquina clam, *Donax variabilis* Say, 1822 (Bivalvia: Donacidae). I. Intertidal density, distribution and migration. *The Veliger*, 23(3): 230–239.

Mikkelsen, P. S. 1985. A comparison of two Florida populations of the coquina clam, *Donax variabilis* Say, 1822 (Bivalvia: Donacidae). II. Growth rates. *The Veliger*, 27(3): 308–311.

Morrison, J. P. E. 1971. Western Atlantic *Donax*. *Proceedings of the Biological Society of Washington*, 83(48): 545–568.

Narchi, W. 1972. On the biology of *Iphigenia brasiliensis* Lamarck, 1818 (Bivalvia: Donacidae). *Proceedings of the Malacological Society of London*, 40(2): 79–91.

Narchi, W. 1978. Functional anatomy of *Donax hanleyanus* Philippi 1847 (Donacidae—Bivalvia). *Boletim de Zoologia, Universidad de São Paulo*, 3: 121–142.

Passos, F. D., and O. Domaneschi. 2004. Biologia e anatomia functional de *Donax gemmula* Morrison (Bivalvia, Donacidae) do litoral de São Paulo, Brasil. *Revista Brasileira de Zoologia*, 21(4): 1017–1032.

Purchon, R. D. 1963. A note on the biology of *Egeria radiata* Lam. (Bivalvia, Donacidae). *Proceedings of the Malacological Society of London*, 35(6): 251–271.

Salas-Casanova, C., and E. Hergueta. 1990. The functional morphology of the alimentary canal of *Donax venustus* Poli and *D. semistriatus* Poli. Pages 213–222, in: B. Morton, ed., *The Bivalvia, Proceedings of a Memorial Symposium in Honour of Sir Charles Maurice Yonge (1899–1986), Edinburgh, 1986*. Hong Kong University Press, Hong Kong.

Schneider, D. 1982. Predation by Ruddy Turnstones (*Arenaria interpres*) on a polymorphic clam (*Donax variabilis*) at Sanibel Island, Florida. *Bulletin of Marine Science*, 32(1): 341–344.

Simone, L. R. L., and J. R. Dougherty. 2004. Anatomy and systematics of northwestern Atlantic *Donax* (Bivalvia, Veneroidea, Donacidae). In: R. Bieler and P. M. Mikkelsen, eds., *Bivalve Studies in the Florida Keys*, Proceedings of the International Marine Bivalve Workshop, Long Key, Florida, July 2002. *Malacologia*, 46(2): 459–472.

Tiffany, W. J., III. 1971. The tidal migration of *Donax variabilis* Say (Mollusca: Bivalvia). *The Veliger*, 14(1): 82–85.

Wade, B. 1969. Studies on the biology of the West Indian beach clam, *Donax denticulatus* Linné. 3. Functional morphology. *Bulletin of Marine Science*, 19(2): 306–322.

Yonge, C. M. 1949. On the structure and adaptations of the Tellinacea, deposit-feeding Eulamellibranchia. *Philosophical Transactions of the Royal Society of London, Series B, Biological Sciences*, 234(609): 29–76.

# Family Psammobiidae – Sanguin or Sunset Clams

**Classification**
AUTOLAMELLIBRANCHIATA Grobben, 1894
HETEROCONCHIA Hertwig, 1895
Heterodonta Neumayr, 1883
Veneroida H. Adams & A. Adams, 1856
Tellinoidea Blainville, 1814
Psammobiidae J. Fleming, 1828

**Featured species**
*Asaphis deflorata* (Linnaeus, 1758) – **Gaudy Asaphis**

  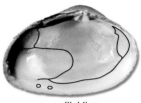

pallial line

Elongated oval, posterior truncated, equilateral or with umbones slightly anterior, solid, with coarse radial ribs crossed by commarginal ridges, scaly posteriorly and anteriorly, variably colored whitish to purple, orange, pink, or yellow, often rayed, often darker at umbones; interior glossy white or bright yellow, with purple stain posteriorly and at hinge, pallial sinus reaching midvalve, turning posteriorly before joining ventral pallial line. North Carolina to Florida, Bermuda, Bahamas, West Indies, Gulf of Mexico, Caribbean Central America, South America (to Brazil). Indo-West Pacific *Asaphis violascens* (Forsskål, 1775) is very similar and differs mainly in characters of the alimentary tract. Length 40 mm (to 67 mm). Note: Also known as Gaudy Sanguin.

*Asaphis deflorata* **is common** at only one location in the Florida Keys, where it lives in gravelly sand among intertidal rocks.

**The shells of *Asaphis deflorata*** show remarkably bright and variable colors, prompting its common name, the "gaudy" *Asaphis*. Caribbean shells tend to be overall more colorful than those from Florida.

# Family description

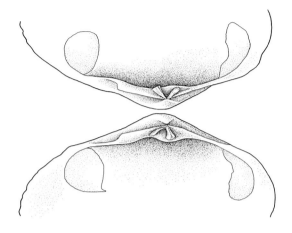

The psammobiid shell is small to medium-sized (to 200 mm), thin-walled to solid, quadrangular to oval, and posteriorly truncate. The shell is EQUIVALVE OR INEQUIVALVE (right valve more convex), compressed to inflated, and gaping (narrowly to widely) posteriorly and in some cases also anteriorly. The shell is EQUILATERAL TO INEQUILATERAL (umbones anterior or posterior), with OPISTHO- or ORTHOGYRATE UMBONES. Shell microstructure is ARAGONITIC and three-layered, with a composite PRISMATIC outer layer, a CROSSED LAMELLAR middle layer, and a COMPLEX CROSSED LAMELLAR inner layer, the last confined within the pallial line. TUBULES are absent. Exteriorly psammobiids are often brightly colored with radiating bands (hence the common name "sunset clams"), covered by a conspicuous, adherent or dehiscent, PERIOSTRACUM. Sculpture is smooth or radial or with weak incised or oblique commarginal lirae. LUNULE and ESCUTCHEON are absent (a very narrow lunule is present in some *Asaphis* spp.). Interiorly the shell is non-NACREOUS. The PALLIAL LINE has a deep, rounded SINUS. The inner shell margins are smooth in adults (can be denticulate in juveniles). The HINGE PLATE is narrow and HETERODONT, with two (bifid in some species) CARDINAL TEETH in each valve; anterior and posterior LATERAL TEETH are absent. The LIGAMENT is large, PARIVINCULAR, OPISTHODETIC, and set on strong NYMPHS.

The animal is slightly HETEROMYARIAN (anterior ADDUCTOR MUSCLE narrower/smaller); anterior and posterior pedal retractor muscles are present. Pedal protractor mus-

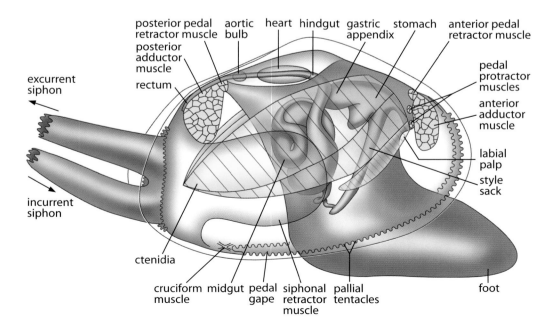

cles are present (two on each side in *Asaphis deflorata*) or absent (e.g., in *Asaphis dichotoma* (Anton, 1839)). Pedal elevator muscles are absent or very small. The MANTLE margins are largely not fused ventrally, with a large anteroventral pedal gape. Posterior EXCURRENT and INCURRENT SIPHONS are separate and highly extendable, with a CRUCIFORM MUSCLE at their base, and can be brightly colored at the tips (yellow in *A. dichotoma*). A specialized, open, sensory organ accompanies the cruciform muscle. HYPOBRANCHIAL GLANDS have not been reported. The FOOT is large, laterally compressed, wedge-shaped, often anteriorly pointed, and in some species has a BYSSAL GROOVE; a BYSSUS is absent in the adult.

The LABIAL PALPS are small to large and trigonal. The CTENIDIA are EULAMELLIBRANCH (SYNAPTORHABDIC) and HETERORHABDIC. The outer demibranchs are smaller than the inner demibranchs. The tips of the inner demibranchs are not inserted or fused into the distal oral groove of the palps (CATEGORY III association). Incurrent and excurrent water flows are posterior. The STOMACH is TYPE IV (in *Heterodonax*) or V, with a posterodorsal, in some cases lobed, gastric appendix; the MIDGUT is tightly coiled and greatly dilated posterior to the stomach. The HINDGUT passes through the ventricle of the heart, and leads to a sessile rectum. An AORTIC BULB is present. Psammobiids are GONOCHORISTIC and produce planktonic VELIGER larvae that in their late stages have a distinctively trigonal shell with an extended anterior end. The nervous system is not concentrated. STATOCYSTS are present (reported for *Gari*). ABDOMINAL SENSE ORGANS have not been reported.

Psammobiids are marine or estuarine DEPOSIT FEEDERS (or rarely [*Nuttalia*] SUSPENSION FEEDERS), living INFAUNALLY in clean or muddy sand or gravel, often in mangrove or other estuarine habitats. Known predators include fish, which nip the tips of the extended siphons; the siphon tips can be autotomized in some species as an antipredator device, and presumably can regenerate.

The family Psammobiidae is known since the Cretaceous and is represented by ca. 7 living genera and ca. 130 species, distributed worldwide in temperate and tropical regions, in shallow to deep seas. *Asaphis* is utilized for food in many regions, from the Caribbean Sea to the Pacific Ocean.

## References

Berg, C. J., Jr., and P. Alatalo. 1985. Biology of the tropical bivalve *Asaphis deflorata* (Linné, 1758). *Bulletin of Marine Science*, 37(3): 827–838.

Bloomer, H. H. 1911. On the anatomy of the British species of the genus *Psammobia*. *Proceedings of the Malacological Society of London*, 9(4): 231–239.

Coan, E. V. 1973. The northwest American Psammobiidae. *The Veliger*, 16(1): 40–57.

Coan, E. V. 2002. Recent eastern Pacific species of *Sanguinolaria* and *Psammotella* (Bivalvia: Psammobiidae). *The Nautilus*, 116(1): 1–12.

Domaneschi, O. 1992. Anatomia functional de *Gari solida* (Gray, 1828) (Bivalvia: Psammobiidae) do litoral de Dichato, Chile. *Boletim de Zoologia, Universidade de São Paulo*, 15: 41–80.

Domaneschi, O., and E. K. Shea. 2004. Shell morphometry of western Atlantic and Indo-West Pacific *Asaphis*; functional morphology and ecological aspects of *A. deflorata* from Florida Keys, USA (Bivalvia: Psammobiidae). In: R. Bieler and P. M. Mikkelsen, eds., *Bivalve Studies in the Florida Keys*, Proceedings of the International Marine Bivalve Workshop, Long Key, Florida, July 2002. *Malacologia*, 46(2): 249–275.

Frenkiel, L. 1979. L'organe sensoriel du muscle cruciforme des Tellinacea: importance systématique chez les Psammobiidae. *The Journal of Molluscan Studies*, 45(2): 231–237.

Graham, A. 1934. The structure and relationships of lamellibranches possessing a cruciform muscle. *Proceedings of the Royal Society of Edinburgh*, 54: 158–187.

Narchi, W. 1980. A comparative study of the functional morphology of *Caecella chinensis* Deshayes, 1855 and *Asaphis dichotoma* (Anton 1839) from Ma Shi Chau, Hong Kong. Pages 253–276, in: B. Morton, ed., *The Malacofauna of Hong Kong and Southern China*. Hong Kong University Press, Hong Kong.

Narchi, W., and O. Domaneschi. 1993. Functional morphology of *Heterodonax bimaculatus* (Linné, 1758) (Bivalvia: Psammobiidae). *American Malacological Bulletin*, 10(2): 139–152.

Pohlo, R. H. 1972. Feeding and associated morphology in *Sanguinolaria nuttallii*. *The Veliger*, 14(3): 298–301.

Sasaki, K., M. Kudo, and K. Ito. 1999. Structures of the siphons of the bivalve *Nuttallia olivacea* (Tellinacea, Psammobiidae) and changes of their states under extended conditions. *Fisheries Science*, 65(6): 839–843.

Willan, R. C. 1993. Taxonomic revision of the family Psammobiidae (Bivalvia: Tellinoidea) in the Australian and New Zealand region. *Records of the Australian Museum, Supplement*, 18: 1–132.

Yonge, C. M. 1949. On the structure and adaptations of the Tellinacea, deposit-feeding Eulamellibranchia. *Philosophical Transactions of the Royal Society of London, Series B, Biological Sciences*, 234(609): 29–76.

### *Gari circe* (Mörch, 1876) – **Western Atlantic Gari**

Elongated quadrangular, posterior obliquely truncate and broader than anterior end, equilateral, fragile, smooth and glossy with faint radial scratches (stronger posteriorly on larger shells) and commarginal ridges, whitish or pink with speckles and vague light brown rays, opaque radial rays at umbones, with orange periostracum; interior glossy white. Florida Keys, Bahamas, West Indies. Length 25 mm (to 40 mm). Formerly in *Psammobia*.

### *Heterodonax bimaculatus* (Linnaeus, 1758) – **Falsebean**

Rounded trigonal, posterior truncate, smooth with fine commarginal growth lines stronger on posterior slope, variably colored from purple to reddish orange, occasionally rayed or with radial rows of purplish dots; interior glossy and more vibrantly colored than exterior, pallial sinus extending past midvalve. Florida, Bermuda, Bahamas, West Indies, Gulf of Mexico, Caribbean Central America, South America (to Brazil). Length 20 mm. Note: The right valve of this specimen has been drilled by a predatory gastropod. Also known as False Donax.

### *Sanguinolaria sanguinolenta* (Gmelin, 1791) – **Atlantic Sanguin**

Elongated oval, posterior narrowed, smooth, glossy white, flushed with pink exteriorly and interiorly, umbones bright red; pallial sinus with conspicuous hump middorsally, dropping vertically to join ventral pallial line. Florida, West Indies, Gulf of Mexico, South America (to Brazil). Length 55 mm.

*Asaphis deflorata* **inhabits** gravelly sand at the edge of quiet bay localities, such as this one at Crawl Key.

*Gari circe*

*Heterodonax bimaculatus*

*Sanguinolaria sanguinolenta*

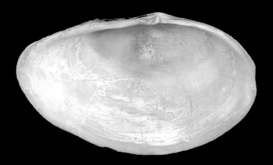

# Family Semelidae – Semele Clams

**Classification**
AUTOLAMELLIBRANCHIATA Grobben, 1894
HETEROCONCHIA Hertwig, 1895
Heterodonta Neumayr, 1883
Veneroida H. Adams & A. Adams, 1856
Tellinoidea Blainville, 1814
Semelidae Stoliczka, 1870 [1825]

**Featured species**
*Semele proficua* (Pulteney, 1799) – **White Atlantic Semele**

  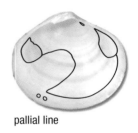

pallial line

Oval to circular, solid, lunule narrow and sunken, smooth with commarginal ridges and crowded radial threads (most evident in juveniles), white flushed at umbones with pale orange or yellow, occasionally with pink or purplish rays; interior glossy yellowish, rarely speckled with purple or pink, and often pitted, pallial sinus rounded and ascending. North Carolina to Florida, Bermuda, Bahamas, West Indies, Gulf of Mexico, Caribbean Central America, South America (to Argentina). Length 30 mm (to 36 mm). Compare *Semele purpurascens*, which is generally more brightly colored and has smooth interspaces (see p. 355 and scanning electron micrograph, p. 352).

# Family description

The semelid shell is very small (*Ervilia*) to medium-sized (to 120 mm), thin-walled to solid, and rounded oval to lenticular, often with a slight posterior flexure toward the right, with the anterior end rounded and the posterior end truncated, rounded, or ROSTRATE. The shell is EQUIVALVE, compressed, and can be gaping posteriorly. The shell is usually EQUILATERAL, with PROSO- or ORTHOGYRATE UMBONES. Shell microstructure is ARAGONITIC and three-layered, with a composite PRISMATIC outer layer, a CROSSED LAMELLAR middle layer, and a COMPLEX CROSSED LAMELLAR or HOMOGENOUS inner layer, the last confined within the pallial line. TUBULES are apparently absent. Exteriorly semelids are often brightly colored (e.g., pink, reddish, orange, or purple), covered by thin, adherent PERIOSTRACUM. Sculpture is smooth to strongly commarginal, in some species with radial or

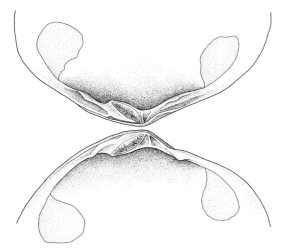

oblique striae. LUNULE and a weak ESCUTCHEON can be present or absent. Interiorly the shell is non-NACRE-OUS. The PALLIAL LINE has a deep, rounded SINUS. The inner shell margins are usually smooth. The HINGE PLATE is narrow to wide and HETERODONT, with one or two CARDINAL TEETH in each valve; anterior and posterior LATERAL TEETH are usually present (absent in Scrobulariinae). The LIGAMENT is PARIVINCULAR and OPISTHODETIC; an internal portion (RESILIUM) sits on an elongated, oblique to vertical RESILIFER.

The animal is slightly HETEROMYARIAN (anterior or posterior ADDUCTOR MUSCLE slightly smaller); anterior and posterior pedal retractor muscles are present. Pedal protractor muscles insert on the posteroventral surface of the anterior adductor muscle. Small pedal elevator muscles insert dorsally in some species (absent in others, e.g., *Semele casali* Doello-Jurado, 1949). The MANTLE margins are largely not fused ventrally, with a large anteroventral pedal gape. Posterior EXCURRENT and INCURRENT SIPHONS are long (up to three to four times the shell length), separate, and independently mobile, with a CRUCIFORM MUSCLE at their base. A specialized, open, sensory organ accompanies the cruciform muscle. HYPOBRANCHIAL GLANDS have not been reported. The FOOT is large, laterally compressed, and anteriorly pointed; a BYSSAL GROOVE and BYSSUS are usually absent in the adult (*Ervilia* has remnants of a byssal pore and gland).

The LABIAL PALPS are small (*Ervilia*) to large and narrowly trigonal. The CTENIDIA are EULAMELLIBRANCH (SYNAPTORHABDIC) and HOMORHABDIC (e.g., *Scrobicularia*) or HET-

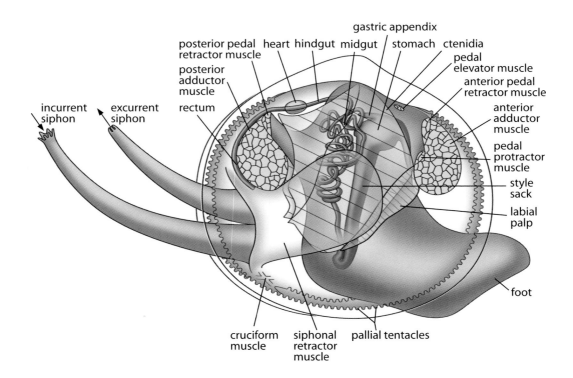

ERORHABDIC (e.g., *Semele*). The outer demibranchs lack ascending lamellae and are up-turned. The tips of the inner demibranchs are not inserted or fused into the distal oral groove of the palps (CATEGORY III association), or (in *Semele*) are inserted and fused into the groove (CATEGORY II). Incurrent and excurrent water flows are posterior. The STOMACH is TYPE V, with a posterodorsal, in some cases lobed, gastric appendix; the MIDGUT is usually highly and tightly coiled (not coiled in *Ervilia*) posterior to the stomach. The HINDGUT passes through the ventricle of the heart, and leads to a sessile rectum. An AORTIC BULB has been reported in some species. Semelids are GONOCHORISTIC and produce planktonic VELIGER larvae that in their late stage have a distinctively trigonal shell with an extended anterior end. The nervous system is not concentrated. STATOCYSTS have been reported in adult *Scrobicularia* and *Ervilia*. ABDOMINAL SENSE ORGANS have not been reported.

Semelids are marine DEPOSIT and SUSPENSION FEEDERS (in combination), INFAUNAL in sandy or muddy substrata, lying on the left side with the siphons bent to the right to extend toward the sediment surface (thus requiring the posterior shell flexure). The slightly trumpet-shaped aperture of the incurrent siphon can be extended above the sediment surface for deposit feeding. According to one author, a long, coiled midgut (as found in semelids) is correlated with the need to consolidate fecal pellets from a diet heavy in ingested mud. Semelids are capable of rapid reburrowing if uncovered.

The family Semelidae is known since the Eocene and is represented by 10 living genera and ca. 65 species, distributed worldwide.

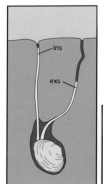
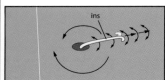

**Independently mobile siphons** of semelids (here in the Peppery Furrow Shell [*Scrobicularia plana* (Da Costa, 1778)], of northern Europe) create separate burrows in the sediment. The incurrent siphon harvests organic material from the surface sediment around its burrow opening.

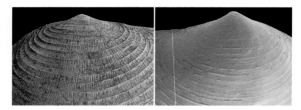

**Scanning electron micrographs** of *Semele proficua* (left) and *Semele purpurascens* show differences in shell sculpture that help distinguish even the smallest specimens. *Semele proficua* has dense radial striae in the interspaces of the commarginal ribs, whereas *S. purpurascens* is smooth.

## References

Boss, K. J. 1972. The genus *Semele* in the western Atlantic (Semelidae; Bivalvia). *Johnsonia*, 5(49): 1–32.

Coan, E. V. 1988. Recent eastern Pacific species of the bivalve genus *Semele*. *The Veliger*, 31(1–2): 1–42.

Davis, J. D. 1973. Systematics and distribution of western Atlantic *Ervilia* (Pelecypoda: Mesodesmatidae) with notes on living *Ervilia subcancellata*. *The Veliger*, 15(4): 307–313, 3 pls.

Domaneschi, O. 1995. A comparative study of the functional morphology of *Semele purpurascens* (Gmelin, 1791) and *Semele proficua* (Pulteney, 1799) (Bivalvia: Semelidae). *The Veliger*, 38(4): 323–342.

Graham, A. 1934. The structure and relationships of lamellibranchs possessing a cruciform muscle. *Proceedings of the Royal Society of Edinburgh*, 54: 158–187.

Grave, B. H. 1927. The natural history of *Cumingia tellinoides*. *Biological Bulletin*, 53(3): 208–219.

Hughes, R. N. 1969. A study of feeding in *Scrobicularia plana*. *Journal of the Marine Biological Association of the United Kingdom*, 49(3): 805–823.

Morton, B. 1990. The biology and functional morphology of *Ervilia castanea* (Bivalvia: Tellinacea) from the Azores. *Açoreana*, Supplement (*The Marine Fauna and Flora of the Azores. Proceedings of the First Internatonal Workshop of Malacology, São Miguel, Azores, 1988*): 75–96.

Morton, B., and P. H. Scott. 1990. Relocation of *Ervilia* Turton, 1822 (Bivalvia) from the Mesodesmatidae (Mesodesmatoidea) to the Semelidae (Tellinoidea). *The Veliger*, 33(3): 299–304.

Narchi, W., and O. Domaneschi. 1977. *Semele casali* Doello-Jurado, 1949 (Mollusca—Bivalvia) in the Brazilian littoral. *Studies on Neotropical Fauna and Environment*, 12(4): 263–272.

Russell-Hunter, W. D., and J. S. Tashiro. 1985. Life-habits and infaunal posture of *Cumingia tellinoides* (Tellinacea, Semelidae): an example of evolutionary parallelism. *The Veliger*, 27(3): 253–260.

Schröder, O. 1916. Beiträge zur Anatomie von *Amphidesma solidum*. *Jenaische Zeitschrift für Naturwissenschaften*, 54(1): 101–132.

Trueman, E. R. 1953. The structure of the ligament of the Semelidae. *Proceedings of the Malacological Society of London*, 30(1–2): 30–36.

Yonge, C. M. 1949. On the structure and adaptations of the Tellinacea, deposit-feeding Eulamellibranchia. *Philosophical Transactions of the Royal Society of London, Series B, Biological Sciences*, 234(609): 29–76.

### *Semele bellastriata* (Conrad, 1837) – **Cancellate Semele**

Oval, solid, lunule narrow and sunken, strongly cancellate, white and spotted or stained with orange-brown; interior glossy white, cream or suffused with mauve or violet. North Carolina to Florida, Bermuda, Bahamas, West Indies, Gulf of Mexico, Caribbean Central America, South America (Suriname, Brazil). Length 8 mm (to 24 mm).

### *Semele purpurascens* (Gmelin, 1791) – **Purplish Semele**

Oval, thin-walled, lunule narrow and sunken, smooth with fine oblique striae (juveniles smooth, without radial threads), gray or cream with purple or orange markings; interior glossy and suffused with purple, brown, or orange. North Carolina to Florida, Bermuda, West Indies, Gulf of Mexico, Caribbean Central America, South America (to Uruguay). Length 30 mm (to 34 mm). Compare *Semele proficua*, which is generally lighter in color and has radially striated interspaces (see p. 350 and scanning electron micrograph, p. 352).

### *Abra aequalis* (Say, 1822) – **Common Atlantic Abra**

Rounded trigonal, posterior blunt, fragile, lunule barely impressed, glossy smooth, translucent white exteriorly and interiorly, exterior slightly iridescent, with thin, yellowish periostracum. Maryland to Florida, Bermuda, West Indies, Gulf of Mexico, Caribbean Central America, South America (to Brazil). Length 7 mm (to 13 mm). Compare *Abra lioica*, which is slightly smaller (as an adult) and has a golden prodissoconch.

### *Abra lioica* (Dall, 1881) – **Smooth Abra**

Rounded trigonal, posterior blunt, fragile, lunule barely impressed, glossy smooth, translucent white exteriorly and interiorly, with opaque white patches umbonally, prodissoconch large and golden. Massachusetts to Florida, Bermuda, Bahamas, West Indies, Gulf of Mexico, South America (Brazil to Argentina). Length 6 mm (to 8 mm). Compare *Abra aequalis*, which reaches larger adult size and is white throughout.

### *Abra longicallus americana* Verrill & Bush, 1898 – **Long Abra**

Elongated oval, posterior pointed, fragile, lunule barely impressed, glossy smooth, translucent white exteriorly and interiorly. New Jersey to North Carolina, Florida Keys, West Indies. Length 18 mm (to 20 mm).

### *Semelina nuculoides* (Conrad, 1841) – **Little Nut Semele**

Elongated oval, umbones well posterior, solid, lunule poorly developed, with regular, low, flat commarginal ridges, somewhat coarser posteriorly where there can also be fine radial threads, two or three coarse commarginal grooves marking growth stages, white, occasionally with yellow-orange tint or radial rays; interior glossy white. North Carolina to Florida, Bahamas, West Indies, Gulf of Mexico, Caribbean Central America, South America (to Brazil). Length 5 mm (to 6 mm). Formerly in *Semele*. Compare species of Nuculidae, which have taxodont teeth and lack a pallial sinus.

*Semele bellastriata*

*Semele purpurascens*

*Abra aequalis*

*Abra lioica*

*Abra longicallus americana*

*Semelina nuculoides*

### *Cumingia coarctata* G. B. Sowerby I, 1833 – **Southern Cumingia**

Rounded trigonal, often distorted, fragile, lunule small and smooth, with coarse, erect commarginal ridges and fine radial threads, white to yellow-white; interior glossy white, with large chondrophores, pallial sinus reaching past midvalve. Florida, Bermuda, Bahamas, West Indies, Caribbean Central America, South America (to Brazil). Length 6 mm (to 16 mm). Compare *Cumingia vanhyningi*, which is more elongate, rostrate, and smooth. Note: Living *Cumingia coarctata* (here from bayside of Big Coppitt Key) show bright pink gills through the translucent shell. Also known as Contracted Semele.

### *Cumingia vanhyningi* Rehder, 1939 – **Van Hyning's Cumingia**

Rounded trigonal, posterior pointed, fragile, lunule small and smooth, with low, dense, relatively smooth commarginal ribs and radial scratches, translucent white; interior glossy white, with large chondrophores, pallial sinus long and flat. Florida Keys, Bahamas, Gulf of Mexico, Caribbean Central America. Length 10 mm (to 20 mm). Formerly considered a subspecies of *C. tellinoides* (Conrad, 1830). Compare *Cumingia coarctata*, which is less elongate and has coarse, erect commarginal ridges.

### *Ervilia subcancellata* E. A. Smith, 1885 – **Subcancellate Ervilia**

Elongated trigonal, posterior distinctly rostrate, solid, lunule pitlike, smooth with radiating striae often over entire shell, especially prominent posteriorly, white exteriorly and interiorly, pallial sinus broad, not quite reaching midvalve. Florida Keys, Bermuda, West Indies, Caribbean Central America, South America (Brazil), also St. Helena. Length 5 mm.

### *Ervilia nitens* (Montagu, 1808) – **Shining Ervilia**

Elongated trigonal, posteriorly rounded, fragile, lunule pitlike, smooth with commarginal ridges crossed posterodorsally by radial striae, glossy white, occasionally tinted with pink concentrated at umbones; interior glossy white, pallial sinus broad, reaching about midvalve. Florida, Bermuda, Bahamas, West Indies, Caribbean Central America, South America (to Uruguay). Length 5 mm (to 12 mm). Note: This species was originally described from Dunbar, Scotland, apparently from western Atlantic ship ballast material.

### *Ervilia concentrica* (Holmes, 1858) – **Concentric Ervilia**

Elongated trigonal, posterior rounded, solid, lunule pitlike, with coarse, dense commarginal ridges crossed by radial striae posteriorly, opaque chalky white; interior glossy white, pallial sinus broad, reaching about midvalve. North Carolina to Florida, Bermuda, Bahamas, West Indies, Gulf of Mexico, Caribbean Central America, South America (Brazil). Length 3 mm (to 10 mm).

*Cumingia coarctata*

*Cumingia vanhyningi*

*Ervilia subcancellata*

*Ervilia nitens*

*Ervilia concentrica*

# Family Solecurtidae — Tagelus or Razor Clams

### Classification
AUTOLAMELLIBRANCHIATA Grobben, 1894
HETEROCONCHIA Hertwig, 1895
Heterodonta Neumayr, 1883
Veneroida H. Adams & A. Adams, 1856
Tellinoidea Blainville, 1814
Solecurtidae d'Orbigny, 1846

**Featured species**
*Tagelus divisus* (Spengler, 1794) – **Purplish Tagelus**

pallial line

Elongated oval, dorsal margin broadly arched, smooth, translucent white tinged with purple, with radial purplish streak at center; interior with central radial ridge. Massachusetts to Florida, Bermuda, Bahamas, West Indies, Gulf of Mexico, Caribbean Central America, South America (to Brazil). Length 20 mm (to 38 mm).

# Family description

The solecurtid shell is small to medium-sized (to 110 mm), thin-walled, and narrowly elongated quadrangular to oval, with rounded or truncated ends. The shell is EQUIVALVE, compressed, and widely gaping anteriorly and posteriorly. The shell is usually EQUILATERAL, with ORTHOGYRATE UMBONES. Shell microstructure is ARAGONITIC and three-layered, with a composite PRISMATIC outer layer, a CROSSED LAMELLAR middle layer, and a COMPLEX CROSSED LAMELLAR inner layer, the latter confined within the pallial line. TUBULES have not been reported. Exteriorly solecurtids are covered by thin, conspicuous or eroded PERIOSTRACUM. Sculpture is smooth or with weak, oblique or DIVARICATING incised lines. LUNULE and ESCUTCHEON are absent. Interiorly the shell is non-NACREOUS. The PALLIAL LINE has a shallow to deep SINUS. The inner shell margins are smooth. The HINGE PLATE is weak and HETERODONT, with one to three, usually bifid, CARDINAL TEETH in each valve; anterior and posterior LATERAL TEETH are usually absent. The LIGAMENT is

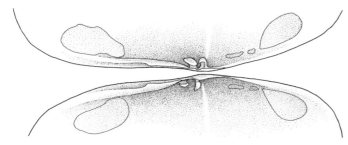

PARIVINCULAR, OPISTHODETIC, and set on strong NYMPHS.

The animal greatly exceeds the size of the shell and cannot be fully withdrawn in many species. It is HETEROMYARIAN (anterior ADDUCTOR MUSCLE smaller); anterior and posterior pedal retractor and pedal elevator muscles are present. Pedal protractor muscles are short and close to the anterior retractors. The siphonal retractor muscles are oval. The MANTLE margins are fused ventrally, with a large anteroventral pedal gape. Posterior EXCURRENT and INCURRENT SIPHONS are long, separate, and independently mobile, with a CRUCIFORM MUSCLE at their base; they are annulated and capable of autotomy in *Solecurtus*. A specialized, open or closed (in *Tagelus*), sensory organ accompanies the cruciform muscle. The MANTLE CAVITY can extend posteriorly beyond the shell when the animal is fully expanded. HYPOBRANCHIAL GLANDS have not been reported. The FOOT is large and laterally compressed; a BYSSAL GROOVE and BYSSUS are absent in the adult (a rudimentary byssal apparatus has been shown in *Tagelus dombeii* (Lamarck, 1818)).

The LABIAL PALPS are small to large and narrowly trigonal. The CTENIDIA are EULAMELLIBRANCH (SYNAPTORHABDIC) and HETERORHABDIC. The outer demibranchs are smaller than the inner demibranchs. The gills are united with the SIPHONAL SEPTUM, separating INFRA- and SUPRABRANCHIAL CHAMBERS. The tips of the inner demibranchs are not inserted or fused into the distal oral groove of the palps (CATEGORY III association). Incurrent and excurrent water flows are posterior. The STOMACH is TYPE V, in some species with a posterodorsal, in some cases lobed, gastric appendix; the MIDGUT is not coiled. The HINDGUT passes through the ventricle of the heart, and leads to a sessile rectum. An AORTIC BULB has been reported in some species. Solecurtids are GONOCHORISTIC and produce planktonic VELIGER larvae that in their late stage have a distinctively trigonal shell with an extended anterior end. The nervous system is not concentrated. STATOCYSTS have

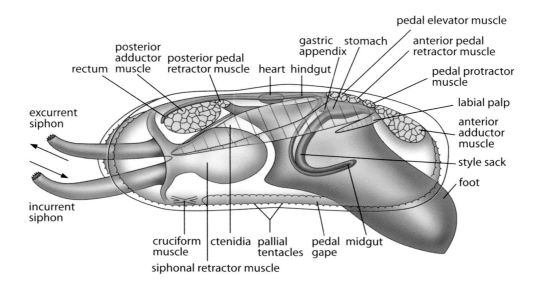

**In living *Solecurtus sulcatus*** (Dunker, 1862) from Queensland, Australia, the body greatly exceeds the size of the shell and cannot be fully withdrawn. When perturbed, the animal can automize its long annulated siphons. Note the prominent cruciform muscle at the base of the siphons in the specimen at right.

been reported in adult *Tagelus dombeii*. ABDOMINAL SENSE ORGANS have not been reported.

Solecurtids are marine SUSPENSION or DEPOSIT FEEDERS that inhabit more or less permanent, occasionally deep, vertical burrows in clean substrata. *Solecurtus strigilatus* (Linnaeus, 1758) of the Mediterranean Sea has been found to burrow rapidly as an escape response using a water jet expelled from the mantle cavity to fluidize the substratum; its burrow is obliquely Y-shaped (with the shell in the oblique stem, and one siphon in each arm of the "Y") and the walls are strengthened by mucus.

The family Solecurtidae is known since the Cretaceous and is represented by 3 living genera and ca. 40 species, distributed in tropical and warmer maritime regions. Some species, such as the Chilean "navajuela" (*Tagelus dombeii*), are harvested by subsistence and commercial fisheries. As softshell clam (Myidae) catches have dropped in Maryland, hydraulic clamming dredges have increasingly targeted the solecurtid *Tagelus plebeius*, which is marketed as bait for eel and crab-pot fisheries.

# References

Bloomer, H. H. 1903a. The anatomy of certain species of *Ceratisolen* and *Solecurtus*. *The Journal of Malacology*, 10(2): 31–40, pl. 2.

Bloomer, H. H. 1903b. The anatomy of *Pharella orientalis*, Dunker and *Tagelus rufus*, Spengler. *The Journal of Malacology*, 10(4): 114–121, pl. 10.

Bloomer, H. H. 1907. On the anatomy of *Tagelus gibbus* and *T. divisus*. *Proceedings of the Malacological Society of London*, 7: 218–223, pl. 19. Correction [to this paper]. *Proceedings of the Malacological Society of London*, 7(5): 260.

Bloomer, H. H. 1912. On the anatomy of species of *Cultellus* and *Azor*. *Proceedings of the Malacological Society of London*, 10(1): 5–10, pl. 1.

Bromley, R. G., and U. Asgaard. 1990. *Solecurtus strigilatus*: a jet-propelled burrowing bivalve. Pages 313–320, in: B. Morton, ed., *The Bivalvia, Proceedings of a Memorial Symposium in Honour of Sir Charles Maurice Yonge (1899–1986), Edinburgh, 1986*. Hong Kong University Press, Hong Kong.

Chanley, P. E., and M. Castagna. 1971. Larval development of the stout razor clam *Tagelus plebeius* Solander (Solecurtidae: Bivalvia). *Chesapeake Science*, 12(3): 167–172.

Fraser, T. H. 1967. Contributions to the biology of *Tagelus divisus* (Tellinacea: Pelecypoda) in Biscayne Bay, Florida. *Bulletin of Marine Science*, 17(1): 111–132.

Frenkiel, L. 1979. L'organe sensoriel du muscle cruciforme des Tellinacea: importance systématique chez les Psammobiidae. *The Journal of Molluscan Studies*, 45(2): 231–237.

Ghosh, E. 1920. Taxonomic studies of the soft parts of the Solenidae. *Records of the Indian Museum*, 19(2): 47–78.

Graham, A. 1934. The structure and relationships of lamellibranchs possessing a cruciform muscle. *Proceedings of the Royal Society of Edinburgh*, 54: 158–187.

Hoffmann, F. 1914. Beiträge zur Anatomie und Histologie von *Tagelus dombeyi* (Lamarck). *Jenaische Zeitschrift für Naturwissenschaft*, 52(4): 521–566, pls. 12–14.

Holland, A. F., and J. M. Dean. 1977. The biology of the stout razor clam *Tagelus plebeius*: I. Animal–sediment relationships, feeding mechanism, and community biology. *Chesapeake Science*, 18(1): 58–66.

Pohlo, R. H. 1973. Feeding and associated functional morphology in *Tagelus californianus* and *Florimetis obesa* (Bivalvia: Tellinacea). *Malacologia*, 12(1): 1–11.

Villarroel, M., and J. Stuardo. 1977. Observationes sobre la morfología general, musculatura y aparato digestivo en *Tagelus (Tagelus) dombeii* y *T. (T.) longisinuatus* (Tellinacea: Solecurtidae). *Malacologia*, 16(2): 333–352.

Yonge, C. M. 1949. On the structure and adaptations of the Tellinacea, deposit-feeding Eulamellibranchia. *Philosophical Transactions of the Royal Society of London, Series B, Biological Sciences*, 234(609): 29–76.

Yonge, C. M. 1952. Studies on Pacific coast mollusks IV. Observations on *Siliqua patula* Dixon and on evolution within the Solenidae. *University of California Publications in Zoölogy*, 55(9): 421–438.

### *Solecurtus cumingianus* (Dunker, 1861) – **Corrugated Razor Clam**

Elongated oval, dorsal margin straight, with irregular commarginal growth lines crossed by wavy oblique incised lines (absent anteriorly), white exteriorly and interiorly. North Carolina to Florida, Bahamas, West Indies, Gulf of Mexico, Caribbean Central America, South America (Colombia, Brazil). Length 30 mm (to 76 mm).

### *Tagelus plebeius* (Lightfoot, 1786) – **Stout Tagelus**

Elongated oval to quadrangular, dorsal margin straight, smooth with irregular commarginal growth lines, white with thick, olive green to brownish yellow periostracum; interior with large, bulbous callus underlying ligament. Massachusetts to Florida, West Indies, Gulf of Mexico, Caribbean Central America, South America (to Argentina), also West Africa. Length 50 mm (to 95 mm).

**A salt marsh** in eastern Florida typifies the habitat of *Tagelus plebeius*.

*Solecurtus cumingianus*

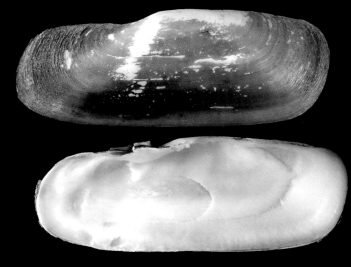

*Tagelus plebeius*

# Family Pharidae – Razor and Jackknife Clams

### Classification
AUTOLAMELLIBRANCHIATA Grobben, 1894
HETEROCONCHIA Hertwig, 1895
Heterodonta Neumayr, 1883
Veneroida H. Adams & A. Adams, 1856
Solenoidea Lamarck, 1809
Pharidae H. Adams & A. Adams, 1856

**Featured species**
*Ensis minor* Dall, 1900 – **Minor Jackknife**

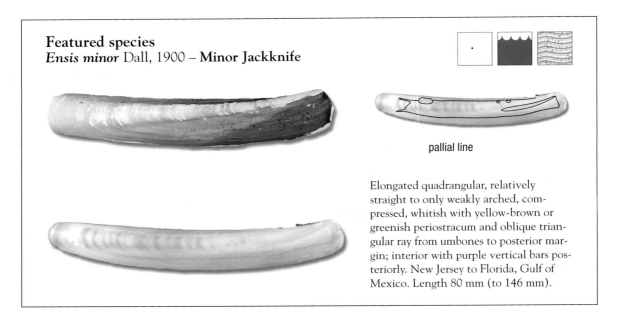

pallial line

Elongated quadrangular, relatively straight to only weakly arched, compressed, whitish with yellow-brown or greenish periostracum and oblique triangular ray from umbones to posterior margin; interior with purple vertical bars posteriorly. New Jersey to Florida, Gulf of Mexico. Length 80 mm (to 146 mm).

## Family description

The pharid shell is small to medium-sized (to 230 mm), thin-walled, fragile to rather solid, and longer than high, narrowly quadrangular, with rounded ends (or truncate in *Ensis*), curved or straight, with a pronounced oblique dividing line between anteroventral and posterodorsal regions. The shell is EQUIVALVE, compressed to cylindrical, and widely gaping anteriorly and posteriorly. The shell is markedly INEQUILATERAL (umbones at or near the anterior end), with ORTHOGYRATE UMBONES. Shell microstructure is ARAGONITIC and two-layered, with a CROSSED LAMELLAR or HOMOGENOUS outer layer and a COMPLEX CROSSED LAMELLAR or (usually) homogenous inner layer, the latter confined within the pallial line. TUBULES are apparently absent. Exteriorly pharids are covered by thin to thick, glossy, brownish PERIOSTRACUM. Sculpture is smooth with commarginal growth lines only. LUNULE and ESCUTCHEON are absent. Interiorly the shell is non-NACREOUS, often reinforced with a flattened calcareous rib emanating from beneath the umbones and to which the anterior pedal retractor muscles attach. The PALLIAL LINE has a short to

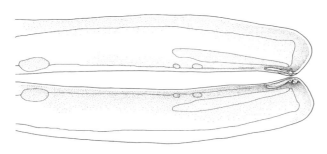

moderately deep SINUS. The inner shell margins are smooth. The HINGE PLATE is weak and HETERODONT; the left valve has two vertical and two more or less horizontal CARDINAL TEETH (the two middle teeth can be fused), whereas the right valve has one vertical and one subhorizontal tooth; anterior and posterior LATERAL TEETH are absent. (Members of the closely related family Solenidae [not represented in the Florida Keys] have only one cardinal tooth in each valve). The LIGAMENT is strong, PARIVINCULAR, OPISTHODETIC, and set on long, narrow NYMPHS. A secondary ligament of fused periostracum unites the valves dorsally.

The animal is HETEROMYARIAN (posterior ADDUCTOR MUSCLE smaller), with the posterior adductor muscle united with the pallial line in some species; anterior and posterior pedal retractor muscles are present, the anterior of which can have two insertions. Pedal elevator and protractor muscles have not been reported, except in *Sinonovacula,* which has anterior and posterior pedal protractor muscles. The MANTLE margins are fused ventrally, with a large anterior and anteroventral pedal gape, and in some species a tentacled FOURTH PALLIAL APERTURE (present in *Ensis* and *Pharus;* secondarily derived from the pedal gape and occasionally confluent with it; not homologous with the fourth pallial aperture of other bivalves). A CRUCIFORM MUSCLE is absent. Posterior EXCURRENT and INCURRENT SIPHONS are short to long, naked, and separate or united; the smaller excurrent siphon has a terminal cone (VALVULAR MEMBRANE). HYPOBRANCHIAL GLANDS have not been reported. The FOOT is large, long and narrow, laterally compressed, obliquely truncate, and with a terminal bulb; a BYSSAL GROOVE and BYSSUS are absent in the adult.

The LABIAL PALPS are medium-sized to large, and are extended by a long, unridged oral groove to the mouth. The CTENIDIA are EULAMELLIBRANCH (SYNAPTORHABDIC), HOMORHABDIC or HETERORHABDIC, and posteriorly placed. The tips of the inner demibranchs are inserted into and fused with the distal oral groove of the palps (CATEGORY II association; *Ensis directus* Conrad, 1843), or are not inserted (or fused) (CATEGORY III, in Novaculininae). Incurrent and excurrent water flows are posterior. The STOMACH is TYPE V; the MIDGUT is coiled. The HINDGUT passes through the ventricle of the heart, and leads to a freely hanging rectum. An AORTIC BULB is present; hemoglobin has been reported in some species. Pharids are GONOCHORISTIC (as reported for *Siliqua*) and produce planktonic

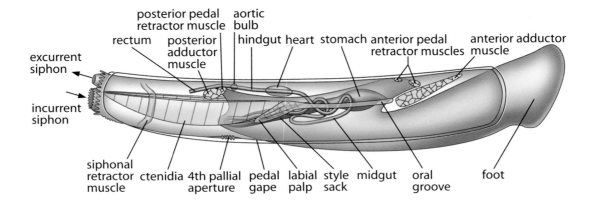

VELIGER larvae. The nervous system is not concentrated. STATOCYSTS and ABDOMINAL SENSE ORGANS have not been reported.

Pharids are marine or estuarine SUSPENSION FEEDERS, modified for rapid INFAUNAL burrowing in soft, unstable sediment, including sand and mud, from the intertidal zone, mangrove communities, and continental shelves. They construct more or less permanent vertical or oblique burrows in which they can ascend and descend. Rapid burrowing is achieved in some by expanding and contracting the piston like foot using hemocoelic pressure; propulsion in *Ensis minor* can be effected by adducting the shells to expel water from the anterior pedal gape. Predators include boring naticid gastropods, nemertine worms, fish, and shorebirds, for which they form important food items. Some species are known to leap out of their burrows to escape predators, swimming for short distances, then reburrowing rapidly.

The family Pharidae is known since the Cretaceous and is represented by 13 living genera and ca. 65 species, distributed worldwide mainly in maritime regions. The larger species are consumed by humans in many countries; *Siliqua patula* (Dixon, 1789), for instance, is harvested extensively in the Pacific Northwest by commercial and sport fisheries. *Sinonovacula constricta* (Lamarck, 1818) is an important cultured species in Asia, increasingly used in polycultures with penaeid shrimp and tilapia. In the late 1970s, *Ensis americanus* (Gould, 1870) was introduced into the North Sea, apparently by larval stages transported in bilge water, and is rapidly extending its range along the European mainland and British coastlines. In addition to its possible impact on the native ecology, the large V-shaped empty shells often cause considerable damage to fishing nets.

# References

Armonies, W., and K. Reise. 1999. On the population development of the introduced razor clam *Ensis americanus* near the Island of Sylt (North Sea). *Helgoländer Meeresuntersuchungen*, 52: 291–300.

Bloomer, H. H. 1903a. The anatomy of certain species of *Ceratisolen* and *Solecurtus*. *The Journal of Malacology*, 10(2): 31–40, pl. 2.

Bloomer, H. H. 1903b. The anatomy of *Pharella orientalis*, Dunker and *Tagelus rufus*, Spengler. *The Journal of Malacology*, 10(4): 114–121, pl. 10.

Cosel, R. von. 1990. An introduction to the razor shells (Bivalvia: Solenacea). Pages 283–311, in: B. Morton, ed., *The Bivalvia, Proceedings of a Memorial Symposium in Honour of Sir Charles Maurice Yonge (1899–1986), Edinburgh, 1986*. Hong Kong University Press, Hong Kong.

Cosel, R. von. 1993. The razor shells of the eastern Atlantic. Part 1: Solenidae and Pharidae I (Bivalvia: Solenacea). *Archiv für Molluskenkunde*, 122(Zilch-Festschrift): 207–321.

Ghosh, E. 1920. Taxonomic studies of the soft parts of the Solenidae. *Records of the Indian Museum*, 19(2): 47–78.

McMahon, R. F., and C. O. McMahon. 1983. Leaping and swimming as predator escape responses in the jackknife clam, *Ensis minor* Dall (Bivalvia: Pharellidae). *The Nautilus*, 97(2): 55–58.

Morton, B. 1984. The functional morphology of *Sinonovacula constricta* with a discussion on the taxonomic status of the Novaculininae (Bivalvia). *Journal of Zoology*, 202: 299–325.

Owen, G. 1959. Observations on the Solenacea with reasons for excluding the family Glaucomyidae. *Philosophical Transactions of the Royal Society of London, Series B, Biological Sciences*, 242(687): 59–97.

Severijns, N. 2002. Distribution of the American jack-knife clam *Ensis directus* (Conrad, 1843) in Europe 23 years after its introduction. *Gloria Maris*, 40(4–5): 61–111.

Yonge, C. M. 1952. Studies on Pacific coast mollusks IV. Observations on *Siliqua patula* Dickson and on evolution of the Solenidae. *University of California Publications in Zoölogy*, 55(9): 421–438.

# Family Mactridae – Surf or Trough Clams

**Classification**
AUTOLAMELLIBRANCHIATA Grobben, 1894
HETEROCONCHIA Hertwig, 1895
Heterodonta Neumayr, 1883
Veneroida H. Adams & A. Adams, 1856
Mactroidea Lamarck, 1809
Mactridae Lamarck, 1809

## Featured species
*Mactrotoma fragilis* (Gmelin, 1791) – **Fragile Surf Clam**

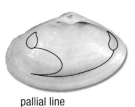

pallial line

Rounded trigonal, posterior truncate, thin-walled, smooth with two radial ridges from umbo to posteroventral margin (one close to dorsal margin), white with yellowish gray periostracum most evident on posterior ridge; interior white. North Carolina to Florida, Bahamas, West Indies, Gulf of Mexico, Caribbean Central America, South America (to Uruguay). Length 35 mm (to 94 mm). Formerly in *Mactra*.

The siphons of *Mactrotoma fragilis*, as seen here in a juvenile from Newfound Harbor in the Lower Florida Keys, are long and united to the tips.

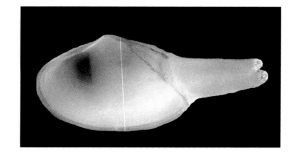

## Family description

The mactrid shell is small to large (to 225 mm, for *Spisula solidissima* (Dillwyn, 1817), the largest western Atlantic bivalve), thin-walled to solid, and oval to trigonal to transversely elongated, with the posterodorsal area marked by a high fringe or keel. The shell is EQUIVALVE, inflated, and usually narrowly gaping posteriorly, often also anteriorly. The shell is EQUILATERAL, with PROSOGYRATE UMBONES. Shell microstructure is ARAGONITIC and two-

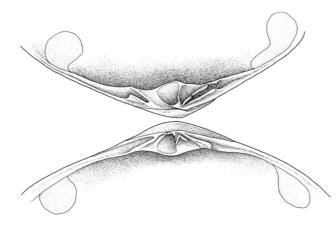

layered, with a CROSSED LAMELLAR outer layer and COMPLEX CROSSED LAMELLAR inner layer. TUBULES are apparently absent. Exteriorly mactrids are covered by thick, often glossy, fibrous, dehiscent, brownish PERIOSTRACUM that is eroded in larger individuals and covers the ventral surface of the free mantle margins and the siphons. Sculpture is smooth or with weak to well-developed commarginal ridges; a few species have prominent radial ribs. The LUNULE is poorly defined in *Anatina*, but usually absent; the ESCUTCHEON is always absent. Interiorly the shell is non-NACREOUS. The PALLIAL LINE has a deep SINUS. The inner shell margins are smooth. The HINGE PLATE is strong and HETERODONT, with three CARDINAL TEETH in the left valve and two in the right valve; anterior and posterior LATERAL TEETH are present and can be transversely ridged. All teeth can be more or less obsolete. The LIGAMENT is small, PARIVINCULAR, OPISTHODETIC, and set on NYMPHS; an internal portion (RESILIUM) is large, trigonal, and set on large oblique CHONDROPHORES posterior to the cardinal teeth; a secondary external ligament of fused periostracum can be present, uniting the valves along the dorsal hinge margin.

The animal is ISOMYARIAN or HETEROMYARIAN (anterior ADDUCTOR MUSCLE more elongated in cross section in *Raeta*); anterior and posterior pedal retractor muscles are present. Pedal elevator and protractor muscles have been recorded in *Mulinia*. The MANTLE margins are fused ventrally, with a large anteroventral pedal gape, and in some species a FOURTH PALLIAL APERTURE ventral to the incurrent siphon. Posterior EXCURRENT and INCURRENT SIPHONS are united and usually sheathed with periostracum; the usually equally sized excurrent siphon has a terminal cone (VALVULAR MEMBRANE). The siphons are fully retractable into the shell in some, not so in others. Lamellar sense organs can be present at the base or the ventral inside of the incurrent siphon. Various waste canal

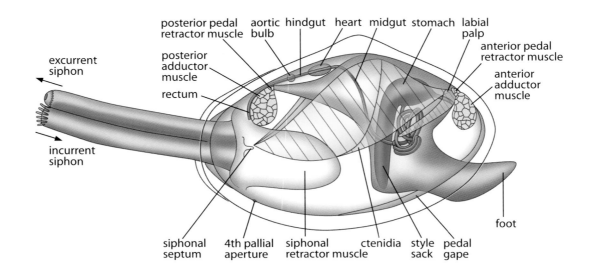

structures occur posteriorly on the inner ventral mantle margins to remove particles from the MANTLE CAVITY. HYPOBRANCHIAL GLANDS have not been reported. The FOOT is usually large (small in *Raeta*), laterally compressed, and heeled; a BYSSAL GROOVE and BYSSUS are absent in the adult.

The LABIAL PALPS are large (very large in *Mactrinula*) and trigonal or elongated. The CTENIDIA are EULAMELLIBRANCH (SYNAPTORHABDIC) and usually HOMORHABDIC, rarely HETERORHABDIC. The outer demibranchs are smaller than the inner demibranchs. The tips of the inner demibranchs are not inserted or fused into the distal oral groove of the palps (CATEGORY III association). The gills are united behind the foot with a SIPHONAL SEPTUM (a continuation of the diaphragm formed by the gill attachments), separating INFRA- and SUPRABRANCHIAL CHAMBERS. Incurrent and excurrent water flows are posterior. The STOMACH is TYPE V; the MIDGUT is long and often complexly coiled. The HINDGUT passes through the ventricle of the heart, and leads to a sessile rectum. An AORTIC BULB is present. Mactrids are GONOCHORISTIC and produce planktonic VELIGER larvae. The nervous system is not concentrated. STATOCYSTS have been reported in adult *Mulinia*. ABDOMINAL SENSE ORGANS are absent.

Mactrids are marine or estuarine SUSPENSION FEEDERS, and INFAUNAL in areas of shifting sand or sandy mud. Most burrow rapidly using the powerful foot in conjunction with jets of water created by rapidly closing the shell valves. Some species have a wide range of salinity tolerance; *Rangia cuneata* (G. B. Sowerby I, 1831) inhabits the outer surf zone of Gulf of Mexico beaches but can tolerate entirely fresh waters. Some species host symbiotic pea crabs (Pinnotheridae) in their mantle cavities; others harbor symbiotic chemautotrophic bacteria in the gills. Predators include fish, crustaceans, boring gastropods, and water birds.

The family Mactridae is known since the Cretaceous and is represented by 19 living genera and ca. 150 species, distributed worldwide in relatively shallow waters. Mactrids are important components of the world's fisheries, with a few species (e.g., *Spisula solidissima*) sustaining commercial hydraulic clam-dredge fisheries in North America. Larger species were used historically as food and scraping tools by natives in North America and Australia.

## References

Barnes, P. A. G., and B. Morton. 1997. The functional morphology of *Mactrinula reevesii* (Bivalvia: Mactroidea) in Hong Kong: adaptations for a deposit-feeding lifestyle. *Journal of Zoology*, 241(1): 13–34.

Cargnelli, L. M., S. J. Griesbach, D. B. Packer, and E. Weissberger. 1999. *Atlantic Surfclam*, Spisula solidissima, *Life History and Habitat Characteristics*. NOAA Technical Memorandum NMFS-NE-142, v + 13 pp. Online version http://www.nefsc.noaa.gov/nefsc/publications/tm/tm142/tm142.pdf, last accessed 10 December 2006.

Fischer, R. 1915. Über die Anatomie von *Mactra (Mulinia) coquimbana* Philippi. *Jenaische Zeitschrift für Naturwissenschaft*, 53(Neue Folge 46)(4): 597–662.

Gentile, A., comp. 2005. *Species fact sheet:* Spisula solidissima *(Dillwyn, 1817)*. *Species Identification and Data Programme (SIDP)*. FAO-FIGIS, 2 pp. Online version http://www.fao.org/figis/servlet/species?fid=3538, last accessed 01 November 2005.

Harry, H. W. 1969. Anatomical notes on the mactrid bivalve, *Raeta plicatella* Lamarck, 1818, with a review of the genus *Raeta* and related genera. *The Veliger*, 12(1): 1–23.

Kellogg, J. L. 1915. Ciliary mechanisms of lamellibranchs with description of anatomy. *Journal of Morphology*, 26: 625–701.

LaSalle, M. W., and A. A. de la Cruz. 1985. *Species Profiles: Life Histories and Environmental Requirements of Coastal Fishes And Invertebrates (Gulf of Mexico)—Common Rangia*. United States Fish and Wildlife Service Biological Report 82(11.31), TR EL-82-4, 16 pp. Online version http://www.nwrc.usgs.gov/wdb/pub/0145.pdf, last accessed 07 November 2005.

Pearce, J. B. 1966. On *Pinnixa faba* and *Pinnixa littoralis* (Decapoda: Pinnotheridae) symbiotic with the clam, *Tresus capax* (Pelecypoda: Mactridae). Pages 565–589, in: H. Barnes, ed., *Some Contemporary Studies in Marine Science*. George Allen & Unwin, London.

Ropes, J. W., and A. S. Merrill, 1966. The burrowing activities of the surf clam. *Underwater Naturalist*, 3(4): 11–17.

Saul, L. R. 1973. Evidence for the origin of the Mactridae (Bivalvia) in the Cretaceous. *University of California Publications in Geological Science*, 97: 1–51.

Walker, R. L., and F. X. O'Beirn. 1996. Embryonic and larval development of *Spisula solidissima similis* (Say, 1822) (Bivalvia: Mactridae). *The Veliger*, 39(1): 60–64.

Yonge, C. M. 1948. Cleansing mechanisms and the function of the fourth pallial aperture in *Spisula subtruncata* (da Costa) and *Lutraria lutraria* (L.). *Journal of the Marine Biological Association of the United Kingdom*, 27: 585–596.

Yonge, C. M. 1982. Ligamental structure in Mactracea and Myacea (Mollusca: Bivalvia). *Journal of the Marine Biological Association of the United Kingdom*, 62(1): 171–186.

### *Anatina anatina* (Spengler, 1802) – **Smooth Duck Clam**

Rounded trigonal, posterior with distinct, low radial rib behind which shell gapes, fragile, smooth with irregularly wavy sculpture just in front of posterior keel, translucent white to tan; interior white. North Carolina to Florida, West Indies, Gulf of Mexico, Caribbean Central America, South America (to Brazil). Length 60 mm.

### *Mulinia lateralis* (Say, 1822) – **Dwarf Surf Clam**

Rounded trigonal, with distinct radial ridge near posterior end, solid, smooth, white to cream; interior white. Eastern Canada to Florida, West Indies, Gulf of Mexico, Caribbean Central America. Length 12 mm (to 15 mm).

### *Raeta plicatella* (Lamarck, 1818) – **Channeled Duck Clam**

Rounded trigonal, posterior narrowed and pointed, fragile, with smooth, rounded commarginal ridges and very fine, crinkly radial threads, white exteriorly and interiorly; interior with commarginal grooves corresponding to external ridges. New Jersey to Florida, West Indies, Gulf of Mexico, Caribbean Central America, South America (to Argentina). Length 60 mm (to 70 mm).

### *Spisula raveneli* (Conrad, 1832) – **Southern Surf Clam**

Rounded trigonal, solid, smooth, yellowish white with yellowish brown periostracum; interior white. Massachusetts to Florida, Gulf of Mexico. Length 60 mm (to 125 mm). Syn. *similis* auctt. non Say, 1822. Formerly considered a subspecies of the northern *S. solidissima* (Dillwyn, 1817), of which *similis* has been determined to be a pathologic form from New Jersey.

**The commercial Atlantic Surf Clam**, *Spisula solidissima*, lives in sandy substrata from eastern Canada to Cape Hatteras (North Carolina) in depths from the surf zone to 128 m. It has been recorded as large as 226 mm and 31 years of age, and thus ranks as the largest western Atlantic bivalve.

**A dredge haul** on a commercial clam dredger harvesting surf clams (*Spisula solidissima*). Since the 1930s, this species has been the target of a commercial industry off New Jersey and the Delmarva Peninsula in depths from 18 to 36 m. A management plan adopted in 1991, including individual, transferable fishing quotas, has stabilized previously declining annual landings.

*Anatina anatina*

*Mulinia lateralis*

*Raeta plicatella*

*Spisula raveneli*

*Spisula solidissima*

# Family Dreissenidae – False Mussels

**Classification**
AUTOLAMELLIBRANCHIATA Grobben, 1894
HETEROCONCHIA Hertwig, 1895
Heterodonta Neumayr, 1883
Veneroida H. Adams & A. Adams, 1856
Dreissenoidea J. E. Gray, 1840
Dreissenidae J. E. Gray, 1840

**Featured species**
*Mytilopsis leucophaeata* (Conrad, 1831) – **Dark False Mussel**

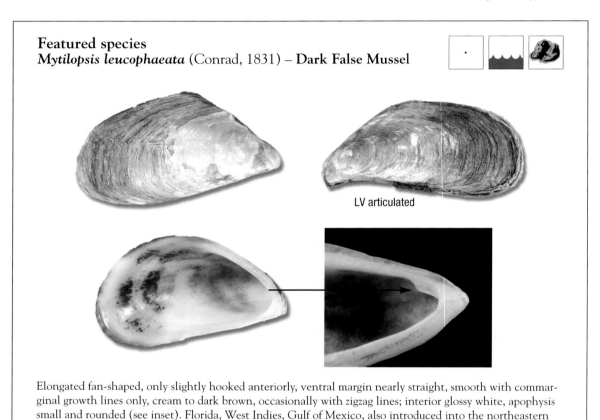

LV articulated

Elongated fan-shaped, only slightly hooked anteriorly, ventral margin nearly straight, smooth with commarginal growth lines only, cream to dark brown, occasionally with zigzag lines; interior glossy white, apophysis small and rounded (see inset). Florida, West Indies, Gulf of Mexico, also introduced into the northeastern United States (including the Hudson River) and Europe. Length 13 mm.

# Family description

The dreissenid shell is small to medium-sized (usually <100 mm), thin-walled, and mussel-shaped (MYTILIFORM) to quadrangular, pointed and hooked anteriorly, rounded posteriorly. The ventral margin is almost straight, flattened to concave, often bordered by an obtuse keel. The shell is INEQUIVALVE (right valve slightly larger), moderately inflated,

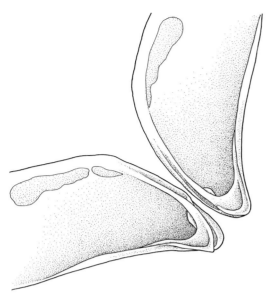

and narrowly gaping anteriorly. The shell is INEQUILATERAL (umbones anterior or terminal), with PROSOGYRATE UMBONES. Shell microstructure is ARAGONITIC and two-layered, with a CROSSED LAMELLAR outer layer and COMPLEX CROSSED LAMELLAR inner layer. TUBULES are present in the outer layer only of some individuals of some species. Exteriorly dreissenids are drably colored in shades of brown and gray, covered by thick, brownish PERIOSTRACUM that is two-layered in *Dreissena*. Sculpture is smooth or restricted to coarse growth lines. LUNULE and ESCUTCHEON are absent. Interiorly the shell is non-NACREOUS and whitish. The PALLIAL LINE is ENTIRE. The inner shell margins are smooth. The HINGE PLATE is EDENTATE in adults; the UMBONAL CAVITY is equipped with a myophoral SEPTUM and a toothlike APOPHYSIS. The LIGAMENT is somewhat sunken, PARIVINCULAR, and OPISTHODETIC; a secondary external ligament of fused periostracum unites the valves anteriorly and posteriorly.

The animal is HETEROMYARIAN (posterior ADDUCTOR MUSCLE irregularly rounded in cross section; anterior adductor smaller), with the anterior adductor inserting at the pointed anterior end between the myophoral septa. The large posterior pedal retractor muscles insert along the dorsal margin near the posterior adductor muscle; the smaller anterior pedal retractor muscles insert near or on the myophoral septum. Pedal elevator and protractor muscles have not been reported. The MANTLE margins are extensively fused ventrally, with a small anteroventral pedal gape. Posterior EXCURRENT and INCURRENT SIPHONS are separate; however, siphonal retractor muscles are absent, hence so is a PALLIAL SINUS. Hypobranchial glands have not been reported. The FOOT is small, digitiform, and has a BYSSAL GROOVE; the adult is byssate.

The LABIAL PALPS are small to large. The CTENIDIA are EULAMELLIBRANCH (SYNAPTORHABDIC) and HOMORHABDIC, with the tips of the inner demibranchs inserted and

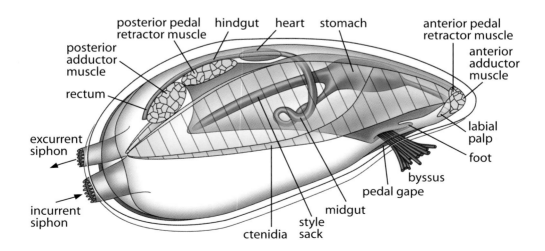

fused into the distal oral groove of the palps (CATEGORY II association). The outer demi-branchs are smaller than the inner demibranchs. Incurrent and excurrent water flows are posterior. The STOMACH is TYPE V; the MIDGUT is coiled. The HINDGUT passes through the ventricle of the heart, and leads to a sessile rectum. Dreissenids are GONOCHORISTIC and produce planktonic VELIGER larvae. The nervous system is not concentrated. STATOCYSTS (with STATOLITHS) have been reported in adult *Dreissena*. ABDOMINAL SENSE ORGANS have not been reported.

Dreissenids are estuarine or freshwater SUSPENSION FEEDERS, and EPIFAUNAL on hard substrata. The invasive freshwater Zebra Mussel (*Dreissena polymorpha* (Pallas, 1771)) is a member of this family. Zebra mussels are native to the Black and Caspian seas and were introduced into western Europe through canals and inland waterways during the nineteenth century. They were discovered in the Great Lakes in 1985, introduced via ballast water from commercial ship traffic, and have since spread throughout the Mississippi River and other major drainages in the eastern United States. Their physiology has been discussed as ideally suited to biological invasion.

The family Dreissenidae is known since the Eocene and is represented by 3 living genera and ca. 10 species, distributed worldwide in slowly moving or still, fresh and low-salinity waters (<13 ppt) down to ca. 50 m. In the Florida Keys, *Mytilopsis* species have been recorded only recently from Florida Bay and certain boating canals (see p. 8), perhaps indicative of increased freshwater input from the Everglades and recreational boat traffic in recent years.

---

**Mytilopsis sallei (Récluz, 1849) – Santo Domingo False Mussel**

Elongated fan-shaped, hooked anteriorly, ventral margin concave, smooth with commarginal growth lines only, cream to dark brown, occasionally with zigzag lines; interior glossy, mottled cream to bluish gray, apophysis large and hook-shaped (see inset). Florida, Gulf of Mexico, Caribbean Central America, South America (to Venezuela), introduced via the Panama Canal into the Indo-Pacific Ocean where it is spreading widely and causing severe fouling problems. Length 12 mm.

**Freshwater pearl mussels** in the United States are severely imperiled by environmental change (damming, pollution, etc.) and by competition from several invasive species. Introduced dreissenid species are particularly effective competitors for food and space. The introduced Zebra Mussel (*Dreissena polymorpha*) is known to smother freshwater mussels (here the Fatmucket [*Lampsilis siliquoidea* (Barnes, 1823)] from Ohio) by overwhelming coverage of its valves. In addition, Zebra Mussels inflict serious economic damage by clogging power plant intake pipes and attaching to boat hulls, docks, and buoys. *Dreissena polymorpha* has rapidly expanded in the continental United States since its introduction into the Great Lakes in the 1980s, but has not yet been recorded in Florida.

### References

D'Itri, F. M., ed. 1997. *Zebra Mussels and Aquatic Nuisance Species*. Ann Arbor Press, Chelsea, Michigan, ix + 638 pp.

Marelli, D. C., and S. Gray. 1983. Conchological redescriptions of *Mytilopsis sallei* and *Mytilopsis leucophaeta* of the brackish western Atlantic (Bivalvia: Dreissenidae). *The Veliger*, 25(3): 185–193, 1 pl.

Morton, B. S. 1969. Studies on the biology of *Dreissena polymorpha* Pall. 1. General anatomy and morphology. *Proceedings of the Malacological Society of London*, 38(4): 301–321.

Nalepa, T. F., and D. W. Schloesser, eds. 1993. *Zebra Mussels: Biology, Impacts, and Control*. Lewis Publishers, Boca Raton, Florida, 810 pp.

Pathy, D. A., and G. L. Mackie. 1993. Comparative shell morphology of *Dreissena polymorpha*, *Mytilopsis leucophaeata*, and the "quagga" mussel (Bivalvia: Dreissenidae) in North America. *Canadian Journal of Zoology*, 71: 1012–1023.

Therriault, T. W., M. F. Docker, I. M. Orlova, D. D. Heath, and H. J. MacIsaac. 2004. Molecular resolution of the family Dreissenidae (Mollusca: Bivalvia) with emphasis on Ponto-Caspian species, including first report of *Mytilopsis leucophaeata* in the Black Sea basin. *Molecular Phylogenetics and Evolution*, 30(3): 479–489.

Yonge, C. M., and J. I. Campbell. 1968. On the heteromyarian condition in the Bivalvia with reference to *Dreissena polymorpha* and certain Mytilacea. *Transactions of the Royal Society of Edinburgh*, 68(2): 21–43.

# Family Myidae – Soft-Shell or Sand Gaper Clams

**Classification**
AUTOLAMELLIBRANCHIATA Grobben, 1894
Heteroconchia Hertwig, 1895
Heterodonta Neumayr, 1883
Myoida Stoliczka, 1870
Myoidea Lamarck, 1809
Myidae Lamarck, 1809

**Featured species**
*Sphenia fragilis* (H. Adams & A. Adams, 1854) –
**Antillean Sphenia**

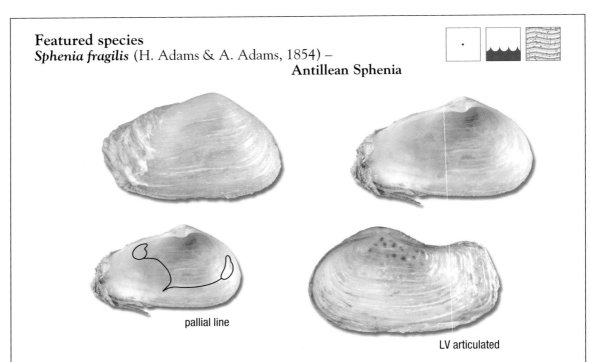

pallial line

LV articulated

Elongated quadrangular but irregular, posterior rostrum truncate, commonly twisted and often extended by periostracum, smooth with fine to moderate commarginal ridges, chalky white with yellowish periostracum (mainly ventrally); interior glossy white. Georgia to Florida, West Indies, Gulf of Mexico, Caribbean Central America, South America (to Uruguay), also eastern Pacific. Length 7 mm (to 15 mm). Syn. *antillensis* Dall & Simpson, 1901. Note: *Sphenia dubia* (H. C. Lea, 1843), from the Miocene of Virginia, might be an earlier name for this species.

## Family description

The myid shell is small to medium-sized (to 180 mm), thin-walled, and elongated to oval or quadrangular, with the posterior angled or sharply truncated. In nestling species (e.g., *Sphenia*), adult shape can be highly variable. The shell is slightly INEQUIVALVE (right valve larger), often somewhat inflated, and widely gaping posteriorly (also anteriorly in

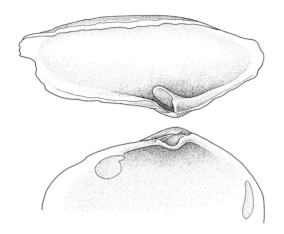

some). The shell is INEQUILATERAL (umbones anterior), with PROSOGYRATE UMBONES. Shell microstructure is ARAGONITIC and three-layered, with a HOMOGENOUS outer layer, a CROSSED LAMELLAR middle layer, and a COMPLEX CROSSED LAMELLAR or homogenous inner layer, the last confined within the pallial line. TUBULES are apparently absent. Exteriorly myids are chalky white and covered by a thick, dehiscent, yellow to brown PERIOSTRACUM that is adherent only at the margins; the periostracum can extend beyond the posterior end in nestling species (e.g., *Sphenia*). Sculpture is limited to commarginal growth lines, rarely with radial ribbing (in *Cryptomya*). LUNULE and ESCUTCHEON are absent. Interiorly the shell is non-NACREOUS and chalky or glossy white. The pallial line is frequently discontinuous, usually with a deep PALLIAL SINUS (shallow in *Sphenia*; absent in *Cryptomya*). The inner shell margin is smooth. The HINGE PLATE is EDENTATE (DESMODONT) in adults or very rarely with a small anterior tooth in the right valve (in *Sphenia*). The LIGAMENT is fully internal (RESILIUM) and extends vertically from the upper surface of the large, projecting CHONDROPHORE (which shows various species-specific ridges that act as functional hinge teeth) in the left valve to a smaller RESILIFER in the UMBONAL CAVITY of the right valve; a secondary external ligament of fused periostracum unites the valves along the dorsal hinge margin.

The animal is ISOMYARIAN OR HETEROMYARIAN (posterior ADDUCTOR MUSCLE smaller and/or less elongated in cross section); pedal retractor muscles are present. Pedal elevator and protractor muscles are absent. The MANTLE margins are extensively fused ventrally, with a small anteroventral pedal opening. PALLIAL MUCUS GLANDS line both sides of the

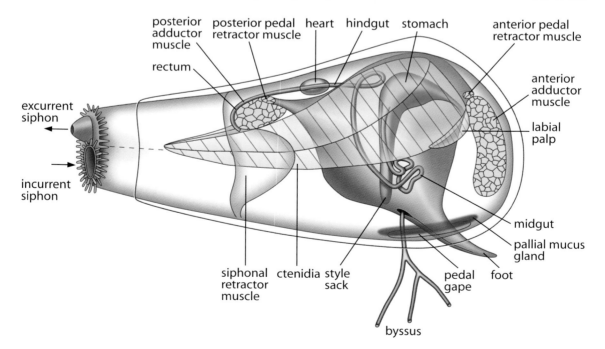

pedal gape. Posterior EXCURRENT and INCURRENT SIPHONS are usually long (short in *Cryptomya* and *Sphenia*), united into a single tube with an internal partition and two openings, and covered by periostracum; they cannot be fully retracted into the shell, and in some taxa (e.g., *Mya*) have light-sensitive areas (but not PALLIAL EYES). The excurrent siphon has a terminal cone (VALVULAR MEMBRANE), and both siphons are surrounded by a common outer ring of tentacles. HYPOBRANCHIAL GLANDS have not been reported. The FOOT is small to well developed, with a BYSSAL GROOVE, and in some cases (e.g., in *Sphenia*) produces a BYSSUS in the adult.

The LABIAL PALPS are small to large. The CTENIDIA are EULAMELLIBRANCH (SYNAPTORHABDIC) and HOMORHABDIC. The tips of the inner demibranchs are not inserted into (or fused with) the distal oral groove of the palps (CATEGORY III association). The outer demibranchs are smaller than the inner demibranchs, and they are united to the SIPHONAL SEPTUM, separating INFRA- and SUPRABRANCHIAL CHAMBERS. Incurrent and excurrent water flows are posterior. The STOMACH is TYPE V, usually with a very large STYLE SACK; the MIDGUT is coiled. The HINDGUT passes through the ventricle of the heart, and leads to a sessile rectum. Myids are GONOCHORISTIC and produce planktonic VELIGER larvae; rare records of HERMAPHRODITIC individuals of *Mya arenaria* Linnaeus, 1758, have been noted. The nervous system is not concentrated. STATOCYSTS (with STATOLITHS) have been reported in adult *Mya*. ABDOMINAL SENSE ORGANS have not been reported.

Myids are marine SUSPENSION FEEDERS, primarily in shallow waters, usually deeply INFAUNAL in soft subtidal sandy mud. *Sphenia* is a byssate EPIFAUNAL nestler in various habitats such as rock crevices, or among mytilid byssi, branching algae, bryozoans, or colonial ascidians. *Cryptomya* and *Paramya* live commensally in crustacean or echiuroid burrows. *Platyodon* burrows mechanically into packed clay or mudstone, using water pressure in the MANTLE CAVITY (rather than hydrostatic pressure in the blood) to expand and contract its body. Natural predators of myids include carnivorous gastropods, horseshoe crabs, lobsters, crabs, sea stars, bottom-feeding fish, and birds.

The family Myidae is known since the Tertiary and is represented by 6 living genera and ca. 25 species, distributed worldwide but best represented in the Northern Hemisphere. The edible soft-shelled or "steamer" clam, *Mya arenaria*, has been harvested commercially on the tidal flats of New England since the mid-1800s and still provides the third most important commercial clam fishery in the United States.

***Mya arenaria*** is the commercially harvested Soft-Shell Clam. Today's primary source is the state of Maine, where they are harvested with hand rakes, which results in less habitat impact and less unwanted bycatch than clam dredging.

## References

Abraham, B. J., and P. L. Dillon. 1986. *Species Profiles: Life Histories And Environmental Requirements of Coastal Fishes And Invertebrates (Mid-Atlantic)—Softshell Clam*. U.S. Fish and Wildlife Service Biological Report 82(11.68). U.S. Army Corps of Engineers, TR EL-82-4, 18 pp. Online version http://www.nwrc.usgs.gov/wdb/pub/0129.pdf, last accessed 03 December 2006.

Adams, A. 1851. Monograph of *Sphaenia*, a genus of lamellibranchiate Mollusca. *Proceedings of the Zoological Society of London for 1850*, 18(206): 86–89, pl. 10.

Bernard, F. R. 1979. Identification of the living *Mya* (Bivalvia: Myoida). *Venus*, 38(3): 185–204.

Coan, E. V. 1999. The eastern Pacific species of *Sphenia* (Bivalvia: Myidae). *The Nautilus*, 113(4): 103–120.

Dow, R. L., and D. E. Wallace. 1961. *The Soft-Shell Clam Industry of Maine*. United States Fish and Wildlife Service Circular 110: vi + 36 pp.

Foster, R. W. 1946. The genus *Mya* in the western Atlantic. *Johnsonia*, 2(20): 29–35.

Hanks, R. W. 1963. *The Soft-Shell Clam*. United States Fish and Wildlife Service Circular 162: 16 pp.

Narchi, W. and O. Domaneschi. 1993. The functional anatomy of *Sphenia antillensis* Dall & Simpson, 1901 (Bivalvia: Myidae). *Journal of Molluscan Studies*, 59(2): 195–210.

Yonge, C. M. 1923. Studies on the comparative physiology of digestion. I. The mechanism of feeding, digestion, and assimilation in the lamellibranch *Mya*. *The British Journal of Experimental Biology*, 1(1): 15–63.

Yonge, C. M. 1951a. Studies on Pacific coast mollusks. I. On the structure and adaptations of *Cryptomya californica* (Conrad). *University of California Publications in Zoölogy*, 55(6–8): 395–400.

Yonge, C. M. 1951b. Studies on Pacific coast mollusks. II. Structure and adaptations for rock boring in *Platyodon cancellatus* (Conrad). *University of California Publications in Zoölogy*, 55(6–8): 401–407.

Yonge, C. M. 1982. Ligamental structure in Mactracea and Myacea (Mollusca: Bivalvia). *Journal of the Marine Biological Association of the United Kingdom*, 62(1): 171–186.

# Family Corbulidae – Basket Clams

**Classification**
AUTOLAMELLIBRANCHIATA Grobben, 1894
HETEROCONCHIA Hertwig, 1895
Heterodonta Neumayr, 1883
Myoida Stoliczka, 1870
Myoidea Lamarck, 1809
Corbulidae Lamarck, 1818

**Featured species**
*Varicorbula limatula* (Conrad, 1846) – **Oval Varicorbula**

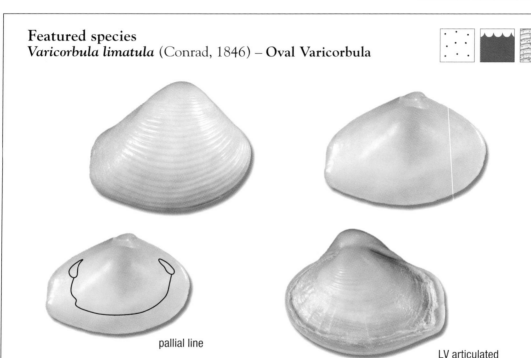

pallial line

LV articulated

Rounded trigonal, bluntly truncated posteriorly, strongly inequivalve, right valve inflated and thick, left valve smaller, flatter and thinner, right valve with coarse, rounded commarginal ridges and weak posterior keel, left valve with radial threads; white to pale yellow, brownish periostracum mainly on left valve, forming foliations and radial lines and overlapping margin; interior white with light yellow blush and rose dorsal margins, pallial sinus shallow, margin smooth, right valve with peripheral groove paralleling margin into which margin of left valve rests when closed. Virginia to Florida, Bahamas, West Indies, Gulf of Mexico, Caribbean Central America. Length 7 mm (to 10 mm). Formerly known as *V. operculata* (Philippi, 1848) [name of uncertain status]. Syn. *disparilis* d'Orbigny, 1853.

**These specimens of *Varicorbula limatula*** were collected on the "tickle chain" of a trawl off the Dry Tortugas.

# Family description

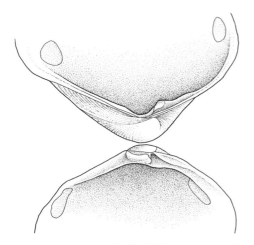

The corbulid shell is small to medium-sized (to ca. 35 mm), solid, trigonal to elongated trigonal, rounded anteriorly, and frequently posteriorly ROSTRATE. The shell is usually markedly INEQUIVALVE (right valve larger and more convex, and overlapping left valve ventrally), often inflated, and not markedly gaping. The smaller left valve lacks calcification at the ventral margin and this periostracal edge fits flexibly into and against the more fully calcified right valve. The shell is usually INEQUILATERAL (umbones slightly anterior), with PROSOGYRATE UMBONES. Shell microstructure is ARAGONITIC and two-layered, with a CROSSED LAMELLAR outer layer and a COMPLEX CROSSED LAMELLAR inner layer, the latter confined within the pallial line. TUBULES are apparently absent. An unusual arrangement of organic conchiolin layers, alternating with layers of aragonite, is believed to deter shell-drilling gastropods. Exteriorly corbulids are covered by a thin to thick PERIOSTRACUM that is most prominent on the smaller left valve especially at the ventral edge (interpreted as exposed conchiolin layers by some authors). At the rostral tip, periostracum is strengthened by calcification, likely to protect the siphons. Sculpture is smooth or commarginal, and can have strong radial ridges on the smaller left valve (thus differing between the two valves). LUNULE and ESCUTCHEON are absent. Interiorly the shell is non-NACREOUS. The PALLIAL LINE is weakly sinuate posteriorly, occasionally with a shallow SINUS. The inner shell margin is smooth. The HINGE PLATE is weak and HETERODONT, with

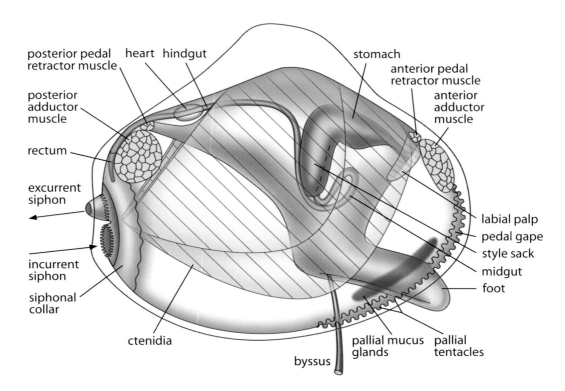

a single cardinal tooth in the right valve, a corresponding socket in the left valve, and absent or (rarely) weak LATERAL TEETH. The LIGAMENT is internal (RESILIUM) and extends more or less vertically from a dorsally tilted CHONDROPHORE in the left valve to a "sunken," ventrally tilted socket or pit on the hinge plate of the right valve; a secondary external ligament of fused periostracum can be present, uniting the valves along the dorsal hinge margin.

The animal is ISOMYARIAN or HETEROMYARIAN (anterior ADDUCTOR MUSCLE often slightly smaller); pedal retractor muscles are present. Pedal elevator and protractor muscles are absent. The MANTLE margins are fused ventrally, with a small to medium-sized anteroventral pedal gape. PALLIAL MUCUS GLANDS line both sides of the pedal gape (reported in *Varicorbula*). Posterior EXCURRENT and INCURRENT SIPHONS are short, united, and wholly retractile; the excurrent siphon is tubular with a terminal cone (VALVULAR MEMBRANE). The siphons can be surrounded by a common outer ring of tentacles or by a thickened, pigmented "collar." HYPOBRANCHIAL GLANDS have not been reported. The FOOT is digitiform or laterally compressed, with a BYSSAL GROOVE, and usually producing a small (occasionally single-stranded) BYSSUS in the adult.

The LABIAL PALPS are small to medium-sized. The CTENIDIA are EULAMELLIBRANCH (SYNAPTORHABDIC) and either HOMORHABDIC or HETERORHABDIC. The tips of the inner demibranchs are not inserted into (or fused with) the distal oral groove of the palps (CATEGORY III association). The outer demibranchs are smaller than the inner demibranchs, and they are united to the SIPHONAL SEPTUM, separating INFRA- and SUPRABRANCHIAL CHAMBERS. Incurrent and excurrent water flows are posterior. The STOMACH is TYPE V; the MIDGUT is loosely coiled. The HINDGUT passes through the ventricle of the heart, and leads to a sessile or freely hanging rectum. Corbulids are GONOCHORISTIC and produce planktonic VELIGER larvae. The nervous system is not concentrated. STATOCYSTS and ABDOMINAL SENSE ORGANS have not been reported.

*Varicorbula philippii* (E. A. Smith, 1885) – **Philippi's Corbula**

LV articulated

Rounded trigonal, bluntly truncated posteriorly, strongly inequivalve, right valve inflated and thick, left valve smaller, flatter and thinner, right valve with coarse rounded commarginal ridges and strong posterior keel, left valve with radial threads; whitish with thick brown periostracum; interior white flushed with orange-brown. North Carolina to Florida, Bermuda, West Indies, Gulf of Mexico, Caribbean Central America, South America (Brazil). Length 7 mm.

Corbulids are marine or estuarine SUSPENSION FEEDERS, and usually shallowly INFAU-NAL or EPIFAUNAL, in muddy sand or shell hash. The burrowing process uses water pressure in the MANTLE CAVITY (rather than hydrostatic pressure in the blood) to alternately expand and contract the body. Stomach contents include diatoms, bacteria, and organic debris from the surface sediment. Corbulids can occur in large numbers, and are highly tolerant of environmental degradation or greatly fluctuating to anoxic conditions (e.g., in harbors), perhaps associated with their tightly closing valves. Some deeper-water species can be so abundant in the benthic fauna that early authors remarked of the European species *Varicorbula gibba* (Olivi, 1792), "its extreme prevalence is a subject of almost petulant complaint from the habitual dredger." The latter species has been introduced to southeastern Australian waters where it competes for food with a commercially important local scallop.

The family Corbulidae is known since the Jurassic and is represented by 8 living genera and ca. 100 species, distributed worldwide in shallow water to the continental shelves. Some species of the genus *Caryocorbula* require shell morphometrics to distinguish one from another.

## References

Anderson, L. C., A. Aronowsky, and P. D. Roopnarine. 2005. Detecting and interpreting morphologic constraint in the fossil record (abstract). *[Program and Abstracts]*, *38th Annual Western Society of Malacologists, 71st Annual American Malacological Society, Asilomar, Pacific Grove, CA, June 26th–30th 2005*, p. 14.

Coan, E. V. 2002. The eastern Pacific Recent species of the Corbulidae (Bivalvia). *Malacologia*, 44(1): 47–105.

Harper, E. M. 1994. Are conchiolin sheets in corbulid bivalves primarily defensive? *Palaeontology*, 37(3): 551–578.

Mikkelsen, P. M., and R. Bieler. 2001. *Varicorbula* (Bivalvia: Corbulidae) of the western Atlantic: taxonomy, anatomy, life habits, and distribution. *The Veliger*, 44(3): 271–293.

Morton, B. 1990. The biology and fuctional morphology of *Corbula crassa* (Bivalvia: Corbulidae) with special reference to shell structure and formation. Pages 1055–1073, in: B. Morton, ed., *The Marine Flora and Fauna of Hong Kong and Southern China II, volume 3*. Hong Kong University Press, Hong Kong.

Talman, S. G., and M. J. Keough. 2001. Impact of an exotic clam, *Corbula gibba*, on the commercial scallop *Pecten fumatus* in Port Phillip Bay, south-east Australia: evidence of resource-restricted growth in a subtidal environment. *Marine Ecology Progress Series*, 221: 135–143.

Vokes, H. E. 1945. Supraspecific groups of the pelecypod family Corbulidae. *Bulletin of the American Museum of Natural History*, 86(1): 1–32, pls. 1–4.

Yonge, C. M. 1946. On the habits and adaptations of *Aloidis* (*Corbula*) *gibba*. *Journal of the Marine Biological Association of the United Kingdom*, 26(3): 258–276.

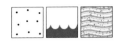

*Caryocorbula caribaea* (d'Orbigny, 1853) – **Caribbean Corbula**

Rounded trigonal, posterior rostrate often extended by calcified periostracum, strongly inequivalve, with irregular commarginal ridges; white; interior white flushed with orange-brown. Massachusetts to Florida, West Indies, Gulf of Mexico, Caribbean Central America, South America (Venezuela, Suriname, Brazil, Uruguay, Argentina). Length 7 mm (to 9 mm). Syn. *barrattiana* C. B. Adams, 1852, and *swiftiana* C. B. Adams, 1852 (see note on p. 385).

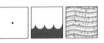

*Caryocorbula chittyana* (C. B. Adams, 1852) – **Snubnose Corbula**

Rounded trigonal, posterior strongly rostrate and obliquely truncate, strongly inequivalve, with commarginal ridges, often defining two distinct growth stages; white to yellowish or brownish, margins with thick, dark brown periostracum; interior white. North Carolina to Florida, West Indies, Gulf of Mexico, South America (Colombia). Length 7 mm (to 9 mm).

*Caryocorbula contracta* (Say, 1822) – **Contracted Corbula**

Rounded trigonal, posterior rostrate and pointed, with radial ridge from umbones to posteroventral corner, slightly inequivalve, with numerous poorly defined commarginal ridges extending over posterior ridge; white to dull gray; interior white to pinkish. Massachusetts to Florida, West Indies, Gulf of Mexico, Caribbean Central America, South America (to Venezuela, Uruguay). Length 6 mm (to 12 mm).

*Caryocorbula cymella* (Dall, 1881) – **Wavy Corbula**

Rounded trigonal, posterior rostrate and pointed, slightly inequivalve, with coarse commarginal ridges, smooth umbonally; white with opaque white mottlings; interior white flushed with pink or orange-brown. Florida Keys, West Indies, South America (Brazil). Length 6 mm (to 14 mm).

*Caryocorbula dietziana* (C. B. Adams, 1852) – **Dietz's Rose Corbula**

Quadrangular, posterior rostrate and pointed, umbones flattened with strong posterior angle, strongly inequivalve, with few coarse, rounded commarginal ridges (often defining two distinct growth stages) and microscopic threads; whitish with ventral margins flushed or rayed with carmine-rose; interior pinkish. North Carolina to Florida, West Indies, Gulf of Mexico, Caribbean Central America, South America (to Brazil). Length 13 mm (to 15 mm).

*Juliacorbula aequivalvis* (Philippi, 1836) – **Equivalve Corbula**

Rounded trigonal, posterior rostrate and pointed, nearly equivalve, with strong commarginal ridges (often defining two distinct growth stages); white or cream exteriorly and interiorly. Florida, West Indies, Caribbean Central America, South America (to Brazil). Length 7 mm (to 11 mm). Syn. *cubaniana* d'Orbigny, 1853.

artyculated

*Caryocorbula caribaea*

articulated

*Caryocorbula chittyana*

articulated

*Caryocorbula contracta*

articulated

*Caryocorbula cymella*

articulated

*Caryocorbula dietziana*

articulated

*Juliacorbula aequivalvis*

# Family Pholadidae Lamarck, 1809 – Piddock Clams

**Classification**
AUTOLAMELLIBRANCHIATA Grobben, 1894
HETEROCONCHIA Hertwig, 1895
Heterodonta Neumayr, 1883
Myoida Stoliczka, 1870
Pholadoidea Lamarck, 1809
Pholadidae Lamarck, 1809

**Featured species**
*Martesia striata* (Linnaeus, 1758) – **Striate Piddock**

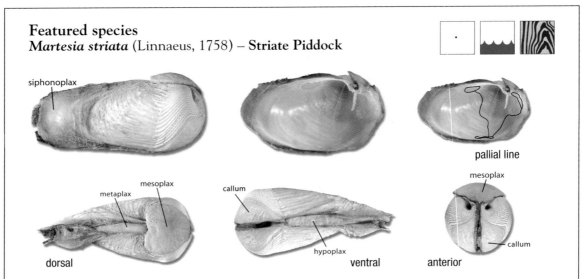

Elongated oval but highly variable, posterior narrowed, with callum, circular to arrow-shaped mesoplax, and long, narrow, anteriorly pointed metaplax and hypoplax, with denticulate commarginal ridges anterior to oblique radial sulcus, smooth ridges posterior to sulcus, whitish with brownish periostracum at margins; interior with oblique raised ridge duplicating external sulcus, apophyses long, narrow and fragile. North Carolina to Florida, Bermuda, Gulf of Mexico, Caribbean Central America, South America (Venezuela, Brazil), also eastern Pacific, Indo-Pacific. Length 20 mm (to 44 mm). Compare *Martesia cuneiformis*, which has a heart-shaped mesoplax with a median groove, and a posteriorly divided metaplax and hypoplax.

# Family description

The pholadid shell is small to large (to 200 mm), thin-walled to solid, and globular to oval to elongated, in part dependent on the nature of the substratum into which it bores. The anteroventral margin of the shell is often emarginated, exposing a pedal gape that later closes in some taxa. The shell is EQUIVALVE or INEQUIVALVE (in Jouannetiinae, see following), moderately inflated, and usually gaping (broadly posteriorly, variably anteriorly). The valves either cover the entire body or the visceral mass alone, leaving the siphons exposed; the body is wormlike in Xylophagainae, with a tiny shell that approaches that of a tere-

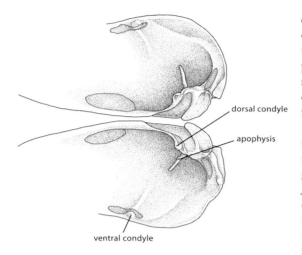

dorsal condyle

apophysis

ventral condyle

dinid. The anteroventral pedal gape can be large and circular or slitlike, and is closed in the adults of some taxa by two semicircular calcareous or periostracal plates (CALLUM; in Martesiinae [equal halves] and some Jouannetiinae [left portion larger]). This closure completely seals the anterior end of the shell except for a small pore in the periostracum at the junction (probably for water circulation) and effectively removes the roughened anterior shell edge from functioning as a boring organ; as a result, boring and growth cease with deposition of the callum. In at least *Jouannetia*, the callum is laid down external to, thus covering, the periostracum. The shell is INEQUILATERAL (umbones anterior), with PROSOGYRATE UMBONES. The anterodorsal margin is more or less reflected (UMBONAL REFLECTION). In addition to the callum, one or more of the following ACCESSORY PLATES are present (no one species has all six plates). The PROTOPLAX (in some Pholadinae) is a wide, single or divided, periostracal (*Cyrtopleura*) or calcareous plate covering the anterodorsal margin and protecting the exposed anterior adductor muscle on the umbonal reflection. Also protecting the anterior adductor muscle is the calcareous MESOPLAX, frequently infolded and occasionally in several sections; this is the most common accessory plate in the family, and the only one present in at least some members of all four currently used subfamilies (Pholadinae, Jouannetiinae, Martesiinae, and Xylophagainae). The dorsal METAPLAX and ventral HYPOPLAX (each in some Martesiinae) are elongated, usually calcareous, and extend along the dorsal or ventral shell margin, respectively. A posterior extension of the shell margin, the periostracal or calcareous SIPHONOPLAX (in Jouannetiinae [on the right valve only] and some Martesiinae), protects the siphons. The burrow can also be equipped with a tube composed of feces and/or pseudofecal particles cemented with mucus;

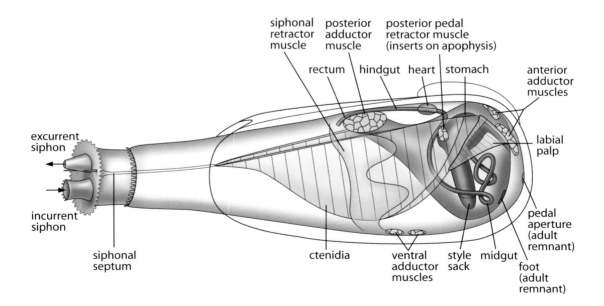

siphonal retractor muscle — posterior adductor muscle — posterior pedal retractor muscle (inserts on apophysis)

rectum — hindgut — heart — stomach — anterior adductor muscles

excurrent siphon

incurrent siphon

siphonal septum

labial palp

pedal aperture (adult remnant)

foot (adult remnant)

ctenidia — ventral adductor muscles — style sack — midgut

if fused to the siphonoplax (in, e.g., *Pholadidea*), it is called a SIPHONAL TUBE, whereas if fitting over the posterior end of the shell (sometimes far anteriorly; in, e.g., *Xylophaga*), it is called a CHIMNEY. Shell microstructure is ARAGONITIC and two- or three-layered, with a simple PRISMATIC or HOMOGENOUS outer layer (absent in most pholadids), a CROSSED LAMELLAR middle layer, and a COMPLEX CROSSED LAMELLAR or homogenous inner layer. The callum is prismatic (recorded only in *Jouannetia*). TUBULES are apparently absent. Exteriorly pholadids are whitish, covered by an inconspicuous or thin yellowish PERIOSTRACUM (two-layered in at least *Jouannetia*) that darkens with age and exposure to substratum and that also sheaths the siphons. Surface sculpture is complex, often heavily reticulate, with spinose or imbricated commarginal ridges, often segregated into distinct anterior and posterior zones separated by a strong radial sulcus from the umbo to the ventral margin. LUNULE and ESCUTCHEON are absent. Interiorly the shell is non-NACREOUS, in thin-shelled species with ridges and pits reflecting the external sculpture. Posteriorly a buttress or shelf is present in *Jouannetia* that supports the posterior adductor and pedal retractor muscles. Dorsal and/or ventral knoblike CONDYLES serve as fulcrums upon which the valves rock during the boring process; a strong internal rib often extends dorsally from the ventral condyle reflecting the external radial sulcus. One or two MYOPHORES called APOPHYSES (absent in Jouannetiinae and Xylophagainae), extending below each umbo, are attachment points for pedal retractor muscles. The PALLIAL LINE is thick, with a broad, deep PALLIAL SINUS. The inner shell margins are smooth but can be scalloped by the external ribs. The HINGE PLATE is irregular, weak, and EDENTATE in adults, comprised solely of the articulating condyles. The LIGAMENT is internal (RESILIUM), weak, OPISTHODETIC, and set on a small RESILIFER; one author considers this an external ligament that merely appears internal because of the sunken umbones. A secondary external ligament of fused periostracum unites the valves along the dorsal hinge margin, incorporating the calcareous metaplax; similarly, fused periostracum incorporating the calcareous hypoplax unites the ventral margins of the shell.

The animal is TRIMYARIAN, with (1) a small anterior ADDUCTOR MUSCLE that can be divided into several parts (single in Xylophagainae), each attaching exteriorly to the umbonal reflection and protected in adults by various accessory plates and in juveniles by a periostracal CEPHALIC HOOD; (2) a large posterior adductor muscle along the posterodorsal margin; and (3) a ventral adductor muscle (in some cases bilobed) attached near the ventral condyle. Anterior and posterior pedal retractor muscles in Jouannetiinae insert in typical bivalve positions near their respective adductor muscles. Only one pair of pedal retractor muscles (the posterior?) has been reported in other subfamilies, usually inserting upon the apophyses (or inserting on the shell near the posterior adductor muscle in Xylophagainae, in which apophyses are absent). Pedal elevator and protractor muscles have not been reported. The large siphonal retractor muscles are bilobed in some cases; the siphons are not retractable in some genera (e.g., *Zirphaea* and *Umitakea*). The MANTLE margins are extensively fused, with a small anteroventral pedal opening that closes in adults of taxa that close the pedal gape with a callum. Posterior EXCURRENT and INCURRENT SIPHONS are long (extending four to five times the shell length in living *Barnea*), equal or unequal in length (in, e.g., *Xylophaga* the shorter excurrent siphon prevents complete expulsion of fecal material from the burrow, which in turn promotes chimney formation), separate or partly to fully united, and naked or covered by a periostracal sheath; PALLETS (as in Teredinidae) are absent and the animal is usually capable of completely retracting the siphons. Luminescent organs are present in the mantle tissues of several species of *Pholas* and *Barnea*. HYPOBRANCHIAL GLANDS have not been reported. The FOOT is short and truncate, terminally suckerlike (for securing a position in the burrow during boring), and usually has a BYSSAL GROOVE, but does not produce a BYSSUS in the adult. The foot and pedal retractor muscles degenerate in the adults of callum-forming species.

The LABIAL PALPS are large, elongated and trigonal (in rock-boring species) or small (in most wood-borers). The CTENIDIA are EULAMELLIBRANCH (SYNAPTORHABDIC), (usually) HO-MORHABDIC or HETERORHABDIC, and typically extend into the siphons far beyond the posterior adductor muscle. The outer demibranchs are equal to or smaller than the inner demibranchs, and are absent in Xylophagainae. The tips of the inner demibranchs are inserted and fused into the distal oral groove of the palps (CATEGORY II association). The gills of *Pholas* have been shown to be capable of absorption of dissolved organic matter in addition to respiration and particle retention. Incurrent and excurrent water flows are posterior. The STOMACH is TYPE V, with a wood-storing appendix or CAECUM in the wood-eating Xylophagainae. The MIDGUT can be extensively coiled. The HINDGUT passes through the ventricle of the heart, and leads to a sessile or freely hanging rectum. Pholadids are PROTANDRIC HERMAPHRODITES (although one species of *Barnea* might be GONOCHORISTIC), and either are ovoviparous or produce planktonic VELIGER larvae that are often brooded for short to long periods, probably in the burrow (rather than the SUPRABRANCHIAL CHAMBER). Fertilization can be external or internal and self-fertilization is possible although unconfirmed; in *Xylophaga*, sperm are maintained in females in paired SEMINAL VESICLES (present in both sexes), having been stored there during the male phase along with secretions from a glandular ACCESSORY GENITAL OR-GAN on the posteroventral surface of the posterior adductor muscle. Juveniles have been recorded byssally attached to the exterior mantle or shells of the adult. Fairly rapid growth rates have been reported for several members of this family (e.g., 5.7 mm/month for *Martesia cuneiformis*). The nervous system is somewhat concentrated, with the pedal ganglia close to the cerebropleural ganglia, and a transverse connective (between the right and left cerebro-pleural–visceral connectives just anterior to the visceral ganglia) that can be swollen into an accessory visceral ganglion. STATOCYSTS (with STATOLITHS) have been reported in adult *Pholas* and *Zirphaea*. ABDOMINAL SENSE ORGANS have not been reported.

Pholadids are marine or estuarine (rarely in freshwater), and generally inhabitants of shallow water (except the deepwater Xylophagainae). Most are SUSPENSION FEEDERS, using dead (not living) wood or another hard substratum as a habitat but not food; only xylopha-gaines actually consume wood in addition to suspension feeding (these, the shipworms [Tere-dinidae], and a few crustaceans [*Sphaeroma* and *Limnoria*] are the only known marine wood-digesting invertebrates). Many form monospecific colonies in shallow water, i.e., only one pholad species will colonize a given wood panel; however, this is not so for deepwater xy-lophagaines, which are known to cohabit wood panels with one to four other congenerics. Pholadid burrows are generally short (or long in *Xylophaga*, up to eight times the shell length) and are usually not lined with calcium carbonate (except in *Xyloredo*). The pholadid shell and musculature are highly modified for mechanical abrasion of solid substrata, including clay, mud, soft rock (e.g., shale and limestone), shell, wood, and other plant products (e.g., seeds and nuts); they are also recorded as burrowed into styrofoam, lead sheathing, and various plastics including polyvinyl chloride (causing failure of a submerged electrical cable in at least one case). Mantle glands that appear to secrete substances involved in boring have been identified surrounding the pedal gape in *Barnea* and *Jouannetia*, suggesting that chemical dissolution (or at least softening) of the substratum also occurs in some species. During the boring cycle (extensively studied in *Zirphaea*, *Barnea*, *Martesia*, and *Xylophaga*), (1) the foot, adhering to the walls by suction and mucus, draws the shell to the base of the burrow (and possibly also secretes burrow-softening agents); (2) the base is abraded by the rough anterior shell edge, as a result of consecutive contractions of the posterior and anterior adductor muscles (the former muscle spreading the valves anteriorly, the latter closing them, with the shell rocking on the internal condyles); and (3) the animal rotates slightly clock-wise or counterclockwise in the burrow, and the cycle is repeated, producing a burrow that is circular in cross section. Rasped particles collect in the MANTLE CAVITY and in most species are collected by cilia and removed as PSEUDOFECES through the incurrent siphon; at least some of these particles are consumed by xylophagaines (which bore exclusively in wood). In *Martesia,* probably the most damaging pholadid to human-made structures, boring ceases when the burrow is complete (dependent upon crowding and nature of the substratum), after which the pedal gape closes, a callum forms, and the foot with its muscles atrophies. Because most pholadids do not ingest wood, antiboring coatings on wood are less effective than they are against shipworms (Teredinidae); as a result, pholadids probably cause greater total damage to human-made structures. Predators include sea stars, which feed on pholads in situ by everting their stomachs to digest the bivalves within their burrows.

The family Pholadidae is known since the Carboniferous; fossil burrows can often be distinguished from those of borers in other bivalve families. The family is represented by 18 living genera and ca. 100 species, mainly in shallow temperate and tropical waters. Several species have been transported worldwide in flotsam by ocean currents or in the wooden hulls of seagoing vessels. Because of their many accessory shell plates, pholadids were once classified as "Multivalva" along with barnacles and chitons. The ACCESSORY PLATES are often necessary for positive species identification, and in the past isolated accessory plates have been named as new species of limpets. Species records are sparse for the Florida Keys, because of the infrequency that collectors sample this specialized habitat, and are restricted mainly to isolated dead shells washed ashore or as remains in driftwood. Some of the larger-bodied species have been used historically for human food; *Cyrtopleura costata* has been experimentally raised in mariculture for this purpose.

**Part of a wooden piling** shows extensive damage due to burrowing *Martesia* clams.

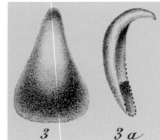

**Before its true identity was realized,** an isolated apophysis of *Cyrtopleura costata* was described as a new species of limpet-shaped gastropod (*Capulus schreevei* Conrad, 1869) from South Carolina. Its unusual shape nevertheless qualified the identification as provisional—"it is not sufficiently perfect to classify without some doubt of its generic character."

### References

Ansell, A. D., and N. B. Nair. 1969. The mechanisms of boring in *Martesia striata* Linné (Bivalvia: Pholadidae) and *Xylophaga dorsalis* Turton (Bivalvia: Xylophaginidae). *Proceedings of the Royal Society of London, Series B, Biological Sciences,* 174(1034): 123–133.

Bartsch, P., and H. A. Rehder. 1945. The west Atlantic boring mollusks of the genus *Martesia. Smithsonian Miscellaneous Collections,* 104(11): 1–16, 3 pls.

Boyle, P. J., and R. D. Turner. 1976. The larval development of the wood boring piddock *Martesia striata* (L.) (Mollusca: Bivalvia: Pholadidae). *Journal of Experimental Marine Biology and Ecology,* 22(1): 55–68.

Chanley, P. E. 1965. Larval development of a boring clam, *Barnea truncata*. *Chesapeake Science*, 6(3): 162–166.

Culliney, J. L., and R. D. Turner. 1976. Larval development of the deep-water wood boring bivalve, *Xylophaga atlantica* Richards (Mollusca, Bivalvia, Pholadidae). *Ophelia*, 15(2): 149–161.

Dall, W. H. 1889. Notes on the anatomy of *Pholas (Barnea) costata* Linne, and *Zirphaea crispata* Linne. *Proceedings of the Academy of Natural Sciences of Philadelphia*, 41(2): 274–276.

Egger, E. 1887. *Jouannetia Cumingii* Sow. Eine morphologische Untersuchung. *Arbeiten aus dem Zoologisch–Zootomischen Institut in Würzburg*, 8: 129–199, pls. 8–11.

Evseev, G. A. 1993. Anatomy of *Barnea japonica* (Bivalvia: Pholadidae). *Ruthenica*, 3(1): 31–49.

Fischer, P. 1858. Études sur les pholades. *Journal de Conchyliologie*, 7(1): 47–58.

Gustafson, R. G., R. L. Creswell, T. R. Jacobsen, and D. E. Vaughan. 1991. Larval biology and mariculture of the angelwing clam, *Cyrtopleura costata*. *Aquaculture*, 95(3–4): 257–279.

Haderlie, E. C. 1980. Sea star predation on rock-boring bivalves. *The Veliger*, 22(4): 400.

Hoagland, K. E., and R. D. Turner. 1981. Evolution and adaptive radiation of wood-boring bivalves (Pholadacea). *Malacologia*, 21(1–2): 111–148. [Data matrix published separately: Hoagland, K. E. 1983. Characters, character states, and taxa used in multivariate analysis of the Pholadacea. *Tryonia*, 8: 51 pp.]

Knudsen, J. 1961. The bathyal and abyssal *Xylophaga* (Pholadidae, Bivalvia). *Galathea Report*, 5: 163–209.

Mann, R. 1988. The physiology of marine wood borers of the families Teredinidae and Pholadidae. Pages 440–452, in: M.-F. Thompson, R. Sarojini, and R. Nagabhushanam, eds., *Marine Biodeterioration, Advanced Techniques Applicable to the Indian Ocean*. Oxford & IBH Publishing Company, New Delhi, India.

Mann, R., and S. M. Gallager. 1984. Physiology of the wood boring mollusc *Martesia cuneiformis* Say. *The Biological Bulletin*, 166(1): 167–177.

Morton, B. 1985. A pallial boring gland in *Barnea manilensis* (Bivalvia: Pholadidae)? Pages 191–197, in: B. S. Morton and D. Dudgeon, eds., *The Malacofauna of Hong Kong and southern China, II, volume 1*. Hong Kong University, Hong Kong.

Morton, B. 1986. The biology and functional morphology of the coral-boring *Jouannetia cumingii* (Bivalvia: Pholadacea). *Journal of Zoology, Series A*, 208(3): 339–366.

Nair, N. B., and A. D. Ansell. 1968. The mechanism of boring in *Zirphaea crispata* (L.) (Bivalvia: Pholadidae). *Proceedings of the Royal Society of London, Series B, Biological Sciences*, 170(1019): 155–173.

Purchon, R. D. 1941. On the biology and relationships of the lamellibranch *Xylophaga dorsalis* (Turton). *Journal of the Marine Biological Association of the United Kingdom*, 25(1): 1–39.

Purchon, R. D. 1956. A note on the biology of *Martesia striata* L. (Lamellibranchia). *Proceedings of the Zoological Society of London*, 126: 245–258.

Srinivasan, V. V. 1961. Ciliary currents and associated organs of *Martesia fragilis*, a wood boring pholad of Madras. *Journal of the Marine Biological Association of India*, 2(2): 186–193.

Srinivasan, V. V. 1963. Further notes on the anatomy of *Martesia fragilis*. *Journal of the Marine Biological Association of India*, 4(1): 1–9.

Thorne, J. 1983. A fresh look at the eulamellibranch gill of *Pholas dactylus* Linne. In: A. Bebbington, ed., *Proceedings of the Second Franco–British Symposium on Molluscs*. *Journal of Molluscan Studies, Supplement*, 12A: 225.

Turner, R. D. 1954. The family Pholadidae in the western Atlantic and the eastern Pacific. Part I—Pholadinae. *Johnsonia*, 3(33): 1–63.

Turner, R. D. 1955. The family Pholadidae in the western Atlantic and the eastern Pacific. Part II—Martesiinae, Jouannetiinae and Xylophaginae. *Johnsonia*, 3(34): 65–160.

Turner, R. D. 1968. The Xylophagainae and Teredinidae—a study in contrasts. *The American Malacological Union, Annual Reports, for 1967*, pp. 46–48.

Turner, R. D. 1972. *Xyloredo*, a new teredinid-like abyssal wood-borer (Mollusca, Pholadidae, Xylophagainae). *Breviora*, 397: 1–19.

Turner, R. D., and A. C. Johnson. 1971. Biology of marine wood-boring molluscs. Pages 259–301, in: E. B. G. Jones and S. K. Eltringham, eds., *Marine Borers, Fungi and Fouling Organisms of Wood*. Organization for Economic Cooperation and Development, Paris.

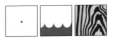

### *Martesia cuneiformis* (Say, 1822) – **Wedge Piddock**

Elongated oval but highly variable, posterior narrowed, with callum, heart-shaped mesoplax with median longitudinal groove, posteriorly divided metaplax and hypoplax, with denticulate commarginal ridges anterior to oblique radial sulcus, smooth ridges posterior to sulcus, whitish with brownish periostracum at margins; interior with oblique, raised ridge duplicating external sulcus, apophyses long, thin and fragile. North Carolina to Florida, West Indies, Gulf of Mexico, Caribbean Central America, South America (to Brazil), also eastern Pacific. Length 13 mm (to 20 mm). Compare *Martesia striata* (p. 388), which has a circular to arrow-shaped mesoplax without a median groove, and an undivided metaplax and hypoplax.

### *Cyrtopleura costata* (Linnaeus, 1758) – **Angelwing**

Elongated oval, with chitinous trigonal protoplax, transverse butterfly-shaped calcareous mesoplax, with ca. 30 beaded radial ribs that are scalelike anteriorly, plus fine commarginal growth lines, pure white, rarely with pink commarginal bands; interior reflecting external color, surface ribbed and pitted reflecting external sculpture, apophyses broad, concave and trigonal. Massachusetts to Florida, West Indies, Gulf of Mexico, Caribbean Central America, South America (Suriname, Brazil). Length 150 mm (to 180 mm).

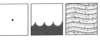

### *Barnea truncata* (Say, 1822) – **Fallen Angelwing**

Elongated oval to quadrangular, anterior pointed, with only one accessory plate (lanceolate calcareous protoplax), with beaded commarginal and radial ribs, white; interior glossy white, apophyses long, narrow and curved. Eastern Canada to Florida, Gulf of Mexico, South America (to Brazil), also West Africa. Length 40 mm (to 70 mm). Note: Also known as Atlantic Mud-Piddock.

### *Pholas campechiensis* Gmelin, 1791 – **Campeche Angelwing**

Elongated oval, umbonal reflection supported by ca. 12 vertical shelly ribs, with three accessory plates (calcareous protoplax that is broadly lanceolate and divided into two pieces longitudinally, transverse butterfly-shaped and calcareous mesoplax, and elongated narrow metaplax), with beaded commarginal and radial ribs, white; interior white and glazed, apophyses short, broad, and ridged on free ends. North Carolina to Florida, West Indies, Gulf of Mexico, Caribbean Central America, South America (to Uruguay). Length 30 mm (to 110 mm).

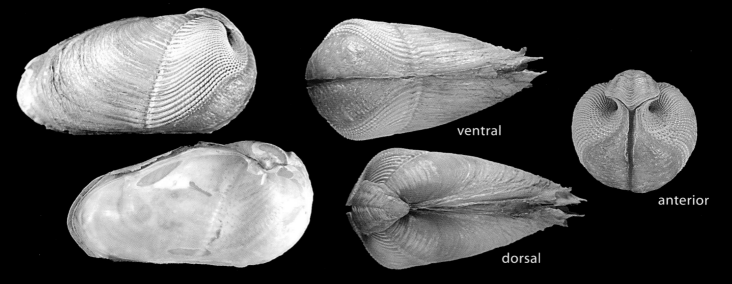

ventral

anterior

dorsal

*Martesia cuneiformis*

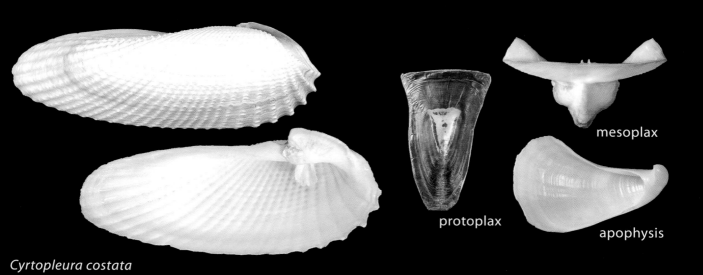

protoplax

mesoplax

apophysis

*Cyrtopleura costata*

*Barnea truncata*

*Pholas campechiensis*

# Family Teredinidae – Shipworms

**Classification**
AUTOLAMELLIBRANCHIATA Grobben, 1894
HETEROCONCHIA Hertwig, 1895
Heterodonta Neumayr, 1883
Myoida Stoliczka, 1870
Pholadoidea Lamarck, 1809
Teredinidae Rafinesque, 1815

**Featured species**
*Teredo clappi* Bartsch, 1923 – Clapp's Shipworm

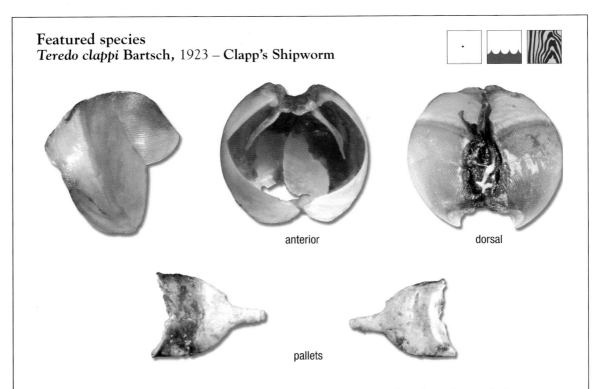

anterior

dorsal

pallets

Dorsoventrally elongated and auriculate, with median sulcus and fine regular ridges, white with thin outer covering of purplish brown or golden periostracum; pallets unsegmented paddle-shaped, red-brown, shallowly excavated medially with short lateral horns, occasionally with medial cleft. Florida Keys, Bermuda, Bahamas, West Indies; also Indo-Pacific. Length (shell) 5 mm. Note: The valves and pallets photographed are of the HOLOTYPE specimen of *Teredo clappi*.

# Family description

The teredinid shell is small (to ca. 5 mm), covering only the anteriormost part of a much larger, flexible, wormlike body, which in turn creates a shell-lined burrow in wood both for protection and (to a greater or lesser extent depending on species) for food. The AD-

VENTITIOUS tube of the mud-dwelling Indo-Pacific Stove-Pipe Clam (*Kuphus polythalamia* (Linnaeus, 1767)), can reach more than 1 m in length and 60 mm in diameter (larger than any other living bivalve except the Giant Clam (*Tridacna gigas*), although not technically a shell). The teredinid shell is thin-walled and globular or hemispherical, and segregated into three distinct zones (the median disk, and anterior and posterior AURICLES) by radial sulci. It is EQUIVALVE, inflated, and widely gaping anteroventrally and posteriorly. It is INEQUILATERAL (umbones anterior), with PROSOGYRATE UMBONES and an UMBONAL REFLECTION. ACCESSORY PLATES are absent, but the siphonal tips are equipped with PALLETS. These calcareous structures (a SYNAPOMORPHY of the family) are either solid and paddle- or shovel-shaped (Teredininae and Kuphinae) or composed of multiple internested cone-shaped segments (Bankiinae), and can have periostracal caps or awns (pointed lateral processes). Either type, when retracted, blocks the burrow entrance; they function both in protecting the soft body and in establishing positive pressure in the MANTLE CAVITY necessary for boring. The pallets usually include more taxonomically useful characters than the shell in teredinids, and are often essential for species-level identification. Shell microstructure is ARAGONITIC and two-layered, with a CROSSED LAMELLAR outer layer and a COMPLEX CROSSED LAMELLAR inner layer. TUBULES are apparently absent. The burrow lining is also aragonitic. Exteriorly teredinid shells are whitish, covered by an inconspicuous PERIOSTRACUM. Periostracum can also sheath the tips of the siphons, extending beyond the surface of the wood in some cases. Sculpture is commarginal with numerous, closely spaced, finely denticulate ridges on the anterior auricle and (often at right angles to the former) the anterior edge of the disk. LUNULE and ESCUTCHEON are absent. Interiorly the shell is non-NACREOUS. Dorsal and/or ventral knoblike CONDYLES serve as fulcra upon which the valves rock during the boring process; a strong internal rib often extends dorsally from the ventral condyle reflecting the external radial sulcus. Elongated MYOPHORES (APOPHYSES) extend from the interior umbones and serve as attachment points for the pedal retractor muscles. The PALLIAL LINE is ENTIRE. The inner shell margins are smooth. The HINGE PLATE is weak and EDENTATE in adults, and comprised solely of the articulating condyles. The internal LIGAMENT (RESILIUM) is weak, OPISTHODETIC, and set on a small RESILIFER.

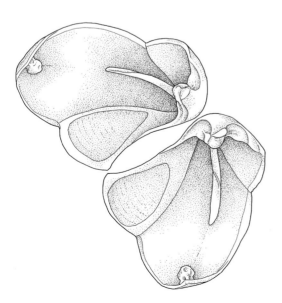

Although the shipworm body is extremely modified through elongation and shell reduction, its organ systems and their relative positions are quite recognizably bivalvan. The animal is TRIMYARIAN, with (1) a small anterior ADDUCTOR MUSCLE that attaches exteriorly to the umbonal reflection, and is protected by the CEPHALIC HOOD (an extension of the mantle); (2) a large posterior adductor muscle along the posterodorsal margin (small in *Kuphus*); and (3) a ventral adductor muscle attached near or in place of the ventral condyle (not reported in many accounts; present in at least *Psiloteredo* and *Nototeredo*). A single pair of PEDAL RETRACTOR MUSCLES (the posterior?) insert on the apophyses. Pedal elevator and protractor muscles are apparently absent. The MANTLE is tubular, with extensively fused margins, leaving a small anterior pedal opening. The glandular walls of long POSTVALVULAR (i.e., posterior to the shell) PALLIAL EXTENSION secrete the calcareous bur-

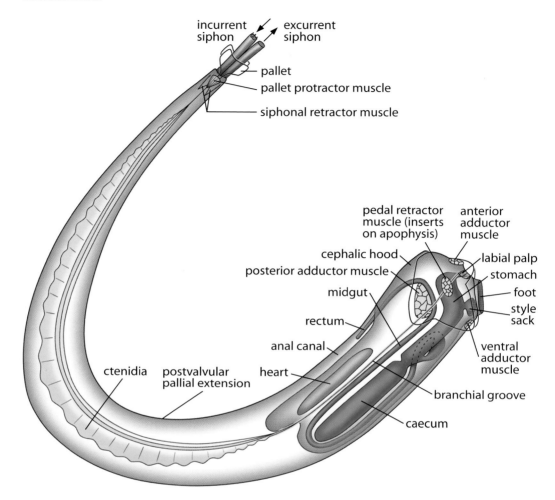

row lining onto which the posterior siphonal and pallial retractor muscles insert; this is the only place where the shipworm is attached to its burrow. Most of the major organ systems of the shipworm lie posterior to the shell and posterior adductor muscle (unlike in more typical bivalves) within the MANTLE CAVITY of this extension. Posterior EXCURRENT and INCURRENT SIPHONS are short to relatively long, partially or fully separate, and protrude from the burrow opening when the animal is active. The siphons and associated pallets are equipped with protractor, retractor, and adductor muscles. The mantle walls in *Kuphus* are exceptionally thick and in all species are somewhat thicker around the base of the siphons. HYPOBRANCHIAL GLANDS have not been reported. The FOOT is short and truncate, terminally suckerlike (for securing a position in the burrow during excavation); a BYSSAL GROOVE and BYSSUS are absent in the adult.

The LABIAL PALPS are small to medium-sized, narrow, and with low folds. The CTENIDIA are EULAMELLIBRANCH (SYNAPTORHABDIC), HOMORHABDIC, and harbor symbiotic gram-negative, nitrogen-fixing, CELLULASE-producing bacteria (several bacterium species have been isolated from shipworms). The bacterial aggregation occupies the interfilamentar spaces of the gill lamellae and is collectively called the ORGAN OF DE-SHAYES; ducts in the afferent branchial vessels empty into the esophagus. The gills are not inserted into (or fused with) the distal oral groove of the palps (CATEGORY III association). The outer demibranchs are very small or absent. The inner demibranchs are broad or narrow, and usually extend within the postvalvular pallial extension from the pericardium to the siphons; a branchial (food) groove on the outer surface of the visceral mass provides a pathway for filtered particles to a small, isolated group of gill filaments near the labial palps, and in turn, the palps and mouth. The gills of *Teredora* and *Uperotus* are longer, extending from the mouth to the siphons, and lack a branchial groove on the visceral mass. The right and left ascending lamellae of the demibranchs are fused and unite with the SIPHONAL SEPTUM, separating INFRA- and SUPRABRANCHIAL CHAMBERS. Incurrent and excurrent water flows are posterior. The STOMACH is either globular or elongated, and a modified TYPE V, with a small to large accessory posterior wood-storing appendix or CAECUM (absent in Kuphinae) with tissue folds partially dividing the orifice, allowing simultaneous incoming and outgoing streams; the caecum is lined with microvilli, suggesting that it also functions in absorption of nutrients from the contained wood particles. Two types of digestive diverticula, called "normal" and "specialized," are involved, respectively, in suspended- and wood-particle digestion (specialized diverticula are probably a synapomorphy for the family); complex sorting areas and ciliated tracts in the stomach separate the two types of particles and transport them to the appropriate diverticula. The MIDGUT is loosely to tightly coiled. The HINDGUT passes ventral to the ventricle of the heart (or through it in Kuphinae), and leads to a freely hanging, dorsal rectum that empties into the narrow ANAL CANAL (continuous with the suprabranchial chamber, and equipped with a terminal sphincter muscle in some species). The heart and pericardium are greatly elongated, and the kidney is dorsal to them (rather than ventral as in other bivalves). (The teredinid heart and nearby kidney have been rotated 180° around the posterior adductor muscle; as a result the anterior end is posterior and the dorsal side is ventral, therefore the intestine anatomically passes [as is typical for bivalves] dorsal to the ventricle.) Myoglobin has been reported in the adductor muscles of some teredinids. The gonad is large and parallels the gills well into the postvalvular extension. Teredinids are almost always PROTANDRIC, rarely SIMULTANEOUS HERMAPHRODITES. One species of *Zachsia* is GONOCHORISTIC with dwarf males that live in females in lateral mantle pouches near the base of the siphons (up to 69 males in one female have been recorded); larvae that settle on uninhabited substratum develop into females, whereas those that settle on substratum already colonized by females become parasitic males. Fertilization is either external (eggs and sperm dispersed) or internal (dispersed sperm drawn into female MANTLE CAVITY), and self-fertilization also has been reported. Copulation occurs in *Bankia* (in one of the few instances of this in Bivalvia), during which the excurrent siphon of a male is inserted into the incurrent siphon of a nearby female, after which sperm are transferred. Teredinids produce either planktonic VELIGER larvae or retain them for part or all of the larval period in brood pouches in the gills. Settling in planktonic veligers can be delayed if wood is unavailable, but metamorphosis and sexual maturation occur rapidly (in all species studied, regardless of developmental type) once boring has commenced. The adult nervous system is somewhat concentrated, with the pedal ganglia (in the dorsal foot), together with the cerebropleural ganglia and connectives, forming a tight nerve ring around the esophagus, and the visceral ganglia at the posterior end of the visceral mass, i.e., well posterior of the typical position near the posterior adductor muscle. STATOCYSTS (with STATOLITHS) have been reported in adult *Teredo*. ABDOMINAL SENSE ORGANS have not been reported.

Teredinids are marine or estuarine (rarely freshwater) SUSPENSION FEEDERS during early life stages and also as adults (to various extents depending on species; those that rely more exclusively on wood have smaller palps, smaller gills, and larger caeca; the opposite is true of species depending more on suspended particles). The large-bodied *Kuphus* is unusual morphologically (see above) and ecologically, mainly suspension feeding, and burrowing in mangrove mud rather than wood (although evidence suggests that juveniles initially bore into wood). *Zachsia zenkewitchi* Bulatoff & Rjabtschikoff, 1933, from Japan bores exclusively into living seagrass rhizomes. However, typical adult teredinids bore into, ingest, and digest wood and other plant products (either living and dead). Mechanical boring (studied most thoroughly in *Teredo navalis* Linnaeus, 1758) is achieved by abrasion of the wood surface using the highly modified shells and musculature. The foot and cephalic hood act to hold the anterior end of the body against the burrow end. The anterior adductor muscle then contracts, bringing the shell valves together; the posterior adductor muscle contracts next (while the anterior adductor relaxes), spreading the valves anteriorly (pivoting on the internal condyles), forcing the ridged anterior auricles to scrape against the wood. Repeated cycles rasp wood from the burrow like a file, producing ingestible food particles (taken into the mantle cavity) while also extending the tube. Except at the base where rasping occurs, the burrow is lined by smooth calcium carbonate, varying in thickness (according to species and type of wood) and occasionally extending beyond the surface of the wood. The burrow can be relatively straight or can twist and curve as necessary around obstacles (including the burrows of other shipworms) in the wood. Chemical etching of the tube lining is possible around the siphons when the need to enlarge the opening occurs. Individuals bore and ingest throughout their lives or until no further space is available, at which point they cap their burrows anteriorly and live at subsistence levels (on suspended particles and/or stored wood) or perish. As stated by Hoagland & Turner (1981: 140), teredinids are unusual in that they "destroy their own substrate," which is already unstable through waterlogging and decay. Shipworms are opportunistic and well suited to these limitations because of high fecundity (*Teredo navalis* spawns three or four times per season), short generation times, early maturation, high tolerance to crowding (e.g., very dense colonies result in smaller adult sizes), and broad physiological tolerances to temperature and salinity extremes. Storage of vast amounts of glycogen in the mantle, muscles, and gills allows shipworms to survive under adverse conditions for prolonged periods. Teredinid predators include fish, which nip at the exposed siphon tips; polychaetes and other invertebrates have also been reported as predators, but probably only scavenge on unhealthy or dead shipworms.

The family Teredinidae is known since the Cretaceous (these are tubes in fossilized wood; the oldest fossilized shells and pallets are from the Paleocene) and is represented by 15 living genera and ca. 70 species, distributed worldwide mainly in intertidal and shallow waters. Their specialized habitat results in undersampling in many biodiversity surveys, including that in the Florida Keys. Shipworms or "teredos" are widely distributed by drifting wood and (at least in the past) by commercial shipping. The economic impact of shipworm damage (to, e.g., wooden piers, ships, and lobster traps), especially in tropical regions, has historically been high. The ancient Greek texts of Pliny the Elder and Homer considered them a plague. Various antiboring measures (e.g., creosote and other coatings, sheathing with metal or plastic, and layering with tar and felt) have been used effectively in recent times to deter boring by shipworms. However, teredinids also play a major positive role in ecosystems as recyclers of wood in the sea. These unusual bivalves were considered by Linnaeus in 1758 (along with scaphopods) to be polychaete tube worms (overlooking a 1733 anatomical monograph that already showed *Teredo navalis* to be a mollusk). Teredinids are used as a food item by Australian Aboriginal peoples.

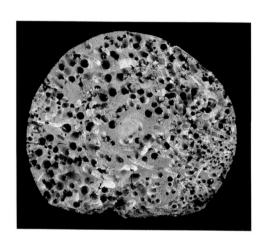

**The museum label** of this riddled piling section reads "wood bored by *Teredo navalis* in twelve months."

## References

Bartsch, P. 1922. A monograph of the American shipworms. *Bulletin of the United States National Museum*, 122: 51 pp., 37 pls.

Bulatoff, G. A., and P. I. Rjabtschikoff. 1933. Eine neue Gattung aus der Familie der Teredinidae aus dem Japanischen Meer. *Zoologischer Anzeiger*, 104: 165–176.

Coe, W. R. 1933. Sexual phases in *Teredo*. *The Biological Bulletin*, 65(2): 283–303.

Coe, W. R. 1941. Sexual phases in wood-boring mollusks. *The Biological Bulletin*, 81(2): 168–176.

Eckelbarger, K. J., and D. J. Reish. 1972. A first report of self-fertilization in the wood-boring Teredinidae (Mollusca: Bivalvia). *Bulletin of the Southern California Academy of Sciences*, 71(1): 48–50.

Fuller, S. C., Y.-P. Hu, R. A. Lutz, and M. Castagna. 1989. Shell and pallet morphology in early developmental stages of *Teredo navalis* Linné (Bivalvia: Teredinidae). *The Nautilus*, 103(1): 24–35.

Grave, B. H. 1928. Natural history of shipworm, *Teredo navalis*, at Woods Hole, Massachusetts. *The Biological Bulletin*, 55(4): 260–282.

Hoagland, K. E., and R. D. Turner. 1981. Evolution and adaptive radiation of wood-boring bivalves (Pholadacea). *Malacologia*, 21(1–2): 111–148. [Data matrix published separately: Hoagland, K. E. 1983. Characters, character states, and taxa used in multivariate analysis of the Pholadacea. *Tryonia*, 8: 51 pp.]

Lazier, E. L. 1924. Morphology of the digestive tract of *Teredo navalis*. *University of California Publications in Zoology*, 22(14): 455–474, pls. 21–24.

Lopes, S.G.B.C., O. Domaneschi, D. T. de Moraes, M. Morita, and G. de L. C. Meserani. 2000. Functional anatomy of the digestive system of *Neoteredo reynei* (Bartsch, 1920) and *Psiloteredo healdi* (Bartsch, 1931) (Bivalvia: Teredinidae). Pages 257–271, in: E. M. Harper, J. D. Taylor, and J. A. Crame, eds., *The Evolutionary Biology of the Bivalvia*. Geological Society of London, Special Publication 177.

Lopes, S.G.B.C., W. Narchi, and O. Domaneschi. 1998. Digestive tract and functional anatomy of the stomach of *Nausitora fusticula* (Jeffreys, 1860) (Bivalvia: Teredinidae). *The Veliger*, 41(4): 351–365.

Mann, R. 1988. The physiology of marine wood borers of the families Teredinidae and Pholadidae. Pages 440–452, in: M.-F. Thompson, R. Sarojini, and R. Nagabhushanam, eds., *Marine Biodeterioration, Advanced Techniques applicable to the Indian Ocean*. Oxford & IBH Publishing Company, New Delhi, India.

Morton, B. 1970. The functional anatomy of the organs of feeding and digestion of *Teredo navalis* Linnaeus and *Lyrodus pedicellatus* (Quatrefages). *Proceedings of the Malacological Society of London*, 39(2–3): 151–167.

Morton, B. 1978. Feeding and digestion in shipworms. *Oceanography and Marine Biology, An Annual Review*, 16: 107–144.

Morton, B. 1985. Tube formation in the Bivalvia—csövek kialakulása kagylónál. *Soosiana*, 13: 11–26.

Nair, N. B., and M. Saraswathy. 1971. The biology of wood-boring teredinid molluscs. *Advances in Marine Biology*, 9: 335–509.

Purchon, R. D. 1941. On the biology and relationships of the lamellibranch *Xylophaga dorsalis* (Turton). *Journal of the Marine Biological Association of the United Kingdom*, 25(1): 1–39.

Saraswathy, M., and N. B. Nair. 1971. Observations on the structure of the shipworms, *Nausitoria hedleyi*, *Teredo furcifera* and *Teredora princesae* (Bivalvia: Teredinidae). *Transactions of the Royal Society of Edinburgh*, 68(14): 507–566.

Savazzi, E. 1982. Adaptations to tube dwelling in the Bivalvia. *Lethaia*, 15(3): 275–297.

Sigerfoos, C. P. 1908. Natural history, organization, and late development of the Teredinidae, or ship-worms. *Bulletin of the Bureau of Fisheries*, 27: 191–231, pls. 7–21.

Turner, R. D. 1966. *A Survey and Illustrated Catalogue of the Teredinidae*. Museum of Comparative Zoology, Harvard University, Cambridge, Massachusetts, 265 pp., 64 pls.

Turner, R. D., and A. C. Johnson. 1971. Biology of marine wood-boring molluscs. Pages 259–301, in: E. B. G. Jones, and S. K. Eltringham, eds., *Marine Borers, Fungi and Fouling Organisms of Wood*. Organization for Economic Cooperation and Development, Paris.

Turner, R. D., and Y. M. Yakovlev. 1983. Dwarf males in the Teredinidae (Bivalvia, Pholadacea). *Science*, 219(4588): 1077–1078.

Yonge, C. M. 1926. Protandry in *Teredo Norvegica*. *Quarterly Journal of Microscopical Science*, 70(3) (new series 279): 391–394.

### *Bankia carinata* (Gray, 1827) – **Carinate Shipworm**

Shell similar to *Teredo clappi*; pallets segmented, cones shallow-cupped, with blunt, smooth edges, very crowded distally and covered by periostracal caps. North Carolina to Florida, West Indies, Gulf of Mexico, South America (to Brazil), also western Europe, Indian Ocean, and Indo-Pacific. Length (shell) 5 mm.

### *Bankia fimbriatula* Moll & Roch, 1931 – **Fringed Shipworm**

Shell similar to *Teredo clappi*; pallets segmented, with long, serrated lateral awns, inner margins of deeply cupped cones with comblike serrations. North Carolina to Florida, West Indies, Gulf of Mexico, South America (Colombia, Brazil), also Pacific Panama and northern Europe. Length (shell) 4 mm.

### *Lyrodus pedicellatus* (de Quatrefages, 1849) – **Black-tipped Shipworm**

Shell similar to *Teredo clappi*; pallets unsegmented with oval calcareous bases, terminally deeply excavated or blunt, with brown-black periostracal cap and lateral horns. Massachusetts to Florida, West Indies, Gulf of Mexico, Caribbean Central America, South America (Brazil), also eastern Pacific, Hawaiian Islands, western Europe, Australia, Indian Ocean, South Africa, and Indo-Pacfic. Length (shell) 4 mm (to 10 mm).

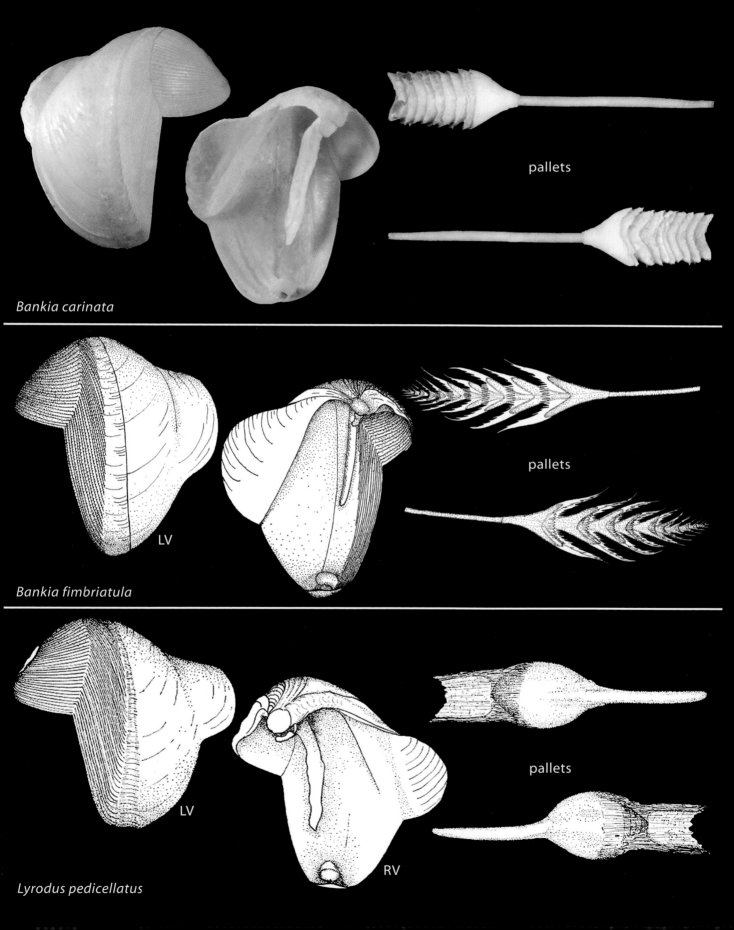

*Bankia carinata*

pallets

*Bankia fimbriatula*

LV

pallets

*Lyrodus pedicellatus*

LV

RV

pallets

### *Nototeredo knoxi* (Bartsch, 1917) – **Knox's Shipworm**

Shell similar to *Teredo clappi*; pallets unsegmented, with short stalks, blades soft, friable and flattened oval, inner face of closely packed segments separated by thin layers of periostracum, entire blade covered by pale periostracum extending distally as a border. North Carolina to Florida, West Indies, Caribbean Central America, South America (Brazil); also western Africa. Length (shell) 5–7 mm. Note: Two views show the anterior and posterior ends of a preserved intact animal.

### *Teredo bartschi* Clapp, 1923 – **Bartsch's Shipworm**

Shell similar to *Teredo clappi*; pallets unsegmented, paddle-shaped, deeply excavated medially with long lateral horns, with periostracum on distal half only, with long stalks. North Carolina to Florida, Bermuda, Gulf of Mexico, South America (Brazil to Uruguay), also western Europe, Hawaiian Islands, Indian Ocean, and Australia, and introduced to numerous locations in tropical and temperate North Pacific. Length 4 mm (to 15 mm). Note: The valves and pallets illustrated are the HOLOTYPE specimen of *Teredo bartschi*; the apophyses are missing from the specimen.

### *Teredo somersi* Clapp, 1924 – **Somers' Shipworm**

Shell similar to *Teredo clappi*; pallets unsegmented, paddle-shaped, solid, heavy, shallowly excavated medially with short lateral horns, with horn- to red-brown–colored periostracum on distal half only. Florida Keys, Bermuda; also South Africa. Length 2 mm. Note: The valve and pallets illustrated are the LECTOTYPE specimen of *Teredo somersi*.

### *Teredora malleolus* (Turton, 1822) – **Malleated Shipworm**

Shell similar to *Teredo clappi*, except posterior auricle small, high and set at right angle to dorsoventral axis; pallets unsegmented, paddle-shaped, oval, with smooth blade with prominent thumbnaillike, commarginally lamellated depression, other side concave with central elevation, stalks short. Florida, Bermuda, West Indies, Gulf of Mexico, also temperate Atlantic and western Europe. Length 10 mm.

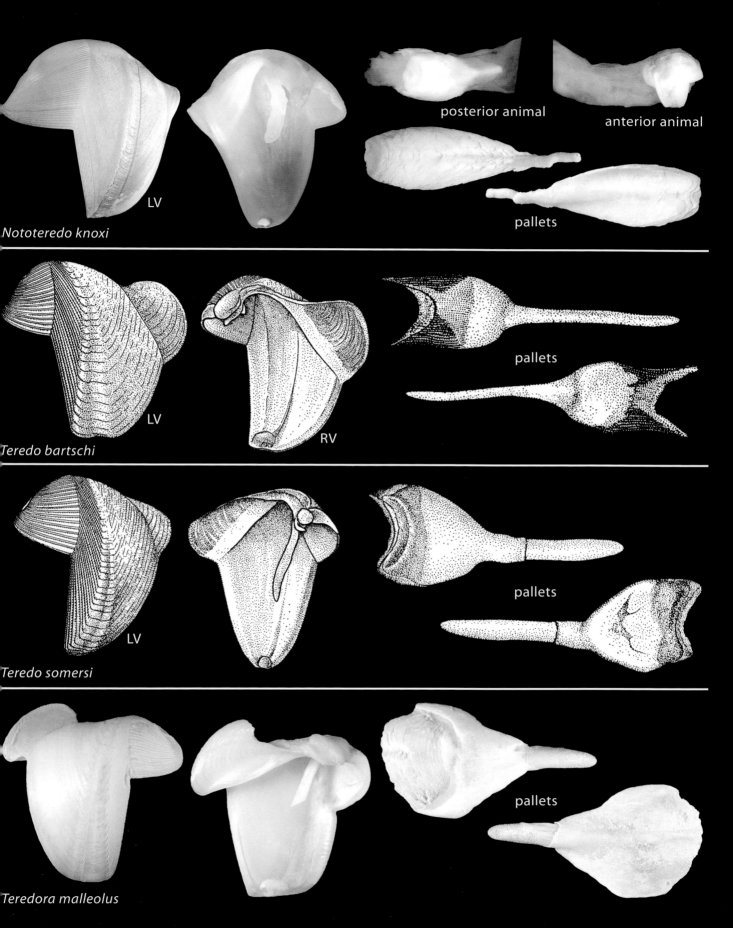

*Nototeredo knoxi*

posterior animal

anterior animal

pallets

LV

*Teredo bartschi*

LV

RV

pallets

*Teredo somersi*

LV

pallets

*Teredora malleolus*

pallets

# Acknowledgments

This work is part of an ongoing investigation of the molluscan diversity of the Florida Keys. This multiyear biotic inventory research project would not have been possible without the generous core support provided by the Comer Science and Education Foundation. Bivalve research in our laboratories and associated student support were made possible under National Science Foundation (NSF) PEET grant DEB-9978119. Over the course of the project, supplemental financial support was provided the Bertha LeBus Charitable Trust, FMNH's Womens Board, Harbor Branch Oceanographic Institution (Ft. Pierce, Florida), Delaware Museum of Natural History, AMNH's Proctor-Old-Sage Malacology Fund, FMNH's Seymour Persky's Zoology Fund honoring his sister, Zoologist Kaylia Katz, and FMNH's Zoology Department's Marshall Field Fund. We gratefully acknowledge the Florida Keys National Marine Sanctuary (FKNMS) for assistance with collecting permits and general support, and the many colleagues and staff members who have provided access to their museum collections and/or assistance in collecting and processing data.

The surveys and collections were supported by permits from the Florida Keys National Marine Sanctuary (080-98, 2000-036, 2002-078, 2002-079, and 2005-011); in John Pennekamp Coral Reef State Park under Florida Department of Environmental Protection permit 5-01-22; in the vicinity of Pigeon Key (on National Register of Historic Places) under the auspices of the Pigeon Key Foundation; in the Dry Tortugas National Park under collecting permits DRTO-19970030 and 2002-SCI-0005; in Long Key State Park (Long Key, Florida Keys) under Florida Department of Environmental Protection permit 5-02-43; and in Key West National Wildlife Refuge (near Sand Key, Florida Keys) under United States Fish and Wildlife Service permit 41580-01-07. Additional collecting was sanctioned under Florida Fish and Wildlife Conservation Commission permit 99S-024 to affiliates of The Bailey-Matthews Shell Museum (Sanibel, Florida) and permit 01S-056 (as well as annual permits for prior years of this study) to affiliates of the Smithsonian Marine Station (Ft. Pierce, Florida; logistic support by former director Mary E. Rice and staff is much appreciated).

Numerous colleagues and friends joined our fieldwork over the years. We specifically would like to thank Timothy Collins, Timothy Rawlings, Roberto Cipriani, Deirdre Gonsalves-Jackson, Cecelia Miles, Louise Crowley, Isabella Kappner, Daniel Miller, Tom Frankovitch, Homer and Anne Rhode, Ed Yastrow, Jim Culter, Jochen Gerber, Petra Sierwald, Anke Bieler, Jay Cordeiro, Lynn Funkhouser, the Smithsonian Marine Station at Fort Pierce, Mote Marine Laboratory's Center for Tropical Research (Summerland Key, Florida), Keys Marine Laboratory (Long Key, Florida), and the captains and crews of R/V *Eugenie Clark* (Mote Marine Laboratory, Sarasota, Florida), R/V *Bellows* (Florida Institute of Oceanography), F/V *Strange Bru* (Raymond Baiz, Florida Keys), and R/V *Coral Reef II* (Shedd Aquarium, Chicago, Illinois) for collecting assistance. Data gathering from other museum collections was facilitated by Gary Rosenberg (ANSP); José Leal and Tina Petrikas (BMSM); Charles Sturm (CMNH); Timothy Pearce, Elizabeth Shea, Leslie Skibinski, and Albert Chadwick (DMNH); Adam Baldinger and Kenneth J. Boss (MCZ); Katrin Schniebs (MTD); Nancy Voss (UMML); Jerry Harasewych and Paul Greenhall (USNM); Ronald Jansen (Senckenberg Museum, Frankfurt); and Matthias Glaubrecht and Lothar Maitas (ZMB). Bernard Metivier (Museum National d'Histoire Naturelle, Paris) and Kathie Way (BMNH) provided type specimen information and loans. Richard

E. Petit, as so often before, helped us disentangle obscure literature references. Philippe Bouchet (Museum National d'Histoire Naturelle, Paris) and Eugene Coan (Palo Alto, California) conferred about family-level taxonomy of bivalves. Individual photographs were kindly provided by John D. Taylor (BMNH), Robert Myers (Coral Graphics, Davie, Florida), John Reed (Harbor Branch Oceanographic Institution, Ft. Pierce), Ilya Tëmkin (AMNH), Neil Bourne (Pacific Biological Station, Nanaimo, British Columbia, Canada), Osmar Domaneschi (Universidade de São Paolo), Richard Willan (Northern Territory Museum of Arts and Sciences, Darwin, Australia), and Lynn Funkhouser (Chicago, Illinois). Several participants of our bivalve workshops in Florida (2002) and Thailand (2005), and our bivalve systematics symposium in Perth, Australia (2004), provided taxonomic insight; here we thank especially John Taylor, Emily Glover, Richard Willan, Graham Oliver, Paul Valentich-Scott, Luiz Simone, Thomas Waller, and Ilya Tëmkin for comments that helped to improve the manuscript. Henk Dijkstra kindly offered advice on taxonomic issues with Pectinidae and Propeamussiidae, and Harry G. Lee (Jacksonville) shared data on bivalves from northern Florida. Luiz Simone (then FMNH) generated anatomical data for many featured species. Steve Thurston (AMNH) rendered the transparent clams with artistic excellence, as did Lisa Kanellos (FMNH) the hinge drawings and various other graphics in this book. Sean Bober (FMNH) generated the GIS map. Over the years numerous people assisted with sorting bulk sediment samples, with data entry, with labeling, and with rehousing, including Leah Becker, Jay Biederman, Chris Boyko, Cheryl Breedlove, Claudia Capitini, Patricia Conway, Jochen Gerber, Marilyn Gerstenhaber, Richard Guzik, Michelynn Hassert, Susan Hewett, Janeen Jones, Isabella Kappner, Brett Kubricht, Armand Littman, Tony Marinello, Juri Miyamae, Laura Porro, Rebecca Price, James Pulizzi, Cynthia Rivera, Erin Roche, Carrie Seltzer, Rebecca Shell, Julia Sigwart, Ilya Tëmkin, Arend Thorp, Sezgi Ulucam, Christine Vittoe, David Walker, Elsa Whitmore, and Joe Zich. Roy Larimer (Microptics, Inc., Richmond, Virginia) offered much help and advice on photographic technique using the photographic system he designed. Martin Pryzdia (FMNH) handled the extensive loan traffic between FMNH, AMNH, and other institutions. Marla Coppolino (AMNH) assisted with scanning electron microscopy, literature research, and many other essential tasks, and Janeen Jones (FMNH) masterfully kept track of the thousands of specimens that we collected, sorted, identified, photographed, reidentified, and (very occasionally, of course) temporarily misplaced.

We thank the staff at Princeton University Press, especially Executive Editor Robert Kirk, who also ushered the tome through the peer review process, and Barbara Clauson, copy editor (Bluestem Editorial Services), who caught mistakes and inconsistencies. Special thanks are due to Eugene Coan, Gary Rosenberg, and Paul Valentich-Scott, whose helpful comments improved the content and readability of this volume.

# A Note About Shell Collecting

Shell collecting is a very popular and very educational hobby. Much of what we know today about past conditions of the Florida Keys fauna was learned from well-documented specimens (i.e., those with accurate locality and date information) collected by private enthusiasts and that ultimately found their way into the permanent holdings of formal museums. South Florida is extremely popular with collectors and the Keys are no exception. The natural resources in this region are stressed by a multitude of factors. Indiscriminate collecting, particularly of living mollusks, only threatens to perpetuate the degradation of local populations and habitats. Removal of individual shells (which in case of small-shelled specimens literally form the local "sand") and occasional harvesting of individual living specimens do little harm to local habitats and populations—as long as this collecting is done in an environmentally prudent way (we suggest following the spirit of the conservation resolution and code of ethics endorsed by the Conchologists of America, http://www.conchologistsofamerica.org).

Examination of our historic data has shown that individual shell collecting has not markedly impacted the Keys bivalve fauna (this is in contrast to our data on gastropods; large-bodied species of several "collectable" snail groups are disappearing from accessible habitats). Large-scale collecting activities, on the other hand, have been cause for concern. These include commercial harvesting of certain species such as Flame Scallops (now regulated) and large-scale removal of "live rock" (now forbidden) for the aquarium trade.

Shell collecting in the Keys must follow local rules and regulations of which there is a surprising diversity. Collection of living bivalves requires a Recreational Fishing License issued by the state of Florida, which is readily obtainable locally or via the Internet. Various special rules apply for the taking of individual bivalve species such as Bay Scallops and certain Flame Scallops. Several areas that fall under the jurisdiction of the Florida Keys National Marine Sanctuary are designated as Ecological Reserve or Sanctuary Preservation Areas. Most of these are protecting popular shallow reefs that now enforce higher levels of protection and forbid the taking of ANY animal, dead or alive. A few other places within the FKNMS are designated as research-only Special Use areas and cannot be entered without special permit. In addition, various other local, state, and federal parks and refuges have their own conservation rules, not all of which are locally posted or intuitive. For instance, at the time of this writing the Florida Keys National Wildlife Refuges (which include the large marine areas of the Great White Heron and Key West Refuges) explicitly allow fishing while stating that "disturbing, injuring or removing of plants and shells is illegal." By contrast, Bahia Honda State Park (which also allows fishing) forbids the taking of "live shells." In any case, we strongly advise obtaining locally available maps and contacting local authorities to inquire about such rules BEFORE collecting in the area.

# A Note on Species Names Introduced by d'Orbigny

Numerous bivalve species known from Florida Keys waters were named by Alcide d'Orbigny (1842–1853) in Sagra's work on the natural history of Cuba. The publication dates of this important work have never been fully established. The date for the bivalve plates and captions has long been generally accepted as 1842, but was recently established by G. Rosenberg (pers. comm., May 2006) to be considerably later, perhaps as late as April 1853 and almost certainly not earlier than 1851. We have adopted an 1853 date herein, affecting 27 species-level taxa covered in this book. For 19 of these, this has resulted in a simple date change from 1842 to 1853, without nomenclatural implications. However, the period of 1842–1853 was one of active publishing in malacology and several of these species were formally introduced, before 1853, in other works by d'Orbigny, Philippi, Conrad, and C. B. Adams. The date change of Sagra's work thus resulted in various priority issues that in some cases threatened nomenclatural stability.

Two species also were described, with the same names, by d'Orbigny in the *Voyage dans l'Amérique Méridionale* (1845 in 1834-1847) and the names are thus preserved. A few others had to change: *Polymesoda floridana* (Conrad, 1846) has priority over *P. maritima* (d'Orbigny, 1853), *Anomalocardia cuneimeris* (Conrad, 1846) has priority over *A. auberiana* (d'Orbigny, 1853), and *Varicorbula limatula* (Conrad, 1846) has priority over *V. disparilis* (d'Orbigny, 1853). In all three cases, the senior names had been used as valid names in the past. In the case of *Caryocorbula caribaea* (d'Orbigny, 1853), we noted synonyms dating from C. B. Adams (1852), but have not changed the species name because of the taxonomic instability in this family (in great need of revision) plus the possibility that the ongoing research into the Sagra publication might prove an earlier date in 1852 or even 1851.

Three well-established names of common western Atlantic *Lithophaga* species, *L. antillarum*, *L. bisulcata*, and *L. nigra*, were threatened by earlier unused names introduced by Philippi. For these cases we are invoking Article 23.9, Reversal of Precedence, of the ICZN (1999) to maintain prevailing usage:

(1) *Lithophaga antillarum* (d'Orbigny, 1853), introduced as *Lithodomus antillarum*, is a junior homonym of *Modiola antillarum* Philippi, 1847, and a junior synonym of *Modiola corrugata* Philippi, 1846. To our knowledge, the two senior names have not been used as valid names after 1899 (Art. 23.9.1.1). The conditions of Art. 29.9.1.2 are also met (see following list) and we consider *Lithodomus antillarum* d'Orbigny, 1853 a nomen protectum. The following is a selection of 30 works that have used the name *Lithophaga antillarum* as its presumed valid name (thus fulfilling the conditions of Art. 23.9.1.2): Abbott (1968), Abbott (1974), Abbott & Morris (1994), Bieler & Mikkelsen (2004b), Diaz Merlano & Puyana Hegedus (1994), Emerson & Jacobson (1976), Espinosa & Juarrero (1989), Espinosa et al. (1994), Garcia (1979), Humfrey (1975), Kissling (1977), Lyons & Quinn (1995), Mikkelsen & Bieler (2000), Mikkelsen & Bieler (2004), Moore (1961), Nowell-Usticke (1959), Odé (1979), Porter (1975), Redfern (2001), Rios (1970), Rios (1975), Rios (1994), Romero et al. (2002), Turgeon et al. (1998), Turner & Boss (1962), Valentich-Scott & Dinesen (2004), Vokes & Vokes (1984), Warmke & Abbott (1961), Weber (1961), Zischke (1977).

(2) *Lithophaga bisulcata* (d'Orbigny, 1853), introduced as *Lithodomus bisulcatus*, is a junior homonym of *Modiola appendiculata* Philippi, 1846. To our knowledge, *Modiola ap-*

*pendiculata* Philippi, 1846, has not been used a valid name after 1899 (Art. 23.9.1.1). The only located reference is by Mazÿck (1913: 22), who listed a "*Lithophaga appendiculata* Rav.*" in his *Catalog of Mollusca of South Carolina*, accepting it on the authority of Ravenel's collection catalog as edited by Gibbes (Ravenel, 1874: 55). Gibbes had listed a *Lithodomus appendiculatus* Ravenel. The given author's name is a probable printer's error (perhaps having slid down from the previous line where no author is given), but neither Gibbes nor Mazÿck referred to Philippi's work. Mazÿck marked the species entry with an asterisk, indicating that the name was merely listed on somebody else's authority (and thus not necessarily considered valid by the author). The conditions of Art. 29.9.1.2 are met (see following list) and we consider *Lithodomus bisulcatus* d'Orbigny, 1853 a nomen protectum. The following is a selection of 30 works that have used the name *Lithophaga bisulcata* as its presumed valid name (thus fulfilling the conditions of Art. 23.9.1.2): Abbott (1974), Andrews (1994), Beauperthuy (1967), Bieler & Mikkelsen (2004b), Diaz Merlano & Puyana Hegedus (1994), Emerson & Jacobson (1976), Espinosa & Juarrero (1989), Espinosa et al. (1994), Garcia (1979), Humfrey (1975), Lyons & Quinn (1995), Macsotay & Campos Villarroel (2001), Matthews & Rios (1967), McCloskey (1970), Mikkelsen & Bieler (2000), Mikkelsen & Bieler (2004), Moore (1961), Nowell-Usticke (1959), Odé (1979), Pointier & Lamy (1998), Porter (1975), Redfern (2001), Rios (1994), Romero et al. (2002), Turgeon et al. (1998), Turner & Boss (1962), Valentich-Scott & Dinesen (2004), Vokes & Vokes (1984), Warmke & Abbott (1961), Weber (1961).

(3) *Lithophaga nigra* (d'Orbigny, 1853), introduced as *Lithodomus niger*, is a junior synonym of *Modiola antillarum* Philippi, 1847, of *Modiola caribaea* Philippi, 1847, and of *Lithophaga crenulata* Dunker, 1849. To our knowledge, the three senior names have not been used as valid names after 1899 (Art. 23.9.1.1). The conditions of Art. 29.9.1.2 are also met (see following list) and we consider *Lithodomus niger* d'Orbigny, 1853 a nomen protectum. The following is a selection of 30 works that have used the name *Lithophaga nigra* as its presumed valid name (thus fulfilling the conditions of Art. 23.9.1.2): Abbott (1968), Abbott (1974), Abbott & Morris (1994), Beauperthuy (1967), Bieler & Mikkelsen (2004b), Emerson & Jacobson (1976), Espinosa & Juarrero (1989), Espinosa et al. (1994), Garcia (1979), Humfrey (1975), Kleemann (1983), Kleemann (1984), Lyons & Quinn (1995), Macsotay & Campos Villarroel (2001), Metivier (1967), Mikkelsen & Bieler (2000), Mikkelsen & Bieler (2004), Nowell-Usticke (1959), Odé (1979), Pointier & Lamy (1998), Redfern (2001), Rios (1970), Rios (1975), Rios (1994), Romero et al. (2002), Turgeon et al. (1998), Turner & Boss (1962), Vokes & Vokes (1984), Warmke & Abbott (1961), Weber (1961).

# Illustrated Glossary of Bivalve Terms ——————

**Note:** SMALL CAPITALS **font is used in family descriptions for terms defined here.**

ABDOMINAL SENSE ORGANS

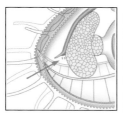

small paired swellings near the anus, on the ventral surface of the adductor muscle and receiving innervation from the visceral ganglion, of unknown function but showing characteristics of both chemo- and mechanoreceptors (perhaps related to, as proposed, detection of vibrations or of excurrent water flow); known as a SYNAPOMORPHY of Pteriomorphia (also present here in arcoids (Arcidae, Noetiidae, Glycymerididae, Limopsidae), mytiloids (Mytilidae), pterioids (Pteriidae, Isognomonidae, Malleidae, Ostreidae, Gryphaeidae, Pinnidae), limoids (Limidae), pectinoids (Pectinidae, Propeamussiidae, Spondylidae, Plicatulidae, Anomiidae); also known as pallial organ (Ostreidae) (figure: Pteriidae).

ACCESSORY FOOT

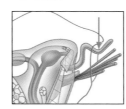

an elongated process on the FOOT of Malleidae, used to clean the INFRABRANCHIAL CHAMBER (figure: Malleidae).

ACCESSORY GENITAL ORGAN

pendulous glandular organ on the posteroventral surface of the posterior adductor muscle in male Pholadidae; secretions from this organ are added to sperm before storage in the SEMINAL VESICLE.

ACCESSORY HEART

enlarged blood vessels in the MANTLE lobes of the SUPRABRANCHIAL CHAMBER that pulsate and deliver blood to the auricles (in Ostreidae).

ACCESSORY PLATE

secondary calcareous or chitinous structure found in Pholadidae to protect the soft body or serve as an attachment for internal muscles; see also APOPHYSIS, CALLUM, HYPOPLAX, MESOPLAX, METAPLAX, PROTOPLAX, SIPHONOPLAX.

ADDUCTOR MUSCLE

one of usually two large muscles (one anterior, one posterior) that contract to close the shell and maintain it in that condition; the position of each muscle is usually clearly marked on the shell interior as an adductor muscle scar or impression (figure: *Mercenaria mercenaria*).

ADVENTITIOUS

of or belonging to a structure formed in an unusual place; applied in bivalve biology to the shell igloos or tubes of Gastrochaenidae (*Eufistulina*) and Teredinidae (*Kuphus*) that are formed in soft sediment.

ALIVINCULAR

type of dorsal LIGAMENT that is usually AM-PHIDETIC in position (typically trigonal in shape), with a central fibrous portion (RESILIUM) bordered by anterior and posterior lamellar layers, and occurring here in Limopsidae, Philobryidae, some Mytilidae, Pteriidae, Malleidae, Ostreidae, Gryphaeidae, Limidae, Pectinidae, Propeamussiidae, Spondylidae, Plicatulidae, Anomiidae, and Crassatellidae.

AMPHIDETIC

type of dorsal LIGAMENT that occurs both on the anterior and posterior sides of the UMBONES; see also PROSODETIC, OPISTHODETIC (figure: *Tucetona pectinata*).

AMYLASE

starch-digesting enzyme found in the CRYSTALLINE STYLE of most mollusks, including many bivalves.

anal aperture or siphon

see EXCURRENT APERTURE OR SIPHON.

ANAL CANAL

long, narrow, dorsal channel in Teredinidae, which terminates in a sphincter muscle in some species, into which the rectum empties and that is continuous with the SUPRABRANCHIAL CHAMBER and EXCURRENT SIPHON (figure: Teredinidae).

anal flap

see ANAL FUNNEL.

ANAL FUNNEL

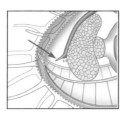 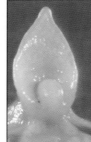

earlike membranous structure protruding from the tip of the rectum at about a right angle to the surface of the posterior adductor and enclosing the anal opening at its base; occurring here in Arcidae, Pteriidae, Isognomonidae, Malleidae, some Ostreidae, and Pinnidae; also called anal papilla or anal flap (figures: Pteriidae, *Pinctada imbricata*).

anal papilla

see ANAL FUNNEL.

ANAL TENTACLES

enlarged pallial tentacles on the posterodorsal margin of some Limidae.

anisomyarian

see HETEROMYARIAN.

ANTERIOR

head end.

AORTIC BULB 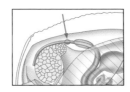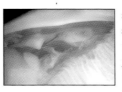 a muscular, spongy, pendulous structure on the ventral side of the posterior aorta and hindgut just posterior to the heart (anterior to the posterior adductor muscle), which serves to prevent rupture of the heart when the siphons and foot retract suddenly, forcing hemolymph backward into the posterior aorta; occurring here in some Pandoridae, Gastrochaenidae, some Cardiidae, Veneridae, some Tellinidae, Psammobiidae, some Semelidae, some Solecurtidae, Pharidae, and Mactridae (figures: Veneridae, *Mercenaria mercenaria*).

APERTURE an opening.

APOPHYSIS (pl. apophyses)  shelly structure to which pedal retractor muscles attach (more generally called MYO-PHORE); extending interiorly below the UM-BONES and occurring here in Teredinidae, Pholadidae, and Dreissenidae (figure: *Cyrtopleura costata*).

ARAGONITE (adj. aragonitic) a form of calcium carbonate found in molluscan shells and other hard parts; see also CALCITE, which differs from aragonite in certain characters of crystallization and density.

ARTICULATED with both valves joined together in living position.

AURICLE 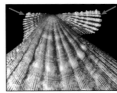 (1) earlike extension of the dorsal HINGE line, usually present both anterior and posterior of the UMBO, most characteristic of the family Pectinidae, also occurring here in Limidae, Malleidae, Propeamussiidae, Pteriidae, some Arcidae, some Philobryidae, and some Plicatulidae (figure: *Nodipecten fragosus*); (2) earlike extension of the posterior shell margin in Teredinidae; (3) chamber of the heart, usually paired in bivalves.

beak see UMBO.

BIFID 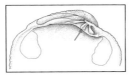 divided into two parts by a groove, usually in reference to HINGE TEETH (figure: *Periglypta listeri*).

branchial membrane/septum see SIPHONAL SEPTUM.

branchial sieve see SEPTIBRANCH.

branchial siphon see INCURRENT SIPHON.

BUCCAL FUNNEL

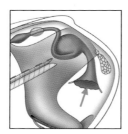

expanded trumpet-shaped mouth, occurring here in Verticordiidae (figured), Poromyidae, and Cuspidari-idae.

BYSSAL FASCIOLE

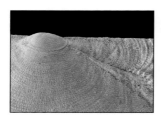

filled-in track of the BYSSAL NOTCH retained on the external shell surface as the shell increases in size (figure: *Similipecten nanus*).

byssal foramen      see BYSSAL NOTCH.

byssal groove      see BYSSUS.

BYSSAL NOTCH

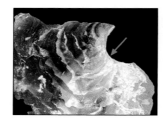

rounded or trigonal embayment in the shell margin for passage of the BYSSUS when the shell is closed; called BYSSAL FORAMEN in Anomiidae (figure: *Isognomon bicolor*).

BYSSUS (pl. byssi)

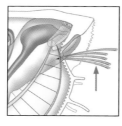 

elastic fibers (or calcified, in Anomiidae) secreted by a gland in the FOOT, exiting through a ventral BYSSAL GROOVE, and used to anchor the animal to a hard substratum (figures: Pteri-idae, *Pteria colymbus*).

CAECUM (pl. caeca)

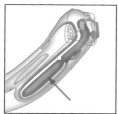

a blind pouch on an anatomical structure, such as the STOMACH or MANTLE; see also TUBULE (figure: Tere-dinidae).

CALCITE (adj. calcitic)      a form of calcium carbonate found in molluscan shells and other hard parts; see also ARAGONITE, which differs from calcite in certain characters of crys-tallization and density.

CALLUM

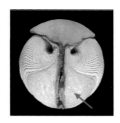

accessory shell or periostracal plate closing the anterior gape in adults of Pholadidae (figure: *Martesia striata*).

calymma

see PERICALYMMA.

CANCELLATE

external shell sculpture consisting of COMMARGINAL and radial elements crossing to form a grid- or netlike pattern; also called reticulate (figure: *Semele bellastriata*).

CARDINAL AREA

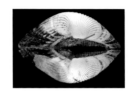

flattened area between the UMBONES dorsal to the HINGE TEETH; also called FOSSETTE (figure: *Noetia ponderosa*).

CARDINAL TEETH

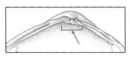

HINGE TEETH originating below the UMBO (figure: *Polymesoda floridana*).

CELLULASE

enzyme that facilitates the reduction of plant material by catalyzing the hydrolysis of cellulose.

cephalic eye

see EYE.

CEPHALIC HOOD

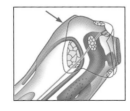

a covering over the external ADDUCTOR MUSCLES inserted on the UMBONAL REFLECTION in Pholadidae (where it is composed of periostracum, found only in juveniles, and replaced by one or more ACCESSORY PLATES in adults) and in Teredinidae (where it is formed by an extension of the MANTLE) (figure: Teredinidae).

CHALKY DEPOSITS

irregular, porous, white deposits of unknown cause or function on the interior shell surface (in Ostreidae and Gryphaeidae).

CHIMNEY

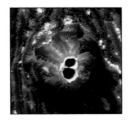

an elongated tube of solid material, here either (1) composed of feces and/or pseudofecal particles cemented with mucus, fitting over the posterior end of the shell, sometimes far anteriorly, in members of Pholadidae; or (2) an extension of the calcareous burrow lining in Gastrochaenidae, that projects above the surface of the substratum (figure: Gastrochaenidae).

CHOMATA

elongated, sinuous, and/or pustular interlocking ridges on the COMMISSURAL SHELF on each side of the LIGAMENT, ranging from simple tubercles and sockets to a complex anastomosing network (in Ostreidae, Gryphaeidae); those of Gryphaeidae are elongate and wormlike and are called VERMICULAR CHOMATA (figure: *Hyotissa mcgintyi*).

CHONDROPHORE

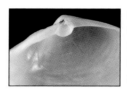

a type of RESILIFER, commonly spoon-shaped and (in some species) buttressed and projecting below the HINGE for insertion of the RESILIUM, especially characteristic of Mactridae and Myidae (figure: *Anatina anatima*).

CILIATED DISKS

ciliated pads that conjoin in hook and loop fastener fashion to connect two soft-tissue surfaces, especially here in PROTOBRANCH or ELEUTHERORHABDIC CTENIDIA.

CLASPER SPINES

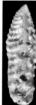

shelly outgrowths of the external valve surface that wrap around and secure the bivalve to a narrow solid substratum (in Ostreidae: Lophinae) (figure: *Dendostrea frons*).

claspers

see CLASPER SPINES.

COMMARGINAL

sculpture parallel to the shell margin, also called concentric (figure: *Astarte crenata subequilatera*).

COMMISSURAL SHELF

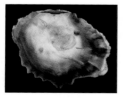

a demarcated shell margin defined interiorly by an angle with the main body of the shell (in Ostreidae, Gryphaeidae) (figure: *Hyotissa hyotis*).

compensation sack

see SEPTAL VALVULA.

complex crossed lamellar

see CROSSED LAMELLAR.

concentric

see COMMARGINAL.

CONDYLE

knoblike internal structure (usually one dorsal, one ventral on each valve) at the valve margins in Pholadidae and Teredinidae, which in pairs form pivoting points upon which the valves articulate during the boring process (figure: *Martesia striata*).

CROSSED LAMELLAR

shell microstructural variety consisting of adjacent aggregations of numerous, parallel, elongated, aragonitic elements arranged in two predominant directions relative to the shell margin; the units in COMPLEX CROSSED LAMELLAR microstructure are arranged in three or more directions.

CRUCIFORM MUSCLE

X-shaped muscle at the base of the INCURRENT SIPHON in Tellinoidea (and recognized as a SYNAPOMORPHY of that superfamily) that absorbs the physical strain experienced when the SIPHONS extend and retract; associated with paired sensory organs thought to monitor water quality.

CRURA (pl. crurae)

variously used term for (1) a raised ridge resembling a tooth (in, e.g., Pandoridae); or (2) an opposed internal shell rib in Pectinoidea (figure: *Caribachlamys sentis*).

CRURUM

the specialized stalked RESILIFER of Anomiidae.

crypt

see IGLOO.

CRYSTALLINE STYLE

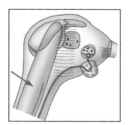

rod-shaped inclusion in the STOMACH of many mollusks, including most bivalves, that includes the starch-digesting enzyme AMYLASE (possibly also CELLULASE for reduction of phytoplankton) and that reduces the size of ingested food particles through rotating abrasion against the gastric shield (and chitinous lining if present); produced by secretory tissues in the STYLE SACK; believed homologous with the PROTOstyle of Protobranchia, which does not include amylase; see also STOMACH, PROTOSTYLE.

ctenidial eye

see EYE.

CTENIDIUM (pl. ctenidia)

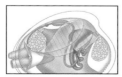

gills, or organ of respiration and in most cases also of food-gathering in bivalves; usually suspended by tissue or cilial junctions from the dorsal region of the animal and typically comprising a W-shaped, doubly folded lamella on each side of the visceral mass; each side typically consisting of an inner and outer demibranch (holobranch =

entire gill), each of these with an ascending and descending lamella and often with ciliated food grooves at the distal edges and/or at the junction of the two demibranchs, also with or without interlamellar junctions connecting the two lamella of one demibranch; see also eulamellibranch, filibranch, heterorhabdic, homorhabdic, protobranch, pseudolamellibranch, plicate, septibranch, synaptorhabdic (figure: Mercenaria mercenaria).

CTENOLIUM

comblike row of denticles along the ventral edge of the BYSSAL NOTCH in Pectinidae; also called pectinidium or pectineum (figure: *Caribachlamys sentis*).

CUPULES

multiple segments of an IGLOO cemented superficially to hard substratum by *Cucurbitula* (Gastrochaenidae), creating an artichoke- or pine-cone–like external appearance.

CYCLODONT

HINGE type characterized by a small or absent hinge plate, and teeth that essentially emerge from hinge margin, occurring here in Cardiidae.

DEHISCENT

splitting, referring here to PERIOSTRACUM that is not adherent.

demibranch

see CTENIDIUM.

DEPOSIT FEEDING

feeding type of some bivalves during which organic particles are harvested (by either the SIPHONS or PALP PROBOSCIDES) from the surface or near-surface sediments; see also SUSPENSION FEEDING.

DESMODONT

HINGE type characterized by very reduced or absent teeth, in some cases replaced by accessory ridges along the valve margins, occurring here in Gastrochaenidae and Myidae.

DIAPHRAGM

ridge separating shell and siphonal chambers in the burrow or IGLOO of Gastrochaenidae; also called a SEPTUM.

DIGESTIVE GLAND

highly branched organ surrounding the stomach within the visceral mass, into which edible food particles pass (through ducts from the stomach) and where digestion occurs; often darkly colored in living specimens, visible as a dark spot through the gills or a thin shell; in earlier literature, often called liver, midgut gland, or hepatopancreas.

DIMYARIAN

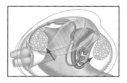

having both anterior and posterior ADDUCTOR MUSCLES (figure: *Mercenaria mercenaria*).

dioecious

see GONOCHORISTIC.

DIRECT DEVELOPMENT

larval development type characterized by lack of a free-swimming stage (VELIGER) that can be absent entirely or passed entirely within an egg capsule or brooding chamber.

DIVARICATING

branching, usually in reference to external radial sculpture (figure: *Ctenoides sanctipauli*).

DORSAL

the HINGE side of a bivalve, opposite of ventral.

dorsal ligament

see LIGAMENT.

DUPLIVINCULAR

type of dorsal LIGAMENT that is AMPHI- or OPISTH-ODETIC in position, and has alternating lamellar and fibrous layers, repeated as a series of parallel or oblique bands, giving the appearance of a series of nested chevrons; occurring here in Arcidae, Noetiidae (modified), and Glycymerididae.

DYSODONT

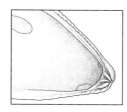

HINGE type featuring small simple denticles close to the UMBONES, characteristic of some members of Mytilidae (figure: *Brachidontes exustus*).

ear

see AURICLE.

EDENTATE

without HINGE TEETH, also called edentulous.

edentulous

see EDENTATE.

ELEUTHERORHABDIC

gill type charcterized by filaments connected by interlocking cilia projecting from CILIATED DISKS; this term does not preclude the existence of tissue junctions in the same gill; see also FILIBRANCH, PSEUDOLAMELLIBRANCH. Ridewood (1903) defined a group Eleutherorhabda based on this character, equivalent to Pelseneer's (1891) group Filibranchia.

ENDOBYSSATE

INFAUNAL with BYSSUS attached to particles or objects below the surface; see also EPIBYSSATE.

ENTIRE

occurring without interruption (break or embayment), usually referring to the PALLIAL LINE, also called simple or integrepalliate (figure: *Codakia orbicularis*).

EPIBYSSATE

EPIFAUNAL with BYSSUS attached to solid substrata; see also ENDOBYSSATE.

EPIFAUNAL

living on top of the sediment, i.e., unburied; also called epibenthic; see also INFAUNAL.

EQUILATERAL

having the UMBO at the center of the dorsal midline of the valve (figure: *Tucetona pectinata*).

EQUIVALVE

having valves that are equal in shape and size (figure: *Periglypta listeri*).

ESCUTCHEON

an oval or spindle-shaped impressed area, often clearly demarcated and of differing sculpture, along the posterodorsal margin of a bivalve (figure: *Periglypta listeri*).

EULAMELLIBRANCH

gill type characterized by a series of narrow, elongated filaments that are interconnected laterally only by tissue junctions (so that the interfilamental spaces are divided into a series of OSTIA or fenestrae), and with or without interlamellar junctions; see also SYNAPTORHABDIC; occurring here in the Limidae and heterodont Carditoida (Crassatellidae, Astartidae, Carditidae, Condylocardiidae), Anomalodesmata (Pandoridae, Lyonsiidae, Periplomatidae, Thraciidae, Verticordiidae), and Veneroida (Lucinidae, Ungulinidae, some Thyasiridae, Chamidae, Lasaeidae, Hiatellidae, Gastrochaenidae, Trapezidae, Sportellidae, Corbiculidae, Cardiidae, Veneridae, Tellinidae, Donacidae, Psammobiidae, Semelidae, Solecurtidae, Pharidae, Mactridae, Dreissenidae, Myidae, Corbulidae, Pholadidae, Teredinidae).

EXCURRENT APERTURE
OR SIPHON

aperture or siphon that controls water outflow from the MANTLE CAVITY; also known as anal, cloacal, or exhalent aperture/siphon (figure: *Mercenaria mercenaria*).

exhalent aperture or siphon      see EXCURRENT APERTURE OR SIPHON.

EXTERNAL LIGAMENT

type of dorsal LIGAMENT that attaches to the dorsally exposed surface (or only slightly sunken on the hinge plate) of the two valves (figure: *Scissula similis*).

EYE  organ of photoreception, existing in a range of structural types (e.g., with or without lens, simple or compound, open-cup–like or closed, etc.) within Bivalvia; (1) CEPHALIC or "true" EYES are innervated by the cerebral ganglion and occur on one or both of the anteriormost gill filaments (also called ctenidial eyes; therefore probably do not function in photoreception); and (2) PALLIAL EYES are innervated by the visceral ganglia, occur on the MANTLE margin or siphon tips (see figure), and function in photoreception; also called ocellus, photoreceptor (figure: *Argopecten gibbus*).

fenestra (pl. fenestrae)      see OSTIUM.

FILIBRANCH  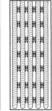 gill type characterized by a series of narrow elongated filaments, usually in the form of two demibranchs per side, which are interconnected laterally by CILIARY JUNCTIONS only (not tissue junctions), and with or without interlamellar junctions; see also ELEUTHERORHABDIC; occurring here in the pteriomorphian Arcoida (Arcidae, Noetiidae, Glycymerididae, Limopsidae, Philobryidae), Mytiloida (Mytilidae), some Pterioida (some Pteriidae, some Isognomonidae, some Malleidae), and Pectinoida (Propeamussiidae, Plicatulidae, Anomiidae).

FILTER FEEDING      feeding type involving the filtering of organic particles from water by the gills, after which appropriately sized particles are transported to the mouth; see also SUSPENSION FEEDING, DEPOSIT FEEDING.

flange      see UMBONAL REFLECTION.

FOOT 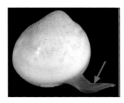 muscular organ at the ventral part of the visceral mass, used by contraction and expansion for locomotion, burrowing, and/or anchoring a bivalve (figure: *Laevicardium serratum*).

FOSSETTE      (1) concave or V-shaped, submarginal, linear support for a SIMPLE dorsal LIGAMENT; or (2) see CARDINAL AREA.

FOURTH PALLIAL APERTURE 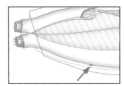 posteroventral opening in the fused MANTLE margin (along the ventral margin or near the base of the INCURRENT SIPHON) of various bivalves, used for rapidly discharging water and PSEUDOFECES from the MANTLE CAVITY when the ADDUCTOR MUSCLES suddenly contract; occurring here in Lyonsiidae, Periplomatidae, Thraciidae, some Hiatellidae, some Pharidae, and some Mactridae; see also PSEUDOFECES (figure: Lyonsiidae).

GAPING

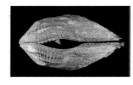

not closing completely, leaving an opening (1) between the valves when the ADDUCTOR MUSCLES are fully contracted (i.e., byssal gape, figure: *Arca imbricata*); or (2) in a partially fused MANTLE (i.e., pedal gape).

gills

see CTENIDIA.

Gland of Deshayes

see ORGAN OF DESHAYES.

GONOCHORISTIC

sexually mature animals in which males and females are separate; also called dioecious.

GRANULOSE

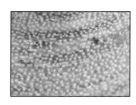

grainy, covered in granules (figure: *Thracia morrisoni*).

GROOVE

an elongated sculptural element that is depressed below the surrounding shell surface; see also RIB, STRIA.

GROWTH LINE

a COMMARGINAL line (usually fine) on the external valve surface that indicates the position of an earlier growth stage.

GUTTER

a curviplanar external support for a SIMPLE dorsal LIGAMENT.

HEIGHT

distance between the UMBO and the VENTRAL margin.

HERMAPHRODITE (adj. hermaphroditic)

sexually mature animal in which male and female gametes are produced by the same individual, either simultaneously or in sequence; see also PROTANDRIC.

HETERODONT

HINGE type characterized by the presence of more than one type of HINGE TEETH (i.e., CARDINAL plus LATERAL TEETH); occurring in most Tertiary and Recent bivalves (figure: *Polymesoda floridana*).

HETEROMYARIAN

with unequally sized ADDUCTOR MUSCLES; in most cases, the posterior adductor muscle is larger than the anterior; also called anisomyarian (figure: Pinnidae).

HETERORHABDIC

gill type characterized by having more than one type of filament, producing a corrugated or plicate surface; see also HOMORHABDIC.

HINDGUT

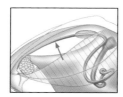

posterior portion of the intestine, here applied to that part external and posterior to the visceral mass (figure: Corbiculidae).

| | |
|---|---|
| HINGE | collective term for the dorsal border of the articulated valves, including the LIGAMENT, HINGE TEETH, and other structures that function to permanently unite the two valves; also called hinge line. |
| hinge line | see HINGE. |
| HINGE PLATE | the flattened interior dorsal area that bears the HINGE TEETH and LIGAMENT; called provinculum in larval bivalves. |

HINGE TEETH 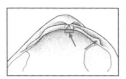 a series of calcified dorsal interlocking teeth and sockets that allow alignment of the valves to be maintained during opening and closing; see also CYCLODONT, DESMODONT, DYSODONT, EDENTATE, HETERODONT, ISODONT, TAXODONT (figure: *Laevicardium serratum*). A tooth numbering system for heterodonts, originally proposed by Bernard (1895), labels hinge teeth according to their position and appearance during ontogeny as follows:

right anterior cardinal tooth – 3a
right middle cardinal tooth – 1
right posterior cardinal tooth – 3b
left anterior cardinal tooth – 2a
left middle cardinal tooth – 2b
left posterior cardinal tooth – 4b
anterior lateral teeth – left AII fitting between right AI and AIII
posterior lateral tooth – left PII fitting between right PI and PIII

| | |
|---|---|
| HIRSUTE | hairy, pertaining here to PERIOSTRACUM that consists of long, shaggy processes. |
| holobranch | see CTENIDIUM. |
| HOLOTYPE | a unique specimen designated to represent the concept of a named species (currently always so designated by the original author in the original description); see also LECTOTYPE, PARATYPE, SYNTYPE. |
| HOMOGENOUS | shell microstructural variety consisting of irregularly shaped but more or less equally sized aragonitic elements that show no regular arrangement. |
| HOMORHABDIC | gill type characterized by having only one type of filament, thus unplicated or smooth; see also HETERORHABDIC. |
| HYALINE ORGANS | MANTLE structures occurring in the subfamily Tridacninae (Cardiidae), equipped with lenses, that transmit light to symbiotic ZOOXANTHELLAE in tissue. |
| HYOTE SPINES | hollow tubular spines that are open distally and along the side, resulting in an ear-shaped opening; occurring here in Ostreidae, Gryphaeidae, ad Chamidae. |
| HYPOBRANCHIAL GLANDS | glands in the SUPRABRANCHIAL CHAMBER of protobranch and a few other bivalves; in protobranchs, the glands function in the consolidation of waste material in the MANTLE CAVITY and in nonprotobranchs, in the nutrition of larvae being incubated in the gills. |

HYPOPLAX

accessory shell plate in Pholadidae that covers the posteroventral margin (figure: *Martesia striata*).

IGLOO

burrow of Gastrochaenidae, especially referring to those built within soft sediment or cemented superficially to hard substratum; also called crypt.

IMBRICATE

composed of scales or scalelike parts overlapping like roof tiles or shingles (figure: *Pinctada imbricata*).

INCURRENT APERTURE
or SIPHON

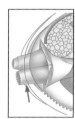

aperture or siphon, usually posterior, rarely anterior, that controls water intake into the MANTLE CAVITY; also known as branchial, pallial, or inhalant aperture/siphon (figure: *Mercenaria mercenaria*).

INEQUILATERAL

having the UMBO closer to one end of the valve than to the center (figure: *Modiolus squamosus*).

INEQUIVALVE

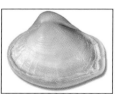

having valves of dissimilar shape or size, i.e., one valve larger than the other (figure: *Varicorbula limatula*).

INFAUNAL

living buried within sediment; see also EPIFAUNAL.

INFRABRANCHIAL CHAMBER

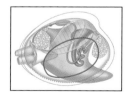

body cavity ventral to the gills; see also SUPRABRANCHIAL CHAMBER (figure: *Mercenaria mercenaria*).

INFRASEPTAL CHAMBER

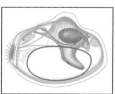

body cavity ventral to the branchial SEPTUM, analogous to infrabranchial chamber in bivalves with nonseptibranch gills; see also SUPRASEPTAL CHAMBER (figure: Poromyidae).

inhalant siphon | see INCURRENT SIPHON.

integrepalliate | see ENTIRE.

INTERLINEAR SCAR

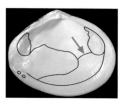

pallial line connecting the anterior end of the sinus with the anterior adductor muscle scar, occurring here in Tellinidae (figure: *Arcopagia Fausta*).

INTERNAL LIGAMENT

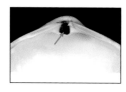

type of LIGAMENT that attaches at or below the level of the HINGE TEETH or (if EDENTATE) near the ventral edge of the shell margin; see also EXTERNAL LIGA-MENT, LIGAMENT, RESILIUM (figure: *Spisula similis*).

intestine | see HINDGUT, MIDGUT.

ISODONT

HINGE type characterized by secondary teeth and sockets, often resembling a ball-and-socket arrange-ment, on either side of a central RESILIUM; occurring here in Spondylidae and Plicatulidae (figure: *Spondy-lus americanus*).

ISOMYARIAN

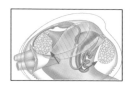

having equally or nearly equally sized ADDUCTOR MUSCLES (figure: *Mercenaria mercenaria*).

LABIAL PALPS

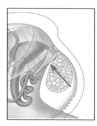

paired lamellae on either side of the mouth, usually with ridged and ciliated opposing surfaces (figure: *Mercenaria mercenaria*). Junctions of the distal oral groove (DOG) of the palps with the ventral tips of the anterior ctenidial fil-aments (ACF) of the inner demibranchs were categorized by Stasek (1963) as:

CATEGORY I – ACF inserted but not fused into the DOG, occurring here in Mytilidae, Astartidae, and some Crassatellidae.

CATEGORY II – ACF inserted into and fused with the DOG, occurring here in some Limidae, Carditidae, Lyonsiidae, Chamidae, Hiatellidae, Cardiidae, Gastrochaenidae, most Veneridae, Pharidae, Dreissenidae, and Pholadidae.

CATEGORY III – ACF not inserted into the DOG (and thus not fused), although the anteroventral margin of the inner demibranchs can be fused to the inner palp lamellae, occurring here in Arcidae, Noetiidae, Limopsi-dae, Philobryidae, Isognomonidae, Malleidae, Ostreidae, Gryphaeidae, Pinnidae, some Limidae, Pectinidae, Propeamussidae, Spondylidae, Pli-catulidae, Anomiidae, some Crassatellidae, Pandoridae, Periplomatidae, Thraciidae, Verticordiidae, Poromyidae, Cuspidariidae, Lucinidae, Ungulinidae, Thyasiridae, Lasaeidae, Trapezidae, some Veneridae, Corbiculidae, Tellinidae, Psammobiidae, Semelidae, Solecurtidae, Mactridae, Myidae, Corbulidae, and Teredinidae.

LACUNA (pl. lacunae)     a space or hemocoel within tissues, serving in place of vessels for the circulation of body fluids; here, the kidneys in Verticordiidae are accompanied by an extensive lateral system of lacunae.

LAMELLA (pl. lamellae)     thin platelike or scalelike structure.

LAMELLIBRANCHIA     Bivalvia.

LATERAL BULBS     elaborations of the LIPS in Lyonsiidae.

LATERAL TEETH

HINGE TEETH in a HETERODONT hinge that are far removed from the UMBO (as opposed to CARDINAL TEETH below the umbo) at the anterior and/or posterior ends, usually oriented parallel to the shell margin and serving to prevent anteroposterior "sliding" or movement of the valves when closed (figure: *Polymesoda floridana*).

LECITHOTROPHIC     larval development type characterized by a free-swimming stage (VELIGER) that carries its own nutritional reserves and does not feed in the plankton.

LECTOTYPE     a single specimen serving the function of a HOLOTYPE, designated by a subsequent author from the members of a syntypic series; the remaining syntypes become PARALECTOTYPES or the equivalents of paratypes by this action; see also HOLOTYPE, PARATYPE, SYNTYPE.

LENGTH     distance between anterior and posterior margins.

LIGAMENT     elastic uncalcified structure, comprised of two layers (lamellar of organic composition, and fibrous of organic material plus aragonite), usually brown to black in color, that connects the two bivalve shells at the HINGE line and functions as a spring to open the valves when the ADDUCTOR MUSCLES relax; either EXTERNAL (dorsal to the hinge teeth, visible, and under tension when the valves are closed) or INTERNAL (ventral to the hinge teeth, not visible, and under compression when the valves are closed) or both.

If external ("dorsal"), the ligament can be (1) SIMPLE, PLANIVINCULAR, PARIVINCULAR, DUPLIVINCULAR, ALIVINCULAR, or MULTIVINCULAR; (2) external or submarginal; (3) OPISTHO-, PROSO-, or AMPHIDETIC; and (4) set on GUTTERS (simple only), FOSSETTES (simple only), NYMPHS, or PSEUDO-NYMPHS (see also these terms).

If internal, the ligament is usually called a RESILIUM, and attaches to a RESILIFER or CHONDROPHORE (see also these terms).

ligamental pit     see RESILIFER.

lips     outgrowths of the buccal dermis (integument surrounding the mouth), which are usually simple flaps dorsal and ventral of the mouth, but hypertrophy into arborescent or ruffled structures in Pectinidae, Limidae, and Spondylidae; in some limid species, the dorsal and ventral lips are fused, leaving only a series of ostia as mouth openings; hypertrophied lips serve as filters of potential food particles.

LITHODESMA

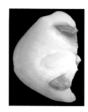

unpaired calcareous structure usually at the center of the inner LIGAMENT layer of the otherwise uncalcified internal ligament (RESILIUM) of some bivalves, especially the Anomalodesmata; in some families (e.g., Periplomatidae, Thraciidae) it is free and bridges the valves adjacent to the CHONDROPHORE; also called calcareous ossicle or osciculum (figure: *Periploma planiusculum*).

LITTORAL

intertidal, or the coastal zone between the highest high-water mark and the lowest low-water mark.

LUNULE

heart- or crescent-shaped impressed external feature, often demarcated and differently sculptured from the main body of the shell, located anterodorsally on the shell of some bivalves (e.g., Veneridae) anterior to the UMBONES (figure: *Periglypta listeri*).

MANTLE

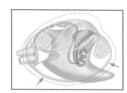

fleshy outer tissue surrounding the organs of a molluscan body and secreting the PERIOSTRACUM and shell; consisting of two lobes in a bivalve, one lining each shell, and at the ventral edge two to four folds that can have different functions or features (figure: *Mercenaria mercenaria*). See also SIPHONS for types of mantle fusion proposed by Yonge (1957).

MANTLE CAVITY

chamber between the MANTLE lobes and interior visceral mass, containing CTENIDIA and other organs; also called pallial cavity.

MARGINAL DENTICLES

teeth or ridges on the free interior valve margins of a nongaping bivalve shell that interlock when the valves are fully closed (figure: *Astarte smithii*).

MARSUPIUM (pl. marsupia)

specialized incubatory chamber for brooded larvae.

MESOPLAX

accessory shell plate in Pholadidae posterior to the PROTOPLAX, both protecting the anterior ADDUCTOR MUSCLE (figure: *Cyrtopleura costata*).

mesosoma

see METASOMA.

METAPLAX

accessory shell plate in Pholadidae extending posteriorly between the two valves from the UMBONES (figure: *Martesia striata*).

METASOMA

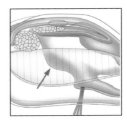

posteroventral sacklike extension of the visceral mass containing the STYLE SACK of the STOMACH and part of the gonad (in Ostreidae and Mytilidae); also called mesosoma, pyloric process (figure: Mytilidae).

MIDGUT

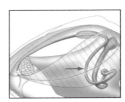

anterior portion of the intestine, here applied to that part within the visceral mass; in text, "coiled" indicates that one or more loops occur (figure: Corbiculidae).

MONOMYARIAN

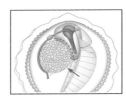

having a single ADDUCTOR MUSCLE or single muscle scar (in some cases formed by the coalesced insertions of several muscles) (figure: Gryphaeidae).

MONOPHYLETIC

(pertaining to a taxonomic group) defined by SYNAPOMORPHIES and including an ancestor and all of its descendants.

mother-of-pearl

see NACRE.

MULTIVINCULAR

type of dorsal LIGAMENT that consists of serially repeated lamellar and fibrous elements (of the ALIVINCULAR type) across the HINGE line, typically OPISTHODETIC in position, and occurring here in Isognomonidae.

MUSCLE SCAR

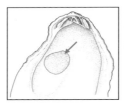

impression on the shell interior that indicates the attachment position of a muscle (figure: *Plicatula gibbosa*).

MYOPHORE

generalized term for shelly structure by which muscles attach to the shell, often in the form of a ridge or shelf; when enlarged, often called SEPTUM or MYOPHORIC RIDGE/flange/buttress; see also APOPHYSIS.

myophoric buttress

see MYOPHORE.

myophoric flange

see MYOPHORE.

myophoric ridge

see MYOPHORE.

MYOSTRACUM

irregularly PRISMATIC aragonite deposited at sites of muscle attachment to the shell.

MYTILIFORM  mussel-shaped (figure: *Ischadium recurvum*).

NACRE (adj. nacreous)  pearly or iridescent form of aragonitic calcium carbonate composed of polygonal or rounded tablets arranged parallel to the depositional surface; present in two forms: sheet (thin sheets of prisms separated by sheets of organic matrix) or lenticular (vertical columns of prisms); also known as mother-of-pearl (figure: *Pinctada imbricata*).

neoteny (adj. neotenous) see PEDOMORPHOSIS.

NEPHROLITH granule produced by the kidney (e.g., calcium-rich in Pinnidae).

NYMPH  a well-defined support for the inner fibrous layer of a more advanced type of dorsal LIGAMENT that is rotated dorsally (thus enhancing the dorsal arching of the ligament); see also LIGAMENT, PSEUDONYMPH (figure: *Periglypta listeri*).

ocellus (pl. ocelli) see EYE.

OPISTHODETIC  type of dorsal LIGAMENT that occurs entirely on the posterior side of the UMBONES; see also AMPHIDETIC, PROSODETIC (figure: *Periglypta listeri*).

OPISTHOGYRATE/
OPISTHOGYROUS 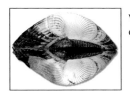 with the UMBONES curling posteriorly; see also PROSOGYRATE, ORTHOGYRATE (figure: *Noetia ponderosa*).

OPISTHOPODIUM expanded heel on the FOOT of some bivalves.

ORBITAL MUSCLES marginal adductor muscles spanning the fused ventral MANTLE folds, acting as accessory adductors; occurring here in Periplomatidae, Hiatellidae, and Trapezidae.

ORGAN OF DESHAYES collective term for the body of nitrogen-fixing, CELLULASE-producing bacteria in the gills of Teredinidae; ducts in the afferent branchial vessels empty into the esophagus; also called Gland of Deshayes.

ORGANS OF WILL pigmented glandular structures of unknown function on the middle and inner MANTLE folds of Pinnidae; they have been shown not to be eyespots, and are possibly associated with production of pigment rays on the shell.

ornament······· see SCULPTURE.

ORTHOGYRATE/ORTHOGYROUS  having the UMBONES curling directly toward one another or ventralward; see also PROSOGYRATE, OPISTHOGYRATE (figure: *Tucetona pectinata*).

osciculum······· see LITHODESMA.

osculum (pl. oscula)······· see SEPTIBRANCH.

ossicle, calcareous······· see LITHODESMA.

OSTIUM (pl. ostia)······· an opening in (1) the ctenidia, created by the intersection of filaments and interfilamental junctions; or (2) the septum, containing gill filaments; see also EULAMELLIBRANCH, SEPTIBRANCH.

otocyst······· see STATOCYST.

PALLETS  calcareous valvelike structures associated with the SIPHONS (Teredinidae) (figure: *Bankia carinata*).

PALLIAL······· pertaining to the MANTLE (= pallium).

pallial cavity······· see MANTLE CAVITY.

pallial eye······· see EYE.

PALLIAL GILLS  a series of ridges, complex folds, or pectinate structures on the inner mantle surface that serve as accessory respiratory organs of Lucinidae, and aid in functionally separating respiratory surfaces from the location of the endosymbionts (figure: Lucinidae).

PALLIAL LINE  impressed line (continuous or discontinuous) on the interior of the shell, usually paralleling the ventral margin between the anterior and posterior ADDUCTOR MUSCLES, that indicates the attachment of the pallial muscles and can include a pallial sinus; see also ENTIRE, PALLIAL SINUS (figure: *Scissula similis*).

PALLIAL MUCUS GLANDS  glands lining the mantle edge near the pedal gape that produce mucus that assists in the removal of pseudofeces; occurring here in Hiatellidae, Myidae, and Corbulidae (figure: Hiatellidae).

PALLIAL ORGAN 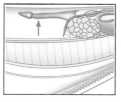 a posterodorsal muscular stalklike structure with a conical glandular head that serves to clear the suprabranchial chamber of shell fragments when the exposed edge of the shell is broken and the mantle withdrawn (an alternative defensive function also has been proposed); a SYNAPOMORPHY of Pinnidae; also called dorsal pallial tentacle (figure: Pinnidae).

pallial rib······· see PALLIAL RIDGE.

PALLIAL RIDGE

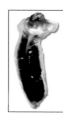

longitudinal median ridge present internally in most species of the family Malleidae, extending from either the VISCERAL RIM or more dorsally from the insertion of the pallial retractor muscle, probably serving to strengthen the thin-walled, elongated shell; also called pallial rib (figure: *Malleus candeanus*).

PALLIAL SINUS

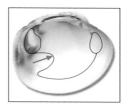

an embayment in the posterior part of the PALLIAL LINE that indicates the attachment of siphonal retractor muscles and demarcates that part of the MANTLE CAVITY into which the SIPHONS can retract; the presence of a sinus in a PALLIAL LINE has been called sinupalliate (figure: *Periglypta listeri*).

pallial siphon            see INCURRENT SIPHON.

PALLIAL VEIL

expanded inner mantle folds that serve to close or partially close an unfused mantle margin when the valves are gaping; occurring here in Pteriidae, Isogno-monidae, Malleidae, Ostreidae, Gryphaeidae, Pinnidae, Limidae, Pectinidae, Propeamussiidae, Spondylidae, Plicatulidae, and Anomiidae; also called velum (figure: Limidae).

palp caecum            see PALP POUCH.

PALP POUCH

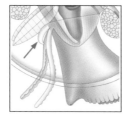

unpaired concave structure at the posterior end of the LABIAL PALPS of Nuculidae, which receives food particles from the PALP PROBOSCIDES before transport to the palps for sorting; also called palp caecum (figure: Nuculidae).

PALP PROBOSCIDES

ciliate tentaculate extensions of the posterior ridge of the palp lamellae in Protobranchia, which are used for DEPOSIT FEEDING (figure: Nuculidae).

palps            see LABIAL PALPS.

paralectotype            see LECTOTYPE.

PARATYPE            one of potentially several specimens examined at the same time as the HOLOTYPE and serving as additional representatives of the species concept (currently usually so designated by the original author in the original description); see also HOLOTYPE, LECTOTYPE, SYNTYPE.

PARIVINCULAR 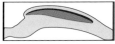 type of dorsal LIGAMENT that is OPISTHODETIC in position, cylindrical in shape, and dorsally arched (in the form of a C-spring), consisting of an elongated lamellar layer running along the HINGE, a long submarginal fibrous layer, and in some cases a ventral shelly layer (LITHODESMA), supported by NYMPHS; occurring here in Solemyidae, Astartidae, Carditidae, Condylocardiidae, Thraciidae, Lucinidae, Ungulinidae, Thyasiridae, Chamidae, Lasaeidae, Hiatellidae, Gastrochaenidae, Trapezidae, Sportellidae, Corbiculidae, Cardiidae, Veneridae, Tellinidae, Donacidae, Psammobiidae, Semelidae, Solecurtidae, Pharidae, Mactridae, and Dreissenidae; see also LITHODESMA, NYMPH.

PEARL a calcareous body of variable color (most familarly white or bluish gray) comprised of concentric layers around a central nucleus, organically produced by a living mollusk (in addition to its normal shell) and highly prized as a gem for its luster; most characteristic of Pteriidae, but also produced by other shelled mollusks (e.g., Veneridae, see p. 309).

PEARL SACK mantle epithelium (usually within the gonad) surrounding a developing cultured pearl and that secretes the nacre.

pectineum see CTENOLIUM.

pectinidium see CTENOLIUM.

PECTINIFORM STAGE early postlarval stage of Spondylidae, characterized by byssal attachment before cementation.

PEDAL GLAND a gland in the foot.

PEDAL ORGAN 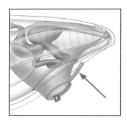 a digitiform process on the foot of Gastrochaenidae that secretes chemicals that assist in boring (figure: Gastrochaenidae).

PEDAL SCAR mark on the internal surface of a burrow or IGLOO of Gastrochaenidae, indicating the position of the foot.

PEDIVELIGER planktonic larval stage subsequent to the veliger but before metamorphosis (settling), at which point the larva has gained use of its foot, and alternates swimming in the water column and crawling on the bottom; see also VELIGER.

PEDOMORPHOSIS evolutionary process by which juvenile or even larval traits are retained in adult (reproductively mature) life stages. There are two alternative processes of pedomorphosis: PROGENESIS, or the acceleration of sexual maturity relative to somatic development, and NEOTENY, or retardation of somatic development with respect to the onset of reproduction.

PERICALYMMA  larval type characteristic of Protobranchia, characterized by a ciliated epithelial covering (calymma) of the larva, which is either discarded or ingested at metamorphosis; see also VELIGER.

PERIOSTRACUM

organic "conchiolin" outer layer of the shell, secreted by the MANTLE, present in various forms from varnishlike to shaggy, also called epidermis by earlier authors (figures: *Fugleria tenera, Eucrassatella speciosa*).

photoreceptor    see EYE.

PLANIVINCULAR

type of dorsal LIGAMENT that is OPISTHODETIC in position, cylindrical in shape, and dorsally arched (in the form of a C-spring), consisting of an elongated lamellar layer running along the HINGE, a long submarginal fibrous layer, and in some cases an innermost shelly plate (LITHODESMA) (all so far same as PARIVINCULAR) but supported only by PSEUDONYMPHS; occurring here in some Mytilidae, Pinnidae, and Poromyidae; see also LITHODESMA, PARIVINCULAR, PSEUDONYMPH.

PLANKTOTROPHIC    larval development type characterized by a free-swimming stage (VELIGER) that feeds in the plankton.

plate, sieve    see SEPTIBRANCH.

PLEUROTHETIC    lying or resting on or cemented by one valve (pertaining only to epifaunal species).

plicate    see HETERORHABDIC.

PLICATE ORGAN    vascularized longitudinal series of transverse folds between each inner demibranch and the visceral mass in Mytilidae, which serves as an accessory respiratory organ.

PORCELLANEOUS    having a glossy appearance, also called non-NACREOUS.

POSTERIOR    anal end.

POSTRECTAL MUSCLE

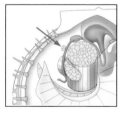

small muscle arching between the two valves dorsal to the HINDGUT in Propeamussidae (figure: Propeamussiidae).

POSTVALVULAR EXTENSION    posteriorly extended mantle, emulating siphons in appearance.

PRE-ORAL GLAND    unpaired excretory gland lying dorsal to the mouth in Pinnidae.

PRISMATIC    shell microstructural variety consisting of parallel columnar (but not strongly interdigitating) prisms of aragonitic or calcitic calcium carbonate.

PRISMATONACREOUS    with shell microstructure consisting wholly or primarily of prismatic and nacreous layers.

PROBING TUBULES — small tubes extending from the anterior end of a Gastrochaenidae burrow, possibly constructed to anticipate features of the substratum ahead of the boring process.

proboscides — see PALP PROBOSCIDES.

PRODISSOCONCH

larval shell, located at the tip of the UMBONES (equivalent to the protoconch in gastropods); in planktonic-developing bivalves, two distinct growth phases (PI and PII) are present, separated by a distinct growth line or change of sculp-ture; in a non–planktonic-developing species, no such transition is visible (figure: *Cratis antillensis*).

progenesis — see PEDOMORPHOSIS.

PROMYAL PASSAGE

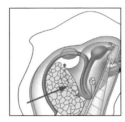

passage either (1) between the posterior ADDUCTOR MUSCLE and visceral mass (figure: Malleidae); or (2) in front of the posterior adductor muscle, lateral to the visceral mass (right side only in Ostreidae, both sides in Gryphaeidae) (these two conditions probably not homologous), which allows excurrent water to pass posteriorly with greater efficiency.

PROSODETIC — type of dorsal LIGAMENT that occurs entirely on the anterior side of the UMBONES; see also AMPHIDETIC, OPISTHODETIC.

PROSOGYRATE/PROSOGYROUS

having the UMBONES curling anteriorly; see also OPISTHOGYRATE, ORTHOGYRATE (figure: *Periglypta listeri*).

PROTANDRIC/PROTANDROUS — a form of HERMAPHRODITISM in which the male phase precedes the female phase during the life cycle of the same individual; see also HERMAPHRODITE.

PROTOBRANCH

gill type characterized by simple, broad, leaflike filaments that are unconnected or loosely connected only by CILIATED DISKS on the filament faces, have chitinous supporting rods, and are used primarily for respiration (rather than feeding); this character forms the basis of the group Protobranchia, recognized by Pelseneer (1891), Ridewood (1903), and by current classifications.

PROTOGYNOUS — a form of HERMAPHRODITISM in which the female phase precedes the male phase during the life cycle of the same individual; see also HERMAPHRODITE.

PROTOPLAX

accessory shell plate in Pholadidae anterior to MESOPLAX, both of which protect the anterior ADDUCTOR MUSCLE (figure: *Cyrtopleura costata*).

PROTOSTYLE

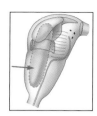

rod-shaped amorphous inclusion in the TYPE I STOMACH of members of Protobranchia that reduces the size of ingested food particles through rotating abrasion against the gastric shield (and chitinous lining if present); produced by secretory tissues in the STYLE SACK and comprised of mucus and ingested food particles; believed homologous with the CRYSTALLINE STYLE of most other mollusks, which also contains the starch-digesting enzyme AMYLASE; see also AMYLASE, CRYSTALLINE STYLE, STOMACH, STYLE SACK.

PROVINCULAR TEETH

larval or early-stage hinge teeth (the hinge plate is called the provinculum at this stage).

PROXIMAL VALVE

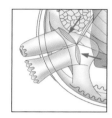

thin tissue flap at the base of a SIPHON that serves to narrow the lumen and thus control water flow into or out of the MANTLE CAVITY; present here in some Mytilidae, some Gastrochaenidae, and Veneridae; also called siphonal membrane (figure: Veneridae).

PSEUDOCTENOLIUM

CTENOLIUM-like structure formed from external sculpture on the edge of the right side of the byssal gape in Pectinidae (some) and Propeamussiidae; see also CTENOLIUM.

PSEUDOFECES

waste particles in the MANTLE CAVITY that must be ejected; usually some consolidation occurs through secretion of mucus and the action of ciliated tracts on the inner mantle surface; ejection of pseudofeces commonly happens through the INCURRENT APERTURE or SIPHON or through the FOURTH PALLIAL APERTURE (if present) by rapid contraction of the ADDUCTOR MUSCLES.

PSEUDOLAMELLIBRANCH

gill type characterized by filaments joined by a combination of ciliary and tissue junctions (the latter "occasional" or "few"); occurring here in some Pteriidae, some Isognomonidae, some Malleidae, Ostreidae, Gryphaeidae, Pinnidae, Pectinidae, Spondylidae, and some Thyasiridae; see also EULAMELLIBRANCH, ELEUTHERORHABDIC, FILIBRANCH.

PSEUDONYMPH

a ridgelike support for a more advanced type of dorsal LIGAMENT that faces the median plane of the shell (is not rotated dorsally; thus not enhancing the dorsal arching of the ligament); see also LIGAMENT, NYMPH.

pyloric process

see METASOMA.

QUADRANGULAR

having four angles, thus four sides, e.g., somewhat square or rectangular in profile (figure: *Hiatella arctica*).

QUENSTEDT MUSCLES

ctenidial (see CTENIDIUM) protractor muscles (in members of Ostreidae).

RADIAL

referring to external sculptural features that originate at the UMBONES and fan outward toward the margins (figure: *Carditamera floridana*).

rectum

end portion of the intestine; either sessile (laying appressed to the posterior adductor muscle) or freely hanging.

RESILIAL TEETH

elevated ridges on a resilifer.

RESILIFER

structure (often a flat or recessed trigonal platform) on each valve, supporting the internal LIGAMENT or RESILIUM; also called ligamental pit; see also CHONDROPHORE (figure: *Semele proficua*).

RESILIUM (pl. resilia)

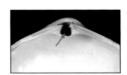

pyramidal enlargement of the LIGAMENT ventral to the HINGE line; also called internal ligament (figure: *Spisula similis*).

reticulate

see CANCELLATE.

RIB

an elongated sculptural element that is raised above the surrounding shell surface; a very fine rib is often called a riblet or thread; also called ridge; see also GROOVE, STRIA.

riblet

see RIB.

ridge

see RIB.

ROSTRUM (adj. rostrate)

elongated or drawn-out posterior extension of a shell, usually enclosing the SIPHONS (figure: *Cuspidaria rostrata*).

SCISSULATE

having sculpture running obliquely across the shell, i.e., not COMMARGINAL or RADIAL.

SCULPTURE

ornament or markings on the surface (usually of a shell) resembling that produced by a carving tool, see also COMMARGINAL, GROOVE, RADIAL, RIB, STRIA.

SEMINAL VESICLE

sperm-storage organ, here present in Pholadidae; also called vesicula seminalis.

SEPTAL VALVULA (pl. valvulae)

thick, muscular, dish- or crescent-shaped posterior extensions of the SEPTUM in Poromyidae; also called compensation sack.

SEPTIBRANCH

gill type characterized by a laterally suspended, muscular, pumping SEPTUM pierced by windowlike OSTIA containing EULAMELLIBRANCH filaments (BRANCHIAL SIEVES in Poromyidae; also called SIEVE PLATES or OSCULA) or their cilia alone (pores in Cuspidariidae); innervation patterns suggest that the septum is mostly ctenidial in origin; occurring here in the heteroconchian heterodont anomalodesmatan Poromyoidea (Poromyidae) and Cuspidarioidea (Cuspidariidae), and functioning solely in respiration; see also SYNAPTORHABDIC. The gills of Verticordiidae are considered "septibranch" by most authors although the gills themselves are separate from the septal ostia, which are perforations in the membrane.

SEPTUM

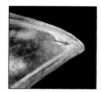

an anatomical partition, either (1) of tissue that separates or assists the gills in separating INFRA- and SUPRABRANCHIAL (or -SEPTAL) CHAMBERS (i.e., body chambers below and above the gills or branchial septum, respectively; figure: Poromyidae); (2) of shell to which pedal retractor muscles attach (see also MYOPHORE; figure: *Mytilopsis sallei*), or (3) of shell separating the shell and siphonal chambers of a burrow or IGLOO of Gastrochaenidae.

shield                         see UMBONAL REFLECTION.

sieve plate                    see SEPTIBRANCH.

SIMPLE LIGAMENT                 type of dorsal LIGAMENT that is arched or planar, external or submarginal, and set on GUTTERS or FOSSETTES; occurring here in Nuculidae, Nuculanidae, Manzanellidae, Yoldiidae, some Arcidae, and some Limopsidae; see also GUTTER, FOSSETTE.

simple pallial line            see ENTIRE.

SINUATE                        of a PALLIAL LINE when a SINUS is present.

sinulapalliate                 see PALLIAL SINUS.

sinus                          see PALLIAL SINUS.

SIPHONAL SEPTUM

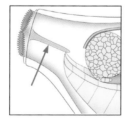

membrane at the base of the INCURRENT/EXCURRENT SIPHONS/APERTURES separating the INFRABRANCHIAL CHAMBER from the SUPRABRANCHIAL CHAMBER; also called branchial membrane/septum (figure: Pandoridae).

SIPHONAL TUBE                  tube composed of feces and/or pseudofecal particles cemented with mucus, which is fused to the SIPHONOPLAX in Pholadidae.

SIPHONOPLAX

accessory shell plate in Pholadidae protecting the SIPHONS (figure: *Martesia striata*).

SIPHONS 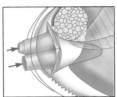 posterior extensions (usually two) of the MANTLE, made tubular by either tissue fusion or (less often) ciliary junctions of the mantle folds, through which water is directed in and out of the body, along with waste products and gametes (figure: *Mercenaria mercenaria*). When the siphons are formed by permanent tissue junctions, a classification proposed by Yonge (1957) is often used to reflect the mantle folds involved in their formation:

TYPE A – fusion of inner folds only (e.g., Tellinidae).

TYPE B – fusion of inner and middle (inner surface only or entire) folds (e.g., Veneridae).

TYPE C – fusion of inner, middle, and outer folds (e.g., Myidae); siphons of this type are at least partly encased in periostracum.

smooth     see HOMORHABDIC.

STATOCONIA     multiple small bodies within a STATOCYST; also called otoconia; see STATOCYST.

STATOCYST     fluid-filled, capsulelike sense organ, open or closed, usually paired, and located near the pedal ganglia (ventral to the posterior adductor muscle) but innervated by the cerebral ganglia, and usually including ciliated "hair" cells and containing a single dense body (STATOLITH) or a number of smaller ones (STATOCONIA); the statolith and/or statoconia interact with the cilia lining the capsule, probably (as has been shown in gastropods and cephalopods) conveying information about orientation to the organism; in INEQUIVALVE bivalves (e.g., *Pecten*) the statocysts are also unequal, with one more complex or larger than the other; usually present in larval bivalves, but often absent in adults; also called otocyst; see also STATOLITH. Morton (1985) defined several types of statocysts in Anomalodesmata (bivalves in other suprafamilial groups have not been so categorized):

TYPE A – capsule penetrated by nerve endings and containing a multicellular statolith (not present in families herein).

TYPE B1 – capsule with ciliated cells as sensory receptors, and containing a single large statolith; occurring here in Pandoridae, Lyonsiidae, Thraciidae, some Verticordiidae, and Poromyidae.

TYPE B2 – capsule with ciliated cells as sensory receptors, and containing both a single large statolith and statoconia; occurring here in Periplomatidae and some Verticordiidae.

TYPE B3 – capsule with ciliated cells as sensory receptors, and containing numerous crystal-like statoconia, one of which can be enlarged into a statolith; occurring here in some Verticordiidae.

TYPE C – small capsule comprised of few cells lined by microvilli, and containing a large oval statolith that does not move freely within the capsule; occurring here in Cuspidariidae.

STATOLITH     single large body within a STATOCYST; also called otolith; see STATOCYST.

STEMPELL'S ORGAN     specialized mechanoreceptor in Nuculidae on the anterior ADDUCTOR MUSCLE that detects its level of contraction.

STOMACH

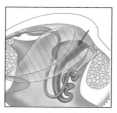

major organ for processing food in bivalves, embedded within in the visceral mass, connecting the esophagus and MIDGUT, and featuring ciliated sorting areas, grooves, and other surfaces, openings to ducts to the digestive diverticula, and in most cases a rotating rodlike structure (see also CRYSTALLINE STYLE, PROTOSTYLE) that abrades against the surface of the gastric shield (and chitinous lining if present) to macerate food particles (figure: *Mercenaria mercenaria*). R. D. Purchon, in a series of seminal papers (1956, 1957, 1958, 1959, 1960, 1963, 1985, 1987; modified by Dinamani, 1967), categorized bivalve stomachs into various morphological types that are still well recognized, and have been used in the past to define various taxonomic units.

TYPE I –

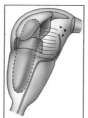

stomach type originally described by Purchon (1956) comprising an elongated bipartite chamber; rounded dorsal region including few (two or three) ducts to the digestive diverticula (a former subclass called Oligosyringia by Purchon, 1963), sorting area on right side, cuticularized region (chitinous girdle) on left side, gastric shield, and dorsal hood; elongated ventral region or STYLE SACK, producing an amorphous PROTOSTYLE, with major typhlosole (and accompanying intestinal groove) either not or only simply entering the dorsal chamber, and tapering to a ventral point from which the MIDGUT emerges; characteristic of the DEPOSIT-FEEDING Protobranchia, here including the Nuculoida (Nuculidae), Nuculanoida (Nuculanidae and Yoldiidae), and in simplified form Solemyoida (Solemyidae and Manzanellidae). Purchon (1959) considered this type of stomach characteristic of a group he called Gastroproteia; Villarroel & Stuardo (1998) further divided it into three subcategories according to details of the sorting areas and typhlosoles.

TYPE II –

greatly simplified stomach type originally described by Purchon (1956) consisting of a large muscular sack, featuring a small to very large esophageal opening, small to absent sorting area, few (two or three) openings (without ducts) to the digestive diverticula (see previous, Oligosyringia), a reduced STYLE SACK (combined or separate from the MIDGUT) producing a short CRYSTALLINE STYLE, with or without a gastric shield and dorsal hood, and with an extensive protective chitinous (scleroproteinaceous) lining; the stomach can be separated from adjacent viscera by a hemocoel, allowing more freedom for crushing action; modified for a carnivorous or scavenging diet, characteristic of the carnivorous "Septibranchia," here including the Poromyoidea (Poromyidae), Cuspidarioidea (Cuspidariidae), and Verticordioidea (Verticordiidae), and of the carnivorous Propeamussiidae. Purchon (1959) considered this type of stomach characteristic of a group he called Gastrodeuteia.

TYPE III –  stomach type originally described by Purchon (1957) comprising a more or less oval dorsal chamber and elongated ventral combined STYLE SACK/MIDGUT, the former chamber featuring a major typhlosole (and accompanying intestinal groove) with a long, slender tongue extending into the food sorting CAECUM, numerous scattered ducts to the digestive diverticula (a former subclass called Polysyringia by Purchon [1963]), a CRYSTALLINE STYLE, and a gastric shield; some species with a ciliated groove linking the apex of the tongue of the major typhlosole with the apex of the dorsal hood; characteristic of filter feeders, here including the pteriomorphian Arcoida (Arcidae, Noetiidae, Glycymerididae, Limopsidae, and Philobryidae), Mytiloida (Mytilidae), and Pterioida (Pteriidae, Isognomonidae, Malleidae, Ostreidae, Gryphaeidae, and Pinnidae). Purchon (1963, 1987) considered this type of stomach characteristic of a group he called Gastrotriteia.

TYPE IV –  stomach type originally described by Purchon (1957, 1958) comprising a more or less oval dorsal chamber and elongated ventral combined or separate STYLE SACK/MIDGUT, the former chamber featuring a major typhlosole (and accompanying intestinal groove) passing across the floor of the stomach (without extension or deviation), and terminating close to the left pouch, with clustered numerous ducts to the digestive diverticula (see previous, Polysyringia; usually in the left pouch [which serves to further anchor the gastric shield], between the pouch and the major typhlosole, and on the right side of the stomach), a CRYSTALLINE STYLE, and a gastric shield; characteristic of SUSPENSION FEEDERS, here including the pteriomorphian Limoida (Limidae) and Pectinoidea (Pectinidae, Spondylidae, Plicatulidae, and Anomiidae), and the heteroconchian heterodont Carditoida (Crassatellidae, Astartidae, Carditidae, and Condylocardiidae), Veneroida (some Psammobiidae), Anomalodesmata (Pandoridae, Lyonsiidae, Periplomatidae, and Thraciidae), and some Veneroida (Lucinidae, Thyasiridae, some Chamidae, Lasaeidae, Hiatellidae, Gastrochaenidae, some Veneridae, and some Donacidae). Purchon (1959, 1987) considered this type of stomach characteristic of a group he called Gastrotetartika, and ancestral to TYPES III and V.

TYPE V – 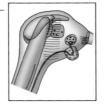 stomach type originally described by Purchon (1960) comprising a more or less oval dorsal chamber and elongated ventral combined or separate STYLE SACK/MIDGUT, the former chamber featuring numerous ducts to the digestive diverticula (see also Polysyringia) concentrated in the left pouch, right and left CAECA, and on the right side of the stomach, a major typhlosole (and accompanying intestinal groove) that is "7-shaped," entering into the right caecum then into the left caecum, a CRYSTALLINE STYLE, and a gastric shield; characteristic of filter feeders, here including most heteroconchian heterodont Veneroida (Ungulinidae, some Chamidae, Trapeziidae, Corbiculidae, Cardiidae, most Veneridae, Tellinidae, some Donacidae, Psammobiidae, Semelidae, Solecurtidae, Pharidae, Mactridae, Dreissenidae, Myidae, Corbulidae, Pholadidae, and Teredinidae). Purchon (1959) considered this type of stomach characteristic of a group he called Gastropempta.

| | |
|---|---|
| STRIA (pl. striae) | general term for a narrow linear furrow or raised sculptural element on a shell surface; also called lines; see also RIB, GROOVE. |
| style | see CRYSTALLINE STYLE, PROTOSTYLE, STYLE SACK. |

STYLE SACK

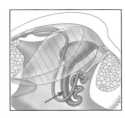

ventral or posterior extension of the STOMACH of many mollusks, including most bivalves, which secretes a rod-shaped inclusion (see also CRYSTALLINE STYLE, PROTOSTYLE) that reduces the size of ingested food particles through rotating abrasion against the gastric shield and, in most mollusks, through the digestive actions of the included enzyme AMYLASE (figure: *Mercenaria mercenaria*).

| | |
|---|---|
| SUBEQUAL | nearly equal. |
| SUBEQUILATERAL | nearly EQUILATERAL. |
| SUBEQUIVALVE | nearly EQUIVALVE. |
| SUBLITTORAL | that part of the coastline immediately below the low-water mark, and which is never exposed. |
| SUBMARGINAL LIGAMENT | type of dorsal LIGAMENT that attaches immediately below the dorsal shell margin on the surface between the two valves. |
| SULCUS | a large fold or groove. |

SUPRABRANCHIAL CHAMBER

body cavity dorsal to the gills; see also INFRABRANCHIAL CHAMBER (figure: *Mercenaria mercenaria*).

SUPRAMYAL SEPTUM

membrane connecting the right and left MANTLE lobes on the dorsal side of the posterior ADDUCTOR MUSCLE; its absence in Malleidae allows the PROMYAL PASSAGE to shunt excurrent water posterodorsally.

SUPRASEPTAL CHAMBER

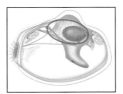

body cavity dorsal to the SEPTUM, analogous to SUPRABRANCHIAL CHAMBER in bivalves with nonseptibranch gills; see also INFRASEPTAL CHAMBER (figure: Poromyidae).

SUSPENSION FEEDING    feeding type of most bivalves during which organic particles are harvested from the water column.

SYNAPOMORPHY    a shared, derived, taxon-defining trait or characteristic.

SYNAPTORHABDIC    gill type characterized by filaments that are joined together only by tissue junctions, so that the interfilamental spaces are divided into a series of OS-TIA or fenestrae; see also EULAMELLIBRANCH, SEPTIBRANCH. Ridewood (1903) defined a group Synaptorhabda based on this character, equivalent to Pelseneer's (1891) Eulamellibranchia plus Septibranchia.

SYNTYPE    one of two or more specimens examined by the original author of a species in the original description, but none of which was uniquely designated as the holotype; the entire lot is called a syntypic series; see also HOLOTYPE, LECTOTYPE, PARATYPE.

TAENIOID MUSCLES    pallial retractor muscles in the fused ventral MANTLE margin of Verticordi-idae (also Pholadomyidae and Parilimyida) that serve to retract the elabo-rate incurrent apparatus during prey capture.

TAXODONT    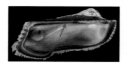 HINGE type characterized by numerous undifferenti-ated similarly sized (usually small) vertical or oblique hinge teeth along the hinge plate, occurring here in Nuculidae, Manzanellidae, Nuculanidae, Yoldiidae, Arcidae, Noetiidae, Glycymerididae, Limopsidae, and Philobryidae (figure: *Arca imbricata*).

teeth    see HINGE TEETH, MARGINAL DENTICLES.

thread    see RIB.

TRIMYARIAN    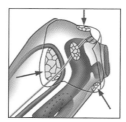 having three ADDUCTOR MUSCLES, occurring here in Pholadidae and Teredinidae (figure: Teredinidae).

TRUNCATE     abruptly cut off or appearing so, usually in reference to a squared-off end of a bivalve shell (figure: *Spengle-ria rostrata*).

TUBULE    in the shell, one of many pores extending from the inner surface through all or some shell layers, containing extensions (CAECA) of the MANTLE that prob-ably secrete a chemical to deter boring organisms (although several other functions have been proposed, and none has been supported conclusively).

TUMID    swollen or bulging.

typhlosole    a ridge, commonly so described in molluscan digestive systems.

UMBO/UMBONE(S)

rounded or pointed extremity of a bivalve shell, often projecting above the HINGE line, that reflects the early growth stage (= oldest part of the shell) and includes the PRODISSOCONCH and adjacent convex area; also called beak (figure: *Dosinia discus*).

UMBONAL CAVITY

space inside the valve within the UMBO and under the HINGE PLATE.

UMBONAL CRACK

crack in the shell of Periplomatidae that occurs naturally at the umbone of each valve (figure: *Periploma margaritaceum*).

UMBONAL REFLECTION

anterodorsal valve margin that is turned up and over the UMBO; also called flange or shield; occurring here in Pholadidae and Teredinidae (figure: *Pholas campechiensis*).

UMBONAL RIDGE

prominent angled or rounded feature on the external shell surface that begins at the UMBO and runs obliquely posteriorly to the ventral border.

UNDULATE

smoothly wavy, usually referring to a shell or MANTLE margin.

VALVE

one half of a bivalve's two shell halves.

valvula, septal

see SEPTAL VALVULA.

VALVULAR MEMBRANE

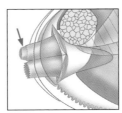

cone-shaped extension on the EXCURRENT SIPHON; occurring here in Hiatellidae, Trapezidae, Veneridae, Pharidae, Mactridae, Myidae, and Corbulidae (figure: *Mercenaria mercenaria*).

VELIGER

planktonic larval type characteristic of most Mollusca, including bivalves, characterized by a ciliated locomotory organ (VELUM) that is either discarded or resorbed at metamorphosis; see also PERICALYMMA.

VELUM

(1) see PALLIAL VEIL; or (2) the locomotory organ of a VELIGER larva.

VENTRAL

the foot-side of a bivalve, opposite of DORSAL.

VENTRICOSE

swollen, usually in the middle after the manner of an animal's belly.

vermicular chomata

see CHOMATA.

vesicula seminalis     see SEMINAL VESICLE.

VISCERAL LOBE     lateral extension of the VISCERAL MASS in Lucinidae, containing part of the reproductive organs.

VISCERAL MASS

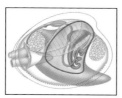

region of the bivalve body containing most of the digestive, excretory, circulatory, and nervous systems that is suspended dorsally between the gills by pedal retractor muscles and that usually terminates ventrally as the foot (figure: *Mercenaria mercenaria*).

VISCERAL RIM

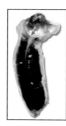

thickened margin of the small internal NACREOUS layer in most Malleidae (and at least one species of Isognomonidae) beyond which the MANTLE and gills can be withdrawn (figure: *Malleus candeanus*).

WAMPUM     beads, usually elongated rectangular in shape, carved by eastern native Americans from the purple-and-white shells of the venerid *Mercenaria mercenaria* (also from gastropod shells of the genus *Busycon*), and used to produce belts, the design of which recorded significant agreements or events (with a higher proportion of purple beads if the agreement was considered very serious or important); it was used as formal currency after contact with colonizing Europeans.

ZOOXANTHELLAE     unicellular algal cells that live symbiotically within the cells of a larger organism (i.e., a bivalve).

# General Literature Cited and Suggested Reading

(see also individual family references at end of each family description)

Abbott, R. T. 1968. *Seashells of North America*. Golden Press, New York, 280 pp.

Abbott, R. T. 1974. *American Seashells: the Marine Mollusca of the Atlantic and Pacific Coasts of North America. Second edition*. Van Nostrand Reinhold, New York: 663 pp., 24 pls.

Abbott, R. T., and P. A. Morris. 1994. *A Field Guide to Shells—Atlantic and Gulf Coasts and the West Indies. Fourth Edition*. The Peterson Field Guide Series. Houghton Mifflin Company, Boston, xxxiii + 350 pp., 74 pls.

Adamkewicz, S. L., M. G. Harasewych, J. Blake, D. Saudek, and C. J. Bult. 1997. A molecular phylogeny of the bivalve mollusks. *Molecular Biology and Evolution*, 14(6): 619–629.

Andrews, J. 1994. *A Field Guide to Shells of the Florida Coast*. Nature Field Guide Series. Gulf Publishing Company, Houston, Texas, xxiii + 182 pp.

Barber, V. C. 1968. The structure of mollusc statocysts, with particular reference to cephalopods. Pages 37–62, in: J. D. Carthy and G. E. Newell, eds., *Invertebrate Receptors. Symposia of the Zoological Society of London*, 23. Academic Press, New York.

Beauperthuy, I. 1967. Los mitilidos de Venezuela (Mollusca: Bivalvia). *Boletin del Instituto Oceanografico de la Universidad de Oriente, Cumana*, 6(1): 7–115.

Bernard, F. 1895. Première note sur le développement et la morphologie de la coquille chez les lamellibranches. *Bulletin de la Société Géologique de France, Série 3*, 23: 104–154.

Bernard, F. R. 1972. Occurrence and function of lip hypertrophy in the Anisomyaria (Mollusca, Bivalvia). *Canadian Journal of Zoology*, 50: 53–57.

Bieler R., and P. M. Mikkelsen. 2003. The cruises of the *Eolis*—John B. Henderson's mollusc collections off the Florida Keys, 1910–1916. *American Malacological Bulletin*, 17(1–2) "2002": 125–140.

Bieler, R., and P. M. Mikkelsen, eds. 2004a. Bivalve studies in the Florida Keys: Proceedings of the International Marine Bivalve Workshop, Long Key, Florida, July 2002. *Malacologia*, 46(2): 241–677.

Bieler, R., and P. M. Mikkelsen, 2004b. Marine bivalves of the Florida Keys: a qualitative faunal analysis based on original collections, museum holdings and literature data. In: R. Bieler and P. M. Mikkelsen, eds., *Bivalve Studies in the Florida Keys*, Proceedings of the International Marine Bivalve Workshop, Long Key, Florida, July 2002. *Malacologia*, 46(2): 503–544.

Bieler, R., and P. M. Mikkelsen. 2006. Bivalvia—a look at the branches. *Zoological Journal of the Linnean Society*, 148:223–235.

Bouchet, P., and J.-P. Rocroi. 2005. Classification and nomenclator of gastropod families. *Malacologia*, 47(1–2): 1–397.

Calkins, W. W. 1878. Catalogue of the marine shells of Florida, with notes and descriptions of several new species. *Proceedings of the Davenport Academy of Natural Sciences*, 2: 232–252, pl. 8.

Campbell, D. C. 2000. Molecular evidence on the evolution of the Bivalvia. Pages 31–46, in: E. M. Harper, J. D. Taylor, and J. A. Crame, eds., *The Evolutionary Biology of the Bivalvia. Geological Society of London, Special Publication*, 177.

Campbell, D. C., K. J. Hoekstra, and J. G. Carter, 1998. 18S ribosomal DNA and evolutionary relationships within the Bivalvia. Pages 75–85, in: P. A. Johnston and J. W. Haggart, eds., *Bivalves: An Eon of Evolution*. University of Calgary Press, Calgary, Alberta Canada.

Carter, J. G. 1980. Environmental and biological controls of bivalve shell mineralogy and microstructure. Pages 69–113, in: D. C. Rhoads and R. A. Lutz, eds., *Skeletal Growth of Aquatic Organisms*. Plenum Publishing Corporation, New York.

Carter, J. G. 1990. Evolutionary significance of shell microstructure in the Palaeotaxodonta, Pteriomorphia and Isofilibranchia (Bivalvia: Mollusca). Pages 135–296, in: J. G. Carter, ed., *Skeletal

*Biomineralization: Patterns, Processes and Evolutionary Trends, volume 1*. Van Nostrand Reinhold, New York.

Carter, J. G., D. C. Campbell, and M. R. Campbell. 2000. Cladistic perspectives on early bivalve evolution. Pages 47–79, in: E. M. Harper, J. D. Taylor, and J. A. Crame, eds., *The Evolutionary Biology of the Bivalvia. Geological Society of London, Special Publication*, 177.

Charles, G. H. 1966. Sense organs (less cephalopods). Pages 455–522, in: K. M. Wilbur and C. M. Yonge, eds., *Physiology of Mollusca, volume 2*. Academic Press, New York.

Clench, W. J., and R. D. Turner. 1950. The western Atlantic marine mollusks described by C. B. Adams. *Occasional Papers on Mollusks, Harvard University*, 1(15): 233–403.

Coan, E. V., P. Valentich-Scott, and F. R. Bernard. 2000. *Bivalve Seashells of Western North America: Marine Bivalve Mollusks from Arctic Alaska to Baja California*. Santa Barbara Museum of Natural History, Santa Barbara, California, viii + 764 pp.

Conrad, T. A. 1846. Catalogue of shells inhabiting Tampa Bay and other parts of the Florida coast. *[Silliman's] American Journal of Science and Arts, Series 2*, 2(6): 393–398.

Cragg, S. M., and J. A. Nott. 1977. The ultrastructure of the statocysts in the pediveliger larvae of *Pecten maximus* (L.) (Bivalvia). *Journal of Experimental Marine Biology and Ecology*, 27: 23–36. [Review of statocysts in bivalve pediveligers.]

Dall, W. H. 1883. On a collection of shells sent from Florida by Mr. Henry Hemphill. *Proceedings of the United States National Museum*, 6(21): 318–342, pl. 10.

Dall, W. H. 1886. Report on the results of dredging, under the supervision of Alexander Agassiz, in the Gulf of Mexico (1877–78) and in the Caribbean Sea (1879–80), by the U.S. Coast Survey Steamer "*Blake*," Lieut.-Commander C. D. Sigsbee, U.S.N., and Commander J. R. Bartlett, U.S.N., commanding. XXIX. Report on the Mollusca. Part I. Brachiopoda and Pelecypoda. *Bulletin of the Museum of Comparative Zoology at Harvard College*, 12: 171–318, pls. 1–9.

Dall, W. H. 1889a. A preliminary catalogue of the shell-bearing marine mollusks and brachiopods of the southeastern coast of the United States, with illustrations of many of the species. *Bulletin of the United States National Museum*, 37: 1–121, 74 pls.

Dall, W. H. 1889b. Report on the results of dredging, under the supervision of Alexander Agassiz, in the Gulf of Mexico (1877–78) and in the Caribbean Sea (1879–80), by the U.S. Coast Survey Steamer "*Blake*," Lieut.-Commander C. D. Sigsbee, U.S.N., and Commander J. R. Bartlett, U.S.N., commanding. XXIX. Report on the Mollusca. Part II. Gastropoda and Scaphopoda [with "Addenda and Corrigenda to Part I, 1886," pp. 433–452]. *Bulletin of the Museum of Comparative Zoology*, 18: 1–492, pls. 10–40.

Dall, W. H. 1903. A preliminary catalogue of the shell-bearing marine mollusks and brachiopods of the southeastern coast of the United States, with illustrations of many of the species. Reprint to which are added twenty-one plates not in the edition of 1889. *Bulletin of the United States National Museum*, 37: 1–232, 95 pls.

Diaz Merlano, J. M., and M. Puyana Hegedus. 1994. *Moluscos del Caribe Colombiano: Un Catálogo Ilustrado*. Colciencias y Fundación Natura Colombia, Santafé de Bogotá, Columbia, 291 pp., 74 pls.

Dinamani, P. 1967. Variation in the stomach structure of the Bivalvia. *Malacologia*, 5(2): 225–268.

d'Orbigny, A. 1834–1847. *Voyage dans l'Amérique Méridionale [. . .] Exécuté Pendant les Années 1826, [. . .] 1833. Volume 5(3), Mollusques*, pp. 1–48, 73–128, pls. 1–2, 9–13, 15–16, 56 [1834]; pp. 49–72, 129–176, pls. 3–8, 17–23, 25, 55 [1835]; pp. 177–184, pls. 14, 24, 26–28, 30–32, 34, 35, 37, 58 [1836]; pls. 33–36 [1836?]; pp. 185–376, pls. 29, 38–52, 57 [1837]; pls. 54, 59–66, 68, 69 [1839]; pp. 377–424, pls. 53, 67, 70, 71 [1840]; pp. 425–488, pls. 72–77, 80 [1841]; pls. 83, 85 [1842]; pl. 84 [1842?]; pp. 529–600 [1845]; pp. 489–528, 601–728 [1846]; pp. 729–758 [1847?]; pls. 78, 79, 81, 82 [1847]; Paris Bertrand, Paris & Levrault, Strasbourg [dates teste C. D. Sherborn & F. J. Griffin. 1934. On the dates of publication of the natural history portions of Alcide d'Orbigny's 'Voyage Amérique Méridionale.' *Annals and Magazine of Natural History, Series 10*, 13(73): 130–134].

d'Orbigny, A. 1842–1853. Mollusques. In: R. de la Sagra, *Histoire Physique, Politique, et Naturelle de l'Ile de Cuba. Volume 2*: pp. 1–112, pls. 10–21? [1842]; pp. 113–128 [1844]; pp. 129–224, pls.

22–25? [1848]; pp. [iv] + 225–380, pls. 26–28? [1853]; Bertrand, Paris [dates teste G. Rosenberg, pers. comm., May 2006].

Dunker, W. 1849. Diagnoses molluscorum novorum. *Zeitschrift für Malakozoologie*, 5(12): 177–186.

Emerson, W. K., and M. K. Jacobson. 1976. *The American Museum of Natural History Guide to Shells; Land, Freshwater, and Marine, From Nova Scotia to Florida.* Alfred A. Knopf, New York, 482 + xviii pp., 47 pls.

Espinosa, J., and A. Juarrero. 1989. Moluscos bivalves del litoral rocoso de Ciudad Habana. *Revista de Investigaciones Marinas*, 10(2): 125–132.

Espinosa, J., J. Ortea, and A. Valdés. 1994. Catálogo de los moluscos bivalvos recientes del Archipiélago cubano. *Avicennia*, 2: 109–129.

Garcia, E. F. 1979. Belize and its molluscan fauna. *Of Sea & Shore*, 10(3): 184–186.

Giribet, G., and D. L. Distel. 2003. Bivalve phylogeny and molecular data. Pages 45–90, in: C. Lydeard and D. R. Lindberg, eds., *Molecular Systematics and Phylogeography of Mollusks*. Smithsonian Books, Washington, D.C.

Giribet, G., and W. Wheeler. 2002. On bivalve phylogeny: a high-level analysis of the Bivalvia (Mollusca) based on combined morphology and DNA sequence data. *Invertebrate Biology*, 121(4): 271–324.

Gray, M. E. S. 1857. *Figures of Molluscous Animals Selected from Various Authors. Etched for the Use of Students by Maria-Emma Gray, Volume 5, Conchifera and Brachiopoda*. Longman & Co., London, 49 pp., pls. 313–381.

Harper, E. M., Taylor, J. D., and J. A. Crame, eds. 2000. *The Evolutionary Biology of the Bivalvia*. Geological Society of London Special Publication 177, vii + 494 pp.

Haszprunar, G. 1983. Comparative analysis of the abdominal sense organs of Pteriomorpha (Bivalvia). *Journal of Molluscan Studies, Supplement*, 12A: 47–50.

Healy, J. M. 1995. Comparative spermatozoal ultrastructure and its taxonomic and phylogenetic significance in the bivalve order Veneroidea. *Mémoires du Museum National d'Histoire Naturelle*, 166: 155–166.

Healy, J. M. 1996. Molluscan sperm ultrastructure: correlation with taxonomic units within the Gastropoda, Cephalopoda and Bivalvia. Pages 99–113, in: J. Taylor, ed., *Origin and Evolutionary Radiation of the Mollusca*. Oxford University Press, Oxford, United Kingdom.

Humfrey, M. 1975. *Sea Shells of the West Indies: A Guide to the Marine Molluscs of the Caribbean*. Taplinger, New York, 351 pp., 32 pls.

ICZN (International Commission on Zoological Nomenclature). 1999. *International Code of Zoological Nomenclature. Fourth edition*. International Trust for Zoological Nomenclature, London. xxix + pp. 1–306.

Johnston, P. A., and J. W. Haggart, eds. 1998. *Bivalves: An Eon of Evolution*. University of Calgary Press, Calgary, Alberta, Canada, xiv + 461 pp.

Jones, C. C. 1979. Anatomy of *Chione cancellata* and some other chionines (Bivalvia: Veneridae). *Malacologia*, 19(1): 157–199.

Kellogg, J. L. 1892. A contribution to our knowledge of the morphology of lamellibranchiate mollusks. *Bulletin of the United States Fish Commission*, 10: 389–450.

Kellogg, J. L. 1903. Feeding habits and growth of *Venus mercenaria*. *Bulletin of the New York State Museum*, 71(10): 3–28.

Kellogg, J. L. 1915. Ciliary mechanisms of lamellibranches with description of anatomy. *Journal of Morphology*, 26: 625–701.

Kissling, D. L. 1977. [Partial list of organisms . . . from examination of patch reefs south of Boca Chica, Newfound Harbor Keys and at Mosquito Banks]. Pages 181–182, in: H. G. Multer, ed., *Field Guide to Some Carbonate Rock Environments—Florida Keys and Western Bahamas. New Edition*. Kendall/Hunt Publishing Company, Dubuque, Iowa, 415 pp. + 10 maps.

Kleemann, K. H. 1983. Catalogue of Recent and fossil *Lithophaga* (Bivalvia). *Journal of Molluscan Studies, Supplement*, 12: 1–46.

Kleemann, K. H. 1984. *Lithophaga* (Bivalvia) from dead coral from the Great Barrier Reef, Australia. *Journal of Molluscan Studies*, 50(3): 192–230.

Kraeuter J. N., and M. Castagna, eds. 2001. *Biology of the Hard Clam*. Elsevier Science, Amsterdam, The Netherlands, xix + 751 pp.

Lang, A. 1900. *Lehrbuch der vergleichenden Anatomie der Wirbellosen Thiere. 1. Mollusca*. G. Fischer, Jena, Germany, viii + 509 pp.

Lee, T., and D. Ó Foighil. 2004. Hidden Floridian biodiversity: mitochondrial and nuclear gene trees reveal four cryptic species within the scorched mussel, *Brachidontes exustus*, species complex. *Molecular Ecology*, 13: 3527–3542.

Lermond, N. W. 1936. *Check List of Florida Marine Shells*. Privately published, Gulfport, Florida, 56 pp.

Lyons, W. G., and J. F. Quinn, Jr. 1995. Appendix J. Marine and terrestrial species and algae: phylum Mollusca. Pages J-10–J-26, in: *Florida Keys National Marine Sanctuary Draft Management Plan/Environmental Impact Statement, Volume III. Appendices*. National Oceanographic and Atmospheric Administration, Silver Spring, Maryland.

Macsotay, O., and R. Campos Villarroel. 2001. *Moluscos Representativos de la Plataforma de Margarita—Venezuela—Descripcion de 24 Especies Nuevas*. Editora Rivolta, Valencia, Venezuela, iii + 280 pp., incl. 32 pls.

Malchus, N. 2004. Constraints in the ligament ontogeny and evolution of pteriomorphian Bivalvia. *Palaeontology*, 47(6): 1539–1574. [Definitions of prodissoconch and ligament types.]

Matthews, H. R., and E. de C. Rios. 1967. Segunda contribuicao ao inventario dos moluscos marinhos do nordeste Brasileiro. *Arquivos da Estação de Biologia Marinha da Universidade Federal do Ceará*, 7(2): 113–121.

Mazÿck, W. G. 1913. Catalog of Mollusca of South Carolina. *Contributions from the Charleston Museum*, 2: xvi + 39 pp.

McCloskey, L. R. 1970. The dynamics of the community associated with a marine scleractinian coral. *Internationale Revue der Gesamten Hydrobiologie*, 55(1): 13–81.

Melvill, J. C. 1880. List of Mollusca obtained in South Carolina and Florida (principally at the island of Key West in 1871–1872[)]. *Journal of Conchology*, 3: 155–173.

Metivier, B. 1967. Résultats scientifiques des campagnes de *La Calypso*. Fasc. VIII. XXXI. Campagne au large des côtes atlantiques de l'Amérique du Sud (1961–1962). Première Partie (suite). Art. 8. Mollusques Lamellibranches: Mytilidae. *Annales de l'Institut Océanographique, Nouvelle Série*, 45(2): 177–181.

Mikkelsen, P. M., and R. Bieler. 2000. Marine bivalves of the Florida Keys: discovered biodiversity. Pages 207–225, in: E. M. Harper, J. D. Taylor, and J. A. Crame, eds., *The Evolutionary Biology of the Bivalvia*. Geological Society of London Special Publication 177.

Mikkelsen, P. M., and R. Bieler, 2004. Critical catalog and annotated bibliography of marine bivalve records for the Florida Keys. In: R. Bieler and P. M. Mikkelsen, eds., *Bivalve Studies in the Florida Keys*, Proceedings of the International Marine Bivalve Workshop, Long Key, Florida, July 2002. *Malacologia*, 46(2): 545–623.

Moore, D. R. 1961. The marine and brackish water Mollusca of the state of Mississippi. *Gulf Research Reports*, 1(1): 1–58.

Morse, E. S. 1919. Observations on living lamellibranches of New England. *Proceedings of the Boston Society of Natural History*, 35(5): 139–196.

Morton, B. 1977. The hypobranchial gland in the Bivalvia. *Canadian Journal of Zoology*, 55(8): 1225–1234.

Morton, B. 1979. A comparison of lip structure and function correlated with other aspects of the functional morphology of *Lima lima*, *Limaria* (*Platilimaria*) *fragilis*, and *Limaria* (*Platilimaria*) *hongkongensis* sp. nov. (Bivalvia: Limacea). *Canadian Journal of Zoology*, 57(4): 728–742.

Morton, B. 1985. Statocyst structure in the Anomalodesmata (Bivalvia). *Journal of Zoology, Series A*, 206: 23–34.

Nowell-Usticke, G. W. 1959. *A Check List of the Marine Shells of St. Croix, U.S. Virgin Islands, with Random Notations*. Lane Press, Burlington, Vermont, 90 pp.

Odé, H. 1979. Distribution and records of the marine Mollusca in the northwest Gulf of Mexico (a continuing monograph) [Mytilidae, Pteriidae, Isognomonidae, Malleidae]. *Texas Conchologist*, 15(3): 69–80.

Pelseneer, P. 1891. Contribution à l'étude des lamellibranches. *Archives de Biologie*, 11: 147–312.

Pelseneer, P. 1911. *Les Lamellibranches de l'Expédition du* Siboga. *Partie Anatomique*. Siboga-Expeditie, Uitkomsten op Zoologisch, Botanisch, Oceanographisch en Geologisch Gebied, Verzameld in de Oost-Indië 1899–1900, aan Boord H. M. Siboga *Onder Commando van Luitenant ter Zee 1e Kl. G. F. Tydeman* (M. Weber, ed.), E. J. Brill, Leiden, The Netherlands, 125 + [ii] pp., 26 pls.

Philippi, R. A. 1846. *Abbildungen und Beschreibungen Neuer oder Wenig Bekannter Conchylien, 2:* Modiola. Theodor Fischer, Kassel, Germany, pp. 147–150, pl. 1.

Philippi, A. R. 1847. Testaceorum novorum centuria [continued]. *Zeitschrift für Malakozoologie*, 4(8): 113–127.

Pointier, J.-P. and D. Lamy. 1998. *Guide des Coquillages des Antilles*. PLB Editions, Guadeloupe, 225 pp.

Porter, H. J. 1975. Record sizes of North Carolina Mollusks. List 4. *North Carolina Shell Club Notes*, 8: 38–44.

Purchon, R. D. 1956. The stomach in the Protobranchia and Septibranchia (Lamellibranchia). *Proceedings of the Zoological Society of London*, 127: 511–525. [Types I and II.]

Purchon, R. D. 1957. The stomach in the Filibranchia and Pseudolamellibranchia. *Proceedings of the Zoological Society of London*, 129: 27–60.

Purchon, R. D. 1958. The stomach in the Eulamellibranchia; stomach type IV. *Proceedings of the Zoological Society of London*, 131: 487–525.

Purchon, R. D. 1959. Phylogenetic classification of the Lamellibranchia, with special reference to the Protobranchia. *Proceedings of the Malacological Society of London*, 33: 224–230.

Purchon, R. D. 1960. The stomach in the Eulamellibranchia; stomach types IV and V. *Proceedings of the Zoological Society of London*, 135: 431–489.

Purchon, R. D. 1963. Phylogenetic classification of the Bivalvia, with species reference to the Septibranchia. *Proceedings of the Malacological Society of London*, 35: 71–80.

Purchon, R. D. 1985. Studies on the internal structure and function of the stomachs of bivalve molluscs: stomach types III, IV and V. Pages 337–361, in: B. S. Morton and D. Dudgeon, eds., *The Malacofauna of Hong Kong and Southern China, II, volume 1*. Hong Kong University, Hong Kong.

Purchon, R. D. 1987. The stomach in the Bivalvia. *Philosophical Transactions of the Royal Society of London, Series B, Biological Sciences*, 316(1177): 183–276.

Quitmyer, I. R., and D. S. Jones. 2000. The over-exploitation of hard clams (*Mercenaria* spp.) from five archaeological sites in the southeastern United States. *The Florida Anthropologist*, 53(2–3): 158–166.

Ravenel, E. 1874. *Catalogue of the Recent and Fossil Shells in the Cabinet of the Late Edmund Ravenel, M.D.* (L. R. Gibbes, ed.). Walker, Evans & Cogswell, Charleston, South Carolina, 67 pp.

Rawitz, B. 1888. Der Mantelrand der Acephalen. Erster Teil. Ostreacea. *Jenaische Zeitschrift für Naturwissenschaft*, 22(3–4): 415–556, pls. 13–18.

Rawitz, B. 1889. Der Mantelrand der Acephalen. II. Teil. Arcacea. Mytilacea. Unionacea. *Jenaische Zeitschrift für Naturwissenschaft*, 24: 549–631, pls. 21–24.

Rawitz, B. 1891. Der Mantelrand der Acephalen. Dritter Teil. Siphoniata. Epicuticulabildung. Allgemeine Betrachtungen. *Jenaische Zeitschrift für Naturwissenschaft*, 27: 1–232, pls. 1–7.

Redfern, C. 2001. *Bahamian Seashells: A Thousand Species from Abaco, Bahamas*. Bahamianseashells.com, Inc., Boca Raton, Florida, ix + 280 pp., 124 pls.

Ridewood, W. G. 1903. On the structure of the gills of the Lamellibranchia. *Philosophical Transactions of the Royal Society of London, Series B, Containing Papers of a Biological Character*, 195: 147–284.

Rios, E. de C. 1970. *Coastal Brazilian Seashells*. Fundação Cidade do Rio Grande, Museo Oceanográfico de Rio Grande, Rio Grande, Rio Grande do Sul, Brazil, 255 pp., 60 pls.

Rios, E. de C. 1975. *Brazilian Marine Mollusks Iconography*. Fundação Universidade do Rio Grande, Centro de Ciencias do Mar, Museu Oceanográfico, Rio Grande, Brazil, 331 pp., 91 pls.

Rios, E. de C. 1994. *Seashells of Brazil. Second edition*. Museu Oceanográfico, Rio Grande, Brazil, 368 pp., 113 pls.

Romero, J. G., H. J. Severeyn, Y. S. Ramírez, R. D. Chávez, and M. López. 2002. *Geukensia demissa*

(Dillwyn, 1817) (Bivalvia: Mytilidae), new genus and species of mussel for Venezuela and the Caribbean. *Boletín del Centro de Investigaciones Biológicas*, 36(3): 231–243.

Schneider, J. A. 2001. Bivalve systematics during the 20th century. *Journal of Paleontology*, 75(6): 1119–1127.

Simpson, C. T. 1887–1889. Contributions to the Mollusca of Florida. *Proceedings of the Davenport Academy of Natural Sciences*, 5: 45–72, 63–72.

Skelton, P. W., and M. J. Benton. 1993. Mollusca: Rostroconchia, Scaphopoda and Bivalvia. Pages 237–263, in: M. J. Benton, ed., *The Fossil Record 2*. Chapman & Hall, London.

Stasek, C. R. 1963. Synopsis and discussion of the association of ctenidia and labial palps in the bivalved Mollusca. *The Veliger*, 6(2): 91–97.

Steiner, G., and S. Hammer. 2000. Molecular phylogeny of the Bivalvia inferred from 18S rRNA sequences with particular reference to the Pteriomorphia. Pages 11–29, in: E. M. Harper, J. D. Taylor, and J. A. Crame, eds., *The Evolutionary Biology of the Bivalvia. Geological Society of London, Special Publication*, 177.

Taylor, J. D., W. J. Kennedy, and A. Hall. 1969. The shell structure and mineralogy of the Bivalvia. Introduction, Nuculacea–Trigonacea. *Bulletin of the British Museum (Natural History), Zoology, Supplement* 3: 1–125.

Taylor, J. D., W. J. Kennedy, and A. Hall. 1973. The shell structure and mineralogy of the Bivalvia. II. Lucinacea–Clavagellacea conclusions. *Bulletin of the British Museum (Natural History), Zoology*, 22(9): 255–294.

Terwilliger, R. C., and N. B. Terwilliger. 1985. Molluscan hemoglobins. *Comparative Biochemistry and Physiology, B, Comparative Biochemistry*, 81B: 255–261.

Thiele, J. 1899. Die abdominalen Sinnesorgane der Lamellibranchier. *Zeitschrift für Wissenschaftliche Zoologie*, 48: 47–59.

Turgeon, D. D., J. F. Quinn, Jr., A. E. Bogan, E. V. Coan, F. G. Hochberg, W. G. Lyons, P. M. Mikkelsen, R. J. Neves, C. F. E. Roper, G. Rosenberg, B. Roth, A. Scheltema, F. G. Thompson, M. Vecchione, and J. D. Williams. 1998. *Common and Scientific Names of Aquatic Invertebrates from the United States and Canada: Mollusks. Second edition.* American Fisheries Society Special Publication 26: ix + 526 pp.

Turner, R. D., and K. J. Boss. 1962. The genus *Lithophaga* in the western Atlantic. *Johnsonia*, 4(41): 81–116.

Ubukata, T. 2003. A theoretical morphologic analysis of bivalve ligaments. *Paleobiology*, 29(3): 369–380.

Valentich-Scott, P., and G. E. Dinesen. 2004. Rock and coral boring Bivalvia (Mollusca) of the Middle Florida Keys, U.S.A. *Malacologia*, 46(2): 339–354.

Villarroel, M., and J. Stuardo. 1998. Protobranchia (Mollusca: Bivalvia) Chilenos recientes y algunos fósiles. *Malacologia*, 40(1–2): 113–229.

Vokes, H. E., and E. H. Vokes. 1984 (1983). Distribution of shallow-water marine Mollusca, Yucatan Peninsula, Mexico. Mesoamerican Ecology Institute, Monograph 1. *Middle American Research Institute, Publication* 54: 183 pp., 50 pls.

Waller, T. R. 1998. Origin of the molluscan class Bivalvia and a phylogeny of major groups. Pages 1–47, in: P. A. Johnston and J. W. Haggart, eds., *Bivalves: An Eon of Evolution*. University of Calgary Press, Calgary, Alberta, Canada.

Warmke, G. L., and R. T. Abbott. 1961. *Caribbean Seashells*. Dover Publications, New York, 348 pp., 44 pls.

Weber, J. A. 1961. Marine shells of Water Island, Virgin Is. *The Nautilus*, 75(2): 55–60.

Yonge, C. M. 1948. Formation of siphons in Lamellibranchia. *Nature*, 161(4084): 198–199.

Yonge, C. M. 1957. Mantle fusion in the Lamellibranchia. *Pubblicazioni della Stazione Zoologica di Napoli*, 29: 151–171.

Yonge, C. M., and T. E. Thompson. 1976. *Living Marine Molluscs*. Collins, London, 288 pp.

Zischke, J. A. 1977. Checklist of macroflora, invertebrates and fishes of Pigeon Key. Pages 27–30, in: H. G. Multer, ed., *Field Guide to Some Carbonate Rock Environments—Florida Keys and Western Bahamas. New Edition.* Kendall/Hunt Publishing Company, Dubuque, Iowa, 415 pp., 10 maps.

# Image Data and Credits

The following listing gives specific specimen and locality data for the photographs and other illustrations in this work. To avoid repetition, the following conventions are used. All photographs are of shell valves unless otherwise noted. All artwork is original unless listed here as from or redrawn after a cited reference figure. All color photography is by R. Bieler, unless otherwise noted. All localities are in Monroe County, Florida Keys, unless otherwise noted. All "FK" stations were collected by the authors and colleagues as part of our Florida Keys Molluscan Diversity Project. Multiple shell measurements are presented as external view first, followed by internal view; these differ if the two views are not taken from a matched pair. Abbreviations include: LV, left valve; RV, right valve; SEM, scanning electron micrograph. Cited collections include: AMNH, American Museum of Natural History (New York, New York); ANSP, Academy of Natural Sciences of Philadelphia (Pennsylvania); BMNH, The Natural History Museum ([British Museum of Natural History] London, United Kingdom); BMSM, Bailey Matthews Shell Museum (Sanibel, Florida); CMNH, Carnegie Museum of Natural History (Pittsburgh, Pennsylvania); DMNH, Delaware Museum of Natural History (Wilmington); FLMNH, Florida Museum of Natural History (University of Florida, Gainesville); FMNH, Field Museum of Natural History (Chicago, Illinois); HBOM, Harbor Branch Oceanographic Museum (Fort Pierce, Florida); HMNS, Houston Museum of Natural Science (Texas); MCZ, Museum of Comparative Zoology (Harvard University, Cambridge, Massachusetts); MTD, Staatliche Naturhistorische Sammlungen Dresden ([Museum für Tierkunde, Dresden] Germany); UMML, Rosenstiel School of Marine and Atmospheric Sciences (University of Miami [University of Miami Marine Laboratory], Florida); USNM, National Museum of Natural History ([United States National Museum] Washington, D.C.); ZMB, Museum für Naturkunde ([Zoologisches Museum Berlin] Berlin, Germany).

## Introduction images

Map, Florida to Brazil: Original.

Satellite image, South Florida: Jacques Descloitres, MODIS Rapid Response Team, NASA/GSFC; image Bahamas A2004024.1600 (cropped).

Satellite image, Lower Florida Keys: Image Science and Analysis Laboratory, NASA–Johnson Space Center, Earth from Space photo STS038-85-103, November 1990.

Satellite image, detail of Florida Bay: Image Science and Analysis Laboratory, NASA–Johnson Space Center, Astronaut Photography of Earth photo ISS005-E-18051, October 2002.

Sandy beach with shells: Loggerhead Key, Dry Tortugas, October 2005.

Hard substrata including bridge: AMNH transparency collection, Ohio Key, looking toward Bahia Honda Bridge, November 1958, G. Raeihle, photographer.

Triggerfish within *Atrina* shell: FK-701, off Dove Key, oceanside of Key Largo, 25°03.01′N, 80°28.16′W, >1 m, June 2003 (not collected, specimen length unrecorded).

Project area map: Original.

*Chama macerophylla* on coral head: FK-692, Hens and Chickens patch reef, oceanside of Plantation Key, 24°56.10′N, 80°32.92′W, ca. 5 m, 08 June 2003 (not collected, specimen length unrecorded).

R/V *Coral Reef II*: At anchor off Garden Key, Dry Tortugas, L. Funkhouser, photographer.

Deepwater dredging: From R/V *Coral Reef II*, off Dry Tortugas, L. Funkhouser, photographer.

Shrimp fleet: AMNH transparency collection, Key West, G. Raeihle, photographer.

*Mytilopsis leucophaeta* on boat: Ramrod Key, on lower transom of trailered boat pulled from boating canal, January 2006.

External shell morphology of *Mercenaria mercenaria*: Original.

Internal shell morphology of *Mercenaria mercenaria*: Original.

Dorsal articulated shell morphology of *Mercenaria mercenaria*: Original.

Anatomy of *Mercenaria mercenaria*: Based on original dissections.

Phylogenetic diagram: after Bieler & Mikkelsen (2006).

## Florida Keys bivalve images

*Nucula proxima*: FMNH 306094, FK-156, NE of Bamboo Key, bayside, 24°45.38′N, 81°00.23′W, 0.5–1.5 m, 27 August 1997 (3.3 mm).

Nuculidae hinge: Based on photographed specimens of *Nucula proxima* (see above).

Nuculidae transparent clam: Based on the anatomy of *Nucula proxima* by Hampson (1971) and of *N. delphinodonta* by Drew (1901).

*Nucula* with brooded eggs: Redrawn after Drew (1899: fig. 8) for *N. delphinodonta*.

*Nucula* in feeding position: Redrawn after Yonge (1939: fig. 2).

*Ennucula aegeensis*: AMNH 191026, Cane Bay, Salt River, St. Croix, Virgin Islands, 1958 (3.4 mm).

*Nucula calcicola*: ANSP 368518, Grand Bahama Island, Bahamas, W end, at hotel, 26°42.25′N, 78°59.83′W (1.8 mm).

*Nucula crenulata*: FMNH 306096, FK-452, 24°34.54′N, 80°57.76′W, 187.7 m, mud, 26 April 2001 (unpaired, 5.1, 4.8 mm).

*Nucula delphinodonta*: AMNH 191082, Lettaue Island, Nova Scotia, July–August 1910 (2.6 mm).

Sampling by pipe scoop: Off Dry Tortugas, aboard R/V *Coral Reef II*, April 2002, L. Funkhouser, photographer.

*Solemya occidentalis*: AMNH 298932, FK-154, Cocoanut Key, 4 nmi due N of Little Duck Key, 24°44.73′N, 81°14.00′W, 0.9–1.5 m, seagrass, sponges, gorgonians, 25 August 1997 (external view, alcohol lot, 7.0 mm); FMNH 311552, FK-442, 24°39.39′N, 80°59.39′W, 26.5 m, mud, 26 April 2001 (internal view, cleaned specimen, 4.0 mm).

*Solemya occidentalis* [living specimen]: AMNH 298933, FK-219, Mud Key Channel, bayside of Big Coppitt Key area, 24°40.52′N, 81°41.71′W, 3.3–3.9 m, very soft anoxic mud, 19 April 1999 (8.0 mm).

Solemyidae hinge: Based on photographed shell specimens of *Solemya occidentalis* (see above).

Solemyidae transparent clam: Based on original dissection of *Solemya occidentalis* (AMNH 298933, FK-219, data as above) and on the anatomy of *Solemya* spp. by Morse (1913) and Reid (1980).

*Solemya velum*: AMNH 115339, York River, Yorktown, Virginia (16.0 mm, measured without periostracum).

*Nucinella woodei*: FMNH 306099, FK-450, 24°35.93′N, 80°58.23′W, 132.3 m, mud, 26 April 2001 (unpaired, 2.3, 2.5 mm max.).

Manzanellidae hinge: Based on photographed specimens of *Nucinella woodei* (see above).

Manzanellidae transparent clam: Based on the anatomy of *Nucinella serrei* by Allen & Sanders (1969).

*Propeleda carpenteri*: FMNH 306103, FK-451, 24°35.25′N, 80°58.00′W, 154.8 m, mud, 26 April 2001 (6.5 mm).

Nuculanidae hinge: Based on photographed specimens of *Propeleda carpenteri* (see above).

Nuculanidae transparent clam: Based on original dissection of *Propeleda carpenteri* (FMNH 302085, FK-504, 24°20.66′N, 81°53.47′W, 209 m, blackish sand with manganese nodules, 25 July 2001).

*Ledella sublevis*: UMML 28:145, 6 mi SE Molasses Reef Light, 219.5 m, 09 July 1950 (unpaired, 4.2, 4.3 mm).

*Nuculana acuta*: FMNH 195998, SE of Dry Tortugas, 91.4–213.4 m, 15 August 1974 (unpaired, 6.1, 5.7 mm).

*Nuculana concentrica*: FMNH 306090, FK-077, NNW of Marquesas Keys, 24°45.00′N, 82°15.00′W, 21.1 m, 22 April 1997 (6.9 mm).

*Nuculana jamaicensis*: Scanned from original description by A. d'Orbigny, 1853, *Histoire Physique, Politique et Naturelle de l'Ile de Cuba*, by R. de la Sagra, Atlas: pl. 26, figs. 30–32.

*Nuculana* cf. *semen*: FMNH 306093, FK-449, 24°36.61'N, 80°58.44'W, 109.1 m, mud, 26 April 2001 (1.9 mm).

*Nuculana verrilliana*: ANSP 102204, USFC 2275 (unpaired, 4.7, 4.8 mm).

*Nuculana vitrea*: Scanned from original description by A. d'Orbigny, 1853, *Histoire Physique, Politique et Naturelle de l'Ile de Cuba*, by R. de la Sagra, Atlas: pl. 26, figs. 27–29.

*Yoldia liorhina*: Scanned from W. H. Dall, 1886, Report on the results of dredging by the United States Coast Survey Steamer "*Blake*." XXIX. Report on the Mollusca. Part I. Brachiopoda and Pelecypoda, *Bulletin of the Museum of Comparative Zoology*, 12(6): pl. 9, figs. 1, 1a.

*Yoldia seminuda* [pallial line]: FMNH 86631, off Venice, California, 183 m, April 1959 (22.0 mm).

Yoldiidae hinge: Based on *Yoldia sapotilla* (Gould, 1841) (FMNH 185614, Massachusetts).

Yoldiidae transparent clam: Based on the anatomy of *Yoldia seminuda* by Stasek (1965, as *ensifera* Dall, 1897), on original dissection of *Y. limatula* (Say, 1831) (AMNH 181451, Massachusetts), and on anatomical accounts of the latter species by Drew (1899) and Yonge (1939).

*Yoldia* in feeding position: *Yoldia limatula*, redrawn after Bender & Davis (1984: fig. 1).

*Barbatia cancellaria*: FMNH 166530, Key Vaca (33.3 mm).

*Barbatia cancellaria* [specimen with periostracum]: FMNH 288836, FK-045, Indian Key Fill, bayside, 24°53.42'N, 80°40.47'W, 0.5–1 m, rocks, *Thalassia/Halodule* seagrass, 20 September 1996 (alcohol lot, 16.4 mm, measured without periostracum).

*Barbatia cancellaria* [living specimens on hard bottom]: FMNH 311608, FK-701, off Dove Key, oceanside of Key Largo, 25°03.01'N, 80°28.16'W, 0.3–0.6 m, sand, silty *Thalassia* seagrass, 12 June 2003 (length not recorded).

Arcidae hinge: Based on photographed specimens of *Barbatia cancellaria* (see above).

Arcidae transparent clam: Based on original dissection of *Barbatia cancellaria* (various Florida Keys locations) and on the anatomy of the same species by Simone & Chichvarkhin (2004).

*Cucullaearca candida*: FMNH 155511, Key Vaca, Gulf of Mexico (53.6 mm); FMNH 183596, Missouri Key (detail of periostracum, 25.9 mm, measured without periostracum).

*Fugleria tenera*: FMNH 311595, FK-583, Dry Tortugas, 24°34.69'N, 82°56.71'W, 12.2 m, large patch reef with surrounding sand plains, 16 April 2002 (17.7 mm, measured without periostracum); FMNH 295616, FK-115, East Washerwoman Shoal, off Key Vaca, 24°40'N, 81°04.3'W, max. 2.7 m, 12 July 1997 (detail of periostracum, alcohol lot, 20.1 mm, measured without periostracum).

*Acar domingensis*: FMNH 183598, Missouri Key (25.1 mm).

*Lunarca ovalis*: FMNH 183609, Bonefish Key, Marathon, 1939 (44.0 mm).

*Arca zebra*: FMNH 183408, Bonefish Key, 1940 (57.2 mm; cleaned specimen, 47.05 mm).

*Arca imbricata*: FMNH 166532, Key Vaca (exterior, interior views, 50.9 mm; articulated ventral view, 47.5 mm).

*Anadara baughmani*: USNM 597375, 12 mi S of Dry Tortugas, 109.7–120.7 m, 05 August 1932 (holotype of *Anadara springeri*, 46.0 mm).

*Anadara floridana*: AMNH 99993, Captiva Island, [Lee County], Florida (65.0 mm).

*Anadara notabilis*: FMNH 183610, Key West (40.7 mm).

*Anadara transversa*: AMNH 28709, Long Key (20.3 mm, larger LV measured).

*Scapharca brasiliana*: USNM 27220, Indian Key (55.6 mm).

*Scapharca chemnitzii*: AMNH 676, off Guanica Playa, Puerto Rico, 5.5 m, 29 June 1915 (unpaired, 27.3, 29.0 mm).

*Bathyarca glomerula*: FMNH 194526, SE of Dry Tortugas, 91.4–213.4 m, 15 August 1974 (unpaired LV, 5.7 mm).

*Bentharca sagrinata*: ANSP 369023, Tamarind, Grand Bahama Island, Bahamas, 26°30.75'N, 78°36.00'W (external view, 17.9 mm); ANSP 367942, Grand Bahama Island, Bahamas, 26°29.75'N, 78°37.25'W (internal view, 11.1 mm).

*Fugleria tenera* [articulated dorsal view]: FMNH 183739, Missouri Key (30.1 mm)

*Scapharca brasiliana* [articulated dorsal view]: FMNH 54695, Grand Isle, Louisiana (35.0 mm).

*Acar domingensis* [articulated dorsal view]: FMNH 183598, Missouri Key (25.1 mm).

*Anadara notabilis* [articulated dorsal view]: FMNH 183611, Little Duck Key (48.3 mm).

*Barbatia cancellaria* [articulated dorsal view]: FMNH 166530, Key Vaca (33.3 mm).

*Arca imbricata* [articulated dorsal view]: FMNH 166532, Key Vaca (articulated dorsal view, 47.5 mm).

*Arca zebra* [living specimen on rock]: FMNH 311563, FK-733, Looe Key, backreef, 24°32.89′N, 81°24.36′W, 1.2-1.8 m, rubble, sand and seagrass, 05 May 2004 (33.3 mm).

*Fugleria tenera* [living specimens on rock]: FMNH 311564, FK-730, Newfound Harbor coral lumps, oceanside, 24°36.89′N, 81°23.63′W, 2.4–2.7 m, *Thalassia* seagrass, sand patches, coral heads, 04 May 2004 (crawling specimen, 32.2 mm).

*Arcopsis adamsi*: FMNH 94728, Key West (12.3 mm); FMNH 295591, FK-244, "Horseshoe" site, bayside of Spanish Harbor Keys, 24°39.32′N, 81°18.22′W, center of quarry, max. 7.0 m, rock wall and sand slope, 05 August 1999 (dorsal ligamental view, alcohol lot, 12.4 mm).

*Arcopsis adamsi* [living specimens on rock]: FMNH 311565, FK-725, "Horseshoe" site, bayside of Spanish Harbor Keys, 24°39.30′N, 81°18.20′W, intertidal and shallow subtidal, rocks/rubble along arms of quarry, 29 April 2004 (each specimen ca. 9–10 mm).

Noetiidae hinge: Based on photographed specimens of *Arcopsis adamsi* (see above).

Noetiidae transparent clam: Based on original dissection of *Arcopsis adamsi* (FMNH 295595, FK-037, "Horseshoe" site, bayside of Spanish Harbor Keys, 24°39.32′N, 81°18.22′W, 0.5-1 m, 11 March 1996; FMNH 311609, FK-725, "Horseshoe" site, bayside of Spanish Harbor Keys, 24°39.30′N, 81°18.20′W, intertidal and shallow subtidal, rocks/rubble along arms of quarry, 29 April 2004; FMNH 311610, FK-015, Indian Key Fill, bayside, 24°53.42′N, 80°40.47′W, rocks, high intertidal and rubble at surf line, 06 July 1995) and on the anatomy of the same species by Oliver & Järnegren (2004).

*Noetia ponderosa*: FMNH 184106, Key West (44.2 mm); FMNH 227403, Marathon, 25 December 1966 (articulated ligamental view, 49.0 mm).

*Tucetona pectinata*: FMNH 150854, Bonefish Key, January 1940 (16.2 mm).

*Tucetona pectinata* [living specimens on hard bottom]: FMNH 311611, FK-729, Marathon, bayside, E of Stirrup Key channel, 24°44.02′N, 81°02.78′W, 1–2 m, sand patches, rubble, *Thalassia* seagrass, 03 May 2004 (length not recorded).

Glycymerididae hinge: Based on photographed specimens of *Tucetona pectinata* (see above).

Glycymerididae transparent clam: Based on original dissection of *Tucetona pectinata* (AMNH 299403, FK-360, coral lumps off Newfound Harbor Keys, off Big Munson Key, 24°36.96′N, 81°23.64′W, 2.7 m, sand, seagrass, patch reef, gorgonians, 11 July 2000) and on the anatomy of the same species by Thomas (1975).

*Glycymeris americana*: AMNH 247486, WNW of Dry Tortugas, 177 m, June 1974 (25.6 mm).

*Glycymeris decussata*: FMNH 301536, FK-662, Sombrero Reef, 5 nmi S of Knights Key, 24°37.62′N, 81°06.53′W, 5.2–7.0 m, sand near coral reef, 05 August 2002 (36.7 mm).

*Glycymeris spectralis*: FMNH 306123, FK-316, off Halfmoon Shoal Light, 24°33.51′N, 82°28.37′W, 6.7 m, clean sand and wreck, 04 July 2000 (unpaired, 25.6, 23.5 mm).

*Glycymeris undata*: UMML 28:1120, Matecumbe Key (14.0 mm, external view); FMNH 82184, Nassau, New Providence, Bahamas, 1948–1958 (11.7 mm, internal view).

*Tucetona subtilis*: FMNH 306109, FK-448, 24°37.28′N, 80°58.67′W, 88.1 m, mud, 26 April 2001 (unpaired, 6.4, 5.4 mm).

*Tucetona* habitat [with *Strombus gigas*]: FK-722, Looe Key back reef, 24°32.89′N, 81°24.36′W, 1.2–2.4 m, rubble, seagrass, sand, 28 April 2004 (*S. gigas* is ca. 25 cm in length).

*Limopsis cristata*: FMNH 306074, FK-452, 24°34.54′N, 80°57.76′W, 187.7 m, mud, 26 April 2001 (articulated external view, 4.0 mm; unpaired internal view, 4.6 mm); FMNH 306077, FK-504, 24°20.66′N, 81°53.47′W, 209 m, blackish sand with manganese nodules, 25 July 2001 (external view with periostracum, 4.0 mm; dorsal view, 3.0 mm).

Limopsidae hinge: Based on photographed specimens of *Limopsis cristata* (see above).

Limopsidae transparent clam: Based on original dissection of *Limopsis cristata* (FMNH 302084, FK-

503, 24°21.42′N, 81°53.45′W, 196 m, blackish sand, 25 July 2001) and on the anatomy of *L. diegensis* Dall, 1908, by Oliver (1981).

*Limopsis* crawling: *Limopsis aurita*, redrawn after Oliver (1981: fig. 12B).

*Limopsis minuta*: Scanned from description by J. G. Jeffreys, 1869, *British Conchology*, 5: pl. 100, fig. 3 [as syn. *L. borealis* Woodward in Jeffreys, 1863]; based on specimens from Scotland and Norway.

*Limopsis aurita*: FMNH 175808, SE of Dry Tortugas, 91.4 m, 15 August 1974 (unpaired, 9.1, 8.2 mm).

*Limopsis sulcata*: USNM 330463, off Carysfort [Reef], 109.7 m, Co. S. 69.2°, U.S. Bureau of Fisheries sta. 2641 (8.9 mm, paired with periostracum; unpaired cleaned valves, each 6.5 mm).

Deep-sea habitat with Blackfin Goosefish (*Lophius gastrophysus* Miranda-Ribeiro, 1915): Miami Terrace escarpment, Straits of Florida, 427 m. Photograph by and courtesy of John Reed, Harbor Branch Oceanographic Institution.

*Cratis antillensis*: HMNS 35522, off Dry Tortugas, 82.3 m, October 1969 (2.5 mm; SEM of prodissoconch and hinge line, RV, 3.3 mm).

*Cratis antillensis* [SEM of prodissoconch and hinge line]: same specimen as shell views (HMNS 35522, RV, 3.3 mm).

Philobryidae hinge: Based on photographed specimens of *Cratis antillensis* (see above).

Philobryidae transparent clam: Based on the anatomy of *Philobrya munita* Finlay, 1930, by Morton (1978).

*Brachidontes exustus*: FMNH 288687, FK-172, Ninefoot Shoal, 24°34.10′N, 81°33.10′W, 9.7 m, patch reef; 20 April 1997 (10.6 mm max.).

Mytilidae hinge: Based on photographed specimens of *Brachidontes exustus* (see above).

Mytilidae transparent clam: Based on original dissection of *Brachidontes exustus* (FMNH 295632, FK-028, Lake Surprise [SW quadrant], Key Largo, 25°10.50′N, 80°22.80′W, feathery *Caulerpa* on mangrove roots, 07 March 1996; FMNH 311612, FK-691, W of Pigeon Key [bayside of Tavernier], 25°03.30′N, 80°30.71′W, 0.6 m, mixed algae on *Thalassia* seagrass, 07 June 2003).

Mytilidae [group on submerged piling]: *Mytilus edulis*, AMNH transparency collection, New York, January 1976, G. Raeihle, photographer (specimen length unrecorded).

*Brachidontes "domingensis"*: FMNH 188293, Lower Matecumbe Key (21.8 max.).

*Brachidontes modiolus*: FMNH 227434, Key West, shallow water, sand, 06 July 1968 (30.8 mm).

*Lioberus castaneus*: AMNH 232628, E end of Missouri Key, bayside, March 1969 (16.9 mm).

*Ischadium recurvum*: FMNH 227439, Key West (29.2 mm).

*Gregariella coralliophaga*: AMNH 293801, Garden Key, Dry Tortugas, August 1947 (14.6 mm).

*Geukensia granosissima*: FMNH 166064, Grassy Key (46.6 mm).

*Geukensia demissa* [living specimens in salt marsh]: AMNH transparency collection, New York, June 1975, G. Raeihle, photographer (seagrass blades ca. 10 mm width).

*Modiolus americanus*: FMNH 183706, Missouri Key (55.8 mm).

*Modiolus squamosus*: AMNH 296316, FK-273, "Horseshoe" site, bayside of Spanish Harbor Key, northernmost (outermost) point of W arm, 24°39.35′N, 81°18.22′W, 0.3–1.2 m, 19 August 1999 (28.4 mm); FMNH 295721, FK-170, Hawk Channel, E of mouth of Tavernier Creek (Plantation Key), 24°58.77′N, 80°30.88′W to 24°58.66′N, 80°30.85′W, 3.3–4.0 M, sand, *Thalassia* seagrass droves, 17 September 1998 (posterior periostracal detail, alcohol lot, 16.0 mm, measured without periostracum, juvenile specimen on *Thalassia* seagrass blades); FMNH 183703, Bradenton Beach, Manatee County, Florida (32.2 mm, measured without periostracum).

*Perna viridis*: AMNH 290101, Tampa Bay, [Pinellas–Hillsborough Counties], Florida, Tampa Electric Power Station, surfaces in condenser cooling water tunnels, 15 July 1999 (alcohol lot, 35.7 mm).

Green Mussel alert card: Photograph courtesy of Florida Sea Grant.

*Musculus lateralis*: FMNH 183678, Lower Matecumbe Key, 1938 (5.5 mm).

*Amygdalum papyrium*: AMNH 184468, N outskirts of Dunedin, Pinellas County, Florida, by seawall, 29 April 1944 (18.2 mm).

*Amygdalum politum*: USNM 64094, near [S of Dry] Tortugas, 619.9 m, R/V *Blake* sta. 43, late 1870s (11.0 mm).

*Amygdalum sagittatum*: USNM 330469, off Carysfort [Reef], 108.7 m, Co. S. 69.2°, U.S. Bureau of Fisheries sta. 2641 (10.8 mm).

*Crenella decussata*: FMNH 306071, FK-443, 24°39.06′N, 80°59.26′W, 36.0 m, shelly sand, 26 April 2001 (2.1 mm max.).

*Dacrydium elegantulum hendersoni*: FMNH 311566, FK-454, 24°33.16′N, 80°57.31′W, 201.2 m, mud, pteropod ooze, 26 April 2001 (3.4 mm).

*Lithophaga antillarum*: FMNH 183716, Bonefish Key, 1946 (105.0 mm).

*Lithophaga aristata*: FMNH 183671, Missouri Key (29.0 mm).

*Lithophaga bisulcata*: FMNH 183635, Bonefish Key, 1938 (46.4 mm).

*Lithophaga nigra*: FMNH 227440, Missouri Key (41.5 mm).

*Botula fusca*: FMNH 183684, Key Vaca, 1939 (35.1 mm).

*Lithophaga bisulcata* [living specimen in rock with *Petricola lapicida*]: FMNH 311567, FK-725, "Horseshoe" site, bayside of Spanish Harbor Keys, 24°39.30′N, 81°18.20′W, intertidal and shallow subtidal, rocks/rubble along arms of quarry, 29 April 2004 (shell inflation 10.8 mm; total length 38.6 mm).

*Pinctada longisquamosa*: FMNH 302080, FK-368, off W shore of Pigeon Key (bayside of Tavernier, Plantation Key), 25°03.30′N, 80°30.71′W, 0.3–1.2 m, *Thalassia/Syringodeum* seagrass, 08 October 2000 (alcohol lot, 40.1 mm, measured including lamellae on LV).

*Pinctada longisquamosa* [SEM, shell microstructure]: AMNH 308118, PMM-1039, Andros, Bahamas, on beach in front of Forfar Field Station, on *Sargassum*, 28 August 2000 (shell 10.7 mm).

*Pinctada longisquamosa* [living juvenile on algae]: FMNH 302081, FK-700, W of Pigeon Key (bayside of Tavernier, Plantation Key), 25°03.29′N, 80°30.69′W, 0.3–0.6 m, seagrass with algae, 12 June 2003 (specimen ca. 10 mm length).

*Pinctada longisquamosa* [living yellow specimen]: FMNH 295709, FK-064, center of Coupon Bight, off Big Pine Key/Newfound Harbor, 24°38.63′N, 81°22.20′W, mud, *Thalassia* seagrass, 1.5 m, 15 April 1997 (23.5 mm).

Pteriidae hinge: Based on photographed specimens of *Pinctada longisquamosa* (see above).

Pteriidae transparent clam: Based on original dissection of *Pinctada longisquamosa* (see Mikkelsen et al., 2004).

*Pinctada maxima* [living specimen from perliculture operation]: FMNH 298890, Vansittart Bay, Kimberley, Western Australia, 07 June 2000 (145 mm).

*Pinctada imbricata*: FMNH 227467, Missouri Key, Ohio–Missouri Channel, in crevice in old coral rocks, 08 March 1987 (57.2 mm).

*Pinctada margaritifera*: FMNH 165779, Florida, 1950 (61.9 mm).

*Pteria colymbus*: FMNH 183297, Missouri Key (56.0 mm).

*Pteria vitrea*: USNM 92492, between Tampa Bay and Dry Tortugas, 51.2 m, sand, U.S. Fish Commission sta. 2410 (25.8 mm).

*Pteria colymbus* [living specimen on gorgonian]: FMNH 311613, FK-689, NE of Dove Key, oceanside of Key Largo, 25°03.05′N, 80°28.22′W, 0.5–1.0 m, hard bottom with silty sand, sponges, gorgonians, 06 June 2003 (length not recorded).

*Pinctada imbricata* [living specimen on hard bottom]: FMNH 311614, FK-701, off Dove Key, oceanside of Key Largo, 25°03.01′N, 80°28.16′W, 0.3–0.6 m, sand, silty *Thalassia* seagrass, 12 June 2003 (48.5 mm).

*Isognomon alatus*: FMNH 183223, Marathon (74.8 mm).

*Isognomon alatus* [living specimens on dock piling]: FK-750, Ramrod Key, St. Martin Lane, 24°39.41′N, 81°24.29′W, seawall, intertidal at low tide, 15 December 2005 (not collected; PVC pipe is ca. 15 cm diameter; largest specimens ca. 6 cm max.).

Isognomonidae hinge: Based on photographed specimens of *Isognomon alatus* (see above).

Isognomonidae transparent clam: Based on original dissection of *Isognomon alatus* (courtesy of Ilya Tëmkin, AMNH).

*Isognomon bicolor*: FMNH 183227, Missouri Key (28.9 mm).

*Isognomon radiatus*: FMNH 183228, Missouri Key (40.1 mm).

*Isognomon bicolor* [living specimens on rock]: FMNH 306131, FK-724, Missouri Key, oceanside,

24°40.48′N, 81°14.35′W, intertidal rocks at low tide, 29 April 2004 (largest specimen 20.4 mm height).

*Isognomon radiatus* [living specimens on rock]: FMNH 306136, FK-730, Newfound Harbor coral lumps, oceanside, 24°36.89′N, 81°23.63′W, 2.4–2.7 m, *Thalassia* seagrass, sand patches, coral heads, 04 May 2004 (largest specimens 46 mm height).

*Isognomon alatus* [living colony on mangrove roots]: AMNH transparency collection, Crawl Key, November 1959, G. Raeihle, photographer.

*Malleus candeanus*: AMNH 293705, Harrington Sound, Trunk Island, Bermuda, embedded in coral, 0.9–1.8 m, August 1964 (57.1 mm max.).

*Malleus candeanus* [living specimen on algae-covered reef]: AMNH 298194, FK-260, Looe Key coral reef, oceanside of Ramrod Key, 24°32.80′N, 81°24.80′W, 7.3–7.6 m, spur and groove reef, 10 August 1999 (alcohol lot, 14.3 mm visible width).

Malleidae hinge: Based on photographed specimens of *Malleus candeanus* (see above).

Malleidae transparent clam: Based on original dissection of *Malleus candeanus* (courtesy of Ilya Tëmkin, AMNH).

*Dendostrea frons*: FMNH 155510, Key West, on gorgonian (47.1 mm max.).

*Dendostrea frons* [living specimen on gorgonian stalk]: FK-692, Hens and Chickens patch reef, oceanside of Plantation Key, 24°56.10′N, 80°32.92′W, max. 6.4 m, 08 June 2003 (not collected, specimen length unrecorded).

Ostreidae hinge: Based on photographed specimens of *Dendostrea frons* (see above).

Ostreidae transparent clam: Based on original dissection of *Dendostrea frons* (AMNH 298005, FK-236, Coffins Patch coral reef, oceanside of Grassy Key, 24°41.08′N, 80°57.47′W, max. 4.9 m, 02 August 1999).

*Crassostrea virginica* [oyster bar]: Cedar Key, Levy County, Florida, P. Mikkelsen, photographer (oyster bar ca. 40 cm height).

*Crassostrea rhizophorae*: FMNH 306124, San Carlos Bay, Sanibel Island, Lee County, Florida, 20 June 1939 (86.5 mm, measured max. of attached valve).

*Crassostrea virginica*: FMNH 311607, FK-193, off Key Largo, bayside, E of Buttonwood Sound, Swash Keys, E shore of unnamed middle key between Whaleback Key and Shell Key, Everglades National Park [dead shells only], 25°07.27′N, 80°28.77′W, clay/sand beach, mangrove roots, 01 April 1999 (10.6 mm max.).

*Ostrea equestris*: FMNH 279417, FK-244, "Horseshoe" site, bayside of Spanish Harbor Keys, 24°39.32′N, 81°18.22′W, center of quarry, max. 7.0 m, rock wall and sand slope, 05 August 1999 (46.9 mm, measured width of cemented group of two pair).

*Teskeyostrea weberi*: FMNH 198077, Big Pine Key, beach stones at tide line, 1972 (26.9 mm).

*Teskeyostrea weberi* [living specimens on rock]: AMNH transparency collection, Grassy Key, November 1959, G. Raeihle, photographer (specimen lengths unrecorded).

*Hyotissa mcgintyi*: USNM 619593, Key West, [Dry] Tortugas area, old iron surface (unspecified), June 1951 (61.0 mm, plus enlarged detail of shell edge).

Gryphaeidae hinge: Based on photographed specimens of *Hyotissa mcgintyi* (see above).

Gryphaeidae transparent clam: Based on original dissection of *Hyotissa mcgintyi* (FMNH 302069, FK-717, wreck of *Thunderbolt*, approx. 6 nmi S of Marathon, 24°39.68′N, 80°57.82′W, 28.9–36.6 m, steel wreck with fouling bivalves, alcyonarians, and hydroids, 19 August 2003) and on the anatomy of *H. hyotis* by Sevilla et al. (1998).

*Hyotissa hyotis*: FMNH 302010, FK-717, same data as previous (178.0 mm).

*Neopycnodonte cochlear*: HBOM 064:1090, Sebastian Pinnacles, [off St. Lucie County, Florida,] 27°50.19′N, 79°57.99′W, 79–101 m, *Oculina* coral rubble, *Johnson-Sea-Link I* dive 1023 (13.5 mm max. of attached valve).

Wreck with scuba diver [A. Bieler]: FK-717, wreck of *Thunderbolt*, same data as previous.

*Pinna carnea*: FMNH 183249, Bonefish Key (264.0 mm).

*Pinna carnea* [living specimen on sand flat]: FMNH 311615, FK-701, off Dove Key, oceanside of Key Largo, 25°03.01′N, 80°28.16′W, 0.3–0.6 m, sand, silty *Thalassia* seagrass, 12 June 2003 (length not recorded).

*Pinna carnea* [SEM of prodissoconch]: FMNH 311597, FK-685, The Rocks (inside patch reef), oceanside off Windley Key, 24°57.25′N, 80°32.88′W, patch reef with surrounding hard bottom covered by thin veneer of sand, many gorgonians, max. 4.2 m, 04 June 2003 (RV, ca. 35 mm, prodissoconch height 0.48 mm).

Pinnidae hinge: Based on photographed specimens of *Pinna carnea* (see above).

Pinnidae transparent clam: Based on the anatomy of *Pinna carnea* by Yonge (1953) and on other accounts of the pinnid anatomy by Grave (1911) and Turner & Rosewater (1958).

*Atrina rigida* [living specimen on sand flat]: Lake Worth, Palm Beach County, Florida, Intracoastal Waterway, under the Blue Heron Bridge, ca. 5 m (specimen width unrecorded). Photograph by and courtesy of Robert F. Myers, Coral Graphics.

*Pinna nobilis* [glove and byssus]: FMNH 2461, knitted glove made from *Pinna nobilis* byssus, Taranto, southern Italy (greatest length 240 mm); FMNH 2460, *Pinna nobilis* byssus, Naples, Italy (greatest length 100 mm).

*Atrina rigida*: FMNH 184076, Little Duck Key (155.0 mm).

*Atrina seminuda*: AMNH 305162, FK-649, Sprigger Bank, bayside of Conch Keys, just W of Everglades National Park border, 24°54.75′N, 80°56.24′W, 0.1–0.9 m, *Thalassia/Syringodeum* seagrass, 27 July 2002 (alcohol lot, external view, 108.9 mm max.); FMNH 151482, Port Isabel, Cameron County, Texas (internal view, 181.0 mm).

*Atrina serrata*: FMNH 184062, Little Duck Key (unpaired, 180.0, 183.0 mm).

*Ctenoides mitis*: FMNH 182929, Key West, 1938 (59.4 mm max.).

*Ctenoides mitis* [living specimen on reef]: AMNH 308072, FK-244, "Horseshoe" site, bayside of Spanish Harbor Keys, 24°39.32′N, 81°18.22′W, center of quarry, max. 7.0 m, rock wall and sand slope, 05 August 1999 (alcohol specimen, length not recorded).

*Ctenoides mitis* [living juvenile]: FMNH 311616, FK-049, "Picnic Island," off Newfound Harbor/Ramrod Key, 24°38.72′N, 81°28.11′W, sand, 21 September 1996 (alcohol specimen, 20 mm).

*Ctenoides mitis* [mantle edge of living specimen]: AMNH 296899, FK-118, Key Vaca, channel on bayside of Stirrup Key, 24°44.19′N, 81°02.92′W, 4.6 m, rock ledges, 18 July 1997 (alcohol specimen, length not recorded).

*Ctenoides mitis* [SEM of prodissoconch]: AMNH 247503, Key Vaca, rocks at edge of tidal inlet, March 1969 (length of prodissoconch 153 μm).

Limidae hinge: Based on photographed specimens of *Ctenoides mitis* (see above).

Limidae transparent clam: Based on original dissection of *Ctenoides mitis* (see Mikkelsen & Bieler, 2003).

*Ctenoides sanctipauli* [scanning electron micrograph, sculpture]: BMSM 2232, Marathon, bayside (paratype, 19.1 mm, with intact ligament) (shell 19.1 mm; width of figure 4.2 mm).

*Ctenoides miamiensis*: UMML 11763, [Dry] Tortugas (10.0 mm max., paratype).

*Ctenoides planulata*: USNM 62246, Barbados, 182.9 m, R/V *Blake*, 1877–1878 (external view, paralectotype, 9.4 mm max.); MCZ 7826, Barbados, 182.9 m, R/V *Blake*, 1877–1878 (internal view, paralectotype, 10.1 mm max.).

*Ctenoides sanctipauli*: MCZ 316561, Sand Key reef, coral blocks, R/V *Eolis* sta. 37, 28 May 1911 (33.0 mm max.).

*Ctenoides scabra*: FMNH 182936, Pigeon Key (76.2 mm max.).

*Divarilima albicoma*: USNM 62250, near Havana, Cuba, U.S. Fish Commission sta. 2322, 210 m (holotype, 8.3 mm).

*Lima caribaea*: FMNH 182935, Missouri Key (40.4 mm max.).

*Limea bronniana*: FMNH 311568, FK-454, 24°33.16′N, 80°57.31′W, 201.2 m, mud, pteropod ooze, 26 April 2001 (unpaired, 3.0, 2.8 mm).

*Limatula confusa*: Scanned from original description by E. A. Smith, 1885, Report on the scientific results of the voyage of H. M. S. *Challenger* during the years 1873–76 . . . Zoology—Vol. XIII, Part XXXV, Report on the Lamellibranchiata, Edinburgh, pl. 24, figs. 6, 6a.

*Limatula setifera*: USNM 62252, near Barbados, *Albatross* sta. 272, 139 m (syntype, unpaired LV, 6.8 mm height).

*Limatula subovata*: FMNH 306117, FK-454, 24°33.16'N, 80°57.31'W, 201.2 m, mud, pteropod ooze, 26 April 2001 (external view, 1.3 mm); FMNH 306116, FK-451, 24°35.25'N, 80°58.00'W, 154.8 m, mud, 26 April 2001 (internal view, 1.5 mm).

*Limaria pellucida*: FMNH 288811, FK-047, Channel marker 50A off Ramrod Key, 24°35.80'N, 81°27.24'W, 4.6 m, rubble and patch reef, 21 September 1996 (17.3 mm max.).

*Limaria pellucida* [living specimens]: AMNH 299641, FK-058, Indian Key Fill, bayside, 24°53.42'N, 80°40.13'W, rocks, silty mud, *Thalassia* seagrass, 0.5–1.0 m, 12 April 1997 (left, 11.3 mm); AMNH 298942, FK-359, American Shoals, 24°31.56'N, 81°31.10'W, *Thalassia/Syringodeum* seagrass with rubble, 3.3–3.6 m, 10 July 2000 (right, 11.7 mm).

*Caribachlamys sentis*: FMNH 183540, Key West (24.6 mm).

*Caribachlamys sentis* [ctenolium]: FMNH 183540, same specimen as shell views.

*Caribachlamys sentis* [living orange specimen on sand]: AMNH transparency collection, Pigeon Key, December 1964, G. Raeihle, photographer (specimen length not recorded).

*Caribachlamys sentis* [living purple specimen on rock]: FMNH 311569, FK-720, Looe Key, backreef, 24°32.89'N, 81°24.36'W, 1.2–2.1 m, rubble, sand, and seagrass, 27 April 2004 (29.3 mm).

Pectinidae hinge: Based on photographed specimens of *Caribachlamys sentis* (see above).

Pectinidae transparent clam: Based on original dissection of *Caribachlamys sentis* (FMNH 311617, FK-686, The Rocks [inside patch reef], oceanside off Windley Key, 24°57.04'N, 80°33.28'W, to 2.4 m, hard bottom with thin layer of sand, many gorgonians, few coral boulders, 04 June 2003).

*Argopecten gibbus* [living specimen]: AMNH transparency collection, Sanibel Island, [Lee County], Florida, March 1975 (specimen length not recorded).

*Caribachlamys mildredae*: AMNH 275487, off Bahia Honda bridge, 3.0 m, 07 July 1984 (28.7 mm).

*Caribachlamys ornata*: AMNH 247258, off Key West, January 1970 (16.4 mm).

*Caribachlamys pellucens*: FMNH 183577, Garden Key, Dry Tortugas, 1941 (36.1 mm).

*Spathochlamys benedicti*: FMNH 170975, SE of Dry Tortugas, 91.4 m, 15 August 1974 (13.1 mm).

*Cryptopecten phrygium*: AMNH 247375, WNW of Dry Tortugas, 192 m, June 1974 (42.0 mm).

*Laevichlamys multisquamata*: AMNH 171330, SW Cuba (33.7 mm).

*Argopecten gibbus*: FMNH 163671, 70 mi off Key West, 1949 (39.4 mm).

*Argopecten irradians*: FMNH 173018, 70 mi off Key West, 1949 (38.1 mm).

*Argopecten nucleus*: FMNH 227448, Little Duck Key, 1955 (16.7 mm).

*Aequipecten muscosus*: FMNH 183554, off Key West, 45.7 m (41.6 mm).

*Aequipecten glyptus*: FMNH 194550, SE of Dry Tortugas, 91.4–213.4 m, 15 August 1974 (30.6 mm).

*Aequipecten lineolaris*: FMNH 77916, Dry Tortugas, 27.4–36.6 m, 1953 (24.7 mm).

*Euvola raveneli*: FMNH 183572, off Key West, 45.7 m (48.2 mm).

*Euvola ziczac*: FMNH 183571, off Key West, 1950 (66.1 mm).

*Euvola chazaliei*: AMNH 179775, off Dry Tortugas, ca. 202 m (26.7 mm).

*Euvola laurentii*: FMNH 302047, Dry Tortugas, 36.6 m, sand, December 2003 (65.1 mm).

*Euvola "papyracea"*: FMNH 77985, Dry Tortugas (79.0 mm).

*Brachtechlamys antillarum*: FMNH 202917, Boca Chica and Matecumbe [Keys], 1938–1955 (17.3 mm).

*Nodipecten fragosus*: FMNH 302048, Dry Tortugas, 33.5 m, sand, March 2004 (71.0 mm); AMNH transparency collection, no data recorded (yellow pair, specimen length unrecorded).

*Caribachlamys sentis* [articulated dorsal view]: FMNH 183540, same specimen as shell views (length at auricles, 14.3 mm).

*Aequipecten muscosus* [articulated dorsal view]: FMNH 183554, same specimen as shell views (length at auricles, 44.3 mm).

*Euvola raveneli* [articulated dorsal view]: FMNH 308627, Dry Tortugas, sand bottom, 30.5 m (37.0 mm; length at auricles, 17.5 mm).

*Similipecten nanus*: FMNH 311618, FK-445, 24°38.37'N, 80°59.02'W, 57.3 m, muddy sand, 26 April 2001 (unpaired RV, 3.4 mm;); FMNH 306068, FK-446, 24°37.10'N, 80°58.90'W, 69.2 m, mud, 26 April 2001 (unpaired LV, 3.1 mm).

*Similipecten nanus* [SEM of prodissoconch]: FMNH 311598, FK-445, 24°38.37'N, 80°59.02'W, 57.3 m, muddy sand, 26 April 2001 (unpaired RV, 2.9 mm).

Propeamussiidae hinge: Based on photographed specimens of *Similipecten nanus* (see above).

Propeamussiidae transparent clam: Based on the anatomy of *Propeamussium lucidum* (Jeffreys, 1879) by Morton & Thurston (1989).

*Hyalopecten strigillatus*: Scanned from W. H. Dall, 1889, A preliminary catalogue of the shell-bearing marine mollusks and brachiopods of the southeastern coast of the United States, with illustrations of many of the species, *Bulletin of the United States National Museum*, 37: pl. 42, fig. 2.

*Cyclopecten thalassinus*: AMNH 191087, off Egmont Key, [Hillsborough County], Florida, 213.4 m (6.6 mm).

*Propeamussium cancellatum*: AMNH 166576, Cardenas, Matanzas, Cuba, 914 m (11.1 mm).

*Parvamussium pourtalesianum*: FMNH 194514, SE of Dry Tortugas, 91.4–213.4 m, 15 August 1974 (9.0 mm).

*Parvamussium sayanum*: FMNH 194541, SE of Dry Tortugas, 91.4–213.4 m, 15 August 1974 (unpaired RV, 10.5 mm); AMNH 166575, no locality data (unpaired LV, 8.4 mm).

*Spondylus americanus*: FMNH 177515, Garden Key, Dry Tortugas (84.9 mm).

*Spondylus americanus* [living specimen, showing veil]: FMNH 302068, FK-717, wreck of *Thunderbolt*, approx. 6 nmi S of Marathon, 24°39.68′N, 80°57.82′W, steel wreck with fouling bivalves, alcyonarians, and hydroids, 28.9–36.6 m, 19 August 2003 (shell width in image 80 mm).

Spondylidae hinge: Based on photographed specimens of *Spondylus americanus* (see above).

Spondylidae transparent clam: Viewed through the lower cemented right valve. Based on original dissection of *Spondylus americanus* (FMNH 302068, FK-717, same data as previous) and on the anatomy of *S. gaederopus* Linnaeus, 1758, by Dakin (1928).

*Spondylus ictericus*: FMNH 177525, Little Duck [Key]–Missouri [Key] bridge, 1955 (55.2 mm).

*Spondylus americanus* [juvenile external view]: BMSM 26237, Bahia Honda Key, after storms (15.4 mm).

*Spondylus ictericus* [juvenile external view]: BMSM 26238, Florida Keys, after storms (17.0 mm).

*Plicatula gibbosa*: FMNH 197534, SE of Dry Tortugas, 91.4–213.4 m, 15 August 1974 (24.3 mm).

*Oculina* coral habitat: *Oculina* Marine Protected Area, E Florida [between Cape Canaveral and Ft. Pierce], 76.2 m. Photograph by and courtesy of John Reed, Harbor Branch Oceanographic Institution.

Plicatulidae hinge: Based on photographed specimens of *Plicatula gibbosa* (see above).

Plicatulidae transparent clam: Based on original dissection of *Plicatula gibbosa* (AMNH 298913, FK-083, Gulf of Mexico, N of Rebecca Shoal, between Marquesas Keys and Dry Tortugas, 24°44.90′N, 82°36.10′W to 24°44.4′N, 82°37.40′W, 28.9 m, 22 April 1997) and on the anatomy of *Plicatula* spp. by Watson (1930).

*Anomia simplex*: FMNH 290343, FK-287, "Horseshoe" site, bayside of Spanish Harbor Keys, inside of (outermost) W arm, 24°39.35′N, 81°18.22′W, shallow subtidal, 10 April 2000 (29.9 mm).

*Anomia simplex* [living specimen on *Mya arenaria* shell]: FMNH 308036, vicinity of Woods Hole, Massachusetts (23 mm max.).

Anomiidae hinge: Based on photographed specimens of *Anomia simplex* (see above).

Anomiidae transparent clam: Viewed through the lower right valve. Based on original dissection of *Anomia simplex* (AMNH 298953, FK-352, "Horseshoe" site [S arm], bayside of Spanish Harbor Keys, 24°39.32′N, 81°18.22′W, to 1.5 m, rubble, 08 July 2000; FMNH 279470, FK-244, "Horseshoe" site, bayside of Spanish Harbor Keys, 24°39.32′N, 81°18.22′W, center of quarry, max. 7.0 m, rock wall and sand slope, 05 August 1999) and on the anatomy of *Anomia ephippium* Linnaeus, 1758, by Gray (1857).

*Pododesmus rudis*: USNM 92585, off [Dry] Tortugas, 43.9 m, sand, U.S. Fish Commission sta. 2413 (external view, 18.6 mm max.; internal view and external LV, 21.9 mm height).

*Eucrassatella speciosa*: FMNH 302046, Dry Tortugas, 30.5 m, sand, November 2003 (42.7 mm).

*Eucrassatella speciosa* [juvenile shells]: FMNH 306080, FK-342, NW corner of Looe Key National Marine Sanctuary, 24°33.43′N, 81°25.94′W, 8.0–9.1 m, seagrass, 07 July 2000 (juvenile LV, both sides, 8.4 mm).

Crassatellidae hinge: Based on photographed specimens of *Eucrassatella speciosa* (see above).

Crassatellidae transparent clam: Based on original dissection of *Eucrassatella speciosa* (AMNH

147695, 10 mi S of Alligator Point, Franklin County, Florida, bell buoy 26, 21.3 m, 06 September 1968) and on accounts of crassatellid anatomy by Harry (1966) and Taylor et al. (2005).

Crassatellidae [living dissected specimen]: *Eucrassatella donacina* (Lamarck, 1818), Esperance Bay, Western Australia, February 2003. Photograph by and courtesy of John Taylor, Natural History Museum, London.

*Crassinella dupliniana*: FMNH 311561, FK-559, S beach of Loggerhead Key, Dry Tortugas, 24°37.79′N, 082°55.40′W, wrack line to 2.1 m, 15 April 2002 (external view, 2.1 mm); FMNH 311562, FK-520, S of Halfmoon Shoal, 24°30.98′N, 82°27.58′W to 24°31.11′N, 82°27.69′W, shell hash, 12.2 m, 26 July 2001 (internal view, 2.3 mm).

*Crassinella lunulata*: FMNH 306072, FK-539, S of Bahia Honda Key, W of Looe Key reef, 24°34.24′N, 81°16.64′W, 30.2–34.1 m, sand and rubble, 28 July 2001 (4.8 mm).

*Crassinella martinicensis*: FMNH 194518, SE of Dry Tortugas, 91.4–213.4 m, 15 August 1974 (2.7 mm).

Lighthouse: Carysfort Reef, off Key Largo, approx. 25°13′N, 80°13′W (lighthouse is 34 m above water).

*Astarte smithii*: USNM 444846, off Key West, Pourtales Plateau, 5-6 mi S of Sand Key, 182.9 m, rocky, R/V *Eolis* sta. 15, 19 April 1910 (5.7 mm).

Astartidae [SEM of periostracum]: *Astarte* "nana," AMNH 247621, Alligator Light, June 1969 (detail of exterior right valve of 8.25 mm total length; field of view 0.7 mm).

Astartidae hinge: Based on photographed specimens of *Astarte smithii* (see above).

Astartidae transparent clam: Based on original dissection of *A. smithii* (AMNH 308090, FK-505, 24°19.88′N, 81°53.50′W, 221.3 m, sand on hard bottom, 25 July 2001) and *A. castanea* (Say, 1822) (AMNH 269891, New Jersey, 1973) and on accounts of *Astarte* anatomy by Saleuddin (1965, 1967).

*Carditamera floridana*: FMNH 311591, FK-629, "Horseshoe" site, bayside of Spanish Harbor Keys, 24°39.30′N, 81°18.20′W, to ca. 1 m, rocks along arms of quarry, 21 and 26 July 2002 (18.2 mm).

*Carditamera floridana* [living juvenile in algae]: FMNH 311619, FK-700, W of Pigeon Key (bayside of Tavernier, Plantation Key), 25°03.29′N, 80°30.69′W, 0.3–0.6 m, seagrass with algae, 12 June 2003 (length not recorded).

Carditidae hinge: Based on photographed specimens of *Carditamera floridana* (see above).

Carditidae transparent clam: Based on original dissection of *Carditamera floridana* (FMNH 288866, FK-188, Blackwater Sound, Florida Bay, E of Bush Point, 25°08.78′N, 80°25.31′W, seagrass, 1.8 m, 14 October 1998).

*Thecalia concamerata* [showing shelly brood pouch]: AMNH 308271, Simonstown, South Africa (11.8 mm).

*Astarte crenata subequilatera*: AMNH 264026, Greenland (22.0 mm).

*Astarte* "nana": AMNH 247621, Alligator Light, June 1969 (8.8 mm).

*Glans dominguensis*: FMNH 311570, FK-445, 24°38.37′N, 80°59.02′W, 57.3 m, muddy sand, 26 April 2001 (unpaired, 4.4, 3.9 mm).

*Pleuromeris tridentata*: FMNH 306104, FK-557, 24°40.09′N, 80°59.60′W, 7.3 m, sand with shells, 31 July 2001 (unpaired, 3.2, 3.6 mm).

*Pteromeris perplana*: FMNH 311571, FK-558, 24°39.76′N, 80°59.50′W, 8.2 m, hard bottom, sand, 31 July 2001 (unpaired, 3.1, 3.8 mm).

*Carditopsis smithii*: FMNH 306065, FK-551, 24°42.23′N, 81°00.23′W, 5.2 m, sandy mud, 31 July 2001 (unpaired, 1.4, 1.3 mm).

*Carditopsis smithii* [SEM of prodissoconch]: FMNH 311599, FK-444, 24°38.72′N, 80°59.14′W, 45.1 m, muddy sand, 26 April 2001 (total shell length 1.16 mm; prodissoconch length 0.4 mm).

Condylocardiidae hinge: Based on photographed specimens of *Carditopsis smithii* (see above).

Condylocardiidae transparent clam: Based on the anatomy of *Condylocardia notoaustralis* Cotton, 1930, by Middelfart (2002).

*Pandora inflata*: AMNH 270220, Egmont Key, [Hillsborough County], Florida, July 1964 (RV, external and internal view, 9.8 mm; articulated view, 12.0 mm); AMNH 144861, N of Key West, 24.4–45.7 m, September 1963 (LV, internal view, 11.4 mm).

Pandoridae hinge: Based on photographed specimens of *Pandora inflata* (see above).

Pandoridae transparent clam: Based on the anatomy of *Pandora gouldiana* Dall, 1886, by Boss & Merrill (1965).

*Pandora gouldiana* [living specimen]: AMNH transparency collection, Orient Point, [Suffolk County,] New York, August 1961, G. Raeihle, photographer (specimen length not recorded).

*Pandora bushiana*: USNM 444670a, 2 mi SE of Fort Jefferson, [Dry] Tortugas, 29.3 m, hard "bubbly" bottom, R/V *Eolis* sta. 33, 08 June 1911 (10.8 mm).

*Pandora glacialis*: ANSP 98665, *Albatross* sta. 2499, USFC (17.5 mm, larger LV measured).

Deepwater habitat with sponge and *Stylaster* coral: Alligator Hump Bioherm, Pourtales Terrace, Florida, 173 m. Photograph by and courtesy of John Reed, Harbor Branch Oceanographic Institution.

*Entodesma beana*: FMNH 227454, Missouri Key, in sponge, 1939 and 1941 (18.7 mm).

*Entodesma beana* [living specimen]: FMNH 295556, FK-138, Sister Creek, Boot Key, 24°41.88′N, 81°05.41′W to 24°41.92′N, 81°05.46′W, 1.8–3.6 m, mangrove-lined creek connecting Boot Harbor with ocean side, mud bottom with seagrass, 10 August 1997 (alcohol lot, 13.7 mm).

Lyonsiidae hinge: Based on photographed specimens of *Entodesma beana* (see above).

Lyonsiidae transparent clam: Based on original dissection of *Entodesma beana* (FMNH 295556, FK-138, Sister Creek, Boot Key, 24°41.88′N, 81°05.41′W to 24°41.92′N, 81°05.46′W, 1.8–3.6 m, mangrove-lined creek connecting Boot Harbor with ocean side, mud bottom with seagrass, 10 August 1997; AMNH 298018, FK-142, Rachel Bank, NNE of Marker 15, bayside of Key Vaca, 24°44.83′N, 81°04.77′W to 24°44.81′N, 81°04.90′W, 1.8–2.1 m, 17 August 1997) and on the anatomy of *Lyonsia californica* Conrad, 1837, by Narchi (1968).

*Lyonsia* in living position: Redrawn after Ansell (1967: 388) for *Lyonsia norvegica* (Gmelin, 1791).

*Lyonsia floridana*: HMNS 46936, Marathon, reef, April 1974 (6.8 mm).

*Lyonsia floridana* [living specimen]: Apalachee Bay, [NW] Florida, P. Mikkelsen, photographer (specimen length not recorded).

*Periploma margaritaceum*: AMNH 191023, Key West (12.9 mm).

*Periploma margaritaceum* [living specimen]: PSM-832, Indian River Lagoon, just inside St. Lucie Inlet, St. Lucie County, Florida, 0.1–0.7 m, subtidal sand, 27 April 1982 (length not recorded).

Periplomatidae hinge: Based on photographed specimens of *Periploma margaritaceum* (see above).

Periplomatidae transparent clam: Based on the anatomy of *Offadesma angasi* (Crosse & Fisher, 1864) by Morton (1981; plus later comments by Rosewater, 1984) and of *Cochlodesma praetenue* (Pulteney, 1799) by Allen (1958).

Periplomatids in living position: Redrawn after Morton (1981: 42) for *Offadesma angasi*.

Lithodesma of *Periploma planiusculum* [three views]: FMNH 149060, San Pedro Bay, [Los Angeles County,] California (shell 61 mm length, max. dimension of lithodesma 8.1 mm).

*Periploma tenerum*: USNM 64066, North Atlantic, seabed (unpaired syntypes?, external view, 9.0 mm, internal view, 14.0 mm).

*Grippina* sp. A: FMNH 311534, FK-454, 24°33.16′N, 80°57.31′W, 201.2 m, mud, pteropod ooze, 26 April 2001 (2.3 mm).

*Grippina* sp. A [SEM of shells and hinges]: FMNH 311600, FK-452, 24°34.54′N, 80°57.76′W, 187.7 m, mud, 26 April 2001 (external view 2.4 mm, internal views 2.35 mm).

Spheniopsidae hinge: Based on photographed specimens of *Grippina* sp. A (see above).

*Grippina* sp. B: FMNH 311574, FK-454, 24°33.16′N, 80°57.31′W, 201.2 m, mud, pteropod ooze, 26 April 2001 (unpaired, 3.1, 3.1 mm).

*Grippina* sp. B [SEM of shells and hinges]: FMNH 311601, FK-504, 24°20.66′N, 81°53.47′W, 209 m, blackish sand with manganese nodules, 25 July 2001 (external view 1.85 mm, internal view 1.95 mm).

*Thracia morrisoni*: USNM 53692, Key West (paired, 14.4 [+ inset], 14.0 mm).

Thraciidae hinge: Based on photographed specimens of *Thracia morrisoni* (see above).

Thraciidae transparent clam: Based on original dissection of *Thracia alfredensis* Bartsch, 1915, *T. diegensis* Dall, 1915, *Asthenothaerus villosior* Carpenter, 1864, and on the anatomies of

"*Sthenothaerus* sp." by Pelseneer (1911), *T. conradi* Couthouy, 1839, by Morse (1919), and *T. meridionalis* E. A. Smith, 1885, by Sartori & Domaneschi (2005).

*Thracia stimpsoni*: Scanned from W. H. Dall, 1890, [Scientific results of explorations by the US Fish Commission steamer *Albatross*.] VII.—Preliminary report on the collection of Mollusca and Brachiopoda obtained in 1887–'88, *Proceedings of the United States National Museum*, 12(773): pl. 13, fig. 2.

*Spinosipella acuticostata*: AMNH 190980, ESE of Key West, 256.0 m (13.6 mm).

Verticordiidae hinge: Based on photographed specimens of *Spinosipella acuticostata* (see above).

Verticordiidae transparent clam: Based on the anatomy of *Verticordia triangularis* Locard, 1898, by Allen & Turner (1974) and of *V.* (*Halicardia*) *perplicata* Dall, 1890, by Bernard (1974).

*Asthenothaerus hemphilli*: FMNH 227474, Missouri Key, sand (11.0 mm).

*Bushia elegans*: USNM 64067, Florida Straits, USFC sta. 2639, 102 m (syntype, 10.8 mm).

*Thracia phaseolina*: FMNH 203008, England (23.4 mm).

*Euciroa elegantissima*: AMNH 248456, Sand Key Light, Key West (33.8 mm).

*Haliris fischeriana*: AMNH 248460, Dry Tortugas (8.5 mm).

*Trigonulina ornata*: FMNH 194505, SE of Dry Tortugas, 91.4–213.4 m, 15 August 1974 (unpaired 2.7, 2.9 mm).

*Poromya granulata*: FMNH 311533, FK-454, 24°33.16′N, 80°57.31′W, 201.2 m, mud, pteropod ooze, 26 April 2001 (unpaired, 4.6, 4.4 mm).

Poromyidae hinge: Based on photographed specimens of *Poromya granulata* (see above).

Poromyidae transparent clam: Based on the anatomy of *Poromya granulata* by Yonge (1928) and Morton (1981).

Poromyid in feeding position: Redrawn after Morton (1981: fig. 4).

*Poromya albida*: USNM 64033, off Havana, Cuba, USFC sta. 2159, 179 m (holotype, unpaired LV, 21.5 mm).

*Poromya rostrata*: BMSM 26240, Sombrero Light, 109.7 m (unpaired, 7.3, 6.4 mm).

Deep-sea habitat with stalked glass sponge: Long Island, Bahamas, deep island slope, 806 m. Photograph by and courtesy of John Reed, Harbor Branch Oceanographic Institution.

*Cuspidaria rostrata*: AMNH 248472, Dry Tortugas (21.4 mm).

Cuspidariidae hinge: Based on photographed specimens of *Cuspidaria rostrata* (see above).

Cuspidariidae transparent clam: Based on the anatomy of *Cuspidaria obesa* (Lovén, 1846) by Allen & Morgan (1981) and of *Cuspidaria* spp. by Pelseneer (1911).

*Cardiomya gemma* [living specimen]: DMNH 201891, PSM-806, Indian River Lagoon, St. Lucie County, Florida, 27°11.40′N, 80°11.10′W, 1–1.3 m, sand bar with *Halodule* seagrass, 26 January 1982 (3.58 mm).

*Cardiomya alternata*: Scanned from original description by A. d'Orbigny, 1853, *Histoire Physique, Politique et Naturelle de l'Ile de Cuba*, by R. de la Sagra, Atlas: pl. 27, figs. 17–20.

*Cardiomya costellata*: AMNH 248468, Florida Straits (11.5 mm).

*Cardiomya gemma*: FMNH 202826, Bradenton Beach, Manatee County, Florida, 1942 (5.1 mm).

*Cardiomya perrostrata*: FMNH 311572, FK-451, 24°35.25′N, 80°58.00′W, 154.8 m, mud, 26 April 2001 (unpaired, 5.2, 3.9 mm).

*Cardiomya ornatissima*: AMNH 296786, FK-085, NNW of East Key, Dry Tortugas, 24°49.20′N, 82°43.90′W, 33.2–33.5 m, thick clay, 22 April 1997 (external view, 11.7 mm); AMNH 248469, Dry Tortugas (internal view, 13.1 mm); scanned from original description by A. d'Orbigny, 1853, *Histoire Physique, Politique et Naturelle de l'Ile de Cuba*, by R. de la Sagra, Atlas: pl. 27, figs. 13–16 (right pair).

*Cardiomya striata*: USNM 460734, off Western Dry Rocks beacon, Pourtales Plateau, 164.6 m, rocky, R/V *Eolis* sta. 319, 27 May 1916 (13.6 mm).

*Cuspidaria obesa*: AMNH 248936, Egmont Key, [Hillsborough County,] Florida, 51.7 m, April 1966, Hicks Collection (6.8 mm).

*Myonera gigantea*: AMNH 248473, Key West, Florida Straits (20.4 mm).

*Myonera lamellifera*: AMNH 294926, Sombrero Light, 137.2 m, July 1952 (5.3 mm).

*Myonera paucistriata*: USNM 460719, [Dry] Tortugas, 2 mi SE of Fort Jefferson, 29 m, hard "bubbly" bottom, R/V *Eolis* sta. 33, 08 June 1911 (unpaired LV, 9.9 mm).

*Plectodon granulatus*: AMNH 248470, Alligator Light (12.0 mm).

*Codakia orbicularis*: FMNH 176528, Little Duck Key (55.3 mm).

*Codakia orbicularis* [living juveniles on sand]: FMNH 311575, FK-730, Newfound Harbor coral lumps, oceanside, 24°36.89′N, 81°23.63′W, 2.4–2.7 m, *Thalassia* seagrass, sand patches, coral heads, 04 May 2004 (smallest specimen, 8.8 mm).

Lucinidae hinge: Based on photographed specimens of *Codakia orbicularis* (see above).

Lucinidae transparent clam: Based on original dissection of *Codakia orbicularis* (FMNH 295647, FK-139, Florida Bay, E of Bethel Bank, 24°43.96′N, 81°07.61′W, 2.4 m, *Halimeda* rubble, seagrass, 10 August 1997; AMNH 298124, FK-153, Molasses Keys, ca. 1.2 nmi S of Seven-Mile Bridge, 24°41.18′N, 81°11.41′W, ironshore/sand beaches, hardwood/mangrove, seagrass, *Halimeda*, corals, sponges, to 0.6 m, 25 August 1997).

Lucinid gills [SEM with bacteriocytes]: Courtesy of John Taylor, Natural History Museum, London.

*Anodontia alba*: FMNH 301628, FK-661, Molasses Keys, S of center of Seven Mile Bridge, N of westernmost island, 24°41.07′N, 81°11.48′W, 0.3–1.8 m, sandy bottom, coral rubble, *Thalassia* seagrass, 04 August 2002 (46.1 mm).

*Anodontia schrammi*: AMNH 263156, Big Pine Key, shallow water (79.5 mm).

*Lucina pensylvanica*: FMNH 176532, Little Duck Key (33.7 mm).

*Phacoides pectinata*: FMNH 288709, Greyhound [now Fiesta] Key, 1964 (43.9 mm).

*Stewartia floridana*: UMML 28:1599, Long Key (35.2 mm).

*Lucinoma filosa*: FMNH 306143, TC-072197-4, Dry Tortugas, S of Rebecca Shoal, 24°17.92′N, 82°32.73′W, 215 m, 21 July 1997 (24.4 mm).

*Ctena orbiculata*: FMNH 202913, Bonefish Key (11.0 mm).

*Ctena pectinella*: FMNH 311576, FK-527, 24°24.34′N, 82°28.05′W, 47.2 m, soft mud, 27 July 2001 (4.3 mm).

*"Parvilucina" costata*: USNM 446469, 2 mi SE of Fort Jefferson, Dry Tortugas, 29.3 m, hard "bubbly" bottom, R/V *Eolis* sta. 33, 08 June 1911 (8.7 mm).

*Callucina keenae*: FMNH 311577, FK-621, "Long Key Artificial Reefs," oceanside of Long Key, 24°44.78′N, 80°50.00′W, 7 m, sand plain with *Thalassia/Syringodeum* seagrass patches, 17 July 2002 (unpaired, 18.2, 16.3 mm).

*Parvilucina crenella*: FMNH 227432, Grassy Key, 1940 (4.6 mm).

*Parvilucina crenella* [living specimens]: FMNH 311602, FK-219, Mud Key Channel [bayside of Big Coppitt Key area], 24°40.52′N, 81°41.71′W, 3.3–3.9 m, very soft anoxic mud, 19 April 1999 (alcohol lot, largest specimen 3.5 mm).

*Cavilinga blanda*: FMNH 311578, FK-385, oceanside of Snake Creek, 24°54.43′N, 80°32.65′W, 6.4 m, clean rippled sand, 15 October 2000 (alcohol lot, 3.9 mm).

*Pleurolucina leucocyma*: FMNH 306087, FK-447, 24°37.64′N, 80°58.79′W, 76.5 m, muddy sand, 26 April 2001 (unpaired, 5.1, 5.6 mm).

*Pleurolucina sombrerensis*: FMNH 195996, SE of Dry Tortugas, 91.4-213.4 m, 15 August 1974 (unpaired, 6.0, 6.2 mm).

*Radiolucina amianta*: FMNH 306088, FK-346, 24°35.09′N, 81°25.91′W to 24°35.12′N, 81°25.89′W, 7.3 m, *Syringodeum* seagrass, muddy sand, 07 July 2000 (4.0 mm).

*Myrtea sagrinata*: USNM 64277, off Bahia Honda, Cuba, R/V *Blake* sta. 21, 525 m (syntype, 6.0 mm).

*Myrteopsis lens*: ANSP 102206, off Martha's Vineyard, [Massachusetts], 1880, USFC sta. 876 (13.0 mm).

*Lucinisca nassula*: FMNH 311596, FK-370, Everglades National Park, E part of Rabbit Key Basin, S of Rabbit Key, bayside of Long Key, 24°58.64′N, 80°50.36′W, 1.2–1.5 m, dense *Thalassia* seagrass, 12 October 2000 (12.8 mm).

*Lucinisca muricata*: AMNH 29548, St. Thomas (14.8 mm).

*Divalinga quadrisulcata*: FMNH 94727, Grassy Key (17.8 mm).

*Divaricella dentata*: FMNH 177584, Bahia Honda, 1942 (29.4 mm).

Mangrove habitat: Lake Surprise, Key Largo, March 1996.

*Diplodonta punctata*: FMNH 176494, Missouri Key (15.0 mm).

Ungulinidae hinge: Based on photographed specimens of *Diplodonta punctata* (see above).

Ungulinidae transparent clam: Based on the anatomy of *Diplodonta punctata* and *Phlyctiderma semiaspera*, both by Allen (1958).

*Phlyctiderma semiaspera* [living specimens]: FMNH 311579, FK-730, Newfound Harbor coral lumps, oceanside, 24°36.89′N, 81°23.63′W, 2.4–2.7 m, *Thalassia* seagrass, sand patches, coral heads, 04 May 2004 (upper left specimen, 8.8 mm).

*Diplodonta notata*: ANSP 228533, N of Mayaguez, Puerto Rico, 19 April 1953 (8.1 mm).

*Diplodonta nucleiformis*: AMNH 292819, Sandy Point, St. Croix, Virgin Islands (unpaired, 8.6, 8.3 mm).

*Phlyctiderma semiaspera*: FMNH 26370, Missouri Key (8.7 mm).

*Phlyctiderma soror*: ANSP 6503, Jamaica (syntype, 11.6 mm).

*Diplodonta punctata* [SEM sculpture x 2]: FMNH 311580, FK-078, Gulf of Mexico, due E of New Ground, NW of Marquesas Keys, 24°40.67′N, 82°16.77′W, 12.2 m, 22 April 1997 (single LV for external sculptural detail, 8.3 mm)

*Diplodonta notata* [SEM sculpture x 2]: FMNH 311581, FK-082, Gulf of Mexico, NNW of New Ground, between Marquesas Keys and Dry Tortugas, 24°48.28′N, 82°28.70′W, 27 m, 22 April 1997 (single LV for external sculptural detail, 6.2 mm)

*Phlyctiderma soror* [SEM sculpture x 2]: FMNH 311582, FK-139, Florida Bay, E of Bethel Bank, 24°43.96′N, 81°07.61′W, 2.4 m, *Halimeda* rubble, seagrass, 10 August 1997 (single RV for external sculptural detail, 3.7 mm)

*Thyasira trisinuata*: FMNH 311583, FK-443, 24°39.06′N, 80°59.26′W, 36.0 m, shelly sand, 26 April 2001 (external view, 3.4 mm); UMML 28:132, 6 mi SE of Molasses Reef Light, 223 m, 09 July 1950 (internal view, 7.6 mm).

Thyasiridae hinge: Based on photographed specimens of *Thyasira trisinuata* (see above).

Thyasiridae transparent clam: Based on the anatomy of *Thyasira trisinuata* by Payne & Allen (1991) and of *T. flexuosa* (Montagu, 1803) by Allen (1958).

Thyasirid in living position: Redrawn after Ockelmann in Oliver & Killeen (2002: 15).

*Axinus grandis*: Scanned from original description by A. E. Verrill, 1885, Third catalogue of Mollusca recently added to the fauna of the New England coast and the adjacent parts of the Atlantic, consisting mostly of deep-sea species, with notes on others previously recorded, *Transactions of the Connecticut Academy*, 6: pl. 44, fig. 22.

*Chama macerophylla*: FMNH 227408, Missouri Key, Ohio [Key]–Missouri [Key] Channel, shallow water, on styrofoam buoy, 08 Mar 1987 (57.6 mm).

*Chama macerophylla* [living specimens on reef]: FMNH 306132, FK-725, "Horseshoe" site, bayside of Spanish Harbor Keys, 24°39.30′N, 81°18.20′W, intertidal and shallow subtidal, rocks/rubble along arms of quarry, 29 April 2004.

Chamidae hinge: Based on photographed specimens of *Chama macerophylla* (see above).

Chamidae transparent clam: Based on original dissection of *Chama macerophylla* (FMNH 295681, FK-117, Key Vaca, channel W of Stirrup Key, bayside, 24°44.19′N, 81°02.92′W, *Thalassia* seagrass, sand, 4.6 m max., 14 July 1997; FMNH 311603, FK-685, The Rocks [inside patch reef], oceanside off Windley Key, 24°57.25′N, 80°32.88′W, patch reef with surrounding hard bottom covered by thin veneer of sand, many gorgonians, max. 4.2 m, 04 June 2003) and on chamid anatomy by Yonge (1967).

*Chama congregata*: FMNH 288725, Missouri Key, intertidal, attached to large loose rocks on sand, 18 July 1970 (23.9 mm max.).

*Chama florida*: AMNH 232616, Bahia Honda beach, after Hurricane Inez, October 1966 (20.7 mm).

*Chama inezae*: HMNS 35492, off Carysfort Reef, 18.3 m, on old steamship wreck (23.1 mm, measured unattached valve).

*Chama radians*: AMNH 244005, off Missouri Key, 1968 (38.9 mm max.).

*Chama sarda*: AMNH 276865, Fowey Light, [Dade County], Florida (21.6 mm, measured max. of attached valve).

*Chama sinuosa*: FMNH 183141, Bonefish Key (39.9 mm, measured max. of attached valve).

*Chama lactuca*: FLMNH 135481, off Dodge Estate, Palm Beach, Palm Beach County, Florida, rocky reef, 45.7 m, 28 July 1950 (19.3 mm max. diameter [attached valve]).

*Arcinella cornuta*: FMNH 163680, Florida Keys, 1948 (30.1 mm); FMNH 311584, FK-527, 24°24.34′N, 82°28.05′W, 47.2 m, soft mud, 27 July 2001 (juvenile LV on adult shell, 2.0 mm).

Coral reef scene: Conch Reef.

*Lasaea adansoni*: FMNH 311621, Key Colony Beach, Marathon, tidal channel, intertidal rocks, 07 October 1978 (alcohol lot, 2.4 mm).

Lasaeidae hinge: Based on photographed specimens of *Lasaea adansoni* (see above).

Lasaeidae transparent clam: Based on the anatomy of *Lasaea adansoni* by Oldfield (1955) and Pelseneer (1911).

*Mysella planulata* [living specimen]: FMNH 311622, FK-052, Indian Key Fill, mile marker 79, bayside, 24°53.42′N, 80°40.47′W, shallow subtidal, mixed *Thalassia*, *Halimeda*, and *Halodule* seagrass, 23 September 1996 (length not recorded).

*Erycina periscopiana*: FMNH 311585, FK-332, off Looe Key, 24°30.28′N, 81°25.90′W to 24°30.37′N, 81°25.83′W, 78.9–80.5 m, mud, 06 July 2000 (external view, 3.7 mm); FMNH 306081, FK-443, 24°39.06′N, 80°59.26′W, 36.0 m, shelly sand, 26 April 2001 (internal view, 2.8 mm).

*Kellia suborbicularis*: AMNH 106020, Taboga Island, Bay of Panama, Pacific Panama, 1931 (7.2 mm).

*Mysella planulata*: FMNH 306085, FK-063, Missouri Key, 24°40.48′N, 81°14.35′W, bayside of Missouri–Ohio Key bridge, 0–1 m, *Thalassia* seagrass, 14 April 1997 (2.2 mm).

*Orobitella floridana*: FMNH 306089, FK-154, Cocoanut Key, 4 nmi due N of Little Duck Key, 24°44.73′N, 81°14.00′W, 0.9–1.5 m, seagrass, sponges, gorgonians, 25 August 1997 (external view, 8.7 mm); FMNH 311593, FK-733, Looe Key, backreef, 24°32.89′N, 81°24.36′W, 1.2–1.8 m, rubble, sand, seagrass, 05 May 2004 (internal view, 10.9 mm).

*Divariscintilla yoyo* [living specimen]: Ft. Pierce Inlet, Indian River Lagoon, St. Lucie County, Florida, 27°28.30′N, 80°17.90′W, intertidal sand, in burrows of mantis shrimp *Lysiosquilla scabricauda* (Lamarck, 1818) (body without foot and tentacles ca. 12 mm).

*Divariscintilla luteocrinita* [living specimen]: same data as previous (body without foot and tentacles ca. 7 mm).

*Divariscintilla octotentaculata* [living specimen]: same data as previous (each body without foot and tentacles ca. 7 mm).

*Hiatella arctica*: FMNH 194525, SE of Dry Tortugas, 91.4–213.4 m, 15 August 1974 (11.7 mm).

*Hiatella* in living position: Redrawn after Hunter (1949: figs. 9 and 11).

Hiatellidae hinge: Based on photographed specimens of *Hiatella arctica* (see above).

Hiatellidae transparent clam: Based on the anatomy of *Hiatella arctica* (as *H. solida* (Sowerby, 1834)) by Narchi (1973) and of *Hiatella* spp. by Yonge (1971).

*Panopea abrupta* [living specimen]: Strait of Georgia area, British Columbia, Canada, late 1970s (specimen length not recorded). Photograph by and courtesy of Neil Bourne, formerly of Canada's Pacific Biological Station.

*Hiatella azaria*: Scanned from W. H. Dall, 1886, Report on the results of dredging by the United States Coast Survey Steamer "*Blake*." XXIX. Report on the Mollusca. Part I. Brachiopoda and Pelecypoda, *Bulletin of the Museum of Comparative Zoology, Harvard University*, 12(6): pl. 4, figs. 9a–b.

*Lamychaena hians*: FMNH 189226, Bonefish Key, 1939 (24.5 mm).

*Lamychaena hians* [ventral view of preserved specimen]: FMNH 295555, FK-236, Coffins Patch coral reef, oceanside of Grassy Key, 24°41.08′N, 80°57.47′W, max. 4.9 m, scuzzy algae covering reef, 02 August 1999 (3.1 mm).

*Lamychaena hians* [living specimen]: AMNH 298016, FK-202, Molasses Reef, 25°00.55′N, 80°22.58′W, 7.3 m, spur and groove reef, 08 April 1999 (shell 15 mm, extended length 55 mm).

Gastrochaenidae hinge: Based on photographed specimens of *Lamychaena hians* (see above).

Gastrochaenidae transparent clam: Based on original dissection of *Lamychaena hians* (FMNH

295554, FK-206, French Reef, 25°02.15′N, 80°21.08′W, 8.5 m, spur and groove reef, rubble, sand plains, 11 April 1999; AMNH 298016, FK-202, Molasses Reef, 25°00.55′N, 80°22.58′W, 7.3 m, spur and groove reef, 08 April 1999) and on the anatomy of the same species by Carter (1978).

*Lamychaena hians* in living position within coral head: Redrawn after Carter (1978: fig. 44).

*Gastrochaena* cf. *ovata*: AMNH 296056, FK-164, The Triangles, off Key Largo, 25°00.58′N, 80°27.47′W, 48.8 m, gorgonian reef, 13 September 1998 (13.4 mm).

*Spengleria rostrata*: FMNH 188359, Bonefish Key and Marathon, 1939 (27.3 mm).

Gastrochaenid burrow opening [chimney in *Diploria*]: FK-692, Hens and Chickens patch reef, oceanside of Plantation Key, 24°56.10′N, 80°32.92′W, max. 6.4 m, 08 June 2003 (not collected or measured, width of double opening approx. 10 mm).

Gastrochaenid burrow opening [chimney in rock with algae]: FK-047, Channel marker 50A off Ramrod Key, 24°35.80′N, 81°27.24′W, 4.6 m, rubble and patch reef, 21 September 1996 (not collected or measured, width of double opening approx. 10 mm).

*Coralliophaga coralliophaga*: FMNH 188801, Garden Key, Dry Tortugas, 1941 (42.8 mm).

*Coralliophaga coralliophaga* [living juvenile]: FMNH 311623, FK-200, Molasses Reef, W of lighthouse, 25°00.55′N, 80°22.58′W, 4.9–6.1 m, spur and groove reef, 07 April 1999 (10.5 mm).

*Coralliophaga coralliophaga* in living position: Redrawn after Morton (1980: 313).

Trapezidae hinge: Based on photographed specimens of *Coralliophaga coralliophaga* (see above).

Trapezidae transparent clam: Based on the anatomy of *Coralliophaga coralliophaga* by Morton (1980).

*Basterotia elliptica*: USNM 447002, 5 mi off N entrance to Key West [harbor], 12.8 m, sand, R/V *Eolis* sta. 30, 03 June 1911 (unpaired, 6.6, 5.4 mm).

Sportellidae hinge: Based on photographed specimens of *Basterotia elliptica* (see above).

Sportellidae transparent clam: Based on the anatomy of *Anisodonta alata* (Powell, 1952) by Ponder (1971).

*Basterotia quadrata*: AMNH 198523, St. Croix, Virgin Islands, W coast (external view, 12.4 mm); FMNH 311586, FK-332, off Looe Key, 24°30.28′N, 81°25.90′W to 24°30.37′N, 81°25.83′W, 78.9–80.5 m, mud, 06 July 2000 (internal view, 8.7 mm).

*Ensitellops protexta*: ANSP 304602, Aransas Causeway, between Aransas Pass and Mustang Island, Texas, September 1958 (unpaired, 6.6, 7.8 mm).

Deep-sea habitat with dugong rib: Marathon Sinkhole, Pourtales Terrace, Florida, bottom of hole, 521 m. Photograph by and courtesy of John Reed, Harbor Branch Oceanographic Institution.

*Polymesoda floridana*: FMNH 188692, Key West, 1938 (24.3 mm).

Ohio Key habitat: FK-723, Ohio Key, landlocked "pond" adjacent to Ohio–Missouri Key bridge, oceanside, 24°40.34′N, 81°14.60′W, dried mud and ankle-deep water, 29 April 2004.

*Polymesoda floridana* [living specimens on cracked mud]: FMNH 311624, FK-723, same locality as previous.

*Polymesoda floridana* [polychromic shells]: FMNH 311624, FK-723, same locality as previous.

Corbiculidae hinge: Based on photographed specimens of *Polymesoda floridana* (see above).

Corbiculidae transparent clam: Based on original dissection of *Polymesoda floridana* (FMNH 288869, FK-056, Ohio Key, landlocked "pond" adjacent to Ohio–Missouri Key bridge, 24°40.34′N, 81°14.60′W, ankle-deep water, 26 September 1996).

*Corbicula fluminea*: FMNH 198565, Okatibbee Lake, Lauderdale County, Mississippi, 21 October 1978 (32.7 mm)

*Laevicardium serratum*: FMNH 227405, Marathon (31.1 mm).

*Laevicardium serratum* [living juvenile]: FMNH 295696, FK-133, Sister Creek (mangrove-lined creek connecting Boot Harbor with oceanside), Boot Key, 24°41.94′N, 81°05.46′W, 3.3–0.9 m, mud bottom with seagrass, 07 August 1997 (alcohol specimen, 16 mm).

Cardiidae hinge: Based on photographed specimens of *Laevicardium serratum* (see above).

Cardiidae transparent clam: Based on original dissection of *Laevicardium serratum* (FMNH 295693, FK-148, Grassy Key Bank, 0.3 nmi S of Marker 11, N of Grassy Key, bayside, 24°47.24′N, 80°57.66′W, sand, rubble patches, seagrass, *Oculina* rubble, *Halimeda* hash, *Penicillus*, 22 August 1997; AMNH 298180, FK-139, Florida Bay, E of Bethel Bank, 24°43.96′N, 81°07.61′W, 2.4 m, *Halimeda* rubble, seagrass, 10 August 1997).

*Dinocardium robustum* [living specimen]: AMNH transparency collection, no locality recorded, photographer unknown (shell approx. 110 mm max.).

Shell Beach with *Microfragum erugatum*: RB-1889, Shell Beach, Western Australia, Australia, 26°12.892′S, 113°46.547′E (length of specimens approx. 10 mm, FMNH 308503).

*Laevicardium mortoni*: AMNH 243061, Middle Torch Key (25.3 mm).

*Laevicardium pictum*: FMNH 164110, Florida Keys, 1950 (18.6 mm).

*Microcardium peramabile*: FMNH 197526, SE of Dry Tortugas, 91.4–213.4 m, 15 August 1974 (unpaired, 5.6, 7.1 mm).

*Microcardium tinctum*: AMNH 144862, N of Key West, 24.4–45.7 m, September 1963 (10.0 mm).

*Tridacna gigas* [living specimen]: Helen Atoll, Palau, lagoon channel slope, 15 m. Photograph by and courtesy of Robert F. Myers, Coral Graphics.

*Trachycardium egmontianum*: FMNH 176760, Little Duck Key. (38.7 mm).

*Dallocardia muricata*: FMNH 176778, Little Duck Key (39.2 mm).

*Acrosterigma magnum*: FMNH 311592, FK-017, Horseshoe Reef, off Key Vaca, 24°39.91′N, 80°59.56′W, max. 8.8 m, patch reef, ledges, sand patches, 08 July 1995 (58.0 mm).

*Dinocardium robustum*: FMNH 176765, Ft. Pierce, St. Lucie County, Florida (61.0 mm height).

*Ctenocardia media*: FMNH 176774, Little Duck Key (28.7 mm).

*Ctenocardia guppyi*: AMNH 295839, FK-142, Rachel Bank, NNE of Marker 15, bayside of Key Vaca, 24°44.83′N, 81°04.77′W to 24°44.81′N, 81°04.90′W, 1.8–2.1 m, 17 August 1997 (4.0 mm).

*Papyridea lata*: FMNH 182759, Missouri Key (29.6 mm).

*Papyridea soleniformis*: FMNH 301980, FK-706, Elbow Reef, E of Coffins Patch (off Crawl Key), 24°41.55′N, 80°56.79′W, 5.8 m, patch reef surrounded by sand flats, 09 August 2003 (47.0 mm).

*Papyridea semisulcata*: AMNH 295946, FK-119, "The Slabs" between "outer patches" and "coral humps" off Marathon, 24°39.53′N, 81°00.90′W, 7.0 m, patch reef, 20 July 1997 (7.4 mm).

*Trigoniocardia antillarum*: FMNH 185745, St. Thomas, West Indies (7.6 mm height).

*Periglypta listeri*: FMNH 302079, FK-703, Coffins Patch coral reef, oceanside of Grassy Key, 24°41.16′N, 80°57.84′W, 2.4–5.5 m, patch reef with sand pockets, 07 August 2003 (external, dorsal, anterior views, 44.0 mm); FMNH 301426, FK-287, "Horseshoe" site, bayside of Spanish Harbor Keys, inside of (outermost) W arm, 24°39.35′N, 81°18.22′W, shallow subtidal, 10 April 2000 (internal view, 66.8 mm).

*Periglypta listeri* [living specimen]: FMNH 295706, FK-117, Key Vaca, channel W of Stirrup Key, bayside, 24°44.19′N, 81°02.92′W, *Thalassia* seagrass, sand, 4.6 m max., 14 July 1997 (56.3 mm).

*Periglypta listeri* [siphons in sand]: FK-244, "Horseshoe" site, bayside of Spanish Harbor Keys, 24°39.32′N, 81°18′13′′W, rocky slope with sand pockets, 05 August 1999 (specimens not collected).

Veneridae hinge [*Periglypta*]: Based on photographed specimens of *Periglypta listeri* (see above).

Veneridae transparent clam: Based on original dissection (see Bieler et al., 2004).

*Petricola lapicida*: FMNH 189224, Bonefish Key, 1939 (30.7 mm).

Veneridae hinge [*Petricola*]: Based on photographed specimens of *Petricola lapicida* (see above).

*Petricola lapicida* [siphons emerging from rock]: FMNH 311587, FK-725, "Horseshoe" site, bayside of Spanish Harbor Keys, 24°39.30′N, 81°18.20′W, intertidal and shallow subtidal, rocks/rubble along arms of quarry, 29 April 2004.

*Petricola* sp. [living specimen]: AMNH 298891, FK-047, Channel marker 50A off Ramrod Key, 24°35.80′N, 81°27.24′W, 4.6 m, rubble and patch reef, 21 September 1996 (29 mm).

*Petricolaria pholadiformis* [living specimens]: AMNH transparency collection, Bayville, Long Island, New York, 1961, G. Raeihle, photographer (specimen lengths not recorded).

*Choristodon robustus* [living specimen]: AMNH transparency collection, locality not recorded, G. Raeihle, photographer (specimen length not recorded).

*Petricolaria pholadiformis* [death assemblage on beach]: AMNH transparency collection, Bayville, Long Island, New York, September 1965, G. Raeihle, photographer (specimen lengths not recorded).

*Mercenaria mercenaria* [dish of prepared steamed clams]: taken at commercial restaurant, Marathon, September 1998, P. M. Mikkelsen, photographer (dish ca. 25 mm diameter).

*Macrocallista maculata*: FMNH 176333, off Key West, 4.6 m (55.6 mm).

*Macrocallista nimbosa*: FMNH 166375, Dry Tortugas, 54.9 m (unpaired, 81.4, 91.7 mm).

*Callista eucymata*: AMNH 129660, Key Largo (unpaired, 26.0, 18.7 mm).

*Mercenaria mercenaria*: AMNH 248834, Matanzas Inlet [15 mi S of St. Augustine], [St. Johns County], Florida, April 1969 (form *notata* Say, 1822, 68.9 mm).

*Mercenaria mercenaria* [pearl]: AMNH transparency collection, locality data not recorded, G. Raeihle, photographer (size not recorded).

*Mercenaria campechiensis*: FMNH 223476, SE end of Sunshine Skyway bridge, S of St. Petersburg, Manatee County, Florida, 27°36.70′N, 82°36.30′W, sand and seagrass beds, shallow (0–1 m), 24 June 1991 (98.0 mm).

*Dosinia discus*: FMNH 185255, Key West (57.0 mm).

*Dosinia elegans*: UMML 28:1192, Long Key (58.4 mm).

*Globivenus rugatina*: ANSP 397860, Bloody Bay, Tobago, 11°18′N, 60°38′W, 1979–1981 (23.1 mm).

*Globivenus rigida*: AMNH 248310, Dry Tortugas (69.0 mm).

*Circomphalus strigillinus*: AMNH 156371, 210 degrees off Sombrero Key Light, 64.0 m, rubble, 29 June 1952 (38.5 mm).

*Cyclinella tenuis*: FMNH 306069, FK-104, NNE of New Ground, between Marquesas Keys and Dry Tortugas, 24°47.30′N, 82°18.30′W, 22.9 m, 25 April 1997 (external view, 9.0 mm); FMNH 306070, FK-133, Sister Creek (mangrove-lined creek connecting Boot Harbor with oceanside), Boot Key, 24°41.94′N, 81°05.46′W, 3.3–0.9 m, mud bottom with seagrass, 07 August 1997 (internal view, 7.9 mm).

*Chione elevata*: FMNH 176349, Bonefish Key (24.6 mm).

*Chione mazyckii*: FMNH 190789, Missouri Key, 1950 (6.6 mm).

*Chione cancellata*: FMNH 311547, Guadeloupe, ca. 2002 (32.8 mm).

*Lirophora clenchi*: AMNH 165871, [off] Port Isabel, [Cameron County,] Texas, 55 m, (26.4 mm).

*Lirophora* cf. *latilirata*: FMNH 176350, off Key West, 1951 (30.8 mm).

*Lirophora paphia*: AMNH 243811, SSW of Dry Tortugas, 76.2 m, 1966 (44.4 mm).

*Chionopsis intapurpurea*: FMNH 121622, S side of Key West, reef (34.1 mm).

*Chionopsis pubera*: FMNH 283524, off N Santos, São Paulo State, Brazil, 50–60 m (69.3 mm).

*Timoclea pygmaea*: AMNH 296555, FK-115, East Washerwoman Shoal, off Key Vaca, 24°40′N, 81°04.3′W, max. 2.7 m, 12 July 1997 (7.6 mm).

*Timoclea grus*: AMNH 296544, FK-085/104, vicinity of Marquesas Keys and Dry Tortugas, 22.9–33.5 m, April 1997 (unpaired, 7.0, 7.1 mm).

*Gouldia cerina*: AMNH 296516, FK-136, E of Bethel Bank, 24°44.80′N, 81°04.70′W to 24°44.67′N, 81°04.71′W, 1.8–2.1 m, sand, sparse seagrass, occasional gorgonians, algae, 07 August 1997 (5.3 mm).

*Anomalocardia cuneimeris*: FMNH 176386, Key West, 1938 (21.4 mm).

*Pitarenus cordatus*: FMNH 77934, Key West, 1950 (paratype, 42.3 mm).

*Pitar albidus*: FMNH 77092, Kingston, Jamaica, 1928 (25.8 mm).

*Pitar circinatus*: FMNH 185156, Nevis, West Indies (29.0 mm).

*Pitar dione*: FMNH 165480, West Indies, 1967 (44.2 mm, measured without spines).

*Pitar fulminatus*: FMNH 26422, Lake Worth, Boynton, [Palm Beach County], Florida (42.1 mm).

*Pitar simpsoni*: FMNH 311594, FK-681, bayside of Tavernier, Plantation Key, off small mangrove island near border of Everglades National Park, 25°01.95′N, 80°31.17′W, 0.6 m, pavement with sand, sparse to thick *Thalassia* seagrass, 12 April 2003 (10.4 mm).

*Pitar simpsoni* [SEM of calcified periostracum]: AMNH 232626, Islamorada, 3 m, July 1962 (LV, 8.2 mm, 2-μm crop).

*Transennella conradina*: FMNH 190835, Grassy Key (5.5 mm).

*Transennella cubaniana*: FMNH 306111, FK-385, oceanside of Snake Creek, 24°54.43′N, 80°32.65′W, 6.4 m, clean rippled sand, 15 October 2000 (4.8 mm).

*Transennella cubaniana* [SEM of posterodorsal shell margin]: FMNH 311604, FK-558, 24°39.76′N, 80°59.50′W, 8.2 m, hard bottom, sand, 31 July 2001 (LV, 5.4 mm).

*Transennella culebrana*: USNM 160064, off Culebra Island, Puerto Rico, 25–33 m (holotype, 7.0 mm).

*Transennella stimpsoni*: USNM 54100, Egmont Key, Tampa Bay, [Hillsborough County], Florida (holotype, unpaired LV, 13.65 mm).

*Gemma gemma*: HMNS 1125, Crawl Key, sand in ledges of old quarry below tideline, 1977 (3.8 mm).

*Parastarte triquetra*: FMNH 288766, RB-1602, Middle Torch Key, N end of road, 24°24'N, 81°24.7'W, bayside, Pine Channel, rock washing, shallow water at low tide, base of small rocks (15–30 cm) on soft sand, 27 June 1993 (1.4 mm).

*Tivela abaconis*: FMNH 190870, Grassy Key, 1939 (6.1 mm).

*Tivela floridana*: FMNH 190791, Grassy Key (6.8 mm).

*Tivela mactroides*: AMNH 306229, Espiritu Santo, Mexico, shallow water, sand (37.7 mm).

*Cooperella atlantica*: FMNH 311588, FK-345, Hawk Channel, 24°34.33' to 34.40'N, 81°25.93' to 25.91'W, 11.9 m, gray shelly mud, 07 July 2000 (4.0 mm).

*Choristodon robustus*: FMNH 188788, Bonefish Key, 1939 (24.6 mm).

*Petricolaria pholadiformis*: FMNH 188665, Key West, 1939 (29.1 mm).

*Scissula similis*: FMNH 177970, Little Duck Key (15.9 mm).

*Scissula similis* [living specimen]: FMNH 311626, FK-730, Newfound Harbor coral lumps, oceanside, 24°36.89'N, 81°23.63'W, 2.4–2.7 m, *Thalassia* seagrass, sand patches, coral heads, 04 May 2004 (15.6 mm).

Tellinidae hinge: Based on photographed specimens of *Scissula similis* (see above).

Tellinidae transparent clam: Based on original dissection of *Scissula similis* (AMNH 298937, FK-233, Old Dan Bank, bayside of Long Key, N of channel marker 2X, 24°49.95'N, 80°49.75'W, *Thalassia* seagrass with *Halimeda* and *Porites* coral, 0.6–1.2 m, 01 August 1999).

Tellinidae in living position: *Macoma balthica*, redrawn after de Goeij & Luttikhuizen (1998: fig. 1).

*Scissula iris*: CMNH 61.13404, Grassy Key (10.2 mm).

*Scissula candeana*: HMNS 35471, Grassy Key, 0.9 m, seagrass, March 1939 (11.4 mm).

*Scissula consobrina*: USNM 421821, [Dry] Tortugas, 08 July 1932 (unpaired, 12.1, 12.4 mm).

*Strigilla carnaria*: FMNH 202818, Bahia Honda Key, 1942 (13.1 mm).

*Strigilla mirabilis*: FMNH 189417, Bahia Honda [Key] (8.8 mm).

*Strigilla pisiformis*: FMNH 191018, Missouri Key (10.6 mm).

*Arcopagia fausta*: FMNH 177483, Little Duck Key, 1940 (83.9 mm).

*Leporimetis intastriata*: FMNH 311627, FK-696, Sand Island reef (near Molasses Reef), 25°01.12'N, 80°22.05'W, max. 6.7 m, patch reef with rubble, 10 June 2003 (46.3 mm).

*Laciolina laevigata*: FMNH 311628, Florida Keys, July 2002 (external view, 44.3 mm); FMNH 311629, FK-559, S beach of Loggerhead Key, Dry Tortugas, 24°37.79'N, 082°55.40'W, wrack line to 2.1 m, 15 April 2002 (internal view, 62.7 mm).

*Laciolina magna*: AMNH 120989, 1.5 mi NW of Key West, 6.1 m, sand, July 1959 (94.6 mm).

*Tellinella listeri*: FMNH 227473, Key West and Little Duck [Key] and Bradenton Beach, [Manatee County], Florida (combined lot, 52.6 mm).

*Tellidora cristata*: BMSM 26239, Key Largo (28.2 mm).

*Tellina radiata*: FMNH 177501, Little Duck Key, 1940 (90.4 mm).

*Phyllodina squamifera*: FMNH 197581, SE of [Dry] Tortugas, 91.4–213.4 m, 15 August 1974 (unpaired, 22.4, 18.8 mm).

*Angulus merus*: FMNH 311630, FK-035, Indian Key Fill, 24°53.42'N, 80°40.47'W, bayside, 1 m, *Thalassia* seagrass, 10 March 1996 (20.1 mm).

*Angulus tampaensis*: FMNH 182765, Key West, 1938 (21.9 mm).

*Angulus paramerus*: ANSP 285183, Bermuda (external view, RV internal view, 13.0 mm); FMNH 311589, FK-342, NW corner of Looe Key National Marine Sanctuary, 24°33.43'N, 81°25.94'W, 8.0–9.1 m, seagrass, 07 July 2000 (internal view, 11.6 mm).

*Angulus agilis*: FMNH 177945, Revere Beach, Suffolk County, Massachusetts, 1938 (13.0 mm).

*Angulus texanus*: FMNH 311632, FK-166, Tavernier, Plantation Key, bayside, 25°03.16'N, 80°29.07'W, 0–1 m, sand, *Thalassia* seagrass, *Acetabularia* on rock jetty, 14 September 1998 (unpaired, 13.8, 13.1 mm).

*Angulus probrinus*: FMNH 306106, FK-515, 24°27.54′N, 82°28.16′W, 18.6 m, muddy sand and shells, 26 July 2001 (alcohol lot, 10.1 mm).

*Angulus sybariticus*: FMNH 311590, FK-553, 24°41.51′N, 81°00.06′W, 11.5 m, sandy mud, 31 July 2001 (external view, 4.6 mm); FMNH 306114, FK-442, 24°39.39′N, 80°59.39′W, 26.5 m, mud, 26 April 2001 (internal view, 3.3 mm).

*Elliptotellina americana*: USNM 461747, off Sand Key, 164.6 m, R/V *Eolis* sta. 337, 1916 (unpaired, 7.3, 8.3 mm).

*Angulus versicolor*: FMNH 309686, FK-219, Mud Key Channel, bayside of Big Coppitt Key area, 24°40.52′N, 81°41.71′W, 3.3–3.9 m, very soft anoxic mud, 19 April 1999 (alcohol lot, 8.8 mm).

*Angulus versicolor* [living specimens]: same specimen as shell views (FMNH 309686, see above, largest specimen 8.8 mm).

*Merisca aequistriata*: FMNH 190931, Missouri Key, 1940 (10.9 mm).

*Merisca cristallina*: AMNH 78457, Corinto, Nicaragua, 1937-1939 (unpaired, 8.6, 15.7 mm).

*Merisca martinicensis*: AMNH 298004, FK-082, NNW of New Ground, between Marquesas Keys and Dry Tortugas, 24°48.28′N, 82°28.70′W, 27.4 m, 22 April 1997 (external view, 7.3 mm); AMNH 298003, FK-085, NNW of East Key, Dry Tortugas, 24°49.20′N, 82°43.90′W, 33.2–33.5 m, thick clay, 22 April 1997 (7.0 mm).

*Eurytellina alternata*: FMNH 151345, Florida Keys (54.2 mm).

*Eurytellina lineata*: FMNH 151457, Key West, 1949 (27.4 mm).

*Eurytellina angulosa*: FMNH 301394, FK-714, Marker 48, oceanside off Key Vaca, 24°41.50′N, 81°01.52′W, 4.9–7.3 m, patch reef with coral, gorgonians, sponges, surrounded by sand and seagrass, 17 August 2003 (56.7 mm).

*Eurytellina punicea*: FMNH 164118, Florida Keys, 1967 (29.8 mm).

*Eurytellina nitens*: FMNH 164969, Key West, 1969 (27.8 mm).

*Macoma cerina*: FMNH 183153, Lower Matecumbe Key (8.3 mm).

*Macoma brevifrons*: FMNH 202849, Missouri Key, 1941 (12.1 mm).

*Macoma constricta*: FMNH 150179, Lost Mans Key [error for Lossman's Key, Monroe County, on the W border of Ten Thousand Islands], Florida (31.8 mm).

*Macoma extenuata*: USNM 438705, off Middle Sambo Reef, with land out of sight, 137.2 m, R/V *Eolis* sta. 197, 24 May 1915 (unpaired, 12.3, 11.0 mm).

*Macoma limula*: USNM 438692, Garden Key, [Dry] Tortugas, 3 mi out from red sea buoy, 27.4 m, hard sand, J. B. Henderson, Jr., R/V *Eolis* sta. 34, 09 June 1911 (unpaired RV, 16.2 mm).

*Macoma pseudomera*: USNM 462267, Garden Key, [Dry] Tortugas, 3 mi out from red sea buoy, 27.4 m, hard sand, R/V *Eolis* sta. 34, 09 June 1911 (unpaired LV, 20.2 mm).

*Macoma tageliformis*: FMNH 177942, off Key West, 1950 (58.4 mm).

*Macoma tenta*: FMNH 170986, SE of Dry Tortugas, 91.4 m, 15 August 1974 (26.7 mm).

*Acorylus gouldii*: FMNH 202875, Missouri Key (8.2 mm).

*Cymatoica hendersoni*: FMNH 306067, FK-344, Hawk Channel, 24°34.05′N, 81°25.91′W to 24°34.09′N, to 81°25.90′W, 11.3 m, sand, shell hash, 07 July 2000 (6.7 mm).

*Donax variabilis*: FMNH 202068, Matecumbe Key (24.9 mm).

*Donax variabilis* [living specimens on beach]: AMNH transparency collection, Shelter Island, Long Island, New York, April 1976, G. Raeihle, photographer (specimen lengths not recorded).

*Donax variabilis* [color variation]: FMNH 208385, Sanibel Island, Lee County, Florida (largest specimen 19.6 mm).

Donacidae hinge: Based on photographed specimens of *Donax variabilis* (see above).

Donacidae transparent clam: Based on original dissection of *Donax variabilis* (AMNH 269835, Panama City Beach, [Bay County], Florida) and on the anatomy of *Donax* spp. by Simone & Dougherty (2004).

*Iphigenia brasiliana*: USNM 27221, Indian Key (42.5 mm).

*Asaphis deflorata*: FMNH 151500, Key West (43.2 mm).

*Asaphis deflorata* [living specimens in situ]: "Horseshoe" site, bayside of Spanish Harbor Keys, inside of (outermost) W arm, 24°39.35′N, 81°18.22′W, shallow subtidal, July 2002. Photograph by and courtesy of Osmar Domaneschi, Universidade de São Paulo, Brazil.

*Asaphis deflorata* [polychromic shells]: AMNH transparency collection, no locality data, W. Sage, photographer (specimen length not recorded).

Psammobiidae hinge: Based on photographed specimens of *Asaphis deflorata* (see above).

Psammobiidae transparent clam: Based on the anatomy of *Asaphis deflorata* by Domaneschi & Shea (2004).

*Gari circe:* FMNH 311632, FK-366, Conch Reef, oceanside of Key Largo, 24°57.37′N, 80°27.44′W, max. 7.9 m, pavement with algae, vertical walls/outcroppings, sand plains, 08 October 2000 (24.5 mm).

*Heterodonax bimaculatus:* FMNH 295264, Florida Keys (20.1 mm).

*Sanguinolaria sanguinolenta:* USNM 6981, Indian Key (55.1 mm).

Gravelly sand habitat with small mangrove: AMNH transparency collection, Crawl Key, G. Raeihle, photographer.

*Semele proficua:* FMNH 177473, Missouri Key (27.6 mm).

Semelidae hinge: Based on photographed specimens of *Semele proficua* (see above).

Semelidae transparent clam: Based on the anatomy of *Semele proficua* by Domaneschi (1995).

Semelids in living position: Redrawn after Yonge & Thompson (1976: fig. 118; of *Scrobicularia plana* (Da Costa, 1778)).

*Semele proficua* [SEM of juvenile sculpture]: FMNH 311605, FK-344, Hawk Channel, 24°34.05′N, 81°25.91′W to 24°34.09′N, 81°25.90′W, 11.3 m, sand, shell hash, 07 July 2000 (RV, total length 12.2 mm).

*Semele purpurascens* [SEM of juvenile sculpture]: FMNH 311606, FK-344, same data as previous (LV, total length 6.6 mm).

*Semele bellastriata:* FMNH 177985, Missouri Key (8.4 mm).

*Semele purpurascens:* FMNH 288708, Missouri Key, sand, 28 July 1973 (30.7 mm).

*Abra aequalis:* FMNH 306066, FK-539, S of Bahia Honda Key, W of Looe Key reef, 24°34.24′N, 81°16.64′W, 30.2–34.1 m, sand and rubble, 28 July 2001 (unpaired, 6.7, 6.9 mm).

*Abra lioica:* USNM 438997, off Key West, due S of Sand Key, 115.2 m, soft or coarse sand, R/V *Eolis* sta. 43, 02 June 1911 (6.1 mm).

*Abra longicallus americana:* USNM 832927, off Norfolk, Virginia, 36°55.50′N, 058°57′12″W, 4631 m (18.5 mm).

*Semelina nuculoides:* FMNH 311633, FK-366, Conch Reef, oceanside of Key Largo, 24°57.37′N, 80°27.44′W, max. 7.9 m, pavement with algae, vertical walls/outcroppings, sand plains, 08 October 2000 (5.0 mm).

*Cumingia coarctata:* FMNH 279123, FK-021, Lake Surprise, Key Largo, NE end of U.S. Rte. 1 causeway across lake, 25°10.9′N, 80°23′W, mangroves at side of road, sediment, algae, rocks, ca. 1.5 m, 09 July 1995 (5.6 mm).

*Cumingia coarctata* [living specimens]: FK-221, W side of Jewfish Basin [bayside of Big Coppitt Key area], tidal channel, 24°39.10′N, 81°43.96′W, 2.7–3.9 m, mangrove detritus, 19 April 1999.

*Cumingia vanhyningi:* FMNH 202916, Bonefish Key (10.4 mm).

*Ervilia subcancellata:* FMNH 202801, Little Duck Key, 1941 (5.1 mm).

*Ervilia nitens:* FMNH 311634, FK-345, Hawk Channel, 24°34.33′ to 34.40′N, 81°25.93′ to 25.91′W, 11.9 m, gray shelly mud, 07 July 2000 (paired, 5.5, 4.5 mm).

*Ervilia concentrica:* FMNH 311635, FK-345, Hawk Channel, 24°34.33′ to 34.40′N, 81°25.93′ to 25.91′W, 11.9 m, gray shelly mud, 07 July 2000 (paired, 3.2, 3.1 mm).

*Tagelus divisus:* FMNH 177617, Missouri Key, 1942 (23.8 mm).

Solecurtidae hinge: Based on photographed specimens of *Tagelus divisus* (see above).

Solecurtidae transparent clam: Based on original dissection of *Tagelus divisus* (AMNH 298940, FK-109, Bonefish Bay, NE side of causeway to Key Colony Beach, 24°43.7′N, 81°01′W, *Halodule/Thalassia* seagrass, mud, 1.5–2.4 m, 10 July 1997) and on the anatomy of *Tagelus divisus* by Bloomer (1907).

*Solecurtus sulcatus* [living animals]: Fraser Island, Queensland, Australia (length not recorded). Photographs by and courtesy of Richard Willan, Northern Territory Museum, Darwin.

*Solecurtus cumingianus*: FMNH 194524, SE of Dry Tortugas, 91.4–213.4 m, 15 August 1974 (27.7 mm).

*Tagelus plebeius*: FMNH 155866, Port Isabel, Cameron County, Texas (53.5 mm).

Salt-marsh habitat: P. M. Mikkelsen transparency collection, Nassau Sound, N of Talbot Island, Duval–Nassau Counties, Florida, 01 August 1981, P. M. Mikkelsen, photographer.

*Ensis minor*: CMNH 61.15, Key West, ca. 1890 (external view, 82.96 mm); FMNH 59670, St. Petersburg, [Pinellas County], Florida, December 1922 (internal view, 64.1 mm).

Pharidae hinge: Based on photographed specimens of *Ensis minor* (see above).

Pharidae transparent clam: Based on original dissection of *Ensis directus* Conrad, 1843 (AMNH 2229, New Brunswick, Canada), and on the anatomy of *E. siliqua* (Linnaeus, 1758) by Morse (1919).

*Mactrotoma fragilis*: FMNH 183845, Little Duck Key (36.1 mm).

*Mactrotoma fragilis* [living juvenile]: FMNH 311636, FK-730, Newfound Harbor coral lumps, oceanside, 24°36.89′N, 81°23.63′W, 2.4–2.7 m, *Thalassia* seagrass, sand patches, coral heads, 04 May 2004 (shell length, 18.6 mm).

Mactridae hinge: Based on photographed specimens of *Mactrotoma fragilis* (see above).

Mactridae transparent clam: Based on original dissection of *Mactrotoma fragilis* (FMNH 311636), living juvenile as above.

*Anatina anatina*: AMNH 269914, Sanibel Island, [Lee County], Florida, 1962 (47.2 mm).

*Mulinia lateralis*: USNM 484947, Lower Matecumbe Key, November 1937 (12.1 mm).

*Raeta plicatella*: AMNH 272574, Key West (59.7 mm).

*Spisula raveneli*: FMNH 83055, Sanibel Island, [Lee County], Florida (62.6 mm).

*Spisula solidissima*: FMNH 126646, Sunken Meadows, Long Island, New York (104 mm).

Commercial clamming operation: NOAA Fisheries Collection slide fish 0676, off Cape May, New Jersey, August 1968. Courtesy of National Oceanic and Atmospheric Administration, United States Department of Commerce.

*Mytilopsis leucophaeata*: FMNH 73206, 2 mi SW Bayou Chinchuba, Lake Pontchartrain, Louisiana, 1.5 m, 27 August 1953 (right exterior articulated view, 12.6 mm; left exterior articulated view, 13.3 mm); AMNH 295808, FK-035, Indian Key Fill, 24°53.42′N, 80°40′28″W, bayside, 1 m, *Thalassia* seagrass, 10 March 1996 (internal view and detail of apophysis, 9.1 mm).

Dreissenidae hinge: Based on photographed specimens of *Mytilopsis leucophaeata* (see above).

Dreissenidae transparent clam: Based on original dissection of *Mytilopsis leucophaeata* (AMNH 268880, Hudson River, Haverstraw Bay, [Rockland–Westchester Counties,] New York, October 1965) and on the anatomy of *Dreissena polymorpha* (Pallas, 1771) by Morton (1969).

*Mytilopsis sallei*: FMNH 53958, Little Pedro Point, E of Port Kaiser, [Jamaica], 14 June 1953 (right exterior articulated view, 11.7 mm; internal view, 10.3 mm).

*Dreissena polymorpha* [group on *Lampsilis siliquoidea*]: FMNH 308068, Lake Erie, Maumee Bay State Park, [town of] Oregon, Ohio, June–July 1990 (group 95 mm max.; largest *Dreissena* in image 18 mm).

*Sphenia fragilis*: AMNH 248691, Sebastian Inlet, [Brevard–Indian River Counties], Florida, in dead barnacles on ark shells, January–February 1973 (7.1 mm, measured on larger RV; LV external articulated view, 5.8 mm).

Myidae hinge: Based on photographed specimens of *Sphenia fragilis* (see above).

Myidae transparent clam: Based on original dissection of *Sphenia fragilis* (AMNH 279166, St. Catherine's Island, Liberty County, Georgia, 15 May 1995) and on the anatomy of the synonymous *S. antillensis* by Narchi & Domaneschi (1993).

*Mya arenaria* [living specimen]: AMNH transparency collection, Bayville, New York, August 1959, G. Raeihle, photographer (specimen length not recorded).

*Varicorbula limatula*: FMNH 306108, FK-106, NNE of New Ground, between Marquesas Keys and Dry Tortugas, 24°47.30′N, 82°18.30′W to 24°48.20′N, 82°19.10′W, 22.9–24.4 m, 25 April 1997 (6.9 mm, articulated view; 7.1 mm, separated valves, measured from larger RV).

*Varicorbula limatula* [living specimens on tickle chain]: same specimens as shell views (FMNH 306108, FK-106, same data as previous).

Corbulidae hinge: Based on photographed specimens of *Varicorbula limatula* (see above).

Corbulidae transparent clam: Based on original dissection (see Mikkelsen & Bieler, 2001).

*Varicorbula philippii*: FMNH 306107, FK-539, S of Bahia Honda Key, W of Looe Key reef, 24°34.24′N, 81°16.64′W, 30.2–34.1 m, sand and rubble, 28 July 2001 (4.4 mm, articulated view; unpaired, 6.8, 4.7 mm).

*Caryocorbula caribaea*: AMNH 109463, Lake Worth, [Palm Beach County], Florida, S inlet, 1939 (articulated view, 7.1 mm; 6.7 mm).

*Caryocorbula chittyana*: USNM 516474, [Dry] Tortugas, mud, 08 July 1932 (7.1 mm).

*Caryocorbula contracta*: HMNS 47344, SSE of Key West, 208.5 m, March 1975 (articulated and exterior views, 5.4 mm; interior view, 5.9 mm).

*Caryocorbula cymella*: FMNH 306064, FK-336, SW corner of Looe Key National Marine Sanctuary, 24°31.59′N, 81°25.94′W, 43.4 m, sandy mud, 06 July 2000 (articulated view, 3.7 mm; unpaired, 6.4, 6.0 mm).

*Caryocorbula dietziana*: HMNS 37224, off Dry Tortugas, 85.3 m, October 1969 (13.1 mm, measured larger LV).

*Juliacorbula aequivalvis*: FMNH 306076, FK-332, off Looe Key, 24°30.28′N, 81°25.90′W to 24°30.37′N, 81°25.83′W, 78.9–80.5 m, mud, 06 July 2000 (unpaired, 4.5, 5.1 mm); AMNH 294744, Santa Rosa Island, [Escambia County], NW Florida, 27.4 m (articulated view, 7.1 mm).

*Martesia striata*: AMNH 248451, Islamorada, cast ashore in nutlike object, March 1977 (external views, 20.1 mm); AMNH 155681, Matecumbe Key (internal view, unpaired, 10.7 mm).

Pholadidae hinge: Based on photographed specimens of *Martesia striata* (see above).

Pholadidae transparent clam: Based on the anatomy of *Martesia fragilis* Verrill & Bush, 1890, by Srinivasan (1961) and on original dissections of M. *striata* (AMNH 293536, Georgia) and of *Martesia* sp. (AMNH 248451, Philippines).

*Martesia* sp. [wood block]: AMNH 314911, no data, J. Biederman, photographer (entire block 78 mm; cropped width 53 mm).

*Capulus schreevei* [apophysis of *Cyrtopleura costata*]: Scanned from original description by T. A. Conrad, 1869, Notes on Recent Mollusca, *American Journal of Conchology*, 5: 105, pl. 13, figs. 3, 3a.

*Martesia cuneiformis*: FMNH 208381, Sanibel Island, Lee County, Florida, 1963–1972 (paired, external, dorsal, ventral, and anterior views, 13.4 mm; unpaired, internal view, 12.0 mm).

*Cyrtopleura costata*: FMNH 183237, Punta Gorda Beach, DeSoto County, Florida, 1949 (external view, pink-banded specimen, 147.9 mm); CMNH 45312, Florida Keys, 13 September 1973 (internal view, 136.2 mm); FMNH 183248, Cedar Keys, [Levy County], Florida (protoplax, 33.0 mm max. length; mesoplax, 29.8 mm max. length; apophysis from LV, 21.0 mm max. length).

*Barnea truncata*: AMNH 307818, N end of Captiva Island, [Lee County], Florida, in peat bank, November 1965 (39.8 mm).

*Pholas campechiensis*: FMNH 26407, North Inlet, Lake Worth, [Palm Beach County], Florida (30.5 mm).

*Teredo clappi*: USNM 348189, Key West (holotype, 4.7 mm, pallet 1.6).

Teredinidae hinge: Based on photographed specimens of *Teredo clappi* (see above).

Teredinidae transparent clam: Based on the anatomy of *Teredo navalis* by Lazier (1924) and a comparative anatomical tabulation (including *T. clappi*) by Turner (1966: table 1).

*Teredo navalis* [bored wood]: FMNH 2873, "wood bored by *Teredo navalis* in twelve months," Pascagoula Harbor, Jackson County, Mississippi, 1893 (cross section of riddled piling, 23.5 cm diameter).

*Bankia carinata*: FMNH 188630, Grassy Key, wood (4.9 mm; pallet, 9.0 mm length).

*Bankia fimbriatula*: Scanned from W. J. Clench and R. D. Turner, 1946, The genus *Bankia* in the western Atlantic, *Johnsonia*, 2(19): pl. 14, figs. 1–4 (3.8 mm; longest pallet 9.5 mm length); based on MCZ specimen, Port au Prince, Haiti. Reprinted with permission of Museum of Comparative Zoology, Harvard University.

*Lyrodus pedicellatus*: Scanned from R. D. Turner, 1966, A Survey and Illustrated Catalogue of the Teredinidae (Mollusca: Bivalvia), Museum of Comparative Zoology, Harvard University, Cambridge, Massachusetts, pl. 1D, figs. 1–4 (4.0 mm, pallet 5.4 mm length); based on USNM 194280,

British Isles. Reprinted with permission of Museum of Comparative Zoology, Harvard University.

*Nototeredo knoxi*: HBOM 064:00989, E end of creek between Card and Barnes Sounds, vicinity of Cormorant and Jew Points, Key Largo, intertidal, submerged branch, 07 October 1978 (alcohol lot, 1.8 mm shell length, pallet 1.3 mm width); ANSP 370075, Mosquito Point, Fleming Road, Grand Bahama Island, Bahamas, 26°37.50′N, 78°54.00′W (LV internal and external, 7.4 mm; isolated pallet, 14.4 mm length).

*Teredo bartschi*: Scanned from R. D. Turner, 1966, *A Survey and Illustrated Catalogue of the Teredinidae (Mollusca: Bivalvia)*, Museum of Comparative Zoology, Harvard University, Cambridge, Massachusetts, pl. 8A, figs. 1–4 (4.1 mm, longest pallet 5.0 mm length); based on the holotype, MCZ 45301, Port Tampa, [Hillsborough County], Florida. Reprinted with permission of Museum of Comparative Zoology, Harvard University.

*Teredo somersi*: Scanned from R. D. Turner, 1966, *A Survey and Illustrated Catalogue of the Teredinidae (Mollusca: Bivalvia)*, Museum of Comparative Zoology, Harvard University, Cambridge, Massachusetts, pl. 7E, figs. 1–4 (1.9 mm, longest pallet 2.6 mm length); based on the lectotype MCZ 45304, Ireland Island, Bermuda. Reprinted with permission of Museum of Comparative Zoology, Harvard University.

*Teredora malleolus*: AMNH 260802, Seaford, Sussex, United Kingdom (10.3 mm; pallet 8.1 mm length).

## Glossary images (except those also used in the main part)

Alivincular ligament: Redrawn after Ubukata (2003: fig. 1C).

Anal funnel [*Pinctada imbricata*]: AMNH 308115, FK-287, "Horseshoe" site, bayside of Spanish Harbor Keys, inside of (outermost) W arm, 24°39.35′N, 81°18.22′W, shallow subtidal, 10 April 2000 (shell 45.68 mm; anal funnel length 2.35 mm). Photograph by and courtesy of Ilya Tëmkin, AMNH.

Aortic bulb [*Mercenaria mercenaria*]: New Jersey seafood market, 05 April 2005, P. M. Mikkelsen, photographer (specimen ca. 75 mm; length of aortic bulb + heart ca. 16.5 mm).

Cruciform muscle: Redrawn after Yonge & Thompson (1976: fig. 117).

Duplivincular ligament: Redrawn after Ubukata (2003: fig. 1K).

Eulamellibranch gill: Redrawn after Lang (1900: fig. 153C).

Filibranch gill: Redrawn after Lang (1900: figs. 153B–C).

Multivincular ligament: Redrawn after Ubukata (2003: fig. 1G).

Parivincular ligament: Redrawn after Ubukata (2003: fig. 1A).

Pericalymma larva: Redrawn after Gustafson & Lutz (1988: figs. 3A and 5C [*Solemya velum*]).

Planivincular ligament: Redrawn after Ubukata (2003: fig. 1B).

Protobranch gill: Redrawn after Lang (1900: fig. 153A).

Septibranch gill: Redrawn after Lang (1900: fig. 153D).

Veliger larva: Redrawn after Galtsoff (1964: fig. 342 [*Crassostrea virginica*]).

# Index

NOTE: Very common terms, which occur in many or most family descriptions (e.g., mantle, stomach, predators), are restricted in this index to figures intended to illustrate the feature and to entries in the Glossary.